Electronic Density Functional Theory

Recent Progress and New Directions

Electronic Density Functional Theory

Recent Progress and New Directions

Edited by

John F. Dobson
Griffith University
Nathan, Queensland, Australia

Giovanni Vignale
University of Missouri–Columbia
Columbia, Missouri

and

Mukunda P. Das
The Australian National University
Canberra, Australia

Plenum Press • New York and London

Library of Congress Cataloging-in-Publication Data

On file

Proceedings of the International Workshop on Electronic Density Functional Theory: Recent Progress and New Directions, held July 14 – 19, 1996, in Nathan, Queensland, Australia

ISBN 0-306-45834-9

A Division of Plenum Publishing Corporation
233 Spring Street, New York, N.Y. 10013

http://www.plenum.com

10 9 8 7 6 5 4 3 2 1

Printed in the United States of America

PREFACE

This book is an outcome of the International Workshop on Electronic Density Functional Theory, held at Griffith University in Brisbane, Australia, in July 1996. Density functional theory, standing as it does at the boundary between the disciplines of physics, chemistry, and materials science, is a great mixer. Invited experts from North America, Europe, and Australia mingled with students from several disciplines, rapidly taking up the informal style for which Australia is famous. A list of participants is given at the end of the book.

Density functional theory (DFT) is a subtle approach to the very difficult problem of predicting the behavior of many interacting particles. A major application is the study of many-electron systems. This was the workshop theme, embracing *inter alia* computational chemistry and condensed matter physics. DFT circumvents the more conceptually straightforward (but more computationally intensive) approach in which one solves the many-body Schrödinger equation. It relies instead on rather delicate considerations involving the electron number density.

For many years the pioneering work of Kohn and Sham (the Local Density Approximation of 1965 and immediate extensions) represented the state of the art in DFT. This approach was widely used for its appealing simplicity and computability, but gave rather modest accuracy. In the last few years there has been a renaissance of interest, quite largely due to the remarkable success of the new generation of gradient functionals whose initiators include invitees to the workshop (Perdew, Parr, Yang).

This is therefore an exciting time for density functional theorists. Major new directions are being explored at the fundamental level. Theoretical considerations once thought esoteric by some are proving to be vital in the construction of practical computational tools for chemists, physicists, and materials scientists.

Practical applications of the theory are growing explosively as many once-sceptical chemists enthusiastically embrace the new functionals, thanks to the much-improved accuracy afforded by theoretical developments such as the generalized gradient approach.

Major new developments described at the workshop involved:

- gradient functionals ;
- functionals for van der Waals (dispersion) energies;
- time-dependent density functional theory and excited states;
- edge effects (the "Airy Gas");
- hybrid functionals mixing traditional and DFT methods;

- the bandgap problem for solids;
- DFT perturbation schemes.

While the emphasis at the workshop was on state-of-the-art developments, a day of introductory lectures was offered, to set the scene for students and newcomers to the field.

The same approach is taken in this book. The first part contains introductory chapters which provide at least minimal background for every invited chapter found in the later parts. Parts II-V are devoted to new research presentations at the forefront of the field. We were very pleased to note the clarity of these invited contributions, and we would like to thank the authors for this.

We feel that this combination of introductory and research chapters, combined with an extensive index, makes for a book with considerable pedagogical value.

We also would like to thank the following for financial and other support, without which we could not have assembled such a field of international talent:

- Australian Federal Department of Industry, Science and Tourism;
- U.S. National Science Foundation;
- Griffith University: Faculty of Science and Technology, School of Science, Vice-Chancellor's Conference Fund.

JFD would also like to thank Pravin Sen, Bernadette Johnston, Hung Le, Brad Dinte and Matthew Daniell for general assistance at the workshop.

John F. Dobson
Giovanni Vignale
Mukunda P. Das

CONTENTS

I. INTRODUCTORY MATERIAL

II. INVITED CHAPTERS ON GROUNDSTATE ENERGY FUNCTIONALS

III. INVITED CHAPTERS ON EXCITED STATES, TIME-DEPENDENT DFT AND THE VAN DER WAALS INTERACTION

IV. INVITED CHAPTERS ON SPECIAL SYSTEMS

V. ABSTRACTS OF POSTER PRESENTATIONS

Part I

INTRODUCTORY MATERIAL

BRIEF INTRODUCTION TO DENSITY FUNCTIONAL THEORY

John F. Dobson[1] and Mukunda P. Das[2]

[1]School of Science, Griffith University
Nathan, Queensland 4111, Australia

[2]Department of Theoretical Physics, Australian National University
Canberra, ACT 0200, Australia

SCOPE OF THIS VOLUME

The physics of many interacting particles is notoriously difficult. The attraction of density functional theory is that it facilitates approximate calculations on many-body systems without requiring the construction of many-body wavefunctions or their equivalent. Instead one extracts the needed information from a one-body quantity, the number density. While density functional methods can be applied to many systems such as classical fluids, nuclei and thermal ensembles, this book is principally devoted to quantum states of many-electron systems. All of the articles are relevant to quantum chemistry and/or condensed matter physics.

BRIEF HISTORY: INTRODUCTION TO THE INVITED CHAPTERS

The following paragraphs contain a very brief and necessarily biased history of the field, with items selected so as to set in context the technical chapters contributed by the workshop invitees. Some specific topics which are not yet well represented in texts are treated separately in later chapters of the present introductory part of this book.

Thomas-Fermi-Slater Approximations

The history of DFT starts with the Thomas-Fermi (TF) theory [1,2] and Dirac's local exchange approximation [3]. These are attempts to simplify the physics of many-Fermion systems, and are defined by the very simple approximations

$$T_{TF} = \int \frac{3}{5}\epsilon_F(n(\mathbf{r}))n(\mathbf{r})\, d^3r = \frac{3}{5}(\frac{\hbar^2}{2m})(3\pi^2)^{2/3}\int n(\mathbf{r})^{5/3}\, d^3r, \quad (1)$$

$$E_x^{LDA} = -\frac{3e^2}{4}(\frac{3}{\pi})^{1/3}\int n(\mathbf{r})^{4/3}\, d^3r \quad (2)$$

Electronic Density Functional Theory: *Recent Progress and New Directions*
Edited by Dobson *et al.*, Plenum Press, New York, 1998

for the kinetic energy T and exchange energy (off-diagonal Hartree-Fock energy) E_x, based on the electron number density $n(\mathbf{r})$. This approach was taken further in the Xα method of Slater [4].

Hohenberg-Kohn Theorems

In the 1960s the theorem of Hohenberg and Kohn ("HK", [5]) showed that the groundstate density $n(\mathbf{r})$ (plus the specification of fermion statistics, the electron mass and electron-electron Coulomb interaction) is sufficient in principle to determine not only the energy in the Hartree-Fock approximation, but the exact many-body energy including all effects beyond Hartree-Fock theory (loosely termed correlation.) The proof proceeded by showing that a given density $n(\mathbf{r})$ can be the nondegenerate groundstate density for at most one external potential $v(\mathbf{r})$. The complete many-body hamiltonian is thus implied by $n(\mathbf{r})$. It follows that not only the groundstate energy but indeed all excited states and their energies are implied by $n(\mathbf{r})$. This theorem gives no insight as to actual methods of calculation, and at any rate it is usually $v(\mathbf{r})$ rather than $n(\mathbf{r})$ that is known (e.g. in chemistry $v(\mathbf{r})$ is provided by the nuclei). However the HK theorem provides confidence that it is sensible to seek prescriptions for many-body physics based on the density rather than the wavefunctions.

Secondly, HK showed that there exists a universal (i.e. $v(\mathbf{r})$-independent) *functional* $F[n]$ such that, for a given external potential $v(\mathbf{r})$, the actual groundstate energy E and density $n(\mathbf{r})$ are obtained by minimizing the energy functional

$$E[n] = F[n] + \int v(\mathbf{r})n(\mathbf{r})d^3r \tag{3}$$

with respect to variations in $n(\mathbf{r})$, subject to the constraint of constant particle number N.

An Aside: Functionals and Their Derivatives

A *function* maps from a number or finite-dimensional vector to a single number or vector: e.g. the density function n maps from the vector $\mathbf{r}$ to the number $n(\mathbf{r})$. By contrast, a *functional* maps from a *function* to a number. For example, the total particle number N is a functional of the density function $n(\mathbf{r})$, given by $N = \int n(\mathbf{r})d^3r$. In general, a functional $F[f]$ is termed "local" if it can be expressed as an integral involving the function $f(\mathbf{r})$ only at one point $\mathbf{r}$ at a time. For example, the Thomas-Fermi KE functional $T_{TF}[n] = \int cn(\mathbf{r})^{5/3}d^3r$ is a local functional, but the Hartree e-e interaction $V^H[n] = \frac{1}{2}\int n(\mathbf{r})n(\mathbf{r}')V_{ee}(\mathbf{r}-\mathbf{r}')d^3rd^3r'$ (where $V_{ee}(R)$ is the coulomb pair interaction) is nonlocal.

In general, the functional derivative $\delta F[n]/\delta n(\mathbf{r})$ is defined so that the first-order change in the functional due to an arbitrary change in its argument $n(\mathbf{r})$ is

$$\delta F = \int \frac{\delta F}{\delta n(\mathbf{r})}\delta n(\mathbf{r})d^3r + O(\delta n^2).$$

More intuitively, we could imagine changing the value of the function $n(\mathbf{r})$ by an amount Δn, in a small region of volume ΔVsurrounding the point $\mathbf{r}$, causing a change ΔF in the functional. Then in the limit of small Δn and ΔV, we have $\delta F[n]/\delta n(\mathbf{r}) = \Delta F/(\Delta n \Delta V)$. An example of a functional derivative is that of the total electron-electron Hartree potential energy defined in the previous paragraph:

$$v_H(\mathbf{r}) = \frac{\delta V^H[n]}{\delta n(\mathbf{r})} = \int n(\mathbf{r}')V_{ee}(\mathbf{r}-\mathbf{r}')d^3r'.$$

Here an extra factor of two has resulted because the change in density at the point $\mathbf{r}$ can alter either the $n(\mathbf{r})$ or the $n(\mathbf{r}')$ term in V^H.

Levy's Constrained Search Approach

Levy [6] used the Rayleigh-Ritz variational principle to give a "constrained-search" realization of the functional F. F is the minimum value of the expectation of (kinetic energy $\hat{T}$ plus electron-electron potential energy $\hat{V}_{ee}$), obtained by searching over all *many-body* wavefunctions Ψ which yield the given density $n(\mathbf{r})$:

$$F[n] = \min_{\Psi \to n} < \Psi \left| \hat{T} + \hat{V}_{ee} \right| \Psi > . \tag{4}$$

A particularly good visualization of this idea is given in the book by Parr and Yang [7].

Kohn-Sham KE, Definition of E_{xc} in Kohn-Sham DFT

Early attempts to approximate the HK functional $F[n]$ used the Thomas Fermi approximation (1) for the kinetic component $< \hat{T} >$. It was soon realized that only very crude answers can be obtained with this local functional for the KE, no matter how sophisticated the approximation for the $< \hat{V}_{ee} >$ component of F. For example, atomic shell structure is missed in this approximation.

Kohn and Sham therefore [8] suggested a highly nonlocal functional giving the major part of the KE. This is the "single-particle" KE, $T_s[n]$, which is the KE of *independent* electrons (i.e. electrons without mutual Coulomb repulsion) in their groundstate, under the action of an external potential such that their groundstate density is $n(\mathbf{r})$. If $\psi_k(\mathbf{r})$ are the one-particle eigenfunctions of this external potential then

$$n(\mathbf{r}) = \sum_{k=1}^{N} |\psi_k(\mathbf{r})|^2 , \tag{5}$$

and

$$T_s[n] = \sum_{k=1}^{N} \int d^3r \frac{\hbar^2}{2m} |\nabla \psi_k(\mathbf{r})|^2 . \tag{6}$$

This is not the true KE $< \hat{T} >$ of an *interacting* system (i.e. one including electron-electron coulomb repulsion) whose groundstate density is $n(\mathbf{r})$, but in the final optimized description it is much closer to $< \hat{T} >$ than the Thomas-Fermi KE is. The energy functional (3) is then written as the sum of large external PE, single-particle KE and Hartree PE terms, plus a remainder E_{xc} which is usually smaller:

$$E = \int v(\mathbf{r})n(\mathbf{r})d^3r + F[n] = \int v(\mathbf{r})n(\mathbf{r})d^3r + T_s[n] + \frac{e^2}{2} \int \frac{n(\mathbf{r})n(\mathbf{r}')}{|\mathbf{r} - \mathbf{r}'|} d^3r d^3r' + E_{xc}[n]. \tag{7}$$

E_{xc} is termed the exchange-correlation energy. The physical effects included in E_{xc} are of both a potential and a kinetic energy character, and comprise the following:

- a negative potential energy (the exchange energy, tending to cancel part of the of the Hartree PE); this arises because wavefunction antisymmetry causes electrons of like spin projection avoid one another

- a negative potential energy, (the correlation potential energy, a further reduction of the Hartree PE) occurring because the Coulomb potential causes electrons of either spin orientation to avoid each other

- a positive correlation contribution to the kinetic energy, an effect of the uncertainty and Pauli principles; mutual avoidance reduces the space available and hence raises the KE .

(We note that the Kohn-Sham decomposition (7) is not the only possible density-functional decomposition of the total energy: for example, the chapter by Engel and Pickett in the present volume exploits a different kinetic energy decomposition in connection with the bandgap problem in periodic solids.)

Minimization of the energy expression (7) with respect to the $\{\psi_k(\mathbf{r})\}$ (equivalent to minimizing with respect to density) gives the now-famous Kohn-Sham equations:

$$[-\frac{\hbar^2}{2m}\nabla^2 + v_{eff}(\mathbf{r})]\psi_k(\mathbf{r}) = \epsilon_k\psi_k(\mathbf{r}). \tag{8}$$

where ϵ_k is a Lagrange multiplier ensuring normalization. The effective one-electron potential contains external, Hartree and exchange-correlation terms:

$$v_{eff}(\mathbf{r}) = v(\mathbf{r}) + e^2\int\frac{n(\mathbf{r}')}{|\mathbf{r}-\mathbf{r}'|}d^3r' + v_{xc}(\mathbf{r}) \tag{9}$$

where

$$v_{xc}(\mathbf{r}) = \frac{\delta E_{xc}}{\delta n(\mathbf{r})}. \tag{10}$$

The KS formalism based on equs (7) - (10) gives the exact groundstate energy and density for non-degenerate systems. A rigorous theory for degenerate groundstates is now also available: it requires the concept of a quantum ensemble but leads to similar Kohn-Sham equations. A good description is given in Dreizler and Gross [9]. Degeneracy in the quantum Hall effect, and the consequent need for ensemble theory, is of the essence in the contribution by Heinonen, Lubin and Johnson in the main section of the present Volume.

As well as being exact, the KS scheme is intrinsically simpler computationally than, for example, the approximate Hartree Fock scheme [10] because the effective potential $v_{eff}(\mathbf{r})$ in (8) is *local:* it only acts on the wavefunction at the point $\mathbf{r}$. Nevertheless the KS potential v_{xc} has, in principle, a nonlocal dependence on the density $n(\mathbf{r}')$.

Local Density Approximation for E_{xc}

The KS scheme does not lead to computational power as it stands, because the difficulty of the many-body problem is still present in the unknown functional $E_{xc}[n]$. To overcome this, Kohn and Sham proposed a local-density approximation (LDA)

$$E_{xc} \approx E_{xc}^{LDA} = \int n(\mathbf{r})\epsilon_{xc}^{\text{hom}}(n(\mathbf{r}))d^3r \tag{11}$$

in which $\epsilon_{xc}^{\text{hom}}$ is the exchange-correlation energy *per particle* in the homogeneous (spatially uniform) electron gas described under the next heading. The functional derivative (see definition above) of E_{xc}^{LDA} is the local approximation to the KS exchange-correlation potential:

$$v_{xc}^{LDA}(\mathbf{r}) \equiv \frac{\delta E_{xc}^{LDA}}{\delta n(\mathbf{r})} = \frac{d}{dn}(n\epsilon_{xc}^{\text{hom}}(n))\bigg|_{n=n(\mathbf{r})}$$

With this approximation, the potential $v_{eff}(\mathbf{r})$ in the KS equation (8) is easily computable. It is local both in the sense that it acts only on the wavefunction at $\mathbf{r}$, and also in the sense that it depends only on the density at $\mathbf{r}$.

Homogeneous Electron Gas (Egas, Infinite Jellium)

This fictitious system, used as a reference in DFT calculations, has a spatially-uniform positively charged background which neutralizes the charge of the uniformly distributed electrons. Its electron number density $n(\mathbf{r}) = n_0$ is position-independent and is traditionally specified by giving the dimensionless number r_s (Wigner-Seitz radius) which is the mean spacing between electrons divided by the Bohr radius $a_B = \hbar^2/(me^2)$:

$$n_0^{-1} = \frac{4}{3}\pi(r_s a_B)^3. \tag{12}$$

High r_s values correspond to low electron densities. Among real systems, the conduction electron cloud in the simple metals comes closest to the ideal condition of homogeneous density. Most metallic electron gases have $2 < r_s < 6$.

Even for an inhomogeneous system (atom, molecule, solid, etc.) it is common to express the electron density at some point $\mathbf{r}$ in terms of a local value of r_s:

$$r_s(\mathbf{r}) = a_B^{-1}[4\pi n(\mathbf{r})/3]^{-1/3}. \tag{13}$$

Description of the homogeneous egas constitutes an insoluble interacting many-body problem, but its geometrical simplicity has facilitated a wide range of approximate and near-exact numerical treatments. The exchange energy per particle (due to antisymmetry of the many-body wavefunction) in a spin-unpolarized electron gas is exactly

$$\epsilon_x^{\mathrm{hom}} = -\frac{c_x}{r_s}$$

where $c_x = \frac{3}{4\pi}(9\pi/4)^{1/3}$ hartree ≈ 0.458 hartree. (Hartree atomic units (a.u.) are obtained by setting $e = \hbar = m_e = 1$. The unit of energy is: 1 hartree = 2 Rydberg $\approx$ 27.2 eV.)

The correlation energy cannot be found exactly even for the homogeneous egas. For the spin-unpolarized gas at metallic densities perhaps the simplest fit, still in wide use where extremely rapid computation is required, is the Wigner interpolation formula [11,12]:

$$\epsilon_c^{Wig} = -\frac{0.44}{r_s + 7.8} \quad hartree.$$

The near-exact quantum Monte-Carlo egas results of Ceperley and Alder [13] for $\epsilon_c(r_s)$ have been parametrized by Vosko, Wilk and Nusair [14] among others (see also p. 179 of [9]).

Advantages of the KS-LDA Formalism

The Kohn-Sham-LDA scheme is given by equs. (7)-(11). Together with fits for $\epsilon_c^{\mathrm{hom}}(n)$, it allows one to use a knowledge of the homogeneous egas to predict properties of the inhomogeneous electron gases occurring in atoms, molecules and solids. With regard to solid-state calculations over the last 25 years, the importance of the KS-LDA computational scheme (and its spin generalization, the LSD: see paragraph below) can hardly be exaggerated. Until very recently no other widely-useable scheme was available for the correlation component of the energy. In particular, the ability to predict electronic groundstate energies and their dependence on ionic positions has been most useful. For a brief review see [15]. In chemical applications the KS-LDA was not so popular because, for small molecules at least, other quantum-chemical methods such as the Configuration Interaction approach gave better accuracy for most quantities, albeit with much greater computational effort. As a rapid method giving total energy changes as a function of ionic position (e.g. for prediction of vibrational frequencies) the KS-LDA was nevertheless useful, often rivalling the other methods. The same cannot be said of electronic excitation and promotion energies, which are generally considered to lie outside the realm of rigorous applicability of groundstate KS theory. (There are exceptions: see below).

Spin Density Functional Theory

A useful extension of the KS approach [16,17] treats separately the densities $n_\uparrow(\mathbf{r})$ and $n_\downarrow(\mathbf{r})$ of electrons with spin projection up and down. Equivalently one can deal with $n(\mathbf{r}) \equiv n_\uparrow(\mathbf{r}) + n_\downarrow(\mathbf{r})$, together with the polarization $\zeta(\mathbf{r}) = (n_\uparrow(\mathbf{r}) - n_\downarrow(\mathbf{r}))/n(\mathbf{r})$. ζ takes values between -1 (fully polarized downwards) and $+1$ (fully polarized upwards). The spin-up and -down densities are generated from spin-up and spin-down KS wavefunctions, $n_\uparrow(\mathbf{r}) = \sum_k |\psi_{k\uparrow}(\mathbf{r})|^2$ and similarly for $n_\downarrow(\mathbf{r})$. A local approximation, the LSD, based on the spin-polarized uniform gas, is available analogously to the KS-LDA approach. The spin interaction with an external magnetic field B is included, so spin magnetism in solids can be treated. LSD also represents a considerable improvement over LDA for atomic and molecular systems with unpaired spins, for which the unpolarized electron gas is clearly not a very good model. This is also related to the problem of orbital self interaction: see the following paragraph.

Orbital Self-Interaction

A difficulty inherent in local density approximations is that they permit unphysical contributions to the energy of a one-electron system (e.g. hydrogen) arising formally from the pair interaction. In fact, of course, there is no pair interaction in the one-body problem. Formally, this manifests itself by the following constraints:

- for a one-body system, the "naive Hartree" energy $\frac{1}{2}\int n(\mathbf{r})n(\mathbf{r}')V_{ee}(\mathbf{r},\mathbf{r}')d^3rd^3r'$ must be exactly cancelled by the exchange energy
- for a one-body system, the correlation energy must be zero.

The same restrictions apply to the corresponding potentials. The LDA violates these conditions, and the satisfaction of such constraints is a nontrivial matter, having major implications not just for one-body systems but for many-body systems whenever orbitals can be spatially localized. This is discussed at length in the introductory chapter "Driving out the self-interaction error" by Perdew and Ernzerhof. The chapter by Temmerman et al. exemplifies the remarkable improvements that can be obtained in the DFT description of solids by taking these considerations into account.

The Meaning of the Kohn-Sham Eigenvalues

For single-particle systems the eigenvalues ϵ_k of the Schrodinger equation give the energies of excited states, so that their differences represent electronic transition energies. For interacting systems the Kohn-Sham Schrödinger-like equation (8) involves eigenvalues ϵ_k which arise as Lagrange multipliers to enforce normalization of the orbitals ψ_k during the variational procedure. It is tempting to treat differences between these KS eigenvalues ϵ_k as transition energies to (or between) interacting electronic excited states, by analogy with the case of a single-particle problem. The KS scheme is however a strictly groundstate theory, so there does not at first appear to be any formal justification for such a usage. This is the case whether or not one uses a local-density approximation for $v_{xc}(\mathbf{r})$. In fact the only KS eigenvalue that can be assigned an unambiguous physical meaning is the highest occupied eigenvalue $\epsilon_{\max}$. By arguments involving the asymptotic spatial decay constants of the true groundstate density and of the density due to the highest occupied KS orbital, one can show that $\epsilon_{\max}$ is the electron removal energy: see p. 149 of [7]. (Very recently some evidence has come to light that KS eigenvalue differences $\epsilon_j - \epsilon_{\max}$ may represent true excitation energies when ϵ_j lies well above $\epsilon_{\max}$. The same is not necessarily true of the LDA approximations to these eigenvalues, however. See below under the heading "Exact KS Potentials").

Despite these theoretical limitations, usage of the KS eigenvalue differences to discuss excitation energies has been a common procedure in practice, and in chemical calculations KS eigenvalue differences often reasonably represent *trends* in excitation energies across the periodic table [15].

Furthermore the KS LDA was for two decades the commonest way to predict electronic Bloch bandstructures of solids with inclusion of the electron-electron interaction. Such $\epsilon(\mathbf{k})$ vs. $\mathbf{k}$ curves, if meaningful, would represent excitation energies for most values of $\mathbf{k}$, with an exception (the case of the fundamental gap) to be discussed presently. This "KS bandstructure" usually has the correct general shape, but the curves tend to be shifted by an almost $\mathbf{k}$-independent amount from the correct values as given by experiment. In particular, the KS bandgap (minimum conduction band energy minus maximum valence band energy $\epsilon^{(2)}(\mathbf{k}_{02}) - \epsilon^{(1)}(\mathbf{k}_{01})$) often seriously underestimates the fundamental bandgap defined in the following paragraph.

For extended interacting systems within the Hartree-Fock approximation, Koopmans' theorem guarantees that each HF eigenvalue represents an energy of removal of an electron, leaving in general an excited HF state. There is no such theorem in general for the KS eigenvalues, though one does have Janak's theorem for fractional occupancies f_i of the KS orbitals in ensemble density functional theory [18]:

$$\frac{\partial E}{\partial f_i} = \epsilon_i. \tag{14}$$

Despite the lack of a Koopmans' theorem, one can nevertheless discuss the bandgap problem rigorously within the KS approach, without having to deal with excited states. The *fundamental band gap* E_g is the difference between the electron addition and removal energies, starting from the insulating groundstate of N electrons, and arriving in both cases at the groundstate of a system with an altered number of electrons. Thus $E_g = [E(N+1) - E(N)] - [E(N) - E(N-1)]$. This involves only groundstate energies and hence is in principle a proper quantity to calculate in the KS scheme. Unfortunately the practical usefulness of the above exact expression for the bandgap is limited by the fact that the xc energy functional is known to have a discontinuous derivative with respect to particle number N, at values of N corresponding to a filled band. In particular, this derivative discontinuity means that the bandgap is *not* given by a difference of exact KS eigenvalues: an explicit discontinuity contribution must be added to the KS gap [19-22]. The discontinuity can be calculated in principle from the electron self-energy, but none of the approximate density functionals in common use incorporates the discontinuity satisfactorily.

The difficulties in identifying the KS $\{\epsilon_i\}$ as band energies have led to the introduction of schemes outside the Kohn-Sham framework. For example, the GW method has had considerable success [23]. It uses a nonlocal and energy-dependent effective potential Σ (mass or self-energy operator), which essentially replaces the local KS potential v_{eff}. (Selfconsistent Kohn-Sham calculations are still useful, typically providing a time-efficient first step in these GW calculations.)

As an alternative to this approach, the chapter in the present volume by Engel and Pickett shows how one can obtain improved bandstructure by modifying the KS kinetic energy term. The resulting scheme remains within the Density Functional framework, but it differs from the Kohn-Sham scheme.

Orbital Magnetic Effects: Static Current Density Functional Theory

The LSD theory does not treat the effects of a magnetic field on the *orbital* motion. To achieve this one needs to deal with current-carrying groundstates. It is necessary to describe the magnetic field via a vector potential $\mathbf{A}(\mathbf{r})$, and considerations of gauge invariance are paramount. This field was opened up by Rasolt and Vignale [24] who discovered that gauge invariance is achieved if $\mathbf{A}_{xc}$ depends only on the vorticity $\nabla \times (\mathbf{j_p}/n)$, where $\mathbf{j_p}$ is the "paramagnetic" current density. Such considerations are clearly vital for a DFT description of high-field phenomena such as the quantum Hall effect: see the chapter by Heinonen and Lubin. In time-dependent density functional theory (see below) it seems the best approximations will also require the inclusion of a vector potential, as described in the chapter by Vignale and Kohn.

Pair Density: Adiabatic connection formula

The pair density $n_2(\mathbf{r},\mathbf{r}')$ is sometimes denoted $< 0\left|\hat{\Psi}^\dagger(\mathbf{r})\hat{\Psi}^\dagger(\mathbf{r}')\hat{\Psi}(\mathbf{r}')\hat{\Psi}(\mathbf{r})\right|0 >$ or $2\Gamma(\mathbf{rr}' \mid \mathbf{rr}')$. It is obtained by integrating the squared many-body wavefunction over all but two space coordinates. Its physical meaning is that $n_2(\mathbf{r},\mathbf{r}')d^3rd^3r'$ is proportional to the probability of finding an electron inside a small region of volume d^3r located at $\mathbf{r}$ and, simultaneously, finding a different electron inside d^3r' located at $\mathbf{r}'$.

Because the electron-electron Coulomb interaction $e^2/\left|\mathbf{r}-\mathbf{r}'\right| \equiv \hat{V}_{ee}$ involves only two particles at a time, knowledge of n_2 determines the groundstate electron-electron potential energy exactly. If the electrons did not interact at all (and were not subject to fermionic wavefunction antisymmetry requirements) the pair distribution would be have the independent-particle value $n_2^{ind} = n(\mathbf{r})n(\mathbf{r}')$. The corresponding pair energy is often termed the Hartree or "classical" interaction energy (in the sense of classical electrostatics, *not* in the sense that $\hbar \to 0$). Thus we can obtain the exchange-correlation (i.e. beyond-Hartree) part of the potential energy exactly as

$$V_{xc} = \frac{1}{2}\int [n_2(\mathbf{r},\mathbf{r}') - n(\mathbf{r})n(\mathbf{r}')]\frac{e^2}{|\mathbf{r}-\mathbf{r}'|}d^3rd^3r' \equiv < \Psi\left|\hat{V}_{ee}\right|\Psi > -E^H \qquad (15)$$

Eq. (15) is not the whole of the xc energy E_{xc}, however, because as noted earlier there is a kinetic energy component in E_{xc}. One can deduce the total xc energy by switching on the strength λ of the Coulomb interaction. The usual way to do this [25,26] is to introduce a λ- dependent external potential V_λ chosen to keep the density constant. By using the Feynman-Hellmann theorem one can then show (see e.g. sec. 7.2 of Ref. [9]) that the xc energy *including the kinetic energy contribution* is given exactly (excluding cases of λ-driven or Mott-type phase transitions) by

$$E_{xc} = \frac{1}{2}\int_0^1 d\lambda \int d^3r \int d^3r' \frac{e^2}{|\mathbf{r}-\mathbf{r}'|}[n_{2\lambda}(\mathbf{r},\mathbf{r}') - n(\mathbf{r})n(\mathbf{r}')] \qquad (16)$$

where $n_{2\lambda}$ is the pair density with reduced Coulomb interaction $\lambda e^2/|\mathbf{r}-\mathbf{r}'|$. One often writes the pair density as

$$n_{2\lambda}(\mathbf{r},\mathbf{r}') = n(\mathbf{r}')(n(\mathbf{r}) + n_{xc\lambda}(\mathbf{r}|\mathbf{r}')) \tag{17}$$

where $n_{xc\lambda}$ is known as the exchange-correlation hole density: it represents the "hole" in the electron density at $\mathbf{r}$ due to the presence of an electron at $\mathbf{r}'$. The xc energy can then be written

$$E_{xc} = \frac{1}{2}\int d^3r \int d^3r' \frac{e^2}{|\mathbf{r}-\mathbf{r}'|} n(\mathbf{r}')\tilde{n}_{xc}(\mathbf{r}|\mathbf{r}') \tag{18}$$

where $\tilde{n}_{xc} = \int_0^1 n_{xc\lambda} d\lambda$ is the coupling-strength-averaged hole. Thus the xc energy can be obtained formally by integrating the coulomb energy of each electron with its hole. Eq. (16), or equivalently (18), is termed the "Adiabatic Connection Formula". Alternative approaches to adiabatic connection are given in the introductory chapter by Burke, in the chapter by Burke, Perdew and Ernzerhof, and in the chapter on the PW91 functional by Burke, Perdew and Wang.

Partial Reasons for Success of LDA in Highly Inhomogeneous Systems

The original paper of Kohn and Sham [8] showed that the LDA, based as it is on uniform-gas data, is a good approximation for systems which, locally, differ only slightly from a uniform gas in the sense that only *slow* variations of the density are allowed: $|\nabla n|/n_0 << k_F(n_0)$. In fact, the LDA has turned out to provide moderate accuracy even for atomic and molecular situations where this conditions is not met. Partial reasons for this unexpected success were advanced by Gunnarsson and Lundqvist [26]. They referred to the adiabatic conection formula (18) and studied both the exact exchange hole $n_x(\mathbf{r}|\mathbf{r}')$ and its LDA approximation $n_x^{LDA}(\mathbf{r}|\mathbf{r}')$ for atoms. The LDA hole is (incorrectly) symmetric about the point $\mathbf{r}=\mathbf{r}'$ and was found to be a poor approximation to the exact hole point-by point: similar behaviour is known to apply to the full hole $n_{xc}^{LDA}(\mathbf{r}|\mathbf{r}')$. Because the Coulomb interaction is isotropic, however, the xc energy E_{xc} is sensitive only to the angularly averaged hole, and much of the error of the LDA hole turns out to disappear upon angular averaging. Secondly, the exact hole satisfies the important *hole normalization conditions*

$$\int n_{xc}(\mathbf{r}|\mathbf{r}')d^3r = -1 \ \forall \ \mathbf{r}', \tag{19}$$

$$\int n_x(\mathbf{r}|\mathbf{r}')d^3r = -1 \ \forall \ \mathbf{r}', \tag{20}$$

$$\int n_c(\mathbf{r}|\mathbf{r}')d^3r = 0 \ \forall \ \mathbf{r}'. \tag{21}$$

These conditions correspond to the fact that, if an electron has been found at $\mathbf{r}'$, then there is one less electron left to find elsewhere (i.e. by a search (integral) over all $\mathbf{r}$). These conditions put a major constraint on the overall size of the hole. The LDA hole satisfies these conditions, which is a major reason for its success.

Furthermore the LDA seems to do better when *system-averaged* than it does point-by-point.

It has also been found [27,28] that the value of the LDA hole at the origin $\mathbf{r}=\mathbf{r}'$ is remarkably accurate in a wide variety of systems.

While these observations partially explain the success of the LDA outside its expected region of validity, an even more important use of such considerations lies in the construction of functionals beyond the LDA: see the following paragraphs.

Gradient and Generalized Gradient Approximations

Since the LDA formula for E_{xc} is formally justified for the case of small density gradients, it seemed natural to seek gradient corrections to E_{xc}^{LDA}, the lowest of which has the form

$$\Delta E_{xc} = \sum_{ss'} \int F_{ss'}(n_\uparrow(\mathbf{r}),n_\downarrow(\mathbf{r}))\nabla n_s(\mathbf{r}) \cdot \nabla n_{s'}(\mathbf{r})d^3r. \tag{22}$$

To pin down $F_{ss'}(n_\uparrow(\mathbf{r}),n_\downarrow(\mathbf{r}))$, microscopic (diagrammatic perturbation) xc energy calculations were performed for an almost-uniform gas with a weak sinusoidal density perturbation. This was by no means easy, but Geldart and Rasolt provided the principal results in 1986 [29]. The resulting xc energy formula is sometimes termed the Gradient Expansion Approximation (GEA). Disappointingly, the GEA does not predict xc energies better than the LDA for strongly-varying systems such as atoms and molecules. Langreth and co-workers [30] explained this poor result in terms of the small-q behavior of the structure factor $S(\mathbf{q})$, a space transform of the pair distribution, and suggested modified functionals to address the problem. A more intuitive explanation of the problem was provided by Perdew and co-workers [31,32], who noted that the GEA xc hole is an improvement over the LDA hole only at short separations: at large separations the GEA hole is poorly damped and oscillatory, its exchange part is often of the wrong sign, and most importantly it violates the xc hole normalization conditions (19) - (21). Perdew *et al* gave a real-space cutoff procedure which remedied these problems for finite systems, generating a short-ranged hole whose angular and system average was much closer to the true hole. The xc energy functional resulting from substitution of this hole into the ACF (Eq. (18)) is of form

$$E_{xc}[n_\uparrow, n_\downarrow] = \int f(n_\uparrow(\mathbf{r}), n_\downarrow(\mathbf{r}), \nabla n_\uparrow(\mathbf{r}), \nabla n_\downarrow(\mathbf{r})) \tag{23}$$

which generalizes (22), being no longer quadratic in gradients. The term "Generalized Gradient Approximation" has been coined [33] for functionals of the form (23).

Functionals of this GGA form have revolutionized quantum chemistry, with the best ones working systematically better than LDA and, in some calculations, approaching the accuracy of traditional quantum chemical (e.g. Configuration Interaction) methods, at a fraction of the cost [34]. Some of the better-known GGAs are described in the next three paragraphs.

The Perdew-Wang 1991 (PW91) GGA functional has been in use for some years and incorporates no free parameters, being entirely determined from uniform-gas properties and exact constraints. It is described partially in [35] and elsewhere, but the first complete derivation of this important functional is given by Burke, Perdew and Wang in the present volume. This functional has been simplified and further refined to give the Perdew-Burke-Ernzerhof (PBE) GGA functional [36].

Becke derived an exchange [37,38] functional ("Becke 88, B88") incorporating the known behavior of the exchange hole at large distances outside a finite system. He also introduced [39] a correlation GGA fitted to atomic data.

By approximate evaluation of the potential energy from a modified Hartree-Fock many-body wavefunction, followed by some fitting to atomic data, Colle and Salvetti [40,41] introduced correlation energy expressions depending on one-electron orbitals. (This class of functional will be discussed further in the chapter "Driving Out the Self-Interaction Error" by Perdew and Ernzerhof.) Lee, Yang and Parr [42] further approximated the Colle-Salvetti functional, obtaining the correlation energy as an explicit functional of the density and its gradient and Laplacian, now generally known as the "LYP" functional. Combinations of one of Becke's exchange functionals and the LYP correlation functional, BLYP for short, have been perhaps the most widely-used class of GGAs in the recent chemical literature.

In the chapter by R. O. Jones, the performance of the LSD and some GGAs is critically examined for some metal and carbon clusters of up to 24 atoms, systems too big for the use of the traditional quantum chemical methods.

Hybrids of GGAs and Exact Exchange - the Current State-of-the-Art in Groundstate DFT Energy Calculations for Chemistry

If one can perform exchange integrals accurately then one has obtained the $\lambda = 0$ value of the integrand in the coupling-constant integration leading to the xc energy (see Eq. (16)). On the other hand, for nonzero λ one has to deal with correlation. Becke [43] noticed that a linear combination of exact (nonlocal) exchange energy and GGA xc energy, $\tilde{E}_{xc} = \alpha E_x^{exact} + (1-\alpha)E_{xc}^{GGA}$, fitted atomic and molecular data better than either exact exchange or pure GGA calculations. This approach is extremely promising from an empirical point of view, and theoretical reasons for its success, based on an examination of the adiabatic-coupling integral, are analyzed in the chapter by Burke, Perdew and Ernzerhof in Part II of this book. Some further analysis of these arguments, based on microscopic calculations, is given in the chapter by Savin, Colonna and Teuler. One may hope that these analyses will lead to even more accurate hybrid energy formulae in future.

Situations where GGAs will not necessarily give a good description

The dispersion (long-ranged van der Waals) interaction between distant sections of neutral condensed matter arises from long-ranged correlated electronic density fluctuations. They correspond to a very long-ranged part of the correlation hole which differs from that in the uniform electron gas. One cannot expect local (LDA/LSD) or quasilocal (GGA) theories to give these phenomena accurately. The chapter by Andersson *et al* and that by Dobson, Dinte and Wang both discuss functionals which do contain the dispersion energy and are (very nonlocal) functionals of the groundstate density. One would wish to retain the enormous progress made in developing GGAs. In this context there is a prospect of using vdW functionals to provide long-ranged corrections to GGA. Recent work by Kohn and Meir, and older work by Kohn and Hanke, not described here [44], contains the promise of allowing such a synthesis.

Another situation not necessarily well described by GGAs is the behavior of electrons in edge regions where the wavefunctions change from propagating to evanescent. The physics, including xc physics, in these region is presumably different from that in the interior of metallic systems with similar density gradients but propagating wavefunctions, on which the GEA and hence the GGA prescription is based. In the chapter by Kohn a new standard reference system, the Airy gas, is introduced to incorporate the physics of "edge" electrons. This is substantially different from the uniform electron gas but, like it, can be investigated by (e.g.) Monte Carlo methods, to produce new approximations to handle edge regions in general. Such considerations are in fact relevant for improvement of the van der Waals functionals described in the chapters by Dobson, Dinte and Wang and by Andersson *et al.*

A more general case where LDA/GGA methods have difficulty is that of a hole (or portion of the hole) centered far from the electron causing the hole. The two classes of situation just described fall into this category. A further example is the "left-right" correlation in stretched H_2^+.

Scaling vs. Coupling-Strength Integration and Perturbation Theory

For some years now, Levy and co-workers [45,46] have systematically explored the properties of the exact exchange and correlation functionals under scaling transformations of the space coordinate $\mathbf{r}$. An important tool in this task is the Levy constrained-search approach (see Eq. (4) above) which provides formal but rather explicit expressions for various components of the DFT energy functionals. As a result, a large number of new conditions have been derived, which place quite severe constraints on the behaviour of any approximate functionals. These constraints have been useful, for example, in the generation of new functionals described in later chapters of this book: Burke, Perdew and Wang (PW91 functional) and Wilson and Ivanov (Wilson-Levy functional). A list of many of these constraints is given in the latter chapter. Some of these constraints can be regarded as generalizations of the known physics of particular systems, but in other cases they are entirely new.

As an example of the former category of constraint, and as a qualitative introduction to this subtle field from a less formal viewpoint, consider the case of "isoelectronic series" - sequences of ions in which the number of electrons N is held fixed while the nuclear charge Z is increased. Examples are the "He series" He, Li^+, Be^{2+}, ... and the "Be series" Be, B^+, C^{2+}, ... As Z increases, dimensional arguments suggest that the radius R of the electron cloud decreases as Z^{-1}, the electron kinetic energy T increases as Z^2, the electron-nuclear potential energy V_{eN} goes as $-Z^2$ and the electron-electron potential energy (the Hartree part) goes as Z. One therefore expects the one-body part of the hamiltonian to be dominant, so that the smaller correlation energy is amenable to a perturbative treatment. For series such as the He series where the groundstate does *not* become degenerate for large Z, perturbation theory shows E_c approaches a constant as $Z \to \infty$. For cases such as the Be series where the single-particle hamiltonian has a degenerate spectrum as $Z \to \infty$, E_c diverges linearly with Z. By contrast, the local density approximation gives $E_c^{LDA} \sim \ln Z$. The physical reason for this discrepancy is that the density increases as Z increases, causing the LDA to sample the correlation energy of very dense ($r_s \to 0$) electron gases where the continuous energy level spacing and long-ranged Coulomb interaction conspire to create a $\ln r_s$ divergence of ϵ. By contrast, in the exact perturbative treatment of the isolelectronic series, the energy gap either remains discrete (e.g. in the the nondegenerate He series) or narrows rapidly with Z (e.g. in the degenerate Li series). (In making these statements we note that, from a perturbation theory point of view, the spacing of levels on the energy denominator is a major determiner of the correlation energy, with small level spacing corresponding to a strong perturbation, i.e. a large correlation energy).

From a general density functional point of view, all of this behavior (and misbehavior of the LDA) is predictable in principle from the sequence of ever higher, more compressed and localized densities encountered as the nuclear charge Z increases. The scaling approach of Levy and coworkers systematizes this sort of behavior by studying the properties of the exact density functional under transformations of coordinates. The type of transformation relevant to the above examples is the *uniform scaling* transformation: given the normalized density $n(\mathbf{r})$ of a finite N-particle system, one creates a new density

$$n_\mu(\mathbf{r}) = \mu^3 n(\mu\mathbf{r}).$$

The prefactor μ^3 ensures that the new density $n_\mu(\mathbf{r})$ is still a normalized density for N

particles. If $\mu > 1$, the density is compressed by this transformation. For example, the new density at the origin is $n_\mu(\mathbf{0}) = \mu^3 n(\mathbf{0})$, higher than before. Levy proved that the exact correlation energy $E_c[n_\mu]$ remains bounded for nondegenerate systems as $\mu \to \infty$. The LDA fails this most basic scaling relation.

In the nondegenerate isoelectronic series discussed above, the correlation energy becomes smaller (relative to the other energies) as the density is compressed (i.e. scaled towards $\mu \to \infty$). This corresponds in loose terms to a weaker electron-electron interaction (i.e. to decreased coupling-strength parameter$\lambda \to 0$ in the adiabatic connection formula (16)). Further, as both of these phenomena set in, perturbation theory becomes applicable to this class of system. The general scaling theory of Levy *et al* shows that this is no coincidence: there is indeed a general and deep connection between coordinate scaling, the coupling constant λ in the adiabatic connection formula, and perturbation theory. In fact, Görling and Levy [47] have developed a special form of perturbation theory specifically adapted to density functional theory. In view of the importance of the adiabatic connection formula in the justification and derivation of density functionals, this approach is currently receiving much attention: see for example the chapters in the present book by Levy, Ivanov and Görling, and by Savin, Colonna and Teuler.

Exact KS Potentials

The Kohn-Sham version of density functional theory involves an exchange- correlation energy $E_{xc}[n]$ and potential $v_{xc}[n]$ which are nonlocal functionals of the groundstate density $n(\mathbf{r})$. The accuracy of the *local density* and *generalized gradient* approximations for a given physical quantity, such as these, within the KS formalism must not be confused with the accuracy or appropriateness of the KS formalism itself.

It has been realized recently that a knowledge of the exact (or highly accurate) groundstate density for some system (e.g. the Hylleraas density of He or the best CI density of a more complicated system) permits, *in practice and not merely in principle*, an essentially exact determination of the KS potential $v_{eff}(\mathbf{r})$ and hence of the KS xc potential $v_{xc}(\mathbf{r})$. The important point is that the passage from $v_{eff}(\mathbf{r})$ to $n(\mathbf{r})$ (see (5) and (8)) involves only a collection of one-body problems and not an interacting many-body problem. The inversion of this problem is also not a true many-body problem, and can be obtained with any desired accuracy by various methods. These include direct variation of $v_{eff}(\mathbf{r})$, and certain iterative procedures: see the chapter by Morrison and Parr in the present book for a bibliography. The results of such calculations show that, even though the LDA gives quite good values of the groundstate energy and density, it can give an *extremely* poor approximation to the exact KS v_{xc} and especially to the KS correlation potential $v_c(\mathbf{r})$. It was already well-known that the asymptotic behavior of v_{xc}^{LDA} in the tunneling tails outside finite systems is qualitatively wrong, because of the phenomenon of unwanted orbital self-interaction: see the chapter "Driving out the self-interaction error" by Perdew and Ernzerhof in the Introductory section of this book. However the LDA prediction is much worse than even this would suggest: for some small systems, $v_c^{LDA}(\mathbf{r})$ was found to have the wrong sign for all $\mathbf{r}$.

Once the numerically exact KS potential v_{eff} is known, one can obtain the numerically exact KS eigenvalues $\{\epsilon_k\}$ from (8). Of course, under the LDA one might expect inaccuracies, but as explained earlier, in general one cannot expect differences of even the *exact* KS eigenvalues to yield exact transition or excitation energies of the many-body system. Only the highest occupied one, $\epsilon_{\max}$, has a guaranteed physical meaning as the chemical potential (ionization potential, electron removal energy: see p. 11 of [7]). It was therefore surprising to many workers that recent calculations of differences of "exact" KS energy eigenvalues yielded very good values for a series of excitation energies of small atoms. This work is briefly described in the chapter by Umrigar *et al.* It seems that the energies in question involve transitions from the highest occupied KS orbital to substantially higher-lying orbitals. A possible explanation lies in the fact that the asymptotic $r \to \infty$ decay of the exact KS potential $v_{eff}(\mathbf{r})$ agrees with that pertaining to a well-defined "quasiparticle amplitude" from many-body theory, up to several terms in the expansion in powers of $1/r$. Thus the energy to add an electron into a high-lying excited state, which samples mainly the outer part of the potential, will be rather accurately given by the appropriate KS ϵ_k. Also, of course, the energy to remove an electron from the highest occupied state, $\epsilon_{\max}$, is correct, so the overall excitation energy will be approximately correct in the present case. This does suggest, however, that low-lying excitations from the highest occupied state will be less accurately described by the exact KS eigenvalue difference. It further suggests that excitations from deeper-lying KS orbitals (corresponding to excitations leaving an excited core) may not be close to any of the many-particle excitation energies, when calculated by the $\epsilon_i^{KS} - \epsilon_j^{KS}$ prescription. There are many issues to be explored here.

Having the exact groundstate density $n(\mathbf{r})$ leads, in practice, to the KS eigenvalues, but in practice it does not yield the exact groundstate energy: for that one would need, in effect, to know the exact exchange-correlation functional $E_{xc}[n]$. The chapter by Morrison and Parr addresses the issue of obtaining an accurate E_{xc} starting from the "exact" $n(\mathbf{r})$, by means involving a functional Taylor series.

Dynamic Phenomena

Even though the groundstate density in principle determines all excited states and hence many dynamic phenomena, the description of time-varying phenomena is not easily incorporated into the groundstate density functional schemes described so far. Of course, to the extent that certain energy differences between the groundstate and many-body excited states can be accurately found within the Kohn-Sham scheme as described in the preceding paragraph, one has the ability to predict the corresponding resonance frequencies under time-varying exciting fields. A much more general way to approach these matters, however, is time-dependent electron density functional theory, including time-dependent Kohn-Sham theory, as described in a separate introductory chapter by Dobson in this book. This field is also the subject of several invited chapters: that by Vignale and Kohn; that by Petersilka, Gossmann and Gross; that by Zaremba and Tso; and that by Rajagopal on effective action.

In many situations the motion of the ions, as well as that of the electrons, is an issue. In the majority of cases the greater mass of the ions permits one to use the Born-Oppenheimer approximation in which the groundstate electronic energy is determined by KS theory at each ionic configuration (here denoted $\mathbf{R}$) with the ions treated as fixed. The resulting electronic groundstate energy $E(\mathbf{R})$ then serves as an effective potential for classical ion dynamics. There are, however, schemes in which the electrons and ions are permitted to move simultaneously. The chapter by Jones includes molecular dynamics (MD) along with electronic density functional theory: this is used to find the minimum-energy geometries, a task which becomes increasingly difficult to achieve by direct search as the size of the system increases, because of the proliferation of local energy minima.

The chapter on "A functional theory of interacting local spins, spin polarized electrons, and ions: half metallic magnets" by Rajagopal, treats another system involving ionic motions over and above electronic degrees of freedom.

Relativistic DFT

There are a few areas of quantum chemistry and condensed matter physics where special relativity is needed. For example, spin-orbit coupling can be regarded as a relativistic phenomenon. The electrons in inner orbitals of heavy atoms move at relativistic speeds. In the case of the Gold atom, for example, the transition from a nonrelativistic description to a relativistic one causes very major changes to many predicted properties including some not normally considered to be associated with the inner shells. The chapter by Engel, Keller and Dreizler discusses these issues, in particular the relativistic description of exchange and correlation, which is shown to combine non-trivially with the more usual relativistic treatment of the kinetic energy.

Miscellaneous Applications

Many of the ideas discussed above can be applied outside the traditional areas of quantum chemistry and electronic structure of solids. Also included in the present volume are

- Some ideas for a density functional treatment of "super-phenomena" (e.g. superconductivity and superfluidity) by considering functionals of both the usual density and one or more anomalous "densities". See the chapter by M. P. Das.
- Ideas about scattering theory based on some of the Workshop discussions on van der Waals interactions: see the chapter by R. K. Nesbet.

REFERENCES

1. L. H. Thomas, Proc. Cambridge Phil. Soc. **23**, 542 (1927).
2. E. Fermi, Z. Phys. **48**, 73 (1928).
3. P. A. M. Dirac, Proc. Cambridge Phil. Soc. **26**, 376 (1930).
4. J. C. Slater, *The Self-Consistent Field for Molecules and Solids* (McGraw-Hill, New York, 1974), Vol. IV.
5. P. Hohenberg and W. Kohn, Phys. Rev. **136**, B864 (1964).
6. M. Levy, Proc. Natl. Acad. Sci. USA **76**, 6062 (1979).
7. R. G. Parr and W. Yang, *Density Functional Theory of Atoms and Molecules* (Clarendon Press, Oxford, 1989).
8. W. Kohn and L. J. Sham, Phys. Rev. **140**, A1133 (1965).
9. R. M. Dreizler and E. K. U. Gross, *Density Functional Theory* (Springer Verlag, Berlin, 1990).
10. A. S. Davydov, *Quantum Mechanics* (Pergamon Press, Oxford, 1965).
11. G. D. Mahan, *Many-Particle Physics* (Plenum, New York, 1981).
12. E. Wigner, Phys. Rev. **46**, 1002 (1934).
13. D. M. Ceperley and B. J. Alder, Phys. Rev. Lett. **45**, 566 (1980).
14. S. Vosko, Wilk, and Nusair, Can. J. Phys. **58**, 1200 (1980).
15. R. O. Jones and O. Gunnarsson, Rev. Mod. Phys. **61**, 689 (1989).
16. U. V. Barth and L. Hedin, J. Phys. C **5**, 1629 (1972).
17. M. M. Pant and A. K. Rajagopal, Sol. State Commun. **10**, 1157 (1972).
18. J. F. Janak, Phys. Rev. B **18**, 7165 (1978): see also J. Callaway and N. H. March, Solid State Physics **38**, 135 (1984), especially p. 169-170.
19. J. P. Perdew, R. G. Parr, M. Levy, and J. L. Balduz, Phys. Rev. Lett. **49**, 1691 (1982).
20. J. P. Perdew and M. Levy, Phys. Rev. Lett. **51**, 1884 (1983).
21. L. J. Sham and M. Schluter, Phys. Rev. Lett. **51**, 1888 (1983).
22. W. Kohn, Phys. Rev. B **33**, 4331 (1986).
23. M. S. Hybertson and S. G. Louie, Phys. Rev. B **34**, 5390 (1986).
24. G. Vignale and M. Rasolt, Phys. Rev. B **37**, 10658 (1988).
25. D. C. Langreth and J. P. Perdew, Sol. State Commun. **17**, 1425 (1975).
26. O. Gunnarsson and B. I. Lundqvist, Phys. Rev. B **13**, 4274 (1976).

27. K. Burke, J. P. Perdew, and D. C. Langreth, Phys. Rev. Lett. **73**, 1283 (1994).
28. J. P. Perdew, M. Ernzerhof, K. Burke, and A. Savin, Int. J. Quantum Chem. **61**, 197 (1997).
29. M. Rasolt and D. J. W. Geldart, Phys. Rev. B **34**, 1325 (1986).
30. D. Langreth and J. Perdew, Solid State Commun. **31**, 567 (1979).
31. J. Perdew, Phys. Rev. Lett. **55**, 1665 (1985).
32. J. P. Perdew, Phys. Rev. Lett. **55**, 2370 (E) (1995).
33. J. Perdew and Y. Wang, Phys. Rev. B **33**, 8800 (1986).
34. J. P. Perdew *et al.* Phys. Rev. B **46**, 6671 (1992).
35. J. P. Perdew, in *Electronic Structure of Solids '91*, edited by P. Ziesche and H. Eschrig (Akademie Verlag, Berlin, 1991).
36. J. P. Perdew, K. Burke, and M. Ernzerhof, Phys. Rev. Lett. **77**, 3865 (1996).
37. A. D. Becke, Phys. Rev. A **38**, 3098 (1988).
38. A. D. Becke and M. R. Roussel, Phys. Rev. A **39**, 3761 (1989).
39. A. D. Becke, J. Chem. Phys. **88**, 1053 (1987).
40. R. Colle and O. Salvetti, Theor. Chim. Acta **37**, 329 (1975).
41. R. Colle and O. Salvetti, Theor. Chim. Acta **53**, 55 (1979).
42. C. Lee, W. Yang, and R. G. Parr, Phys. Rev. B **37**, 785 (1988).
43. A. D. Becke, J. Chem. Phys. **98**, 5648 (1993).
44. W. Kohn and W. Hanke, private communication.
45. M. Levy, Adv. Quantum Chem. **21**, 69 (1990).
46. M. Levy and J. Perdew, Int. J. Quantum. Chem. **49**, 539 (1994).
47. A. Görling and M. Levy, Phys. Rev. B **47**, 13105 (1993).

DIGGING INTO THE EXCHANGE-CORRELATION ENERGY: THE EXCHANGE-CORRELATION HOLE

Kieron Burke

Department of Chemistry, Rutgers University-Camden
315 Penn Street, Camden, NJ 08102

INTRODUCTION

In this chapter, I summarize much of what is known about the exchange-correlation hole in density functional theory. The aim of this summary is to introduce the non-expert to key concepts in electronic density functional theory. For a much broader introduction, see the book by Parr and Yang [1]. For more detail, see the book by Dreizler and Gross [2], and other chapters in the current volume. Most of the references in this chapter are either classic papers in which a fundamental result has been established (and so are well worth reading), or are references to illustrative figures in my own work. I conclude with some simple exercises. Solutions are provided to the odd-numbered problems, while true experts will make short work of the even-numbered ones.

DEFINITIONS AND PHYSICAL INTERPRETATION

Consider the ground state of N electrons, interacting with each other via the Coulomb potential, in some external potential. The Hamiltonian is, in atomic units ($e^2 = \hbar = m = 1$):

$$\hat{H} = \hat{T} + \hat{V}_{ee} + \hat{V}_{ext} = -\frac{1}{2}\sum_{i=1}^{N} \nabla_i^2 + \frac{1}{2}\sum_{i \neq j}^{N} \frac{1}{|\mathbf{r}_i - \mathbf{r}_j|} + \sum_{i=1}^{N} v_{ext}(\mathbf{r}_i). \tag{1}$$

where $\hat{T}$ is the kinetic energy operator, $\hat{V}_{ee}$ is the (two-body) interelectronic Coulomb repulsion, and $\hat{V}_{ext}$ is the external (one-body) potential. The exact density of the system is

$$n(\mathbf{r}) = N \int d^3r_2 \dots d^3r_N \, |\Psi(\mathbf{r}, \mathbf{r}_2, \dots, \mathbf{r}_N)|^2, \tag{2}$$

where Ψ is the ground-state wavefunction, normalized to unity. Thus $n(\mathbf{r})d^3r$ is the probability of finding an electron in a neighborhood d^3r of $\mathbf{r}$. By construction the density is everywhere non-negative, and normalized to N. Because the external potential is a one-body operator, we can then write

$$\langle\Psi|\hat{V}_{ext}|\Psi\rangle = \int d^3r\ n(\mathbf{r})\ v_{ext}(\mathbf{r}). \tag{3}$$

Similarly, the *pair* density is defined to be

$$P(\mathbf{r},\mathbf{r}') = N(N-1)\int d^3r_3 \ldots d^3r_N\ |\Psi(\mathbf{r},\mathbf{r}',\mathbf{r}_3\ldots,\mathbf{r}_N)|^2, \tag{4}$$

and $P(\mathbf{r},\mathbf{r}')d^3rd^3r'$ is the probability of finding an electron in d^3r at $\mathbf{r}$ and a second electron in d^3r' around $\mathbf{r}'$. The pair density is symmetric, non-negative, and is normalized to the number of indistinct pairs of electrons in the system:

$$\int d^3r \int d^3r'\ P(\mathbf{r},\mathbf{r}') = N\ (N-1). \tag{5}$$

Because the interaction between the electrons is two-body, the pair density determines the value of the electron-electron repulsion:

$$\langle\Psi|\hat{V}_{ee}|\Psi\rangle = \frac{1}{2}\int d^3r \int d^3r'\ \frac{P(\mathbf{r},\mathbf{r}')}{|\mathbf{r}-\mathbf{r}'|}. \tag{6}$$

The density can be extracted by integration of the pair density:

$$n(\mathbf{r}) = \frac{1}{N-1}\int d^3r'\ P(\mathbf{r},\mathbf{r}'), \tag{7}$$

which determines the expectation value of the external potential. Thus the value of two of the three operators in Eq. (1) are determined by the pair density alone.

ADIABATIC CONNECTION

To relate the pair density to the exchange-correlation hole, we need to vary the strength of the Coulomb repulsion. The constrained search definition [3,4] of the universal functional of Hohenberg and Kohn [5] is

$$F[n] = \min_{\Psi\to n}\langle\Psi|\hat{T}+\hat{V}_{\rm ee}|\Psi\rangle, \tag{8}$$

where the minimization is over all wavefunctions yielding the given density $n(\mathbf{r})$. This may be generalized to:

$$F_\lambda[n] = \min_{\Psi_\lambda\to n}\langle\Psi_\lambda|\hat{T}+\lambda\hat{V}_{\rm ee}|\Psi_\lambda\rangle, \tag{9}$$

In this minimization, the density is held constant at its physical value by making the external potential λ-dependent, $v_{ext,\lambda}(\mathbf{r})$. The coupling constant λ may be chosen to have any non-negative value. For $\lambda = 1$, we recover the physical wavefunction and energies, while for $\lambda = 0$ we get the non-interacting Kohn-Sham wavefunction, in which just the kinetic energy is minimized. We can then define the exchange-correlation energy as a function of λ via [6]

$$E^\lambda_{\rm XC} = F_\lambda - T_{\rm S} - \lambda U, \tag{10}$$

where $U = \int d^3r\ \int d^3r'\ n(\mathbf{r})\ n(\mathbf{r}')/2|\mathbf{r}-\mathbf{r}'|$ is the Hartree energy. (Note that our definition of $E^\lambda_{\rm XC}$ differs from that of Ref. [6] by a factor of λ.) Then

$$E^{\lambda'}_{\rm XC} = \int_0^{\lambda'} d\lambda\ \frac{dE^\lambda_{\rm XC}}{d\lambda} = \int_0^{\lambda'} d\lambda\ \Big(\langle\Psi_\lambda|\hat{V}_{\rm ee}|\Psi_\lambda\rangle - U\Big), \tag{11}$$

since the λ-dependence of the wavefunction in Eq. (9) does not contribute to the first derivative, by virtue of the variational principle.

COORDINATE SCALING

There is a conceptually simple way to generalize $E_{XC}[n]$ to $E_{XC}^{\lambda}[n]$. From dimensional analysis (briefly restoring the fundamental constants),

$$E_{XC}[n] = \frac{e^2}{a_o} f[a_o^3 n(\mathbf{r}/a_o)] \tag{12}$$

where $f[\nu]$ is a dimensionless functional of a dimensionless density, and $a_o = \hbar^2/me^2$ is the Bohr radius. Now consider making the change $V_{ee} \to \lambda V_{ee}$. In Eq. (12), $e^2 \to \lambda e^2$, $a_o \to a_o/\lambda$, and so the exchange-correlation energy at coupling strength λ is

$$E_{XC}^{\lambda}[n] = \lambda^2 f[n_{1/\lambda}] = \lambda^2 E_{XC}[n_{1/\lambda}], \tag{13}$$

where $n_\gamma(\mathbf{r}) = \gamma^3 n(\gamma \mathbf{r})$. Eq. (13) is the fundamental scaling relation for the exchange-correlation energy [6], showing that a change in coupling constant is simply related to a uniform scaling of the density. In practice, this is easy to implement for an approximate density functional, but difficult in an exact calculation, as the interacting problem must be solved for each $v_{ext,\lambda}(\mathbf{r})$.

Levy and Perdew also showed that

$$E_X[n_\gamma] = \gamma E_X[n], \;\; E_C[n_\gamma] > \gamma E_C[n] \;\; (\gamma > 1), \tag{14}$$

while Levy [7] has shown that $E_C[n_\gamma]$ tends to a negative constant as $\gamma \to \infty$. E_X is the leading contribution in perturbation theory in powers of λ to E_{XC} (see problem 5), which, from Eq. (13), can be thought of either as small λ or as the limiting behavior as the density is uniformly squeezed toward the high density limit. Since $\Psi_{\lambda=0}$ is just the non-interacting Kohn-Sham wavefunction, which we assume to be a single Slater determinant throughout this article, we have the Fock expression

$$E_X = -\frac{1}{2} \sum_{i \neq j} \int d^3r \int d^3r' \, \frac{\phi_i^*(\mathbf{r})\, \phi_j(\mathbf{r})\, \phi_j^*(\mathbf{r}')\, \phi_i(\mathbf{r}')}{|\mathbf{r} - \mathbf{r}'|}, \tag{15}$$

where the ϕ_i are the Kohn-Sham orbitals.

EXCHANGE-CORRELATION HOLE

To construct the exchange-correlation energy from the pair density, we generalize Eq. (4) to

$$P_\lambda(\mathbf{r}, \mathbf{r}') = N(N-1) \int d^3r_3 \ldots d^3r_N \, |\Psi_\lambda(\mathbf{r}, \mathbf{r}', \mathbf{r}_3 \ldots, \mathbf{r}_N)|^2, \tag{16}$$

where $\lambda = 1$ corresponds to the physical value, discussed in the definition section. If the distribution of electrons were totally uncorrelated, i.e., if the probability of finding an electron at one location were independent of finding another elsewhere, then

$$P_\lambda^{\text{uncorr.}}(\mathbf{r}, \mathbf{r}') = n(\mathbf{r})\, n(\mathbf{r}'). \tag{17}$$

This cannot be true everywhere, as that would violate the normalization condition, Eq. (5). The probability distribution for the second electron is altered (often reduced) by the finding of the first electron, and we describe this effect by writing:

$$P_\lambda(\mathbf{r}, \mathbf{r} + \mathbf{u}) = n(\mathbf{r}) \left[n(\mathbf{r} + \mathbf{u}) + n_{XC,\lambda}(\mathbf{r}, \mathbf{r} + \mathbf{u}) \right] \tag{18}$$

where $n_{\text{XC},\lambda}(\mathbf{r}, \mathbf{r}+\mathbf{u})$ is the (often negative) λ-dependent exchange-correlation hole density at separation $\mathbf{u}$ surrounding an electron at $\mathbf{r}$. From the normalization condition Eq. (5), we find

$$\int d^3u \; n_{\text{XC},\lambda}(\mathbf{r}, \mathbf{r}+\mathbf{u}) = -1, \tag{19}$$

i.e., surrounding every point in space, there is a dearth of probability, integrating up to exactly one missing electron. Now, inserting Eq. (18) into Eqs. (6) and (11),

$$\frac{dE_{\text{XC}}^{\lambda}}{d\lambda} = \int \frac{d^3u}{2u} \int d^3r \; n(\mathbf{r}) \; n_{\text{XC},\lambda}(\mathbf{r}, \mathbf{r}+\mathbf{u}) \tag{20}$$

is the λ-dependent integrand in the adiabatic connection formula. Thus, if we define the λ-integrated hole as

$$n_{\text{XC}}(\mathbf{r}, \mathbf{r}+\mathbf{u}) = \int_0^1 d\lambda \; n_{\text{XC},\lambda}(\mathbf{r}, \mathbf{r}+\mathbf{u}), \tag{21}$$

then

$$E_{\text{XC}} = \int d^3u \frac{1}{2u} \int d^3r \; n(\mathbf{r}) \; n_{\text{XC}}(\mathbf{r}, \mathbf{r}+\mathbf{u}), \tag{22}$$

i.e., the Coulomb energy of the electron density with the exchange-correlation hole surrounding it [8]. Note that, in density functional theory, *the* exchange-correlation hole n_{XC} incorporates the coupling constant average of Eq. (21).

We illustrate many of the ideas in this chapter using the uniform electron gas hole, as parametrized by Perdew and Wang [9]. In Fig. 1, we plot this hole, divided

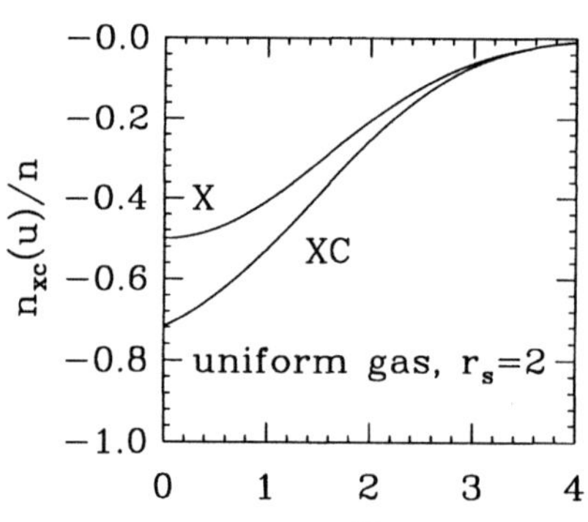

Figure 1. Exchange and exchange-correlation hole densities as a function of separation u for the spin-unpolarized uniform electron gas of $r_s = 2$.

by the density, as a function of separation for $r_s = 2$, where $r_s = (4\pi n/3)^{1/3}$ is the Wigner-Seitz radius, a useful measure of the mean separation between electrons. Note that, when multiplied by the phase factor $4\pi u^2$, this curve integrates to -1, satisfying the normalization condition, Eq. (19).

A third way of conveying the same information is the pair distribution function, defined as

$$g_{\lambda}(\mathbf{r}, \mathbf{r}') = \frac{P_{\lambda}(\mathbf{r}, \mathbf{r}')}{n(\mathbf{r})n(\mathbf{r}')}. \tag{23}$$

Thus g is symmetric, everywhere positive, and normalized so that

$$\int d^3r'\ n(\mathbf{r}')\big(g_\lambda(\mathbf{r},\mathbf{r}') - 1\big) = -1. \tag{24}$$

Because the density is uniform in the case of Fig. 1, this is just a plot of $\bar{g}(u) - 1$, where $\bar{g} = \int_0^1 g_\lambda$.

The pair density (at $\lambda = 1$) is often used in the quantum chemistry literature, because traditional *ab-initio* methods are based on wavefunctions. In that language, $P(\mathbf{r},\mathbf{r}')$ is the reduced diagonal second-order density matrix. The (λ-averaged) exchange-correlation hole is used in density functional theory, because the definition of the exchange-correlation energy includes a kinetic contribution. The pair distribution function is often used in many-body theory [10] and theory of classical liquids. Via the fluctuation-dissipation theorem, it can be related to the equal-time density-density response function (see the chapter "Time-dependent Density Functional Theory" by J. F. Dobson in the introductory section of the present volume). This response function, when frequency decomposed, is proportional to the structure factor probed in scattering experiments. Thus the same basic quantity appears in many guises in several fields.

EXACT CONDITIONS

From its definition in terms of the pair density, we can immediately write down several important conditions the hole satisfies. Since the pair density is symmetric under interchange of $\mathbf{r}$ and $\mathbf{r}'$, we find

$$n_{\text{XC},\lambda}(\mathbf{r},\mathbf{r}+\mathbf{u}) = n_{\text{XC},\lambda}(\mathbf{r}+\mathbf{u},\mathbf{r})\frac{n(\mathbf{r}+\mathbf{u})}{n(\mathbf{r})}. \tag{25}$$

Furthermore, since the pair density is never negative,

$$n_{\text{XC},\lambda}(\mathbf{r},\mathbf{r}+\mathbf{u}) \geq -n(\mathbf{r}+\mathbf{u}), \tag{26}$$

i.e., one cannot dig a hole in the density greater than the density itself. This is reflected by the fact that the curves in Fig. 1 never dip below -1.

The dimensional analysis leading to the scaling relation Eq. (13) can also be used to relate the coupling-constant dependence of the hole to its behavior under scaling:

$$n_{\text{XC},\lambda}([n];\mathbf{r},\mathbf{r}+\mathbf{u}) = \lambda^3\ n_{\text{XC},\lambda=1}([n_{1/\lambda}];\lambda\mathbf{r},\lambda(\mathbf{r}+\mathbf{u})) \tag{27}$$

The exchange (or Fermi) hole is the high-density or weak-coupling limit:

$$n_{\text{X}}(\mathbf{r},\mathbf{r}+\mathbf{u}) = n_{\text{XC},\lambda=0}(\mathbf{r},\mathbf{r}+\mathbf{u}) \tag{28}$$

This is the second curve in Fig. 1. The correlation (or Coulomb) hole is then $n_{\text{C},\lambda} = n_{\text{XC},\lambda} - n_{\text{X}}$. Since the exchange contribution satiates the normalization sum rule

$$\int d^3u\ n_{\text{X}}(\mathbf{r},\mathbf{r}+\mathbf{u}) = -1, \tag{29}$$

the correlation hole satisfies

$$\int d^3u\ n_{\text{C}}(\mathbf{r},\mathbf{r}+\mathbf{u}) = 0. \tag{30}$$

The exchange hole can be written exactly in terms of the one-particle density matrix. For the spin-unpolarized case, this is simply

$$n_X(\mathbf{r}, \mathbf{r}+\mathbf{u}) = -|n_1(\mathbf{r}, \mathbf{r}+\mathbf{u})|^2/(2n(\mathbf{r})), \tag{31}$$

where n_1 is the reduced one-matrix. This is never positive:

$$n_X(\mathbf{r}, \mathbf{r}+\mathbf{u}) \leq 0 \tag{32}$$

For the uniform gas, this can be written explicitly (see problem 3), and was used to construct the exchange curve in Fig. 1. Note that the correlation hole must contain some positive contributions, so that, e.g., for large u, $n_{XC}(\mathbf{r}, \mathbf{r}+\mathbf{u})$ is often positive. This occurs for the hole in Fig. 1, but beyond the range of u plotted there. Note also that this figure neglects the Friedel oscillations at large u.

Small separations

The on-top hole, i.e., the hole at zero separation ($u = 0$) from the electron, can be written exactly in special circumstances. The on-top pair density is just $P_{\lambda=0}(\mathbf{r}, \mathbf{r}) = 2n_\uparrow(\mathbf{r})n_\downarrow(\mathbf{r})$, so that the exchange hole is

$$n_X(\mathbf{r}, \mathbf{r}) = -(n_\uparrow^2(\mathbf{r}) + n_\downarrow^2(\mathbf{r}))/n(\mathbf{r}) \tag{33}$$

For a spin-unpolarized system, $n_X(\mathbf{r}, \mathbf{r}) = -n(\mathbf{r})/2$, just as shown for the exchange hole in Fig. 1. However, there are two other limits with the interaction turned on where we know the on-top hole. In the strong coupling (or low density) limit, and also in the fully spin-polarized case, the electrons completely avoid each other, so that $P_\lambda(\mathbf{r}, \mathbf{r}) = 0$, or

$$n_{XC,\lambda}(\mathbf{r}, \mathbf{r}) \to -n(\mathbf{r}) \ (\lambda \to \infty \text{ or } \zeta(\mathbf{r}) = 1) \tag{34}$$

where $\zeta = n_\uparrow - n_\downarrow$ is the relative spin polarization. One can see that turning on λ in the exchange-correlation hole of Fig. 1 has deepened it relative to $\lambda = 0$. In the low density ($r_s \to \infty$) limit or if $\zeta = 1$, the curve drops to -1 at $u = 0$.

The on-top hole determines more than just the value of the hole at $u = 0$. As $u \to 0$, the Coulomb repulsion between electrons dominates their behavior, leading to a cusp in the correlation hole density at $u = 0$. The strength of this cusp is given by a universal condition [11]

$$n'_{XC,\lambda}(\mathbf{r}, 0) = \lambda\big[n_{XC,\lambda}(\mathbf{r}, 0) + n(\mathbf{r})\big], \tag{35}$$

where $n_{XC,\lambda}(\mathbf{r}, u)$ is the spherical average of the hole at $\mathbf{r}$,

$$n_{XC,\lambda}(\mathbf{r}, u) = \int \frac{d\Omega_{\mathbf{u}}}{4\pi} n_{XC,\lambda}(\mathbf{r}, \mathbf{r}+\mathbf{u}), \tag{36}$$

and the prime in Eq. (35) denotes the derivative with respect to u. Thus the on-top hole sets both the scale of the exchange-correlation hole and its leading behavior as $u \to 0$. This is the origin of the linear behavior of the exchange-correlation curve of Fig. 1, whereas the exchange hole is quadratic in u at small u.

Large separations

We can also write down exact asymptotic expressions for the hole as $u \to \infty$. In any extended system, finding a second electron far from a first electron must be an independent event, so that $g \to 1$. In jellium, the exchange hole decays as $1/u^4$, but so too does the correlation hole, exactly cancelling the exchange hole at large u, yielding a $1/u^5$ behavior overall. For solids, there should also be a $1/u^3$ contribution which averages to zero over a single unit cell [12].

But consider Ψ when one coordinate is far from the nuclei in a Coulombic system. The wavefunction collapses to a product of the distant orbital times the wavefunction of the $N-1$ particle ion, assuming it's not degenerate (see problem 3), i.e., when one electron is far away, the remaining $N-1$ electrons collapse to their ground state, so that

$$\lim_{r\to\infty} P_\lambda(\mathbf{r},\mathbf{r}') = n(\mathbf{r})n_\lambda^{\rm ion}(\mathbf{r}'), \tag{37}$$

where $n_\lambda^{\rm ion}(\mathbf{r})$ is the λ-dependent density of the $N-1$ particle ground-state wavefunction of $v_{ext,\lambda}(\mathbf{r})$. The asymptotic form of the hole is then

$$n_{\rm XC,\lambda}(\mathbf{r},\mathbf{r}+\mathbf{u}) = n(\mathbf{r}+\mathbf{u})\ \left(n_\lambda^{\rm ion}(\mathbf{r})/n(\mathbf{r}) - 1\right)\ (u \to \infty). \tag{38}$$

Note that, contrary to expectation [13], this implies that g does not tend to 1 typically in a finite system as $u \to \infty$ (see Fig. 2 of Ref. [14]). There are even more interesting long-range correlations that occur when the ion's ground state is degenerate (see Figs. 8-10 of Ref. [15]).

APPROXIMATIONS

In this section, we study various approximations to the energy by considering them as approximations for the hole. From Eq. (22), it is clear that a good approximation for the 7-dimensional object $n_{\rm XC,\lambda}(\mathbf{r},\mathbf{r}+\mathbf{u})$ implies a good approximation for $E_{\rm XC}$, but the converse is *not* true. In fact, the underlying shape of the hole cannot be deduced from a given functional approximation for $E_{\rm XC}[n]$. We see below that there are many properties of this 7-dimensional object which standard approximations (e.g., LSD and GGA) do poorly, while still yielding 'good' approximations for $E_{\rm XC}$. The question of which properties of $n_{\rm XC,\lambda}(\mathbf{r},\mathbf{r}+\mathbf{u})$ should be well-approximated to yield a good approximation for $E_{\rm XC}$ is a subtle one.

Local spin density (LSD) approximation

LSD [16] is by far the most successful density functional approximation to date [13]:

$$E_{\rm XC}^{\rm LSD} = \int d^3r\ n(\mathbf{r})\ \epsilon_{\rm XC}^{\rm unif}(n_\uparrow(\mathbf{r}), n_\downarrow(\mathbf{r})), \tag{39}$$

where $\epsilon_{\rm XC}^{\rm unif}(n_\uparrow, n_\downarrow)$ is the accurately known exchange-correlation energy per particle of a uniform gas [17]. It has moderate accuracy for most systems of interest, making errors of order 5-10%. Its most remarkable feature is its reliability, making the same kinds of errors on every system it's applied to. It has been the mainstay of solid-state calculations for many years, because many properties of interest do not require such a drastic rearrangement of bonds as to make LSD inaccurate. On the other hand, it was long avoided by the quantum chemistry community, since reaction energies do require such rearrangements [18].

Eq. (39) suggests the interpretation that LSD works because somehow real systems are close enough to uniform. If that were the case, then the logical extension of LSD, the gradient expansion approximation (GEA) [16], which expands the functional in a Taylor series in gradients of the density [19]:

$$E_{\text{XC}}^{\text{GEA}}[n_\uparrow, n_\downarrow] = E_{\text{XC}}^{\text{LSD}}[n_\uparrow, n_\downarrow] + \sum_{\sigma,\sigma'} \int d^3r\; C_{\sigma\sigma'}\big(n_\uparrow(\mathbf{r}), n_\downarrow(\mathbf{r})\big) \frac{\nabla n_\sigma}{n_\sigma^{2/3}} \cdot \frac{\nabla n_{\sigma'}}{n_{\sigma'}^{2/3}}, \tag{40}$$

would be just as reliable, but more accurate, than LSD for real systems. However, experience has shown that GEA yields *worse* energies than LSD.

We can resolve this paradox by considering the LSD exchange-correlation hole. The LSD approximation to the exchange-correlation hole is

$$n_{\text{XC},\lambda}^{\text{LSD}}(\mathbf{r}, \mathbf{r}+\mathbf{u}) = n(\mathbf{r})\Big(g_\lambda^{\text{unif}}(n_\uparrow(\mathbf{r}), n_\downarrow(\mathbf{r}); u) - 1\Big) \tag{41}$$

where $g_\lambda^{\text{unif}}(n_\uparrow, n_\downarrow; u)$ is the pair distribution function of a uniform gas of spin densities $n_\uparrow$ and $n_\downarrow$. Because LSD approximates the hole by that of another system, it satisfies many exact constraints (see problem 9). Both its exchange and exchange-correlation contributions are correctly normalized, and the exchange hole is never negative. Furthermore, the on-top exchange hole is exact in LSD, the on-top exchange-correlation hole is very accurate [20], and LSD respects the cusp condition. Thus both the scale and the initial slope of the hole are determined by its on-top value, while its shape is determined by the normalization condition. Since the energy and normalization integrals differ simply by a power of u, the LSD energy is usually close to the exact value.

What about the conditions LSD fails to satisfy, such as symmetry and positivity of the pair density? It was shown long ago that in fact LSD does not reproduce *all* details of the hole correctly [21]. In particular, the uniform gas hole is spherical in $\mathbf{u}$, but the true hole may be highly non-spherical [22]. However, the spherically averaged hole usually does compare well with LSD, and the energy only depends on the spherical average. Thus it is most useful to compare only spherically averaged holes when studying LSD.

This argument can be taken a step further, and applied to the system-average in Eq. (22). For r near a nucleus, the Laplacian of the density becomes large, and LSD does not work so well. However, the phase space factor $4\pi r^2$ unweights such a region in the average (see Figs. 1-3 of Ref. [15]). Also, at large distances from finite systems, in the evanescent tail regions, LSD performs poorly, but now the density weighting ensures this makes little contribution. Similar arguments apply to the large u behavior in finite systems, which LSD misses entirely (see the discussion of exact behavior above). These failings do not show up prominently in E_{XC}. For the system-averaged hole, the symmetry condition becomes trivial, while the positivity condition is not very restrictive. Figures 4 and 5 of Ref. [15] show comparisons between exact system-averaged holes and LSD.

Generalized Gradient Approximations (GGA's)

The term generalized gradient approximation is used to denote any functional which uses both the density and its gradient [23-28]:

$$E_{\text{XC}}^{\text{GGA}}[n_\uparrow, n_\downarrow] = \int d^3r\; f\big(n_\uparrow(\mathbf{r}), n_\downarrow(\mathbf{r}), \nabla n_\uparrow, \nabla n_\downarrow\big), \tag{42}$$

We focus here on those GGA's derived using the exchange-correlation hole. The first modern GGA was that of Langreth and Mehl [29], who introduced the idea of truncating the gradient expansion for the hole. This was refined in a sequence of functionals by Perdew (PW86, PW91, PBE) [28]. The details of the real-space cutoff procedure on the hole, which restores the normalization and negativity conditions on the GEA hole, are given in great detail in the chapter by Burke, Perdew and Wang in the present volume. See also Ref. [30]. System-averaged exchange holes are shown in Figs. 9 and 10 of Ref. [31].

We summarize a few results from that chapter to put the GGA in context. The real-space construction restores those exact conditions which LSD satisfies, and a few more besides. It is designed to have the same qualitative properties as LSD, but be somewhat more accurate. The construction applies only to the system and spherically averaged holes, so that even arguments about contributions from specific values of $\mathbf{r}$ do not apply. The LSD on-top hole is built-in as is, while the asymptotic decay with u is different. The long-range decay is cut off by the real-space cutoff procedure, which improves the asymptotic behavior in finite systems, but does not make it exact.

CONCLUSIONS

As we progress in computer power, more sophisticated approximations to the exchange and correlation holes become practical. An early version of a direct construction of these holes is the weighted-density approximation (WDA), which has produced mixed results. The Colle-Salvetti orbital functional, from which the Lee-Yang-Parr functional (LYP [25]) was derived, is a more successful construction: see the chapter "Driving out the self-interaction error" in the Introctuctory section of the present volume. A promising way forward to even greater accuracy is to use the exact exchange hole, calculated within, e.g., the KLI approximation [32], as a starting point for modeling the exchange-correlation hole [33]. However, to retain the reliability of what we know (LSD), new approaches should incorporate at least all the known good features of that approximation, as well as some new ones, but not try to get every last detail correct. This philosophy has led to the successful construction of non-empirical GGA's, and may also lead to their successors.

ACKNOWLEDGMENTS

I am very grateful to John Perdew for three years of invaluable discussions, on which this article is based. I also thank Martin Plihal and Nicola Marzari for critical readings of the manuscript.

EXERCISES

1. Calculate explicitly the exchange and correlation holes in a one-electron system.

2. Repeat the previous problem for $N > 1$.

3. Write an explicit expression for the asymptotic behavior of the exchange hole in a finite system as $u \to \infty$.

4. Develop a functional which incorporates the exact asymptotic exchange-correlation hole.

5. Write down a plausible perturbation expansion for $F_\lambda[n]$ as $\lambda \to 0$.

6. Repeat previous problem for $\lambda \to \infty$.

7. Construct analytically the exact exchange hole of a uniform gas.

8. Redo previous problem for correlation (see Ref. [9] for an approximation).

9. Prove that the LSD hole satisfies the electron-electron cusp condition.

10. Construct a wavefunction with a cusp using a standard quantum chemical program, i.e., by expanding in single-particle states.

ANSWERS TO ODD-NUMBERED PROBLEMS

1. For a one-electron system, $P(\mathbf{r}, \mathbf{r}') = 0$ everywhere, as there is no chance of finding a second electron. Thus

$$n_{\mathrm{XC},\lambda}(\mathbf{r}, \mathbf{r}') = -n(\mathbf{r}'). \tag{43}$$

Since the density is held fixed in the coupling constant integration, $n_{\mathrm{X}}(\mathbf{r}, \mathbf{r}') = -n(\mathbf{r}')$ and $n_{\mathrm{C}}(\mathbf{r}, \mathbf{r}') = 0$. Note, however, that local and semi-local functionals which use the density and its gradient at $\mathbf{r}$ to construct the hole, have difficulty reproducing these simple results. (See Ref. [34] and the chapter "Driving out the self-interaction error", by Perdew and Ernzerhof in the present volume.)

3. Taking $\lambda = 0$ in Eq. (38), we find

$$n_{\mathrm{X}}(\mathbf{r}, \mathbf{r}+\mathbf{u}) = n(\mathbf{r}+\mathbf{u}) \Big(- |\phi_N(\mathbf{r})|^2 / n(\mathbf{r}) \Big) \quad (u \to \infty), \tag{44}$$

where $\phi_N(\mathbf{r})$ is the highest occupied Kohn-Sham orbital.

5. From the same dimensional arguments used to derive Eq. (13), we know $F^\lambda[n] = \lambda^2\ F[n_{1/\lambda}]$. Then, from its definition, we know $T_{\mathrm{S}}[n_\gamma] = \gamma^2 T_{\mathrm{S}}[n]$, we know how E_{X} scales from Eq. (14), and we can write down a plausible expansion for $E_{\mathrm{C}}[n_\gamma]$ as $\gamma \to \infty$ consistent with Eq. (14), yielding

$$F^\lambda[n] = T_{\mathrm{S}}[n] + \lambda\Big(U[n] + E_{\mathrm{X}}[n]\Big) + \lambda^2 E_{\mathrm{C}}^{(2)}[n] + \lambda^3 E_{\mathrm{C}}^{(3)}[n] + \ldots \tag{45}$$

This expansion is correct for finite systems [35,36,37], but logarithmic divergences occur in the uniform gas correlation energy [10].

7. For the uniform gas, the orbitals are simply plane waves, occupied up to the Fermi energy. Evaluation of V_{ee} on this wavefunction, after summing over the occupied states, yields [10] in the spin-unpolarized case

$$n_{\mathrm{X}}^{\mathrm{unif}}(u) = -\frac{9\ n}{2} \left[\frac{\sin(k_{\mathrm{F}} u) - k_{\mathrm{F}} u\ \cos(k_{\mathrm{F}} u)}{(k_{\mathrm{F}} u)^3} \right]^2 \tag{46}$$

where n is the density and $k_{\mathrm{F}} = (3\pi^2 n)$ is the Fermi wavevector.

9. Since the uniform gas is a well-defined system, its hole must satisfy the cusp condition Eq. (35) which, in terms of g, is simply

$$g_\lambda'(0) = \lambda g_\lambda(0). \tag{47}$$

Insertion of the LSD hole of Eq. (41) in Eq. (35) and use of the above implies that the LSD hole does satisfy the cusp condition.

REFERENCES

1. R.G. Parr and W. Yang, *Density Functional Theory of Atoms and Molecules* (Oxford, New York, 1989).
2. R.M. Dreizler and E.K.U. Gross, *Density Functional Theory* (Springer-Verlag, Berlin, 1990).
3. M. Levy, Proc. Natl. Acad. Sci. (U.S.A.) **76**, 6062 (1979).
4. E. H. Lieb, Int. J. Quantum Chem. **24**, 224 (1983).
5. P. Hohenberg and W. Kohn, Phys. Rev. **136**, B 864 (1964).
6. M. Levy and J.P. Perdew, Phys. Rev. A **32**, 2010 (1985).
7. M. Levy, Int. J. Quantum Chem. **S23**, 617 (1989).
8. D.C. Langreth and J.P. Perdew, Solid State Commun. **17**, 1425 (1975).
9. J. P. Perdew and Y. Wang, Phys. Rev. B **46**, 12947 (1992).
10. G.D. Mahan, *Many-Particle Physics*, 2nd ed. (Plenum Press, New York, 1990).
11. J. C. Kimball, Phys. Rev. A **7**, 1648 (1973).
12. D. Vanderbilt and D.C. Langreth, unpublished.
13. R.O. Jones and O. Gunnarsson, Rev. Mod. Phys. **61**, 689 (1989).
14. K. Burke and J. P. Perdew, Int. J. Quantum Chem. **56**, 199 (1995).
15. M. Ernzerhof, J.P. Perdew, and K. Burke, in *Density Functional Theory*, edited by R. Nalewajski (Springer, Berlin, 1996).
16. W. Kohn and L.J. Sham, Phys. Rev. **140**, A 1133 (1965).
17. D. M. Ceperley and B. J. Alder, Phys. Rev. Lett. **45**, 566 (1980).
18. Peter Fulde, *Electron Correlations in Molecules and Solids* (Springer-Verlag, Berlin, 1991).
19. M. Rasolt and D.J.W. Geldart, Phys. Rev. B **34**, 1325 (1986).
20. K. Burke, J. P. Perdew, and M. Levy, in *Modern Density Functional Theory: A Tool for Chemistry*, edited by J. M. Seminario and P. Politzer (Elsevier, Amsterdam, 1995).
21. O. Gunnarsson and B.I. Lundqvist, Phys. Rev. B **13**, 4274 (1976).
22. O. Gunnarsson, M. Jonson, and B. I. Lundqvist, Solid State Commun. **24**, 765 (1977).
23. J.P. Perdew, Phys. Rev. B **33**, 8822 (1986); **34**, 7406 (1986) (E).
24. A.D. Becke, Phys. Rev. A **38**, 3098 (1988).
25. C. Lee, W. Yang, and R.G. Parr, Phys. Rev. B **37**, 785 (1988).
26. J.P. Perdew, in *Electronic Structure of Solids '91*, edited by P. Ziesche and H. Eschrig (Akademie Verlag, Berlin, 1991), page 11.
27. J. P. Perdew, J. A. Chevary, S. H. Vosko, K. A. Jackson, M. R. Pederson, D.J. Singh, and C. Fiolhais, Phys. Rev. B **46**, 6671 (1992); **48**, 4978 (1993) (E).
28. J.P. Perdew, K. Burke, and M. Ernzerhof, Phys. Rev. Lett. **77**, 3865 (1996); **78**, 1396 (1997) (E).
29. D.C. Langreth and M.J. Mehl, Phys. Rev. B **28**, 1809 (1983).
30. J.P. Perdew, K. Burke, and Y. Wang, Phys. Rev. B **54**, 16533 (1996).
31. M. Ernzerhof, K. Burke, and J.P. Perdew, in *Recent developments and applications in density functional theory*, edited by J.M. Seminario (Elsevier, Amsterdam, 1996).
32. J. B. Krieger, Y. Li, and G. J. Iafrate, Phys. Rev. A **45**, 101 (1992).
33. K. Burke, J.P. Perdew, and M. Ernzerhof, Int. J. Quantum Chem. **61**, 287 (1997).
34. J. P. Perdew and A. Zunger, Phys. Rev. B **23**, 5048 (1981).
35. A. Görling and M. Levy, Phys. Rev. B **47**, 13105 (1993).
36. M. Levy, Phys. Rev. A **43**, 4637 (1991).
37. A. Görling and M. Levy, Phys. Rev. A **45**, 1509 (1992).

DRIVING OUT THE SELF-INTERACTION ERROR

John P. Perdew and Matthias Ernzerhof

Department of Physics and Quantum Theory Group, Tulane University
New Orleans, LA 70118

SELF-INTERACTION ERROR

Simple density functionals [1-4] for the exchange-correlation energy typically work well for many-electron systems, but fail to be exact for one-electron systems. This lecture reviews the self-interaction error of standard local and semilocal functionals, and points out just how severe this error can be when the exact exchange-correlation hole is not close to the electron it surrounds, as in "stretched H_2^+". Also discussed are several attempts to eliminate this error, with special emphasis on orbital-functional approximations such as the Perdew-Zunger self-interaction correction (SIC) [5-7] and the Colle-Salvetti functional [8-10]. An alternative approach, the construction of fully-nonlocal explicit density functionals such as the weighted density approximation [11-13], will not be reviewed here.

In Kohn-Sham spin density-functional theory [14], the aim is to predict the spin densities $n_\uparrow(\mathbf{r})$ and $n_\downarrow(\mathbf{r})$ (or the total density $n(\mathbf{r}) = n_\uparrow(\mathbf{r}) + n_\downarrow(\mathbf{r})$) and the energy E for the ground state of a system of N electrons in an external potential $v(\mathbf{r})$. The exact answers can be found in principle by minimizing the energy functional

$$E_v[n_\uparrow, n_\downarrow] = T_s[n_\uparrow, n_\downarrow] + \int d^3r\; n(\mathbf{r})v(\mathbf{r}) + U[n] + E_{\rm XC}[n_\uparrow, n_\downarrow] \tag{1}$$

over all allowed spin densities such that

$$\int d^3r\; n(\mathbf{r}) = N. \tag{2}$$

In Eq. (1), T_s is the ground-state kinetic energy of a system of non-interacting electrons with the given spin densities,

$$U[n] = \frac{1}{2}\int d^3r \int d^3r'\; n(\mathbf{r})n(\mathbf{r}')/|\mathbf{r} - \mathbf{r}'| \tag{3}$$

is the electrostatic self energy of a rigid continuous density distribution $n(\mathbf{r})$, and $E_{\rm XC}$ is the residue or exchange-correlation energy. The probability density to find an electron at position $\mathbf{r}$ is simply $n(\mathbf{r})$. An electron at point $\mathbf{r}$ does not interact with itself, but only with the $N-1$ remaining electrons. The electrostatic self energy $U[n]$ of n however contains such a self interaction. $E_{\rm XC}$ corrects for this error, as well as for the remaining exchange-correlation effects. All equations are expressed in atomic units ($\hbar = m = e^2 = 1$).

T_s can be treated exactly, leading to the Kohn-Sham self-consistent equations

$$T_s = \sum_{\alpha\sigma} f_{\alpha\sigma} \langle \psi_{\alpha\sigma}| - \frac{1}{2}\nabla^2 |\psi_{\alpha\sigma}\rangle, \tag{4}$$

$$n_\sigma(\mathbf{r}) = \sum_{\alpha} f_{\alpha\sigma} |\psi_{\alpha\sigma}(\mathbf{r})|^2, \tag{5}$$

$$\left[-\frac{1}{2}\nabla^2 + v_\sigma^{KS}(\mathbf{r})\right] \psi_{\alpha\sigma}(\mathbf{r}) = \epsilon_{\alpha\sigma}\psi_{\alpha\sigma}(\mathbf{r}), \tag{6}$$

where $\sigma = \uparrow, \downarrow$ and $f_{\alpha\sigma} = \Theta(\mu - \epsilon_{\alpha\sigma})$ is a fermion occupation number (1 for $\epsilon_{\alpha\sigma} < \mu$, 0 for $\epsilon_{\alpha\sigma} > \mu$). The solutions to these equations are the Kohn-Sham single-particle orbitals $\psi_{\alpha\sigma}$ and the corresponding orbital energies $\epsilon_{\alpha\sigma}$. The effective potential in Eq. (6) is

$$v_\sigma^{KS}(\mathbf{r}) = v(\mathbf{r}) + u([n];\mathbf{r}) + v_{\rm XC}^\sigma([n_\uparrow, n_\downarrow];\mathbf{r}). \tag{7}$$

In Eq. (7),

$$u([n];\mathbf{r}) = \frac{\delta U}{\delta n(\mathbf{r})} = \int d^3r'\, n(\mathbf{r}')/|\mathbf{r}' - \mathbf{r}|, \tag{8}$$

$$v_{\rm XC}^\sigma([n_\uparrow, n_\downarrow];\mathbf{r}) = \frac{\delta E_{\rm XC}}{\delta n_\sigma(\mathbf{r})}. \tag{9}$$

The exchange-correlation potential $v_{\rm XC}^\sigma$ ensures that the electron density of the Kohn-Sham determinant of occupied orbitals equals the exact density of the interacting system.

The exchange-correlation energy can be separated into exchange and correlation terms:

$$E_{\rm XC} = E_{\rm X} + E_{\rm C}. \tag{10}$$

The exchange energy is given by the Fock integral

$$E_x = -\frac{1}{2}\sum_\sigma \int d^3r d^3r' |\sum_\alpha f_{\alpha\sigma}\psi_{\alpha\sigma}^*(\mathbf{r})\psi_{\alpha\sigma}(\mathbf{r}')|^2/|\mathbf{r}' - \mathbf{r}|. \tag{11}$$

The density functionals $E_{\rm X}[n_\uparrow, n_\downarrow]$ and $E_{\rm C}[n_\uparrow, n_\downarrow]$ are not known exactly in explicit practical form, except in the simple limit of a one-electron density. Suppose

$$n_\downarrow(\mathbf{r}) = 0, \quad \int d^3r\, n_\uparrow(\mathbf{r}) = 1. \tag{12}$$

Then clearly

$$U[n] + E_{\rm X}[n_\uparrow, 0] = 0, \tag{13}$$

$$E_{\rm C}[n_\uparrow, 0] = 0, \tag{14}$$

$$u([n];\mathbf{r}) + v_{\rm X}^\uparrow([n_\uparrow, 0];\mathbf{r}) = 0, \tag{15}$$

$$v_{\rm C}^\uparrow([n_\uparrow, 0];\mathbf{r}) = 0. \tag{16}$$

Equation (13) means that a single electron does not interact with itself, i.e., the self energy of the electron $U[n]$ is cancelled by self exchange. Similarly a single electron has zero correlation energy (Eq. (14)). Equations (15) and (16) ensure that the electron is moving in the bare external potential $v(\mathbf{r})$, since no many-body effect must be mimicked by the local potential.

The exchange-correlation energy is typically the smallest term in Eq. (1), and is in fact neglected ($E_{\rm XC} = 0$) in the simple Thomas-Fermi density functional theory. The resulting violation of Eqs. (13-16) for $N = 1$ was corrected by Fermi and Amaldi [15],

$$E_{\rm XC}^{FA}[n] = -NU[n/N], \tag{17}$$

but at a heavy price: violation of the size-consistency principle [7], which asserts that the total energy of two well-separated non-overlapping densities should be the sum of the energy of each. The Hartree (H) approximation manages to be both self-interaction free and size-consistent by replacing the average orbital density $n(\mathbf{r})/N$ in Eq. (17) by a set of individual orbital densities

$$n_{\alpha\sigma}(\mathbf{r}) = f_{\alpha\sigma}|\phi_{\alpha\sigma}(\mathbf{r})|^2, \tag{18}$$

$$E_{\rm XC}^{H} = -\sum_{\alpha\sigma} U[n_{\alpha\sigma}]. \tag{19}$$

For any number of electrons, and for reasons now well-understood [16], $E_{\rm XC}$ can usually be approximated reliably by local or semilocal functionals. The local spin density (LSD) approximation [14,17]

$$E_{\rm XC}^{LSD}[n_\uparrow, n_\downarrow] = \int d^3r\ n(\mathbf{r})\epsilon_{\rm XC}(n_\uparrow(\mathbf{r}), n_\downarrow(\mathbf{r})), \tag{20}$$

where $\epsilon_{\rm XC}(n_\uparrow(\mathbf{r}), n_\downarrow(\mathbf{r}))$ is the exchange-correlation energy [18] per electron of an electron gas with uniform spin densities $n_\uparrow$ and $n_\downarrow$, is still widely and successfully applied in solid state physics. The generalized gradient approximation [19-22]

$$E_{\rm XC}^{GGA}[n_\uparrow, n_\downarrow] = \int d^3r\ f(n_\uparrow, n_\downarrow, \nabla n_\uparrow, \nabla n_\downarrow) \tag{21}$$

is currently popular, especially in quantum chemistry. There is a unique LSD but no unique gradient-corrected approximation; several empirical [21,22] and non-empirical [19,20] GGA's are in use. Unfortunately, these local and semilocal approximations necessarily make a self-interaction error, because their restricted functional forms cannot recover Eqs. (13-16) when they are applied to an arbitrary one-electron density.

To illustrate the self-interaction errors of approximate density functionals, we first consider the hydrogen atom. Table 1 lists the exchange energy, the correlation energy, and their sum in the LSD and the PBE [19] GGA. Consistent with a general observation [16], the local and semilocal approximations are better for $E_{\rm XC}$ than for $E_{\rm X}$ or $E_{\rm C}$. LSD and PBE make self-exchange and self-correlation errors, but for the H atom and many other systems these errors are fairly small. In the case of the hydrogen atom, the self-interaction error in $E_{\rm XC}$ carries over directly to the ionization potential of the atom.

Atomization energies of molecules are of special interest in quantum chemistry. Typically, LSD overestimates the energy needed to atomize a molecule, and PBE reduces both the atomization energy and the error substantially [19]. Even the extremely weak binding of the rare-gas dimers is described accurately by PBE [23,24]. In all atoms and most molecules, the exact exchange-correlation hole [25,26] remains close to the electron it surrounds, as it does in an electron gas of uniform or slowly-varying density.

Table 1. Approximate density functionals applied to the hydrogen atom. (All energies are in hartrees.) The self-consistent PBE density has been employed in the calculations. However, the exact density has been employed in the exact calculations shown in parentheses.

	LSD	PBE	exact
E_x	-.264	-.302	-.307(-.3125)
E_c	-.022	-.006	0(0)
E_{xc}	-.286	-.308	-.307(-.3125)

Table 2. Approximate density functionals applied to the atomization of H_2^+. (All energies are in hartrees.) The selfconsistent PBE density and the exact equilibrium bond length have been employed in the calculations. However, the exact density has been employed in the calculations shown in parentheses. The atomized state is the broken-symmetry $(H + H^+)$ state.

	LSD	PBE	exact
ΔE_x	.026	.030	.024(.019)
ΔE_c	.001	.001	0(0)
ΔE_{xc}	.027	.031	.024(.019)

The one-electron molecule H_2^+ poses a challenge to local and semilocal functionals, because its exact exchange-correlation hole is spread out over the entire molecule. Table 2 shows the exchange and correlation contributions to the atomization energy of H_2^+, where the final atomized state is one H atom and one H^+ ion. For this problem, LSD is not so good and PBE is even worse. This error is entirely due to self-interaction, and is absent even in the Hartree approximation.

The better to identify the root of the problem, we next consider the H_2^+ molecule at large bond length. The exact orbital of the H_2^+ molecule at infinite separation (stretched H_2^+) extends over both nuclei and is symmetic. This orbital is degenerate with another orbital of broken-symmetry localized at one of the nuclei. Any normalized linear combination of these two orbitals again has the exact total ground-state energy. However, local and semilocal approximations to E_{XC} give total energies for stretched H_2^+ which are sensitive to the distribution of the single electron among the two nuclei. To illustrate this behavior, Table 3 compares the symmetric H_2^+ ion (denoted by $(H \cdots H)^+$) to $H + H^+$, both at infinite separation.

The LSD and PBE approximations give a realistic description of the broken-symmetry $H + H^+$ system, with the PBE being remarkably accurate for E_{XC}. However, the formally degenerate symmetric $(H \cdots H)^+$ representation of stretched H_2^+ is badly described by both LSD and PBE approximations. The reason for this failure can be traced back to the exchange-correlation hole [26] around an electron. The LSD and PBE holes are always localized around the electron, like the exact hole in $H + H^+$ but unlike the exact hole in $(H \cdots H)^+$.

Worse still, the LSD or PBE total energy of symmetric $(H \ldots H)^+$ is *lower* than the broken-symmetry $H + H^+$ by about 0.1 hartree. Thus there is no spontaneous symmetry breaking, and the LSD or PBE binding energy curve of H_2^+ is disastrously

Table 3. Approximate density functionals applied to stretched H_2^+. (All energies are in hartrees.) The selfconsistent PBE density has been employed in the calculations. However, the exact density has been employed in the calculations shown in parentheses.

	LSD		PBE		exact	
	$H+H^+$	$(H\cdots H)^+$	$H+H^+$	$(H\cdots H)^+$	$H+H^+$	$(H\cdots H)^+$
E_x	-.264	-.234	-.302	-.279	-.307(-.3125)	-.172(-.15625)
E_c	-.022	-.021	-.006	-.004	0(0)	0(0)
E_{xc}	-.286	-.255	-.308	-.283	-.307(-.3125)	-.172(-.15625)

wrong as a consequence of self-interaction error. (Similar binding energy curves for He_2^+ are shown in Ref. [27].)

We have of course chosen a "worst case" example for LSD and PBE. Another less physical "worst case" is a one- or two-electron system with very low average density [28]. In most physical systems, the self-interaction errors of these functionals are a rather small fraction of the total energy or energy difference of interest, but are a much larger fraction of the orbital energy eigenvalue. As an example, Figure 1 displays the "spurious self-interacting potential" $u([n];\mathbf{r}) + v_{\mathrm{XC}}^{\uparrow}([n,0];\mathbf{r})$ for the H atom. Over most of the volume of the atom, this potential is large (about 6 eV out of a total binding energy of 13.6 eV) but approximately constant, so it contaminates the LSD or PBE orbital energy eigenvalue far more than it contaminates the density profile. However, at large distances from the nucleus the spurious self-interacting potential falls slowly to zero (like $1/r$), and so contaminates the long-range behavior of the density.

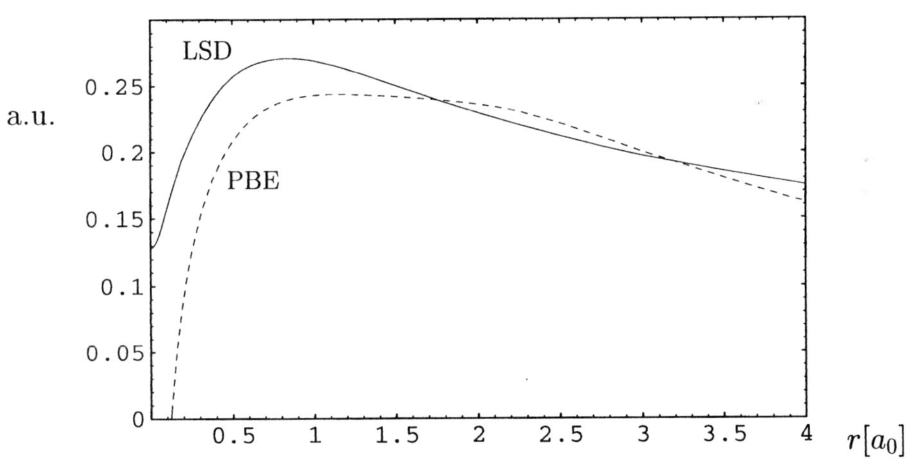

Figure 1. Spurious self-interacting potentials for the hydrogen atom, evaluated on the exact density.

ORBITAL-BY-ORBITAL SELF-INTERACTION CORRECTION

An approximate exchange-correlation energy functional $E_{\mathrm{XC}}^{approx}[n_\uparrow, n_\downarrow]$ which violates Eqs. (13-16) for a one-electron density is not properly self-interaction-free. It can be made so by an orbital-by-orbital self-interaction correction [5,6,7] (SIC) similar to that of Eqs. (18) and (19)

$$E_{\mathrm{X}}^{SIC} = E_{\mathrm{X}}^{approx}[n_{\uparrow}, n_{\downarrow}] - \sum_{\alpha\sigma} \{U[n_{\alpha\sigma}] + E_{\mathrm{X}}^{approx}[n_{\alpha\sigma}, 0]\}, \tag{22}$$

$$E_{\mathrm{C}}^{SIC} = E_{\mathrm{C}}^{approx}[n_{\uparrow}, n_{\downarrow}] - \sum_{\alpha\sigma} E_{\mathrm{C}}^{approx}[n_{\alpha\sigma}, 0]. \tag{23}$$

Note that unoccupied orbitals do not contribute. Lindgren [29] applied Eq. (22) to E_{X}^{LSD}. In 1981, Perdew and Zunger [5] proposed Eqs. (22) and (23) as a general method of self-interaction correction, and tested them for E_{X}^{LSD} and E_{C}^{LSD}. For related work see Refs. [30-37]. More recently, these equations have been applied [38] to E_{XC}^{GGA} as well.

When Eqs. (22) and (23) are applied to the Kohn-Sham orbital densities of an atom, it is typically found that each orbital self-interaction correction to the LSD exchange energy is negative while each correction to the LSD correlation energy is positive, the total correction to E_{XC}^{LSD} being negative. These findings are consistent with the results of the previous section for the H atom, where the SIC correction to LSD and PBE leads to the exact solution. In applications to molecules or solids, it is important *not* to use the Kohn-Sham orbital densities, as this can lead to a size-consistency error [5,7]. The Kohn-Sham orbitals are "molecular" or "symmetry" orbitals for molecules, and "Bloch" orbitals for solids; they can and often do extend over the whole system. Then, as the size of the system grows, the orbitals become more diffuse and the total self-interaction correction tends to zero. To avoid this problem, one must construct a special set of localized SIC orbitals $\phi_{\alpha\sigma}(\mathbf{r})$, to use in Eq. (18) instead of the Kohn-Sham orbitals $\psi_{\alpha\sigma}(\mathbf{r})$.

Minimization of the SIC total energy with respect to variation of the orbitals $\phi_{\alpha\sigma}(\mathbf{r})$, subject only to the constraint of normalization, leads to equations [5] that resemble Eqs. (6) and (7) but with the orbital-independent exchange-correlation potential $v_{\mathrm{XC}}^{\sigma,approx}$ replaced by the orbital-dependent potential

$$v_{\mathrm{XC}}^{\alpha\sigma,SIC}(\mathbf{r}) = v_{\mathrm{XC}}^{\sigma,approx}([n_{\uparrow}, n_{\downarrow}]; \mathbf{r}) - \left\{u([n_{\alpha\sigma}]; \mathbf{r}) + v_{\mathrm{XC}}^{\uparrow,approx}([n_{\alpha\sigma}, 0]; \mathbf{r})\right\}. \tag{24}$$

The selfconsistent SIC orbitals are not strictly orthogonal, but can be made so by orthogonalization [5] or by the introduction of off-diagonal Lagrange multipliers which couple the one-electron Schrödinger equations for different orbitals, or by other means [39]. All symmetries may break as the SIC orbitals localize around atomic centers [40] or in bonds [41].

The SIC effective potential has a much more realistic long-range ($r \to \infty$) behavior [5] than does the LSD or GGA potential. In the latter approximations, $v_{\mathrm{XC}}^{\sigma,approx}(\mathbf{r})$ tends to zero exponentially as $\mathbf{r} \to \infty$. But

$$v_{\mathrm{XC}}^{\alpha\sigma,SIC}(\mathbf{r}) \to -u([n_{\alpha\sigma}]; \mathbf{r}) \to -\frac{1}{r}, \tag{25}$$

the correct long-range behavior of the Kohn-Sham v_{XC}^{σ}. When one of the electrons of the neutral system wanders far away and looks back, it must see a singly-positive ion and not a neutral system; the electron cannot see its own electrostatic potential.

Nevertheless, the appearance of an orbital-dependent potential demonstrates that SIC falls outside the Kohn-Sham orbital scheme. The Hohenberg-Kohn density-functional scheme is broader, and can in a sense encompass [5] SIC.

SIC has been applied to atoms [5,42-47] and atomic ions [48], molecules [39,49-56], solids [40,41,57-72] and model problems [73,74]. Instead of attempting to review these applications, this article will simply list some of the good and bad effects of SIC.

Good effects of SIC include [5]: improved total energies of atoms, improved separate exchange and correlation energies, selfconsistent bound solutions for negative ions, occupied orbital energies that approximate electron removal energies, orbital energy differences that approximate band gaps in solids, intuitively localized orbitals (core states, bonds, lone pairs, etc), and proper dissociation of molecules and solids to separated neutral atoms. LSD and GGA do not always obey this "principle of integer preference" [7], and can dissociate heteronuclear molecules and solids incorrectly to fractionally-charged fragments. In fact, these correct features of SIC led to the discovery of the derivative discontinuity [75,76] of the exact $E_{XC}[n_\uparrow, n_\downarrow]$.

Some of the effects of SIC are best characterized as "neutral". The SIC energy is exact (or nearly exact) for the uniform electron gas at valence-electron densities [57,59]. SIC ionization energies [42] of atoms are neither better nor worse than those of LSD, on average. The SIC exchange-correlation hole around an electron satisfies many of the important exact conditions [26] satisfied by the LSD hole. While the SIC exchange hole violates the negativity constraint, the SIC exchange-correlation hole satisfies a sum rule [5,76] or integral condition not only for integer electron numbers but even (approximately) for fractional electron number. (The symmetric state of stretched H_2^+, described in the first section, can be regarded as two hydrogen ions each with half an electron.)

Now we turn to the bad news about SIC. It can yield "false surface energies" [5,7] for metals. SIC is hard to implement, especially for molecules. It appears that SIC (applied to LSD) does not greatly improve LSD reaction [50] or atomization energies [51], and yields bond lengths that are too short [50-54], associated with vibrational frequencies that are too high [53].

COLLE-SALVETTI FUNCTIONAL

Exact exchange, which is self-interaction free, can now be implemented [77] in Kohn-Sham calculations. The exact exchange energy differs from the Hartree-Fock exchange-energy only by the small difference [78] between Kohn-Sham and Hartree-Fock occupied orbitals. The problem remains to construct an accurate, self-interaction-free correlation energy for use with [79] exact exchange.

The Colle-Salvetti functional [8-10,80] for the correlation energy is of interest here for two reasons: (1) It provides an example of an orbital functional which is automatically free from self-interaction error and invariant under unitary transformation of the occupied orbitals. (2) It underlies the Lee-Yang-Parr (LYP) [22] density functional for the correlation energy. LYP is the most widely-used correlation energy functional in quantum chemistry [81], and is itself free from self-correlation error (but at a high cost, as we will see).

The many-electron wave function $\Psi(\mathbf{r}_1\sigma_1, \mathbf{r}_2\sigma_2, \ldots, \mathbf{r}_N\sigma_N)$ provides the fullest-possible description of the positions $\mathbf{r}_i$ and z-components of spin σ_i for the N electrons. Neglecting correlation, it may usually be represented as a single Slater determinant of Hartree-Fock orbitals.

Since the many-electron Hamiltonian $\hat{H}$ is a sum of one-body and two-body operators, the energy $E = \langle\Psi|\hat{H}|\Psi\rangle$ can be found from two contractions of Ψ. The first is the one-particle density matrix

$$\rho_1(\mathbf{r}, \mathbf{r}') = N \sum_{\sigma_1} \ldots \sum_{\sigma_N} \int d^3r_2 \ldots d^3r_N \\ \times \Psi^*(\mathbf{r}\sigma_1, \mathbf{r}_2\sigma_2, \ldots, \mathbf{r}_N\sigma_N)\Psi(\mathbf{r}'\sigma_1, \mathbf{r}_2\sigma_2, \ldots, \mathbf{r}_N\sigma_N). \tag{26}$$

The electron density is $n(\mathbf{r}) = \rho_1(\mathbf{r},\mathbf{r})$, and $n(\mathbf{r})d^3r$ is the probability of finding an electron in volume element d^3r at position $\mathbf{r}$. Note that

$$\rho_1^{HF}(\mathbf{r},\mathbf{r}') = \sum_{\alpha\sigma} f_{\alpha\sigma}\psi^*_{\alpha\sigma}(\mathbf{r})\psi_{\alpha\sigma}(\mathbf{r}') = \sum_{\sigma} \rho_{1\sigma}^{HF}(\mathbf{r},\mathbf{r}'). \tag{27}$$

The second needed contraction is the pair density

$$\rho_2(\mathbf{r},\mathbf{r}') = N(N-1)\sum_{\sigma_1}\cdots\sum_{\sigma_N}\int d^3r_3\ldots d^3r_N|\Psi(\mathbf{r}\sigma_1,\mathbf{r}'\sigma_2,\mathbf{r}_3\sigma_3,\ldots,r_N\sigma_N)|^2, \tag{28}$$

where $\rho_2(\mathbf{r},\mathbf{r}')d^3rd^3r'$ is the joint probability of finding an electron at d^3r at $\mathbf{r}$ and another in d^3r' at $\mathbf{r}'$. Indeed, the density $n_{\text{XC}}^{\lambda}(\mathbf{r},\mathbf{r}')$ at $\mathbf{r}'$ of the exchange-correlation hole [26] around an electron at $\mathbf{r}$ for coupling constant λ is given by the equation

$$\rho_2(\mathbf{r},\mathbf{r}') = n(\mathbf{r})[n(\mathbf{r}') + n_{\text{XC}}^{\lambda}(\mathbf{r},\mathbf{r}')]. \tag{29}$$

For the Hartree-Fock determinant (essentially $\lambda = 0$),

$$\rho_2^{HF}(\mathbf{r},\mathbf{r}') = n^{HF}(\mathbf{r})n^{HF}(\mathbf{r}') - \sum_{\sigma}|\rho_{1\sigma}^{HF}(\mathbf{r},\mathbf{r}')|^2. \tag{30}$$

In 1975, Colle and Salvetti [8] presented a model for the correlation energy as a functional of the Hartree-Fock one-matrix $\rho_{1\sigma}^{HF}(\mathbf{r},\mathbf{r}')$. The Colle-Salvetti wave function is

$$\Psi^{CS}(\mathbf{r}_1\sigma_1,\mathbf{r}_2\sigma_2,\ldots,\mathbf{r}_N\sigma_N) = \Psi^{HF}(\mathbf{r}_1\sigma_1,\mathbf{r}_2\sigma_2,\ldots,\mathbf{r}_N\sigma_N)\prod_{i<j}[1-\phi(\mathbf{r}_i,\mathbf{r}_j)]. \tag{31}$$

The correlation factor ϕ is a symmetric function of $\mathbf{r}_i$ and $\mathbf{r}_j$, modelled to yield an electron-electron cusp as $\mathbf{r}_j \to \mathbf{r}_i$ [82]. It depends upon $n_\uparrow$ and $n_\downarrow$, and contains one parameter fitted to the correlation energy of the neon atom.

It was then argued that

$$\rho_1^{CS}(\mathbf{r},\mathbf{r}') \approx \rho_1^{HF}(\mathbf{r},\mathbf{r}'), \tag{32}$$

$$\rho_2^{CS}(\mathbf{r},\mathbf{r}') \approx \rho_2^{HF}(\mathbf{r},\mathbf{r}')[1-\phi(\mathbf{r},\mathbf{r}')]^2, \tag{33}$$

making the correlation energy

$$E_{\text{C}}^{CS-I} = \frac{1}{2}\int d^3r \int d^3r\,'\,\rho_2^{HF}(\mathbf{r},\mathbf{r}')\frac{[-2\phi(\mathbf{r},\mathbf{r}')+\phi^2(\mathbf{r},\mathbf{r}')]}{|\mathbf{r}'-\mathbf{r}|}. \tag{34}$$

Eq. (34) is automatically free of self-interaction error, since $\rho_2^{HF}(\mathbf{r},\mathbf{r}')$ vanishes for any one-electron system. Eqs. (32-34) seem to neglect the kinetic-energy contribution to the correlation energy [28], but this is included to some extent by fitting one parameter in ϕ to the actual correlation energy of the neon atom.

Further approximations lead to the simpler result

$$E_{\text{C}}^{CS-II} = \frac{1}{2}\int d^3r\ \rho_2^{HF}(\mathbf{r},\mathbf{r})F(n_\uparrow(\mathbf{r}),n_\downarrow(\mathbf{r});t_s(\mathbf{r})). \tag{35}$$

Here $t_s(\mathbf{r})$ is the non-interacting kinetic energy density at $\mathbf{r}$, which along with $n_\uparrow(\mathbf{r})$ and $n_\downarrow(\mathbf{r})$ controls the small $|\mathbf{r}'-\mathbf{r}|$ behavior of $\rho_2^{HF}(\mathbf{r},\mathbf{r}')$. From Eq. (30),

$$\rho_2^{HF}(\mathbf{r},\mathbf{r}) = [n^{HF}(\mathbf{r})]^2 - [n_\uparrow^{HF}(\mathbf{r})]^2 - [n_\downarrow^{HF}(\mathbf{r})]^2, \tag{36}$$

which vanishes for any fully spin-polarized system (and thus for any one-electron system). By replacing Hartree-Fock with Kohn-Sham orbitals and making a spin-density-gradient expansion for $t_s(\mathbf{r})$, Lee, Yang and Parr [22] converted Eq. (35) into a spin-density functional for the correlation energy.

For the uniform electron gas, the LYP correlation functional is identical to the Colle-Salvetti functional of Eq. (35). McWeeny [83] tested Eq. (34) for the uniform electron gas, and found a realistic correlation energy over the range of valence electron densities. However, Eq. (35) and its LYP counterpart give only about half of the true correlation energy for the uniform gas over this range [16,84]. The problem is partly that Eq. (35) misses parallel-spin correlation. (Note that parallel-spin correlation contributes $\sim$ 25 % to the total correlation energy of the Ne atom [16], and a larger percentage to that of the uniform gas.) Thus, while LYP yields excellent results for many small molecules [81], it must fail for bulk metals and large metallic clusters, and perhaps also for large molecules like C_{20} [85].

CONCLUSIONS

It is easy to describe one-electron systems exactly, and not very hard to construct good approximations for many-electron systems. Ironically, it is not easy at all to do both at the same time. Standard LSD and GGA functionals can make severe errors when the exact exchange-correlation hole is not close to its electron, e.g., "stretched H_2^+". Here we have described two attempts to get one-electron and many-electron limits right by making use of a set of occupied orbitals.

So far, no completely satisfactory solution to this problem has been demonstrated. However, the possibilities inherent in Eq. (34) or related ideas [32,86] remain to be explored. Note that ρ_2^{HF} in Eq. (34) may be replaced by ρ_2^{KS}, the pair density of the Kohn-Sham determinant, and that $\phi(\mathbf{r},\mathbf{r}')$ may be chosen to approximate the coupling-constant-averaged [28,87] correlation hole. Probably $\phi(\mathbf{r},\mathbf{r}')$ should reduce to its LSD form in the $|\mathbf{r}'-\mathbf{r}| \to 0$ and uniform-gas limits, and should satisfy the sum rule on the correlation hole:

$$\int d^3r'\ \rho_2^{KS}(\mathbf{r},\mathbf{r}')[-2\phi(\mathbf{r},\mathbf{r}') + \phi^2(\mathbf{r},\mathbf{r}')] = 0. \tag{37}$$

ACKNOWLEDGMENTS

This work was supported in part by the U.S. National Science Foundation under Grant No.DMR95-21353.

REFERENCES

1. R. M. Dreizler and E. K. U. Gross, *Density Functional Theory* (Springer-Verlag, Berlin, 1990).
2. R. G. Parr and W. Yang, *Density-Functional Theory of Atoms and Molecules* (Oxford University Press, Oxford, 1989).
3. R. O. Jones and O. Gunnarsson, Rev. Mod. Phys. **61**, 689 (1989).
4. *Recent Developments and Applications of Modern Density Functional Theory*, edited by J. M. Seminario, *Theoretical and Computational Chemistry* (Elsevier, Amsterdam, 1996).
5. J. P. Perdew and A. Zunger, Phys. Rev. B **23**, 5048 (1981), and references therein.
6. J. P. Perdew, in *Local Density Approximations in Quantum Chemistry and Solid State Physics*, edited by J. Avery and J. P. Dahl (Plenum, NY, 1984).
7. J. P. Perdew, in *Density Functional Theory of Many-Fermion Systems*, edited by S. B. Trickey, *Advances in Quantum Chemistry* (Academic Press, NY, 1990).

8. R. Colle and O. Salvetti, Theoret. Chim. Acta **37**, 329 (1975).
9. R. Colle and O. Salvetti, Theoret. Chim. Acta **53**, 55 (1979).
10. R. Colle and O. Salvetti, in *Density Matrices and Density Functionals*, edited by R. M. Erdahl and V. H. Smith Jr. (Reidel, Dordrecht, 1987).
11. O. Gunnarsson, M. Jonson, and B. I. Lundqvist, Phys. Rev. B **20**, 3136 (1979), and references therein.
12. J. A. Alonso and N. A. Cordero, in Ref. 4, and references therein.
13. M. Sadd and M. P. Teter, Phys. Rev. B **54**, 13643 (1996).
14. W. Kohn and L. J. Sham, Phys. Rev. **140**, A1133 (1965).
15. E. Fermi and E. Amaldi, Accad. Ital. Rome. **6**, 119 (1934).
16. M. Ernzerhof, J. P. Perdew, and K. Burke, in *Density Functional Theory*, edited by R. F. Nalewajski, *Topics in Current Chemistry* (Springer-Verlag, Berlin, 1996).
17. U. von Barth and L. Hedin, J. Phys. C **5**, 1629 (1972).
18. J. P. Perdew and Y. Wang, Phys. Rev. B **45**, 13244 (1992).
19. J. P. Perdew, K. Burke, and M. Ernzerhof, Phys. Rev. Lett. **77**, 3865 (1996), ibid **78** 1396 (1997)(E).
20. K. Burke, J. P. Perdew, and Y. Wang, chapter in the present volume.
21. A. D. Becke, Phys. Rev. A **38**, 3098 (1988).
22. C. Lee, W. Yang, and R. G. Parr, Phys. Rev. B **37**, 785 (1988).
23. D. C. Patton, D. V. Porezag, and M. R. Pederson, Phys. Rev. B **55**, 7454 (1997).
24. Y. Zhang, W. Pan, and W. Yang, unpublished.
25. M. Ernzerhof, K. Burke, and J. P. Perdew, in *Recent Developments in Density Functional Theory*, edited by J. M. Seminario (Elsevier, Amsterdam, 1996).
26. K. Burke, chapter in the present volume.
27. R. Merkle, A. Savin, and H. Preuss, J. Chem. Phys. **97**, 9216 (1992).
28. K. Burke, J. P. Perdew, and M. Ernzerhof, chapter in the present volume.
29. I. Lindgren, Int. J. Quantum Chem. **4**, 411 (1971).
30. M. S. Gopinathan, Phys. Rev. A **15**, 2135 (1977).
31. H. Stoll, C. M. E. Pavlidou, and H. Preuß, Theoret. Chim. Acta **49**, 143 (1978).
32. J. F. Dobson and J. H. Rose, J. Phys. C **15**, L1183 (1982).
33. P. Cortona, Phys. Rev. A **34**, 769 (1986).
34. S. Manoli and M. A. Whitehead, Phys. Rev. A **38**, 630 (1988).
35. J. F. Dobson, J. Phys. Condens. Matter **4**, 7877 (1992).
36. A. D. Becke, J. Chem. Phys. **104**, 1040 (1996).
37. R. K. Nesbet, unpublished.
38. Y. Li and J. B. Krieger, Phys. Rev. A **41**, 1701 (1990).
39. M. R. Pederson, R. A. Heaton, and C. C. Lin, J. Chem. Phys. **80**, 1972 (1984).
40. S. C. Erwin and C. C. Lin, J. Phys. C **21**, 4285 (1988).
41. Y. Hatsugai and T. Fujiwara, Phys. Rev. B **37**, 1280 (1988).
42. O. Gunnarsson and R. O. Jones, Solid State Commun. **37**, 249 (1981).
43. J. G. Harrison, R. A. Heaton, and C. C. Lin, J. Phys. B **16**, 2079 (1983).
44. M. R. Pederson and C. C. Lin, J. Chem. Phys. **88**, 1807 (1988).
45. M. M. Rieger and P. Vogl, Phys. Rev. A **52**, 282 (1995).
46. J. B. Krieger, J. Chen, Y. Li, and G. Iafrate, Int. J. Quantum Chem. **29**, 79 (1995).
47. J. Chen, J. B. Krieger, Y. Li, and G. J. Iafrate, Phys. Rev. A **54**, 3939 (1996).
48. L. A. Cole and J. P. Perdew, Phys. Rev. A **25**, 1265 (1982).
49. M. R. Pederson, R. A. Heaton, and C. C. Lin, J. Chem. Phys. **82**, 2688 (1985).
50. S. Goedecker and C. Umrigar, Phys. Rev. A **55**, 1765 (1997)
51. J. B. Krieger, J. Chen, and G. J. Iafrate, Bull. Am. Phys. Soc. **41**, 748 (1996), abstracts R19 11-12; J. Chen, J. B. Krieger, and G. J. Iafrate, submitted to Phys. Rev. A.
52. B. G. Johnson, C. A. Gonzales, P. M. W. Gill, and J. A. Pople, Chem. Phys. Lett. **221**, 100 (1994).
53. G. I. Csonka and B. G. Johnson, unpublished.
54. G. I. Csonka, N. A. Nguyen, and I. Kolossvary, J. Comp. Chem. **18**, 1534 (1997).
55. E. S. Fois, J. I. Penman, and P. A. Madden, J. Chem. Phys. **98**, 6352 (1993).
56. M. A. Whitehead, in *Recent Advances in Density Functional Methods, Part II*, edited by D. P. Chong (World Scientific, Singapore, 1997).
57. M. R. Norman, Phys. Rev. B **28**, 3585 (1983).
58. K. Katayama-Yoshida and A. Zunger, Phys. Rev. Lett. **53**, 1256 (1984).
59. M. R. Pederson, R. A. Heaton, and J. G. Harrison, Phys. Rev. B **39**, 1581 (1989).

60. Z. Szotek, W. M. Temmerman, and H. Winter, Solid State Commun. **74**, 1031 (1990).
61. Z. Szotek, W. M. Temmerman, and H. Winter, Phys. Rev. B **47**, 4029 (1993).
62. W. M. Temmerman, Z. Szotek, and H. Winter, Phys. Rev. B **47**, 11533 (1993).
63. Z. Szotek, W. M. Temmerman, and H. Winter, Phys. Rev. Lett. **72**, 1244 (1994); A. Svane, Z. Szotek, W. M. Temmerman, and H. Winter, Solid State Commun. **102**, 423 (1997). S. V. Beiden, W. M. Temmerman, Z. Szotek, and G. A. Gehring, Phys. Rev. Lett., Nov 10, 1997. W. M. Temmerman, A. Svane, Z. Szotek, and H. Winter, chapter in the present volume.
64. D. Vogel, P. Krüger, and J. Pollmann, Phys. Rev. B **52**, 14316 (1995).
65. D. Vogel, P. Krüger, and J. Pollmann, Phys. Rev. B **54**, 5495 (1996).
66. M. M. Rieger and P. Vogl, Phys. Rev. B **52**, 16567 (1995).
67. M. Arai and T. Fujiwara, Phys. Rev. B **51**, 1477 (1995).
68. S. Kobayashi, M. Arai, and T. Fujiwara, Phys. Rev. B **52**, 13718 (1995).
69. A. Svane and O. Gunnarsson, Phys. Rev. Lett. **65**, 1148 (1990).
70. A. Svane, Phys. Rev. Lett. **68**, 1900 (1992).
71. A. Svane, Phys. Rev. Lett. **72**, 1248 (1994).
72. A. Svane, Phys. Rev. B **53**, 4275 (1996).
73. A. Svane and O. Gunnarsson, Phys. Rev. B **37**, 9919 (1988).
74. A. Svane and O. Gunnarsson, Europhys. Lett. **7**, 171 (1988).
75. J. P. Perdew, R. G. Parr, M. Levy, and J. L. Balduz, Phys. Rev. Lett. **49**, 1691 (1982).
76. J. P. Perdew, in *Density Functional Methods in Physics*, edited by R. M. Dreizler and J. da Providencia (Plenum, New York, 1985).
77. J. B. Krieger, Y. Li, and G. J. Iafrate, Phys. Rev. A **45**, 101 (1992).
78. A. Görling and M. Ernzerhof, Phys. Rev. A **51**, 4501 (1995).
79. T. Grabo and E. K. U. Gross, Chem. Phys. Lett. **240**, 141 (1995).
80. E. K. U. Gross, M. Petersilka, and T. Grabo, in *Chemical Applications of Density-Functional Theory*, edited by B. B. Laird, R. B. Ross, and T. Ziegler (ACS Books, Washington, 1996).
81. L. A. Curtiss, K. Raghavachari, P. C. Redfern, and J. A. Pople, J. Chem. Phys. **106**, 1063 (1997).
82. T. Kato, Commun. Pure Appl. Math. **10**, 151 (1957).
83. R. McWeeny, in *The New World of Quantum Chemistry*, edited by B. Pullman and R. Parr (Reidel, Dordrecht, 1976).
84. J. P. Perdew and K. Burke, Int. J. Quantum Chem. **S 57**, 309 (1996).
85. J. C. Grossman, L. Mitas, and K. Raghavachari, Phys. Rev. Lett. **75**, 3870 (1995); ibid. **76**, 1006 (1996)(E).
86. K. Burke, J. P. Perdew, and M. Ernzerhof, Int. J. Quantum Chem. **61**, 287 (1997).
87. D. C. Langreth and J. P. Perdew, Solid State Comm. **17**, 1425 (1975).

TIME-DEPENDENT DENSITY-FUNCTIONAL THEORY

John F. Dobson

School of Science, Griffith University
Nathan, Queensland 4111, Australia

INTRODUCTION

The static Kohn-Sham theory of many-particle systems is basically unsuitable for the description of time-dependent electronic phenomena. (The few exceptions rely on the approximate description of electronic excitation energies, a procedure fraught with difficulties as described in a previous chapter.) In this chapter we introduce the time-dependent density functional theory (TDDFT), which fills this gap.

TDDFT is able to describe phenomena such as the following .

- Linear response to time-dependent external fields. This includes, for example,
  - Optical absorption of atoms at low power [1]
  - Infrared absorption of quantum wells [2,3]
  - Response of metal surfaces to external fields including surface plasmon dispersion [4] and damping and shift of adsorbate resonances [5]
- Nonlinear response to external fields including
  - Second and higher-order nonlinear susceptibilities [6]
  - Optical switching
  - Laser ablation of atoms [7,8]
- Excitation energies (from poles of frequency-dependent linear response: see the chapter by Petersilka *et al*)
- Other manifestations of time-dependent response including new functionals for the van der Waals interaction: see the chapter by Dobson , Dinte and Wang

Ref. [6] gives a recent review of the field, excluding however the very new results presented in the invited chapters of the present volume.

BASIC FORMULATION OF TDDFT

Runge-Gross Theorems (Time Dependent Hohenberg-Kohn)

The basic aim of TDDFT is to describe time-dependent phenomena in interacting many-particle systems entirely in terms of the *time-dependent* one-particle density $n(\mathbf{r},t)$. Following earlier versions restricted to particular situations, a rather general time-dependent generalization of the static Hohenberg-Kohn theorem was given in 1984 by Runge and Gross [9]. Consider a fixed number N of Fermions with a given pair interaction $V_{ee}(\mathbf{r}_1,\mathbf{r}_2)$, *starting at time* t_0 *in a particular fixed many-body state* $|\Psi_0>$. Clearly the only additional information required to specify the many-body Schrödinger equation, and hence all properties of the system for times $t > 0$, is the external potential $v^{ext}(\mathbf{r},t)$. Thus in particular the time-dependent density is determined by $v^{ext}(\mathbf{r},t)$:

$$v^{ext}(\mathbf{r},t) : \ t > 0 \Rightarrow n(\mathbf{r},t) : \ t > 0.$$

Runge and Gross showed, using operator equations of motion, that the reverse is also true: each density can arise from at most one external potential, so that

$$n(\mathbf{r},t) : \ t > 0 \Rightarrow v^{ext}(\mathbf{r},t) : \ t > 0$$

and hence all properties for $t > 0$ are determined by the time-dependent density.

(Actually v^{ext} is only determined up to a time-dependent additive constant $e(t)$, and the many-body wavefunction is ambiguous up to a time-dependent phase $\phi(t)$. This time-dependence, and the implicit dependence on the initial many-body wavefunction, constitute differences from the static Hohenberg-Kohn approach. One often circumvents the dependence on initial state by starting from the many-body groundstate.)

As an intermediate step in their proof, Runge and Gross showed that all properties are determined by the Schrödinger current $\mathbf{j}(\mathbf{r},t) : t > 0$. This leads to a time-dependent *current* density functional theory. In fact, according to the chapter by Vignale and Kohn in the present book, it may be preferable to use current density functional theory because the physics of delayed exchange and correlation is more spatially local in terms of the current than it is in terms of the density. In this chapter we will mainly treat the more traditional density functional approach, however.

For the static case, Hohenberg and Kohn established a minimum principle based on a universal functional $F[n]$ of the density. In the time-dependent case there is no minimum principle, but a stationary principle can be established for the action

$$A([n]) = \int_{t_0}^{t_1} < \Psi([n],t) \left| i\frac{\partial}{\partial t} - \hat{H}(t) \right| \Psi([n],t) > \tag{1}$$

which is a functional of the density $n(\mathbf{r},t)$ since the many-body wavefunction $|\Psi([n],t>$ is determined by $n(\mathbf{r},t)$ as noted above. A is stationary with respect to variations of $n(\mathbf{r},t)$,

$$\frac{\delta A}{\delta n(\mathbf{r},t)} = 0.$$

This action formulation of the time-dependent problem has several advantages, but there are some subtle, as-yet unresolved puzzles concerning the constraint of causality in the second functional derivative of A, which is related to the density-density response function studied further on in the present chapter (see p. 110-111 of Ref. [6]). The chapter by Rajagopal on "Effective Action..." contains some ideas about the resolution of this difficulty, by introducing a more sophisticated definition of the action.

Hydrodynamics - Time-Dependent Thomas-Fermi and Beyond

Analogously with the static Thomas-Fermi theory [10,11] there have been many attempts [12,13,14] to describe the time-dependent behavior of degenerate many-Fermion systems *explicitly* in terms of the density $n(\mathbf{r},t)$. In the time-dependent case one needs also a current $\mathbf{j}(\mathbf{r},t)$. Regardless of the level of theory used, conservation of particle number leads to the exact continuity equation $\partial n/\partial t + \nabla\cdot\mathbf{j} = 0$. The problem of hydrodynamics is to provide an equation of motion for $\mathbf{j}$ also. In principle, for fixed initial many-body wavefunction, the quantity $\partial j_\mu/\partial t = -i < \Psi[n](t) \left|[j_\mu,\hat{H}]\right| \Psi[n](t) >$ is a functional of the density $n(\mathbf{r},t)$ because the many-particle wavefunction Ψ depends, via the Runge-Gross theorem, on $n(\mathbf{r},t)$ (see p. 88 of Ref. [6]) . A more heuristic approach is usually taken, however. The chapter by Zaremba discusses these issues further, and applies quite a sophisticated form of hydrodynamics to the problem of plasmons on quantum wells and metallic films.

Time-Dependent Kohn Sham Scheme

We expect, from experience with the static Kohn-Sham formalism compared with the Thomas-Fermi approach, that a wavefunction-based scheme will more readily provide an accurate description of time-dependent phenomena than a hydrodynamic scheme based explicitly on the density. As was done for $n(\mathbf{r})$ in the original exposition [15] of time-independent Kohn-Sham theory, in the time dependent case we postulate that the density $n(\mathbf{r},t)$ is *non-interacting v-representable.* That is, given a time-dependent density $n(\mathbf{r},t)$ pertaining to *interacting* electrons, we postulate that there exists an external potential $v_s(\mathbf{r},t)$ such that *independent* ($\hat{V}_{ee}=0$) electrons moving under the time-dependent Schrödinger equation with potential $v_s(\mathbf{r},t)$, produce the same density $n(\mathbf{r},t)$. For independent electrons the wavefunction is a determinant of one-body orbitals $\psi_k(\mathbf{r},t)$, and the density is a sum of their squares. Then the exact time-dependent density is reproduced by the following *time-dependent Kohn-Sham equations:*

$$n(\mathbf{r},t) = \sum_{k=1}^{N} |\psi_k(\mathbf{r},t)|^2 \tag{2}$$

$$[i\frac{\partial}{\partial t} + \frac{1}{2}\nabla^2 - v_s(\mathbf{r},t)]\psi_k(\mathbf{r},t) = 0. \tag{3}$$

The $\{\psi_k\}$ are the the time-dependent Kohn-Sham (TDKS) wavefunctions, and v_s is the effective TDKS potential.

As in the groundstate KS theory, it is convenient to separate out from the effective potential $v_s(\mathbf{r},t)$ the part which would formally arise in mean-field theory. In the time-dependent case the appropriate mean-field theory is time-dependent Hartree theory. Thus we write

$$v_s(\mathbf{r},t) = v^{ext}(\mathbf{r},t) + \int V_{ee}(\mathbf{r},\mathbf{r}')n(\mathbf{r}',t)d^3r' + v_{xc}([n],\mathbf{r},t). \tag{4}$$

Eq. (4) is the definition of the time-dependent exchange-correlation potential v_{xc}. The hope is, of course, that the first two terms on the right side of (4) (the terms from TD Hartree theory) include enough of the physics that we can afford to make fairly simple, local approximations for the smaller remainder v_{xc}. Note that

- $v_{xc}(\mathbf{r}, t)$ appears in the effective one-body Schrödinger equation (3) in the combination $v_{xc}(\mathbf{r}, t)\psi_k(\mathbf{r}, t)$: that is, it is local both in space and in time
- the same $v_{xc}(\mathbf{r}, t)$ acts for each k: i.e. v_{xc} is orbital-independent
- $v_{xc}(\mathbf{r},t)$ nevertheless has an unknown, delayed, nonlocal dependence on the density $n(\mathbf{r}', t')$. The fact that v_{xc} depends on n is implied by the square bracket in the notation $v_{xc}([n], \mathbf{r}, t)$.

Note that the first two dot-points mean that the TDKS equations are numerically less demanding than (e.g.) the full TD Hartree-Fock equations which involve a nonlocal, orbital-dependent potential. The essential step needed for applications is to find a simple approximation for the dependence of v_{xc} on n.

TIME-DEPENDENT LOCAL-DENSITY APPROXIMATIONS

Adiabatic LDA

The simplest approximation [16,1] for the dependence of v_{xc} on n is local both in space and in time:

$$v_{xc}([n], \mathbf{r}, t) \approx v_{xc}^{ALDA}(\mathbf{r},t) = v_{xc}^{\text{hom},static}(\bar{n} = n(\mathbf{r},t)). \tag{5}$$

Here,

$$v_{xc}^{\text{hom},static}(\bar{n}) = \frac{d}{d\bar{n}}(\bar{n}\epsilon_{xc}(\bar{n})) \tag{6}$$

is the xc potential of a homogeneous gas of density $\bar{n}$ (see the chapter "Introduction to Density Functional Theory" for a discussion of the xc energy per particle, ϵ_{xc}, of the homogeneous gas). The approximation (5) is known these days as the "Adiabatic Local Density Approximation" (ALDA) - the earlier name "time-dependent LDA" is not specific enough, in view of more sophisticated time-dependent approximations developed subsequently.

The ALDA has been remarkably successful, having provided sensible results for dynamical linear polarizabilities of atoms [17], dispersion relations of metal surface plasmons and various related adsorbate interactions on metals [4], and ionization of atoms in strong laser fields [7].

Beyond the ALDA - Linearized Theory

Some of the phenomena just described occur at high frequencies, so it is remarkable how well the ALDA has fared, considering that it ignores frequency dependence (or equivalently, time delay) and takes static data as input. Some insight into this has come from the work of Gross and Kohn on time dependent linear response. Before describing this we digress to discuss the linear density-density response function in general.

Consider some (generally inhomogeneous) many-electron system, initially in its groundstate with density $n(\mathbf{r})$. When a sufficiently small additional external space- and time-dependent potential v_1^{ext} is applied, in general the electron density at some chosen "viewing" point $(\mathbf{r}, t)$ will become $n(\mathbf{r}) + n_1(\mathbf{r}, t)$ where the density perturbation n_1 is *linear* in the applied potential:

$$n_1(\mathbf{r},t) = \int \bar{\chi}(\mathbf{r}, \mathbf{r}', t - t')v_1^{ext}(\mathbf{r}', t')d\mathbf{r}'dt' + O([v_1^{ext}]^2). \tag{7}$$

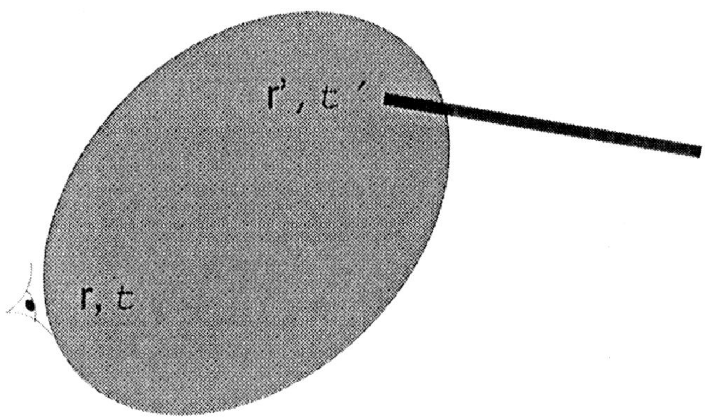

Figure 1. Meaning of $\bar{\chi}(\mathbf{r}, \mathbf{r}\prime, t-t')$: density change at viewpoint $(\mathbf{r}, t)$ (eye sym bol) due to a "poke" (stick symbol) at $(\mathbf{r}\prime, t')$.

Mathematically, the kernel $\bar{\chi}$ in this expression is just the coefficient in the most general linear connection formula between density and potential perturbations. The explicit dependence on $t-t'$, rather than separate dependences on t and t', arises because the groundstate, about which we are perturbing, was time invariant. The unperturbed system was not necessarily spatially invariant (i.e not homogeneous), however, so $\bar{\chi}$ is not a function simply of $\mathbf{r}-\mathbf{r}'$. Physically, $\bar{\chi}(\mathbf{r}, \mathbf{r}', t-t')$ is just the density perturbation at the viewing point ($\mathbf{r}$,t) due to a "unit poke" applied at ($\mathbf{r}'$,t') (i.e. due to an external potential $v_1^{ext}(\mathbf{R}, T) = \delta(t'-T)\delta^3(\mathbf{r}'-\mathbf{R})$). See Figure 1.

If the potential perturbation oscillates in time so that $v_1^{ext}(\mathbf{r}, T) = v_1^{ext}(\mathbf{r})\exp(-i\omega T)$ then (7) becomes $n_1(\mathbf{r}, t) = n_1(\mathbf{r},\omega)\exp(-i\omega t)$ where, to linear order,

$$n_1(\mathbf{r},\omega) = \int \bar{\chi}(\mathbf{r}, \mathbf{r}', \omega)v_1^{ext}(\mathbf{r}', \omega)d\mathbf{r}' \equiv [\bar{\chi} * v_1^{ext}]_{\mathbf{r}}. \tag{8}$$

Here the star represents the generalized spatial convolution integral over $\mathbf{r}'$: $(f*g)(\mathbf{r}) = \int f(\mathbf{r}, \mathbf{r}')g(\mathbf{r}')d^3r'$.

The physics in $\bar{\chi}$ leading to the density change at $(\mathbf{r}, t)$ is as follows:

1. Electrons moving past point $\mathbf{r}'$ at time t' are accelerated by the "poke" (potential) and travel on, some reaching point $\mathbf{r}$ at time t'

2. The changed motion of electrons which experienced the external potential (or have interacted with other electrons which did) and reach points $\mathbf{r}''$ at time t'' causes, via the electron-electron interaction $\hat{V}_{ee}$, further forces on the electrons at $\mathbf{r}$ and hence further changes in density at $\mathbf{r}$.

A special case is that of "independent" electrons which feel a static external potential (giving an inhomogeneous groundstate with one-body wavefunctions and energies $\phi_k(\mathbf{r})$,ϵ_k; $k = 1,..,N$) but suffer no interelectron interaction, $\hat{V}_{ee} = 0$. Then only mechanism 1 contributes to $\bar{\chi}$. The change $\delta\phi_k$ in the one-body wavefunction of each independent electron due to the action of v_1^{ext} can be found from first-order perturbation theory, and the density change at the view point is $n_1(\mathbf{r},t) = \delta(\sum_k \phi_k^*\phi_k)$. By these means one finds a relation of form (8) with $\bar{\chi}$ replaced by the single-particle or

Kohn-Sham response,

$$\chi_{KS}(\mathbf{r},\mathbf{r}',\omega) \equiv \sum_k \sum_{j\neq k} (f_j - f_k) \frac{\phi_k^*(\mathbf{r})\phi_j^*(\mathbf{r}')\phi_k(\mathbf{r}')\phi_j(\mathbf{r})}{\epsilon_j - \epsilon_k - \omega}. \tag{9}$$

where f_k is the Fermi occupation factor. For the case of the homogeneous electron gas, χ_{KS} is the space Fourier transform of the well-known Lindhard function $\chi^{Lindhard}(k,\omega)$ more commonly derived by diagrammatic perturbation theory from the empty-bubble diagram. For independent electrons, then, (8) becomes

$$n_1 = \chi_{KS} * v_1^{ext}. \tag{10}$$

(Many writers [18,6] use the notation χ_s rather than χ_{KS}.)

Now return to the problem of the linear response of *interacting* electrons - i.e., switch $\hat{V}_{ee}$ on again. This interacting problem is described exactly by the TDKS equations (2) and (3). Denoting the selfconsistent groundstate KS density (in the absence of v_1^{ext}) by $n(\mathbf{r})$ we can write the perturbation to the effective potential (4) (due to the effect of v_1^{ext}) as

$$v_{s1}(\mathbf{r},t) = v_1^{ext}(\mathbf{r},t) + \int \frac{n_1(\mathbf{r}',t)}{|\mathbf{r}-\mathbf{r}'|} d^3r' + \int f_{xc}(\mathbf{r},\mathbf{r}',t-t') n_1(\mathbf{r}',t') d^3r' dt' \tag{11}$$

$$\therefore\ v_{s1}(\mathbf{r},\omega) = v_1^{ext}(\mathbf{r},\omega) + \int \frac{n_1(\mathbf{r}',\omega)}{|\mathbf{r}-\mathbf{r}'|} d^3r' + \int f_{xc}(\mathbf{r},\mathbf{r}',\omega) n_1(\mathbf{r}',\omega) d^3r'. \tag{12}$$

Here the *dynamic xc kernel*

$$f_{xc}(\mathbf{r},\mathbf{r}',t-t') = \left.\frac{\delta v_{xc}(\mathbf{r},t)}{\delta n(\mathbf{r}',t')}\right|_{n(\mathbf{r})} \tag{13}$$

is the space-time functional derivative of the effective potential, evaluated at the groundstate density $n(\mathbf{r})$. The term in (11) involving f_{xc} is just the linear change v_{xc1} in the time dependent xc potential v_{xc} defined in (4), due to the effects of the external potential. (For a definition of functional derivatives such as the one in (13) see the chapter "Introduction to Density Functional Theory": here we have generalized to functions of time as well as space.)

The TDKS equations (2) and (3) formally describe independent electrons moving in a total external potential $v_s(\mathbf{r},t)$. The essential observation of Gross and Kohn [18] was that the electronic density therefore responds as though the electrons were independent, and driven by an effective external potential given by (12). Thus using (10) we obtain an exact selfconsistent integral equation for the density perturbation n_1:

$$n_1 = \chi_{KS} * v_{s1} = \chi_{KS} * (v_1^{ext} + V_{ee} * n_1 + f_{xc} * n_1).$$

Writing out the convolutions in full, we find

$$\begin{aligned} n_1(\mathbf{r},\omega) &= \int \chi_{KS}(\mathbf{r},\mathbf{r}',\omega) v_{s1}(\mathbf{r}',\omega) d^3r' \\ &= \int \chi_{KS}(\mathbf{r},\mathbf{r}',\omega)[v_1^{ext}(\mathbf{r}',\omega) + \int (\frac{1}{|\mathbf{r}'-\mathbf{r}''|} + f_{xc}(\mathbf{r}',\mathbf{r}'',\omega)) n_1(\mathbf{r}'',\omega) d^3r''] d^3r'. \end{aligned} \tag{14}$$

Because of the definition (7), we can also write this as an integral equation linking the response $\bar{\chi}$ of interacting electrons to the response χ_{KS} of independent electrons:

$$\bar{\chi} = \chi_{KS} + \chi_{KS} * (V_{ee} + f_{xc}) * \bar{\chi}. \tag{15}$$

This is sometimes referred to as a Dyson-like equation. It can also be regarded as a screening equation because the total effective potential on the right side of Eq. (14) includes the screening effect: electrons move so as to screen out the field felt by other electrons. (Sometimes, near a plasmon resonance, this screening can be negative - i.e. the electron-electron effects enhance the fields).

If f_{xc} is set to zero, (14) or (15) gives the linear time-dependent Hartree (RPA) response, which could alternatively be obtained, after linearization, by the assumption that the time-dependent two-body density is a product of one-body densities. That is, neglect of f_{xc} amounts to neglect of exchange and correlation contributions to the effective field. The xc kernel f_{xc} accounts for these xc effects, and the combination $V_{ee} + f_{xc} \equiv |\mathbf{r} - \mathbf{r}'|^{-1} + f_{xc} \equiv \tilde{V}_{ee}$ can be thought of as an effective pair interaction. The exact screening equation (14) or (15) then has the *form* of a time-dependent Hartree equation with a pair interaction $\tilde{V}_{ee}$.

Gross and Kohn [18] further suggested a doubly local approximation for f_{xc}, which we term here the dynamic LDA:

$$f_{xc}(\mathbf{r}, \mathbf{r}', \omega) \approx f_{xc}^{DLDA} = \delta^3(\mathbf{r} - \mathbf{r}') f_{xc}^{\text{hom}}(\bar{n} = n_0(\mathbf{r}), q = 0, \omega)$$

where $f_{xc}^{\text{hom}}(\bar{n}, q, \omega)$ is the kernel from the homogeneous gas of density $\bar{n}$. For the homogeneous gas, all quantities depend only on differences $|\mathbf{r} - \mathbf{r}'|$, and thus, on space Fourier transformation of (15), the convolution theorem converts all convolutions to simple products. After collecting terms we can thus express (15) as

$$\chi^{\text{hom}}(q, \omega) = \frac{\chi_0(\bar{n}, q, \omega)}{1 - [\frac{4\pi}{q^2} + f_{xc}^{\text{hom}}(\bar{n}, q, \omega)]\chi_0(\bar{n}, q, \omega)} \tag{16}$$

where $\chi_0 = \chi_{KS}^{\text{hom}}$ is the bare density-density response (Lindhard function) of the homogeneous electron gas. Eq. (16) also shows that $-(4\pi)^{-1} q^2 \, f_{xc}^{\text{hom}}(q, \omega) = G(q, \omega)$, where G is the "dynamic local field factor" of the homogeneous gas.

Using the compressibility sum rule for the uniform gas, Gross and Kohn [18] showed that

$$f_{xc}^{\text{hom}}(\bar{n}, q = 0, \omega \to 0) = \frac{d^2}{d\bar{n}^2}(\bar{n}\epsilon_{xc}^{\text{hom}}(\bar{n})) \equiv \frac{d}{d\bar{n}}(v_{xc}^{\text{hom}}(\bar{n})).$$

This value of f_{xc}, if used in (14) for all frequencies, yields the linearized version of the ALDA (see (5) and (6)). Using the third frequency moment sum rule, Gross, Kohn and Iwamoto [19] also obtained exactly the high-frequency kernel $f_{xc}^{\text{hom}}(\bar{n}, q = 0, \omega \to \infty)$ as a more complicated expression involving only static properties of the uniform gas. Using this and other asymptotic properties they proposed an interpolation for the real (frequency-shifting) and imaginary (damping) parts of $f_{xc}^{\text{hom}}(\bar{n}, q = 0, \omega)$ for all frquencies just above the real axis. An examination of this fit (see the Figure in Ref. [19]) reveals the following points:

- $Re(f_{xc})$ starts negative at $\omega = 0$, and approaches a less negative but nonzero value as $\omega \to \infty$.
- $Im(f_{xc})$ starts at zero for $\omega = 0$, has a negative peak and falls again to zero as $\omega \to \infty$.
- For metallic densities, $Re(f_{xc})$ at the plasma frequency is not much different from the zero-frequency value. Only for very low densities (corresponding to $r_s > 5$) does the value at the plasma frequency start to differ widely from the static value. This is basically the reason why the ALDA works well for most inhomogeneous systems.

Here we will term this theory the dynamic LDA (DLDA). There are a few calculations on inhomogeneous systems using the DLDA theory for f_{xc} in inhomogeneous systems. Dobson *et al* [20] investigated multipole plasmons on the jellium model of an aluminium surface, and found only a 3% shift of the plasmon frequency in going from the ALDA (with $f_{xc}(\omega = 0)$) to the Gross-Kohn DLDA (with $f_{xc}(\omega)$). The DLDA did, however, make a 20% increase in plasmon damping compared with the ALDA. More recent calculations [21] on a low-density quantum-well electron gas showed larger effects from the use of the DLDA.

Recently [22] it was found that the DLDA does not satisfy an exact dynamic constraint, the harmonic potential theorem (HPT), which is an extension of the generalized Kohn theorem [23,24,25]. The HPT applies to an inhomogeneous interacting electronic system confined by a harmonic external potential: examples are parabolic quantum wells [3] and certain quantum dots. The theorem guarantees the existence of a mode of oscillation in which the entire many-electron wavefunction moves rigidly with free or driven harmonic motion at arbitrarily large amplitude. To be compatible with this, any theory of xc must predict a rigidly accelerated xc potential when the groundstate density is rigidly acelerated. The ALDA satisfies this condition, unlike the DLDA. This problem was further analyzed by Vignale [26,27] and shown to be related to transformation properties in accelerated reference frames, leading to force sum rules: see also pp 418ff of Ref. [28] and p. 121 of Ref. [6]. This led to a new theory of dynamic exchange and correlation described in full detail for the first time in the chapter by Vignale and Kohn in the present book: see also [29]. The essential feature of the new theory is that, for a description of memory-xc effects which is reasonably local in space, one must go to a *current* density theory involving a *vector* xc potential $\mathbf{a}_{xc}$ rather than a scalar v_{xc}. The linear xc kernel f_{xc} is then replaced by a tensor $\mathbf{f}_{xc}$. The longitudinal or diagonal part of $\mathbf{f}_{xc}$ is the analog of the scalar f_{xc} discussed already in the context of density-density response. The original scalar TD density functional theory would need a very nonlocal kernel $f_{xc}(\mathbf{r}, \mathbf{r}')$ in order to reproduce the longitudinal part of the local tensor kernel $\mathbf{f}_{xc}$ of Vignale and Kohn. This nonlocality can be understood [31] from the linear limit a nonlinear theory in which spatially locality refers to a given fluid element rather than a given point in space. It is not clear whether such a simple interpretation can be found for the transverse part of $\mathbf{f}_{\mathbf{xc}}$, however.

Beyond the ALDA - Nonlinear xc Theory with Memory

As explained above, the frequency dependence of exchange-correlation effects is significant mainly at low electron density. One may expect that nonlinear effects are also stronger in low-density regions, because Pauli stiffness (high KE) at high electron densities causes response to be small and hence presumably describable at the linear level. Thus one might expect that effects beyond the ALDA will be more important in cases of nonlinear response. In fact, xc effects at the 100% level have been obtained in laser ablation of atoms [30], so one might expect frequency dependence to be significant in such cases.

At the level of *linear* response beyond the ALDA, one can either speak of frequency dependent exchange and correlation, $v_{xc} = f_{xc}(\omega) * n_1$, or (by the convolution theorem within time-Fourier transformation) one can speak of *delayed* or *memory* effects, $v_{xc}(t) = \int_{t_0}^{t} f_{xc}(t-t') * n_1(t')dt'$. The ALDA corresponds to a frequency-independent f_{xc}, that is, in the time domain, to an instantanous kernel $f_{xc} \sim \delta(t-t')$ and hence to a lack of memory. In the nonlinear case there is usually no single frequency, so the natural way to speak of these effects is in terms of an *xc potential with memory*.

An initial attempt to describe such nonlinear xc effects with memory was given on pp. 127ff of Ref. [6], but this has had to be modified [31] to conform to the latest theory of linear memory effects as expounded in the chapter by Vignale and Kohn [32].

OTHER USES OF TIME-DEPENDENT DFT

Although the direct study of time-dependent phenomena is the most obvious use for time-dependent DFT, other uses are possible. Two examples follow.

Evaluation of Excitation Energies

By time-dependent perturbation theory, the linear density-density response $\bar{\chi}$ of an inhomogeneous many-electron system as defined in Eq. (8) can be written exactly in terms of the exact interacting many-body groundstate $|0>$ and excited eigenstates $|J>$ of the system, and the corresponding many-body energies E_J:

$$\bar{\chi}(\mathbf{r},\mathbf{r}',\omega)=\sum_J\left[\frac{<J\,|\delta\hat{n}(\mathbf{r})|\,0><0\,|\delta\hat{n}(\mathbf{r}')|\,J>}{E_J-E_0-\hbar\omega}+\frac{<J\,|\delta\hat{n}(\mathbf{r}')|\,0><0\,|\delta\hat{n}(\mathbf{r})|\,J>}{E_J-E_0+\hbar\omega}\right] \tag{17}$$

where $Im(\omega)>0$ and $\delta\hat{n}(\mathbf{r})$ is the number density operator. It follows that $\bar{\chi}$, as a function of frequency, has poles at the many-body excitation energies $E_{J0}=E_J-E_0$ of the interacting system. (This is is contrast to the *static* KS theory where there is no general, exact way to identify excitation energies.) The *time-dependent* Kohn-Sham scheme naturally leads to an evaluation of $\bar{\chi}$ (see e.g. Eq. (15)) starting from χ_{KS}, which by Eq. (9) has poles at differences of the Kohn-Sham eigenvalues ϵ_i: these latter are readily approximated but have no rigorous physical significance. Thus one can hope to find approximate interacting excitation energies via local-density approximations to the time-dependent Kohn-Sham scheme. Gross and co-workers [33] established a scheme based on Eq. (15) leading to an explicit formula correcting the static Kohn-Sham eigenvalue differences $\epsilon_i-\epsilon_j$ toward true excitation energies. A section of the chapter by Petersilka, Gossmann and Gross in the present book is devoted to an improved scheme of this type, giving good values for discrete excitation energies including multiplet splittings. Considering the difficulties with excitation energies and also multiplet effects in the static Kohn Sham theory, this is an important achievement.

Groundstate Correlations From Dynamic Response

Simple approximations for exchange and correlation in time-dependent linear response, as embodied in $\bar{\chi}(\mathbf{r},\mathbf{r}',\omega)$, can also be used to generate more sophisticated approximations for the groundstate xc energy.

As discussed in the chapter "Introduction to density functional theory", the exact groundstate exchange-correlation energy of an arbitrary system is given by the adiabatic connection formula [34,35],

$$E_{xc}=\frac{1}{2}\int_0^1 d\lambda\int d^3rd^3r'\frac{e^2}{|\mathbf{r}-\mathbf{r}'|}\{n_{2\lambda}(\mathbf{r},\mathbf{r}')-n(\mathbf{r})n(\mathbf{r}')\} \tag{18}$$

where the pair density $n_{2\lambda}$ has been evaluated for the groundstate $|\Psi_\lambda>$ with a reduced pair interaction $\lambda e^2\,|\mathbf{r}-\mathbf{r}'|^{-1}$ plus an external potential $v_\lambda(\mathbf{r})$ chosen to keep the density equal to the true ($\lambda=1$) groundstate density. The groundstate density is then

$$n(\mathbf{r})=<\Psi_\lambda|\sum_{i=1}^{N}\delta(\mathbf{r}-\hat{\mathbf{r}}_i)|\Psi_\lambda>\equiv<\Psi_\lambda|\hat{n}(\mathbf{r})|\Psi_\lambda>$$

where $\hat{\mathbf{r}}_i$ is the position operator for the ith particle, and the pair density is

$$\begin{aligned} n_{2\lambda}(\mathbf{r},\mathbf{r}') &= \sum_{i\neq j} < \Psi_\lambda|\delta(\mathbf{r}-\hat{\mathbf{r}}_i)\delta(\mathbf{r}'-\hat{\mathbf{r}}_j)|\Psi_\lambda > \\ &= \sum_i\sum_j < \Psi_\lambda|\delta(\mathbf{r}-\hat{\mathbf{r}}_i)\delta(\mathbf{r}'-\hat{\mathbf{r}}_j)|\Psi_\lambda > - \sum_i < \Psi_\lambda|\delta(\mathbf{r}-\hat{\mathbf{r}}_i)\delta(\mathbf{r}'-\hat{\mathbf{r}}_i)|\Psi_\lambda > \\ &= < \Psi_\lambda|\hat{n}(\mathbf{r})\hat{n}(\mathbf{r}')|\Psi_\lambda > -\delta(\mathbf{r}-\mathbf{r}')n(\mathbf{r}). \end{aligned}$$

It is convenient to define a density fluctuation operator whose expectation is zero:

$$\delta\hat{n}(\mathbf{r}) = \hat{n}(\mathbf{r}) - < \hat{n}(\mathbf{r}) > \equiv \hat{n}(\mathbf{r}) - n(\mathbf{r}).$$

Then

$$\begin{aligned} n_{2\lambda}(\mathbf{r},\mathbf{r}') &= < \Psi_\lambda|(n(\mathbf{r})+\delta\hat{n}(\mathbf{r}))(n(\mathbf{r}')+\delta\hat{n}(\mathbf{r}'))|\Psi_\lambda > -\delta(\mathbf{r}-\mathbf{r}')n(\mathbf{r}) \\ &= n(\mathbf{r})n(\mathbf{r}') + < \Psi_\lambda|\delta\hat{n}(\mathbf{r})\delta\hat{n}(\mathbf{r}')|\Psi_\lambda > -\delta(\mathbf{r}-\mathbf{r}')n(\mathbf{r}) \end{aligned}$$

so (18) becomes

$$E_{xc} = \frac{1}{2}\int_0^1 d\lambda \int d^3r d^3r' \frac{e^2}{|\mathbf{r}-\mathbf{r}'|}\{< \Psi_\lambda|\delta\hat{n}(\mathbf{r})\delta\hat{n}(\mathbf{r}')|\Psi_\lambda > -\delta(\mathbf{r}-\mathbf{r}')n(\mathbf{r})\}. \tag{19}$$

A generalization [36] of the fluctuation-dissipation theorem (FDT) [37] shows that the equal-time fluctuation quantity in (19) can be found from

$$\begin{aligned} &< \Psi_\lambda|\delta\hat{n}(\mathbf{r})\delta\hat{n}(\mathbf{r}') + \delta\hat{n}(\mathbf{r}')\delta\hat{n}(\mathbf{r})|\Psi_\lambda > \equiv C_{n(\mathbf{r})n(\mathbf{r}')}(t=0) \\ &= \frac{1}{2\pi}\int_{-\infty}^{\infty} C_{n(\mathbf{r})n(\mathbf{r}')}(\omega)d\omega \\ &= \frac{1}{2\pi} Im \int_{-\infty}^{\infty} -2\hbar\theta(\omega)[\bar{\chi}_\lambda(\mathbf{r},\mathbf{r}',\omega+i0) + \bar{\chi}_\lambda(\mathbf{r}',\mathbf{r},\omega+i0)]d\omega \\ &= -\frac{\hbar}{\pi}\int_0^{\infty} [\bar{\chi}(\mathbf{r},\mathbf{r}',iu) + \bar{\chi}(\mathbf{r}',\mathbf{r},iu)]du \end{aligned}$$

where the contour along the real frequency axis was deformed to lie up the imaginary frequency axis, where $\bar{\chi}$ can be proven to be real.

It follows from the symmetry of the $\mathbf{r}$ and $\mathbf{r}'$ integrations in (19) that

$$E_{xc} = \frac{1}{2}\int_0^1 d\lambda \int d^3r d^3r' \frac{e^2}{|\mathbf{r}-\mathbf{r}'|}\{-\frac{\hbar}{\pi}\int_0^{\infty} \bar{\chi}_\lambda(\mathbf{r},\mathbf{r}',iu)du - n(\mathbf{r})\delta^3(\mathbf{r}-\mathbf{r}')\}. \tag{20}$$

This form of the adiabatic connection allows one to use a knowledge of time-dependent linear response to construct groundstate correlations. It is worth emphasizing the physical content of the fluctuation-dissipation theorem. The groundstate density - density correlation $< \Psi_\lambda|\delta\hat{n}(\mathbf{r})\delta\hat{n}(\mathbf{r}')|\Psi_\lambda >$ measures the degree to which electrons avoid one another in the groundstate, in the absence of an external perturbation. This quantity is zero in a mean-field (Hartree) description. The FDT relates this to $\bar{\chi}$ which measures how electrons avoid an external probe. * The probe creates a predictable

*The probe referred to here is a unit positive external potential (repulsive delta-function) localized at some point $\mathbf{r}'$ and varying in time as $exp(ut)$. The quantity $\bar{\chi}(\mathbf{r},\mathbf{r}',iu)exp(ut)$ is the linear density perturbation at $\mathbf{r}$ due to this external probe. The typically negative values of the response function $\bar{\chi}$, for $\mathbf{r}$ near to $\mathbf{r}'$, reflect the electrons' tendency to avoid the repulsive potential localized at $\mathbf{r}'$.

distortion ($< \delta\hat{n} > \neq 0$) so that electron-electron interaction appears already at the mean-field (time-dependent Hartree, RPA) level. The nontrivial correlations arising in (20) from this RPA $\bar{\chi}$ are termed RPA correlations. By including f_{xc} in $\bar{\chi}$ via the screening equation (15) we can obtain groundstate correlations in (20) that are beyond the RPA.

This approach is of particular importance in recent attempts to predict van der Waals interactions within a density functional framework. Even the RPA description ($f_{xc} = 0$) leads to an interaction of van der Waals form. One generates an approximate vdW energy directly from the groundstate density by a conserving local-density approximation for χ_{KS}, followed by use of (15) and (20). Inclusion of a carefully constrained local approximation for f_{xc} in the screening equation yields a better vdW approximation. See the chapter by Dobson, Dinte and Wang.

REFERENCES

1. A. Zangwill and P. Soven, Phys. Rev. A **21**, 1561 (1980).
2. J. F. Dobson, Australian J. Phys. **46**, 391 (1993).
3. P. R. Pinsukanjana *et al.* Phys. Rev. B (Rap. Comm.) **46**, 7284 (1992).
4. K.-D. Tsuei *et al.* Surf. Sci. **247**, 302 (1991).
5. J. F. Dobson and G. H. Harris, J. Phys. C: Solid State Phys. **19**, 3971 (1986).
6. E. K. U. Gross, J. F. Dobson, and M. Petersilka, in *Density Functional Theory II*, Vol. 181 of *Topics in Current Chemistry*, edited by R. F. Nalewajski (Springer, Berlin, 1996), Chap. 2, pp. 81 – 172. (Refs. [29] and [31] update this review).
7. C. A. Ullrich, U. J. Gossman, and E. K. U. Gross, Ber. Bunsenges. Phys. Chem. **99**, 488 (1995).
8. C. A. Ullrich, S. Ehrhard, and E. K. U. Gross, in *Super Intense Laser Atom Physics IV*, edited by H. G. Muller (Kluwer, Dordrecht, 1996).
9. E. Runge and E. K. U. Gross, Phys. Rev. Lett. **52**, 997 (1984).
10. L. H. Thomas, Proc. Cambridge Phil. Soc. **23**, 542 (1927).
11. E. Fermi, Z. Phys. **48**, 73 (1928).
12. F. Bloch, Z. Phys. **81**, 363 (1933).
13. G. Barton, Rep. Prog. Phys. **42**, 964 (1979).
14. A. G. Eguiluz, Phys. Rev. B **19**, 1689 (1979).
15. W. Kohn and L. J. Sham, Phys. Rev. **140**, A1133 (1965).
16. T. Ando, Z. Phys. B **26**, 263 (1977).
17. G. D. Mahan and K. R. Subbaswamy, *Local Density Functional Theory of Polarizability* (Plenum, New York, 1990).
18. E. K. U. Gross and W. Kohn, Phys. Rev. Lett. **55**, 2850 (1985).
19. N. Iwamoto and E. K. U. Gross, Phys. Rev. B **35**, 3003 (1987).
20. J. F. Dobson, G. H. Harris, and A. J. O'Connor, J. Phys. Condens. Matt. **2**, 6461 (1988).
21. J. F. Dobson and G. Vignale (unpublished).
22. J. F. Dobson, Phys. Rev. Lett. **73**, 2244 (1994).
23. L. Brey, N. F. Johnson, and B. I. Halperin, Phys. Rev. B **40**, 10647 (1989).
24. S. K. Yip, Phys. Rev. B **43**, 1707 (1991).
25. W. Kohn, Phys. Rev. **123**, 1242 (1961).
26. G. Vignale, Phys. Rev. Lett. **74**, 3233 (1995).
27. G. Vignale, Physics Letts. A **209**, 206 (1995).
28. J. F. Dobson, NATO ASI Series B, Physics **337**, 393 (1995).
29. G. Vignale and W. Kohn, Phys. Rev. Lett. **77**, 2037 (1996).
30. E. K. U. Gross, private communication (1997).
31. J. F. Dobson, M. Bunner, and E. K. U. Gross, Phys. Rev. Lett. **79**, 1905 (1997).
32. G. Vignale and W. Kohn, private communication (1995).
33. M. Petersilka, U. J. Gossmann, and E. K. U. Gross, Phys. Rev. Lett. **76**, 1212 (1996).
34. D. C. Langreth and J. P. Perdew, Sol. State Commun. **17**, 1425 (1975).
35. O. Gunnarsson and B. I. Lundqvist, Phys. Rev. B **13**, 4274 (1976).
36. L. D. Landau and E. Lifshitz, *Statistical Physics* (Addison-Wesley, Reading, Mass., 1969).
37. H. B. Callen and T. A. Welton, Phys. Rev. **83**, 34 (1951).

Part II

INVITED CHAPTERS ON GROUNDSTATE ENERGY FUNCTIONALS

MIXING EXACT EXCHANGE WITH GGA: WHEN TO SAY WHEN

Kieron Burke,[1] John P. Perdew,[2] and Matthias Ernzerhof[2]

[1] Department of Chemistry, Rutgers University-Camden
315 Penn Street, Camden, NJ 08102

[2] Department of Physics and Quantum Theory Group, Tulane University
New Orleans, LA 70118

INTRODUCTION

The local spin density (LSD) approximation [1] has long been the method of choice for solid-state physics calculations. With the advent of generalized gradient approximations (GGAs) [2-7], density functional calculations for bond energies became an inexpensive alternative to traditional *ab-initio* quantum chemical calculations [8]. The recently derived PBE approximation [9] reduces the mean absolute error on a set of 20 small molecules from 31 kcal/mol in LSD to 8 kcal/mol. Both LSD and PBE approximations are non-empirical, in that all their parameters (other than the exchange-correlation energy per electron of a uniform gas) are fundamental constants. PBE is a simplification of the PW91 GGA [5-7].

GGA is not more accurate than LSD for every property of every system, but is usually better for energy differences between more and less inhomogeneous ground states, for reasons that are well-understood [10]. (GGA also provides employment for aspiring young theorists.)

However, Becke [11,12] showed how to do even better, using a mixture of exact exchange and GGA. He reduced errors to about 3 kcal/mol, finding the optimum mixture to lie in the range of 16 -28 % of exact exchange, depending on the choise of GGA. Further empirical parameters were introduced via "B3LYP" [13], arguably the most-used functional in chemistry today.

The present work describes the formal underpinning behind such mixtures in terms of the adiabatic connection, and summarizes recent work in which the validity and accuracy of several hybrid schemes were explored [14-17]. The upshot is a non-empirical prescription for "when to say when." We use atomic units ($e^2 = \hbar = m = 1$) throughout, except for exchange-correlation contributions to atomization energies of molecules, which are in kcal/mol. (1 hartree = 27.21 eV, 1 eV = 627.5 kcal/mol.)

Electronic Density Functional Theory: *Recent Progress and New Directions*
Edited by Dobson *et al.*, Plenum Press, New York, 1998

ADIABATIC CONNECTION

The total ground-state energy of a non-relativistic electronic system is:

$$E = T + V_{ee} + V_{ext}, \tag{1}$$

where T is the kinetic energy, V_{ee} the Coulomb repulsion energy between electrons, and V_{ext} the external potential energy. In the Kohn-Sham scheme [1], in which the wavefunction is non-interacting, the ground-state energy is rewritten as

$$E = T_s + U + V_{ext} + E_{\rm XC}, \tag{2}$$

where T_s is the kinetic energy of the Kohn-Sham wavefunction, and U is the Hartree energy,

$$U = \frac{1}{2}\int d^3r \int d^3r' \, \frac{n(\mathbf{r})\, n(\mathbf{r}')}{|\mathbf{r}-\mathbf{r}'|}. \tag{3}$$

Thus the exchange-correlation energy is

$$E_{\rm XC} = V_{ee} - U + T - T_s, \tag{4}$$

and contains both a potential and a kinetic contribution.

To construct the adiabatic connection, we simply generalize the constrained-search definition of the universal functional [18]

$$F[n] = \min_{\Psi\to n}\langle\Psi|\hat{T}+\hat{V}_{ee}|\Psi\rangle \tag{5}$$

to [19-21]

$$F_\lambda[n] = \min_{\Psi_\lambda\to n}\langle\Psi_\lambda|\hat{T}+\lambda\hat{V}_{ee}|\Psi_\lambda\rangle. \tag{6}$$

Thus Ψ_λ is the ground-state wavefunction of a problem with interelectronic repulsion $\lambda\hat{V}_{ee}$, but which yields n. At $\lambda = 1$, we have the regular interacting wavefunction, while at $\lambda = 0$, we recover the Kohn-Sham non-interacting wavefunction. Note also that the external potential is then λ-dependent, in order to keep the density fixed as λ changes. Inserting this definition into Eq. (4), we have

$$\begin{aligned} E_{\rm XC} &= F_{\lambda=1} - F_{\lambda=0} - U \\ &= \int_0^1 d\lambda \frac{dF_\lambda}{d\lambda} - U \\ &= \int_0^1 d\lambda E_{{\rm XC},\lambda}, \end{aligned} \tag{7}$$

where

$$E_{{\rm XC},\lambda} = \langle\Psi_\lambda|\hat{V}_{ee}|\Psi_\lambda\rangle - U, \tag{8}$$

since only the dependence of the Hamiltonian on λ contributes. Thus we have expressed the exchange-correlation energy in terms of a purely potential contribution, at the price of having to perform an integral over the coupling constant.

Figure 1 shows the adiabatic connection curve for the two-electron Hooke's atom [22], in which the electrons are held to a center by a harmonic potential of radial frequency $\omega = 1/2$. Note that $\lambda = 0$ is the exchange limit, i.e., $E_{{\rm XC},\lambda=0} = E_{\rm X}$. At $\lambda = 1$

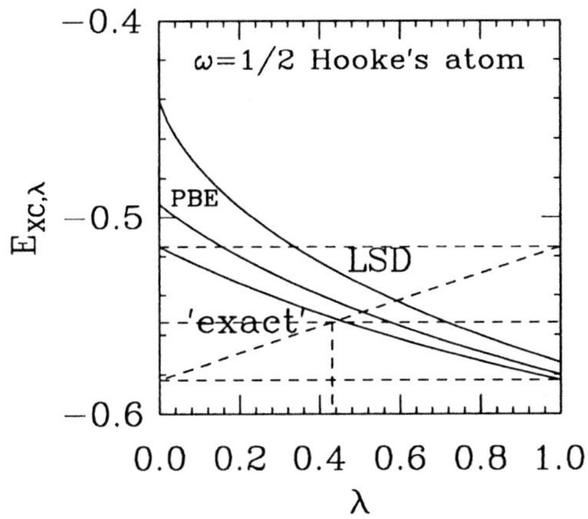

Figure 1. Solid lines: adiabatic connection curves for $\omega = 1/2$ Hooke's atom, within the local density approximation (LSD), the PBE generalized gradient approximation, and "exactly" (see text). The dashed lines show how to construct the correlation parameter b geometrically for the "exact" curve.

we have the fully interacting system, and $E_{XC,\lambda=1}$ is the pure potential contribution to exchange-correlation. The average over λ, which is the area under the curve, gives the actual exchange-correlation energy, $E_{XC} = \int_0^1 d\lambda\ E_{XC,\lambda}$.

Since correlation reduces the energy, these are always ordered:

$$E_X \geq E_{XC} \geq E_{XC,\lambda=1}. \tag{9}$$

Furthermore, since the exchange-correlation energy contains a kinetic component, it is related to the potential contribution via

$$E_{XC} = E_{XC,\lambda=1} + T_C, \tag{10}$$

where $T_C = T - T_s$ is the kinetic contribution to the correlation energy (or the correlation contribution to the kinetic energy).

The 'exact' curve was constructed by fitting three energies known from the exact solution, viz. E_X, E_{XC}, and $E_{XC} - T_C$, to a [1,1]-Padé form in λ [17]. Because it is fitted to the exact values, we expect it to be highly accurate in the range $0 \leq \lambda \leq 1$.

A simple measure of the strength of the correlation in a system is given by the curvature of the adiabatic connection curve in the range $\lambda = 0$ to 1. We write

$$E_{XC} = bE_X + (1 - b)E_{XC,\lambda=1}, \tag{11}$$

and find $0 \leq b \leq 1$ (see problem 1). In fact, for all cases we are aware of, the adiabatic curve is concave upwards, i.e.,

$$\frac{d^2 E_{XC,\lambda}}{d\lambda^2} \geq 0, \tag{12}$$

so that $b \leq 1/2$, as conjectured by Levy [23]. Solving Eq. (11) for b yields

$$b = \frac{T_C}{T_C - E_C}. \tag{13}$$

Table 1. Energies in hartrees and b for the $\omega = 1/2$ Hooke's atom

energy	exact	LSD	PBE
E_X	-0.515	-0.441	-0.493
E_{XC}	-0.554	-0.527	-0.544
$E_{XC,\lambda=1}$	-0.583	-0.574	-0.580
b	0.43	0.35	0.41

For high density (relatively low correlation) systems, the adiabatic curve is almost a straight line, and b is close to 1/2, as in Fig. 1. For low density (relatively strong correlation) systems, the adiabatic curve drops rapidly toward its $\lambda = \infty$ limit, and b is closer to zero. Thus b^{-1} is a measure of the strength of correlation. Figure 2 demonstrates this behavior for $\omega = 0.019$, a low-density analytic solution to Hooke's atom, where $b = 0.12$. The numbers for the $\omega = 1/2$ solution are tabulated in Table I.

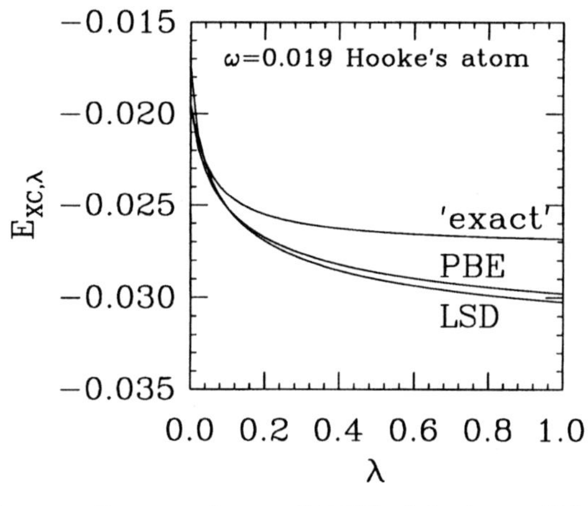

Figure 2. Adiabatic connection curve for $\omega = 0.019$ Hooke's atom, within the local density approximation (LSD), the PBE generalized gradient approximation, and "exactly" (see text).

A simple geometric interpretation can be given to b, and is illustrated for the "exact" curve of Fig. 1 by the dashed lines. Draw a horizontal line at E_{XC} for the adiabatic curve, so that the net area between the horizontal line and the curve is zero. Next draw a backwards diagonal from the point $(\lambda = 0, E_{XC,\lambda=1})$ to the point $(\lambda = 1, E_{XC,\lambda=0})$. The value of λ where these two lines intersect is b.

We conclude this section with a working definition of the nebulous concept of "static" correlation. By static correlation we will mean correlation with little kinetic contribution. From Eq. (13), a system with strong static correlation will have b close to zero. For total energies, this is usually associated with low densities. But for atomization energies of molecules, there is strong static correlation even at typical valence densities.

Although it is barely visible in Fig. 1, note that the LSD approximation has an incorrect logarithmic divergence in its slope as $\lambda \to 0$. This singular behavior is due to the long-range nature of the Coulomb interaction in the extended uniform gas system, and contributes to the LSD overestimate of static correlation in finite systems. For a finite system with non-degenerate Kohn-Sham determinant this slope should be finite, and $E_{\text{XC},\lambda}$ may be written as [23]

$$E_{\text{XC},\lambda} = E_{\text{X}} + \lambda E_{\text{C}}^{(2)} + \dots \ (\lambda \to 0). \tag{14}$$

The real-space construction of a GGA corrects this problem [7], but the PW91 parametrization of its results does not [24]. The recent PBE [9], derived in a simpler fashion, recovers this correct behavior.

CORRELATION MAKES THE HOLE GROW DEEPER

In this section, we analyze the behavior of approximate functionals over the range of the adiabatic curve. The λ-dependent curve can be calculated for any approximate density functional by scaling the density [25,26]:

$$E_{\text{XC},\lambda}[n] = \frac{d}{d\lambda}\Big(\lambda^2 E_{\text{XC}}[n(\mathbf{r}/\lambda)/\lambda^3]\Big). \tag{15}$$

In Fig. 1 we show both the LDA and PBE curves alongside the sketch of the exact curve. We see that both approximate functionals improve their performance as λ grows.

This behavior can be understood in very general terms by considering these approximations as approximations for the (system- and spherically-averaged) exchange-correlation hole [27]. As λ is turned on, correlation deepens the on-top hole [15]. Because the hole is normalized to -1, this implies parts of the hole are being drawn in to smaller separations. Since the LDA and PBE approximations work best for small separations, their accuracy in describing the hole (and therefore in predicting the energy) improves as the hole gets deeper and more short-ranged. Thus they work best at the $\lambda = 1$ end. This explains the well-known cancellation of errors between exchange and correlation for approximate functionals. Since the relative error reduces as correlation is turned on, splitting the exchange-correlation energy into an exchange and a correlation contribution produces larger errors in each than in their sum. In Fig. 3, we show percentage errors in LDA (relative to PW91) for closed-shell atoms. We find

$$|\Delta E_{\text{X}}| \geq |\Delta E_{\text{XC}}| \geq |\Delta E_{\text{XC},\lambda=1}|, \tag{16}$$

consistent with our hole arguments.

NON-EMPIRICAL HYBRID CONSTRUCTION

Consider a situation in which we can calculate any approximate functional energy, and know also the 'exact' exchange energy. Here, by 'exact', we mean the Fock integral evaluated on the Kohn-Sham orbitals, and we ignore small errors in these orbitals from their exact counterparts. Thus we get an approximate adiabatic connection curve from an approximate E_{XC} functional via Eq. (15), plus an exact value at just one point along the curve, $\lambda = 0$. How can we best utilize the additional information in that one point?

A convenient way to express the result of any such procedure is to write the exchange-correlation energy as

$$E_{\text{XC}}^{\text{hyb}} = a(E_{\text{X}} - E_{\text{X}}^{\text{DFA}}) + E_{\text{XC}}^{\text{DFA}}, \tag{17}$$

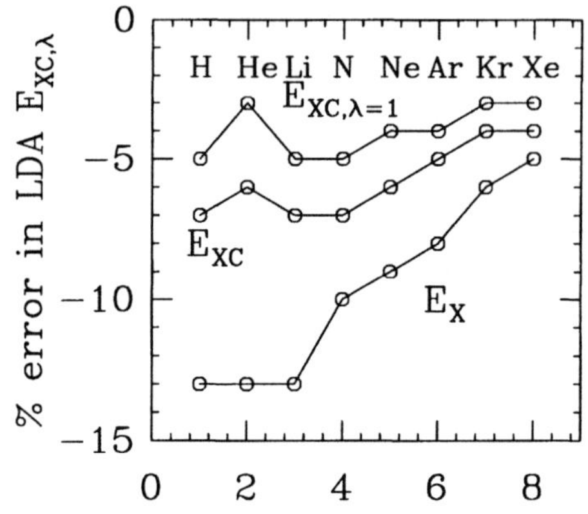

Figure 3. Percent errors made in the three energies discussed in the text for LSD relative to PW91.

where DFA denotes density functional approximation. When the exact exchange energy lies above the DFA exchange-correlation energy, this expression defines a value for $0 \leq a \leq 1$. Becke [12] fitted this form to the atomization energies of many molecules, ionization potentials and proton affinities, finding $0.16 \leq a \leq 0.3$ and reducing the mean error to 2-3 kcal/mol.

Exact exchange plus approximate correlation

A naive solution would be simply to shift the entire approximate curve downward, by the amount of error at $\lambda = 0$. This is equivalent to adding the approximate correlation energy to the exact exchange energy, thereby destroying the cancellation of errors between exchange and correlation mentioned above, and typically producing a larger error than in the original approximation itself. Such a construction ignores the fact that the approximation improves as λ grows.

Half and half

Another crude but more effective scheme is to use a straight line to connect the exact exchange value with the approximation at $\lambda = 1$,

$$E_{\text{XC},\lambda}^{1/2\&1/2} = E_{\text{X}} + \lambda(E_{\text{XC},\lambda=1}^{\text{DFA}} - E_{\text{X}}) \tag{18}$$

which, when integrated over λ, yields

$$E_{\text{XC}}^{1/2\&1/2} = \frac{1}{2}E_{\text{X}} + \frac{1}{2}E_{\text{XC},\lambda=1}^{\text{DFA}}. \tag{19}$$

This model was used by Becke [28], with DFA=LSD, to improve upon LSD results. It has the merit of having a finite slope as $\lambda \to 0$, and becomes exact if $E_{\text{XC},\lambda=1}^{\text{DFA}}$ is exact and the system has high density.

Levy had in fact written an exact half-and-half formula earlier, in the context of the total energy [29]. For the exchange-correlation energy alone, this is

$$E_{\text{XC}} = \frac{1}{2}(E_{\text{XC}} + T_{\text{C}}) + \frac{1}{2}E_{\text{XC},\lambda=1}. \tag{20}$$

The first term has the importance that, *if the exact density is known*, the exchange-correlation potential can be constructed by any of several schemes, and its virial is precisely $E_{\mathrm{XC}} + T_{\mathrm{C}}$. Thus when the exact density is known, an extremely accurate estimate is possible [30], since $E^{\mathrm{DFA}}_{\mathrm{XC},\lambda=1}$ is more accurate than $E^{\mathrm{DFA}}_{\mathrm{XC}}$, and only half of its error appears in Eq. (20).

Two-legged mixing

More recently, we have constructed and tested several non-empirical hybrid schemes. Two of these can be best pictured by mimicking the adiabatic curve by two straight line segments, chosen so that they meet at $\lambda = b$ and $E_{\mathrm{XC},\lambda=b} = E_{\mathrm{XC}}$, i.e., the kink in the curve is at exactly the average. In Fig. 4, the solid lines show both the two-legged representation for the 'exact' and LSD curves for the $\omega = 1/2$ Hooke's atom.

The simplest scheme can be constructed by approximating both b and $E_{\mathrm{XC},\lambda=1}$ in Eq. (11) by their density functional approximations, but using the exact exchange value [31,14]:

$$E^{\mathrm{I}}_{\mathrm{XC}} = b^{\mathrm{DFA}} E_{\mathrm{X}} + (1 - b^{\mathrm{DFA}}) E^{\mathrm{DFA}}_{\mathrm{XC},\lambda=1}. \tag{21}$$

This approximation, applied to LSD, is shown by the line marked I on Fig. 4 and the results listed in Table I. Although it yields almost the exact exchange-correlation

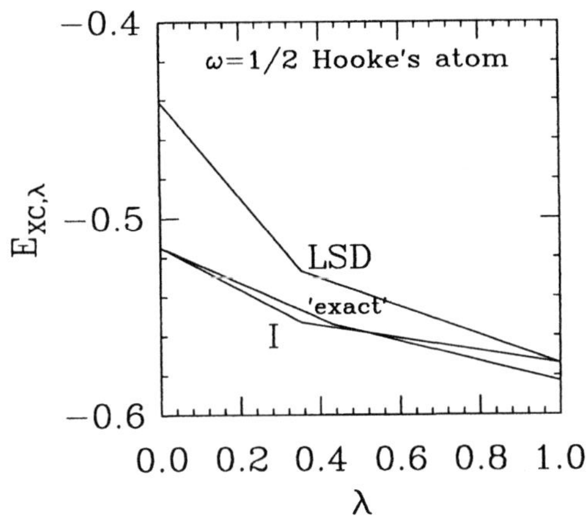

Figure 4. Two-legged line segment representation of adiabatic curves of Fig. 1, and hybrid approximation I (see text).

energy in this case, the agreement is somewhat fortuitous. The value $b^{\mathrm{LSD}} = 0.35$ is too low, due to the excessive static correlation in LSD for this system, but the error is cancelled by the remnant LSD error at $\lambda = 1$. The same fortuitous cancellation often occurs when the hybrid of Eq. (21) is applied to LSD atomization energies of molecules, as shown in Ref. [14].

The atomization energies of molecules are of more practical interest. We apply our reasoning directly to the adiabatic curve of the energy difference between the separated atoms and the molecule. The situation is different from that of Fig. 4, as now $-\Delta E_{\mathrm{X}}$ is always above $-\Delta E^{\mathrm{GGA}}_{\mathrm{XC},\lambda}$. In Fig. 5, we plot both the PW91 curve for the atomization of N_2 and the hybrid I discussed above. Since the exact $-\Delta E_{\mathrm{XC}}$ is about -85 kcal/mol,

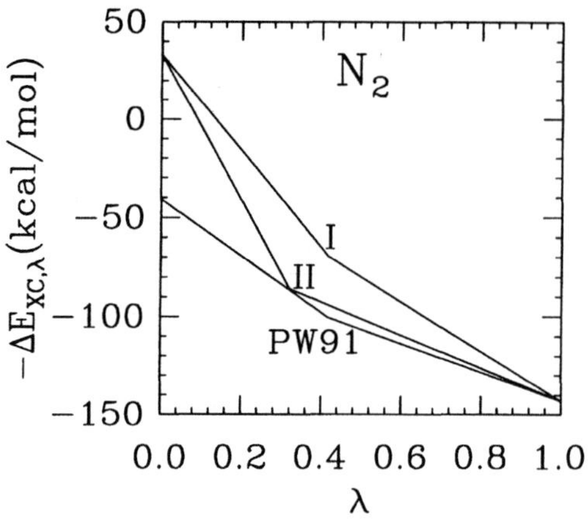

Figure 5. Construction of two-legged line segments representing the adiabatic connection curve for the atomization of N_2. For each case, the average, ΔE_{XC}, is the value at the kink in the curve, $\lambda = b$. The two hybrid curves (I+II) employ the exact ΔE_X at $\lambda = 0$ and the PW91 $\Delta E_{XC,\lambda}$ at $\lambda = 1$, but I uses the PW91 value of b, while II is chosen to be as close to PW91 as possible.

we see that the error in the PW91 curve must be largely coming at the small λ end. We can therefore construct a better hybrid than I, using exactly the same inputs, but now requiring the second line segment to be as close as possible to its GGA counterpart, without actually crossing it. This construction leads to a simple non-empirical hybrid (see Eqs. (15) and (16) of Ref. [15]), and produces the curve marked II in Fig. 5, whose average is -85 kcal/mol. Application of this scheme to many small molecules produces mean errors of about 3-4 kcal/mol, comparable to those of Becke [11,12], and also typically gives a between 0.1 and 0.3. Thus this non-empirical scheme may be viewed as a derivation of Becke's value for a.

Connection to perturbation theory

An alternative construction for the difference between the exact and approximate adiabatic atomization curves provides a heuristic relation between a and n, the approximate order of perturbation theory needed to estimate accurately the atomization energy of molecules. Using the same general idea that the curves should be most similar at the $\lambda = 1$ end, we write [16]

$$E_{XC,\lambda}^{III}(n) = E_{XC,\lambda}^{DFA} + (E_X - E_X^{DFA})(1-\lambda)^{n-1} \tag{22}$$

and choose n as the lowest order in perturbation theory needed to provide a realistic description of the shape of the adiabatic curve. For many small molecules, second order perturbation theory is insufficient, but fourth order is reasonably accurate, so $n = 4$. When we average Eq (22) over λ, we find:

$$E_{XC}^{hyb} = E_{XC}^{DFA} + \frac{1}{n}(E_X - E_X^{DFA}). \tag{23}$$

Thus the need for 4th-order perturbation theory implies $a \approx 1/4$.

An even better hybrid

An additional piece of information about the adiabatic connection curve is provided by the boundary condition [32,33]

$$\lim_{\lambda\to\infty} E_{\rm XC,\lambda} = \text{const.} \tag{24}$$

Since all the schemes discussed so far were designed solely for the range $0 \leq \lambda \leq 1$, none take advantage of this information. On the other hand, the [1/1]-Padé ansatz [17]

$$\Delta E_{\rm XC,\lambda} = a \left[\frac{1+\lambda b}{1+\lambda c}\right] \tag{25}$$

obeys this condition and provides an analytical model for $E_{\rm XC,\lambda}$. The non-empirical parameters in Eq. (25) are obtained with the exact exchange energy, i.e. $E_{\rm XC,\lambda=0}$ and with the value and the derivative with respect to λ of $E^{\rm DFA}_{\rm XC,\lambda}$ at $\lambda = 1$.

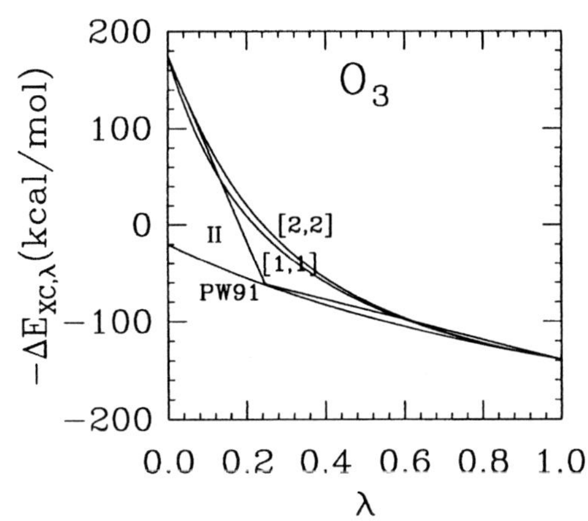

Figure 6. Adiabatic connection curves for O_3 in PW91, two-legged hybrid, [1,1]-Padé, and [2,2]-Padé. The last of these yields essentially the exact atomization energy.

For molecules with weak to moderate correlation, this yields similar results to the two-legged hybrid. But for molecules with very strong static correlation this model leads to more accurate atomization energies [34]. To understand why this is so, consider the atomization of the ozone molecule. In Fig. 6, we plot the adiabatic curves from PW91, the two-legged hybrid, the [1,1]-Padé, and the [2,2]-Padé, discussed below. These yield values of $-\Delta E_{\rm XC}$ of -90, -62, -55, and -50 kcal/mol, respectively, while the exact value is -50 kcal/mol. The two-legged model produces too negative a result because it relies too heavily on the incorrect $\lambda = 0$ information of the PW91 curve.

More sophisticated hybrids

The schemes discussed so far have been limited in how much they can improve upon the DFA result by a paucity of information about the exact adiabatic curve: Just one point at $\lambda = 0$ and general arguments about the nature of the error in the DFA curves. (The silver lining in this cloud is the ease with which they can be implemented – hence the current popularity of such schemes in quantum chemistry.) We can both test the current schemes and hope to improve upon them in cases where they fail by employing more exact information, but at the price of doing a more expensive calculation.

The simplest next step is to use second-order Görling-Levy perturbation theory [35] to find the exact value for the slope of the adiabatic curve at $\lambda = 0$. This information can be incorporated into a [2/2]-Padé model [17] analogous to Eq. (25), but now including the correct slope at $\lambda = 0$. The necessary calculations are technically no more expensive than standard second-order Möller-Plesset perturbation theory. One of us (M. E.) has performed such calculations for small molecules [17]. In most cases, the exact slopes coincide reasonably well with those predicted by the hybrid schemes II, III and Eq. (25). However, in some cases, such as O_3, for which none of the simpler hybrids work so well, the [2/2]-Padé scheme improves the atomization energy significantly. The [2/2]-Padé curve for O_3 is shown in Fig. 6, and yields essentially the exact atomization energy.

CONCLUSIONS

To summarize, we have shown that hybrids of exact exchange and density functional GGA's can be constructed non-empirically. The basic input is the GGA calculation of $E_{XC,\lambda}$ and the exact E_X. Note that some insight into the source of error in the GGA calculation is needed, and that the resulting hybrid only works (i.e., improves on the GGA result) when that insight is correct.

For atoms, Fig. 2 for the low density Hooke's atom indicates a system where the approximate functionals do *worse* at $\lambda = 1$ than $\lambda = 0$. This should also be the case for the $Z \to \infty$ limit of the Be atom [36]. In neither case do we expect our current hybrids to work.

For molecules, our schemes have been developed specifically for atomization energies, where strong non-locality of the exchange hole leads to larger errors at $\lambda = 0$ than at $\lambda = 1$. These schemes will fail when the error does not follow this pattern, e.g., in H_2 where the $\lambda = 0$ end of the adiabatic connection is much better approximated by PW91 and PBE than the $\lambda = 1$ end [37], implying that a more general form of hybrid is required to include these cases. Another difficulty arises when the hybrid is applied directly to energy differences, as this leads to non-unique values for total energies (see problem 3).

The hybrid schemes employed here use only the energy at $\lambda = 0$ (and, in one case, its derivative with respect to λ). However, much more detail at $\lambda = 0$ can easily be calculated, such as the exchange hole of Ref. [27]. Using information from this hole may make possible more sophisticated and accurate hybrid calculations, producing, e.g., a hybrid hole [38].

ACKNOWLEDGMENTS

This work was supported by the National Science Foundation under grant DMR95-21353, and in part by the Deutsche Forschungsgemeinschaft. K.B. thanks A. Becke, U. von Barth, and C. Umrigar for discussions.

EXERCISES

1. Show that b, defined in Eq. (11), is always between 0 and 1.

2. Prove that b is never greater than 1/2.
3. Show that if a, defined by Eq. (17), is system-dependent, then applying the hybrid scheme to energy differences leads to different results than applying it to total energies, and then taking differences. This is a formal shortfall for all the non-empirical hybrids discussed here,
4. Construct a hybrid scheme which overcomes this problem [39].

Answers to odd-numbered problems

1. From the variational principle, one may easily show [25] that

$$T_C \geq 0, \ E_C \leq 0 \tag{26}$$

for all densities. Inserting these inequalities into Eq. (13), we find $0 \leq b \leq 1$.

3. Consider a homogeneous diatomic molecule, A_2. The hybrid I defined in Eq. (21), applied directly to the atomization energy yields:

$$\Delta E^{I}_{XC} = b\Delta E_X + (1-b)\Delta E^{DFA}_{XC,\lambda=1} \tag{27}$$

where $\Delta E = 2E(A) - E(A_2)$, and

$$b = \Delta T_C/(\Delta T_C - \Delta E_C). \tag{28}$$

On the other hand, if we write

$$\Delta E_{XC} = 2E^{I}_{XC}(A) - E^{I}_{XC}(A_2), \tag{29}$$

we find

$$\begin{aligned}\Delta E_{XC} = &\ b(A)2E_X(A) - b(A_2)E_X(A_2)\\ &+ [1-b(A)]2E^{DFA}_{XC,\lambda=1}(A) - [1-b(A_2)]E^{DFA}_{XC,\lambda=1}(A_2)\end{aligned} \tag{30}$$

with $b(A)$ given by Eq. (13) evaluated within the DFA for A, and similarly for $b(A_2)$. Only if all values for b are equal will the two methods yield the same result. The same reasoning applies to all the energy difference hybrids discussed here, and also to the parameter a in Eq. (17).

REFERENCES

1. W. Kohn and L.J. Sham, Phys. Rev. **140**, A 1133 (1965).
2. J.P. Perdew, Phys. Rev. Lett. **55**, 1665 (1985); **55**, 2370 (1985) (E).
3. A.D. Becke, Phys. Rev. A **38**, 3098 (1988).
4. C. Lee, W. Yang, and R.G. Parr, Phys. Rev. B **37**, 785 (1988).
5. J.P. Perdew, in *Electronic Structure of Solids '91*, edited by P. Ziesche and H. Eschrig (Akademie Verlag, Berlin, 1991), page 11.
6. J. P. Perdew, J. A. Chevary, S. H. Vosko, K. A. Jackson, M. R. Pederson, D.J. Singh, and C. Fiolhais, Phys. Rev. B **46**, 6671 (1992); **48**, 4978 (1993) (E).
7. K. Burke, J.P. Perdew, and Y. Wang, *Derivation of a generalized gradient approximation: The PW91 density functional*, chapter in the present volume.
8. *Modern Density Functional Theory: A Tool for Chemistry*, edited by J. M. Seminario and P. Politzer (Elsevier, Amsterdam, 1995).
9. J.P. Perdew, K. Burke, and M. Ernzerhof, Phys. Rev. Lett. **77**, 3865 (1996); ibid. **78**, 1396 (1997)(E).
10. A. Zupan, K. Burke, M. Ernzerhof, and J. P. Perdew, J. Chem. Phys. **106**, 10184 (1997).
11. A.D. Becke, J. Chem. Phys. **98**, 5648 (1993).

12. A. Becke, J. Chem. Phys. **104**, 1040 (1996).
13. L.A. Curtiss, K. Raghavachari, P.C. Redfern, and J.A. Pople, J. Chem. Phys. **106**, 1063 (1997), and references therein.
14. M. Ernzerhof, J.P. Perdew, and K. Burke, Int. J. Quantum Chem. **64**, 285 (1997).
15. K. Burke, M. Ernzerhof, and J.P. Perdew, Chem. Phys. Lett. **265**, 115 (1997).
16. J.P. Perdew, M. Ernzerhof, and K. Burke, J. Chem. Phys. **105**, 9982 (1996).
17. M. Ernzerhof, Chem. Phys. Lett. **263**, 499 (1996).
18. M. Levy, Proc. Natl. Acad. Sci. (U.S.A.) **76**, 6062 (1979).
19. D.C. Langreth and J.P. Perdew, Solid State Commun. **17**, 1425 (1975).
20. O. Gunnarsson and B.I. Lundqvist, Phys. Rev. B **13**, 4274 (1976).
21. J. Harris, Phys. Rev. A **29**, 1648 (1984).
22. M. Taut, J. Phys. A **27**, 1045 (1994).
23. M. Levy, in *Recent Developments and Applications of Modern Density Functional Theory*, edited by J. Seminario (Elsevier, Amsterdam, 1996).
24. C. J. Umrigar and X. Gonze, in *High Performance Computing and its Application to the Physical Sciences*, Proceedings of the Mardi Gras 1993 Conference, edited by D. A. Browne *et al* (World Scientific, Singapore, 1993).
25. M. Levy and J.P. Perdew, Phys. Rev. A **32**, 2010 (1985).
26. M. Levy, N.H. March, and N.C. Handy, J. Chem. Phys. **104**, 1989 (1996).
27. K. Burke, chapter in the present volume.
28. A.D. Becke, J. Chem. Phys. **98**, 1372 (1993).
29. M. Levy and J.A. Goldstein, Phys. Rev. B **35**, 7887 (1987).
30. K. Burke, J.P. Perdew, and M. Levy, Phys. Rev. A **53**, R2915 (1996).
31. O. Gritsenko, R. van Leeuwen, and E.J. Baerends, Int. J. Quantum Chem. S**30**, 163 (1997).
32. E. H. Lieb and S. Oxford, Int. J. Quantum Chem. **19**, 427 (1981).
33. M. Levy and J. P. Perdew, Phys. Rev. B **48**, 11638 (1993).
34. M. Ernzerhof, unpublished results.
35. A. Görling and M. Levy, Phys. Rev. B **47**, 13105 (1993).
36. J.P. Perdew, M. Ernzerhof, K. Burke, and A. Savin, Int. J. Quantum Chem. **61**, 197 (1997).
37. M. Ernzerhof, K. Burke, and J. P. Perdew, in *Recent Developments and Applications of Modern Density Functional Theory*, edited by J. Seminario (Elsevier, Amsterdam, 1996).
38. K. Burke, J.P. Perdew, and M. Ernzerhof, Int. J. Quantum Chem. **61**, 287 (1997).
39. K. Burke, F. Cruz, and K.C. Lam, work in progress.

ADIABATIC COUPLING IN THE HELIUM AND THE BERYLLIUM SERIES

Andreas Savin,[1] François Colonna,[1] and Jean-Marie Teuler[2]

[1]Laboratoire de Chimie Théorique (CNRS)
Université Pierre et Marie Curie
4, Place Jussieu
Case 137
F-75252 Paris, France

[2] CNRS-IDRIS
Batiment 506, B.P. 167
91403 Orsay Cedex, France

INTRODUCTION

Switching on the electron-electron interaction has provided a means for a deeper understanding of the Kohn-Sham formalism [1,2,3,4]. In recent years this adiabatic coupling was also used to produce density functionals (see, *e.g.*, Ref. [5]) among which the so-called hybrid schemes [6] had great success. It was conjectured that the local density approximation works well for values of the coupling constant close to one (cf. [7]). Little is known, however, about the actual behavior of the correlation energy in real systems (see, *e.g.*, [8]). The present paper shows results for simple systems (the helium and the beryllium series) often used as a benchmark in density functional calculations, as they are known not to behave properly in the local density approximation (LDA) [9].

In the following some adiabatic connection formulas will be re-derived, then some calculations will be presented along with a comparison with LDA [10] and the Perdew-Burke-Ernzerhof (PBE) generalized gradient approximation [10].

SOME ADIABATIC CONNECTION FORMULAS

The adiabatic connection provides a direct link between calculations using effective one-particle Hamiltonians and those using the exact one. In density functional theory (DFT) it is natural to use the Kohn-Sham Hamiltonian as the first one. Thus Harris and Jones proposed [2]:

$$\hat{H}^{\lambda} = \hat{H}^{0} + \lambda(\hat{V}_{ee} + \hat{V}_{ext} - \hat{V}_{KS}) \tag{1}$$

Electronic Density Functional Theory: *Recent Progress and New Directions*
Edited by Dobson *et al.*, Plenum Press, New York, 1998

where

$$\hat{H}^0 = \hat{H}^{\lambda=0} = \hat{H}_{KS} = \hat{T} + \hat{V}_{KS}. \tag{2}$$

$\hat{T}$, $\hat{V}_{ee}$, $\hat{V}_{ext}$, $\hat{V}_{KS}$ are the operators for the kinetic energy, for the interaction between the electrons, between the electrons and the external (*e.g.*, nuclear) field, and the Kohn-Sham potential, respectively. λ is the coupling constant, which will be considered to vary between zero and one in this paper. For $\lambda = 0$ ($\lambda = 1$) the Kohn-Sham (the exact) Hamiltonian is obtained.

One of the advantages using the Kohn-Sham Hamiltonian is that it yields the exact density. It is possible to keep this property for all λ by a slight change in the definition of $\hat{H}^\lambda$ [3,4]. * In this paper, we will pursue, however, the adiabatic connection as proposed by Harris and Jones (Eq. (1)). We have the following arguments for doing it:

- The relative simplicity of obtaining $\langle \hat{H}^\lambda \rangle$ not keeping the density fixed, with a standard configuration interaction (CI) program. (In our case we used Molpro [13]).
- It gives the same value for the correlation energy (as defined in DFT).
- The change in the density can be calculated. A program trying to keep the density constant would have a criterion of analyzing the difference between the densities at $\lambda \neq 1$ and at $\lambda = 1$. Such a criterion has been checked, and shown to be satisfied within an accuracy comparable to the error produced by using a finite number of Slater determinants and Gaussian basis functions in the CI calculation. This might not be very surprising, as the densities at $\lambda = 0$ and $\lambda = 1$ are required to be the same.

Let us first define some quantities. The energy at coupling strength λ is:

$$E^\lambda = \langle \Psi^\lambda \mid \hat{H}^\lambda \mid \Psi^\lambda \rangle \tag{3}$$

where Ψ^λ is the antisymmetric minimizing wavefunction. $\Psi^{\lambda=0}$ is – in the absence of degeneracy – a Slater determinant (the Kohn-Sham determinant). The correlation energy (in a density functional context) at coupling strength λ is:

$$\tilde{E}_c^\lambda = \langle \Psi^\lambda \mid \hat{H}^\lambda \mid \Psi^\lambda \rangle - \langle \Psi^{\lambda=0} \mid \hat{H}^\lambda \mid \Psi^{\lambda=0} \rangle. \tag{4}$$

For $\lambda = 1$, $\tilde{E}_c^\lambda$ becomes the usual (density functional) correlation energy, while $\tilde{E}_c^{\lambda=0} = 0$. Keeping the density ($n$) constant for all λ one gets similarly:

$$E_c^\lambda[n] = \langle \Psi_n^\lambda \mid \hat{T} + \lambda \hat{V}_{ee} \mid \Psi_n^\lambda \rangle - \langle \Psi_n^{\lambda=0} \mid \hat{T} + \lambda \hat{V}_{ee} \mid \Psi_n^{\lambda=0} \rangle \tag{5}$$

where Ψ_n^λ is the minimizing antisymmetric wavefunction with the supplementary constraint that it yields n (see, *e.g.*, [14]). † Eq. (5) shows that $E_c^\lambda[n]$ is a universal density functional (i.e. it depends on n alone).

As Ψ_n^λ has to satisfy a supplementary constraint with respect to Ψ^λ we have:

$$\tilde{E}_c^\lambda \leq E_c^\lambda[n] \tag{6}$$

*The Hohenberg-Kohn theorem [12] can be applied to any λ. The correct density is imposed by adding a potential V^λ

†The contribution of the external potential is the same in $\langle \hat{H}^\lambda \rangle$ for all λ and disappears on the r.h.s of Eq. (5).

Let us now show some properties of $\tilde{E}_c^\lambda$. Following Harris and Jones in using the Hellmann-Feynman theorem:

$$\begin{aligned}\frac{\partial \tilde{E}_c^\lambda}{\partial \lambda} &= \langle \Psi^\lambda \mid \frac{\partial \hat{H}^\lambda}{\partial \lambda} \mid \Psi^\lambda \rangle - \langle \Psi^{\lambda=0} \mid \frac{\partial \hat{H}^\lambda}{\partial \lambda} \mid \Psi^{\lambda=0} \rangle \\ &= \langle \Psi^\lambda \mid \hat{V}_{ee} \mid \Psi^\lambda \rangle - \langle \Psi^{\lambda=0} \mid \hat{V}_{ee} \mid \Psi^{\lambda=0} \rangle + \\ &\quad \langle \Psi^\lambda \mid V_{ext} - \hat{V}_{KS} \mid \Psi^\lambda \rangle - \langle \Psi^{\lambda=0} \mid V_{ext} - \hat{V}_{KS} \mid \Psi^{\lambda=0} \rangle \\ &= \langle \Psi^\lambda \mid \hat{V}_{ee} \mid \Psi^\lambda \rangle - \langle \Psi^{\lambda=0} \mid \hat{V}_{ee} \mid \Psi^{\lambda=0} \rangle + \\ &\quad \int (n^\lambda - n)(v_{ext} - v_{KS}) d^3 r. \end{aligned} \tag{7}$$

n^λ is the density obtained from Ψ^λ; the last term on the r.h.s. disappears when $n^\lambda = n$. For obtaining numerical values of $\partial \tilde{E}_c^\lambda / \partial \lambda$ we have used the equivalent form:

$$\frac{\partial \tilde{E}_c^\lambda}{\partial \lambda} = \frac{\partial E^\lambda}{\partial \lambda} - \lim_{\lambda \to 0} \frac{\partial E^\lambda}{\partial \lambda}. \tag{8}$$

Following Levy and Perdew (Eq. (65) of Ref. [15]), one can obtain from the variational principle that:

$$\frac{\partial \tilde{E}_c^\lambda}{\partial \lambda} \leq 0. \tag{9}$$

While the two-body density matrix is needed in order to provide $\partial \tilde{E}_c^\lambda / \partial \lambda$ it can be readily shown that only the first-order density matrix is needed for $\partial (\tilde{E}_c^\lambda / \lambda) / \partial \lambda$ [16]. By applying again the Hellmann-Feynman theorem one gets:

$$\frac{\partial (\tilde{E}_c^\lambda / \lambda)}{\partial \lambda} = -\frac{1}{\lambda^2} [\langle \Psi^\lambda \mid \hat{T} + \hat{V}_{KS} \mid \Psi^\lambda \rangle - \langle \Psi^{\lambda=0} \mid \hat{T} + \hat{V}_{KS} \mid \Psi^{\lambda=0} \rangle] \tag{10}$$

It follows from the preceding equation, and from the fact that $\Psi^{\lambda=0}$ is minimizing $\hat{T}$, that:

$$\frac{\partial (\tilde{E}_c^\lambda / \lambda)}{\partial \lambda} \leq 0. \tag{11}$$

From Eqs. (7), (10) one gets:

$$\begin{aligned}\tilde{E}_c^\lambda &= \int_0^\lambda d\bar{\lambda} [\langle \Psi^{\bar{\lambda}} \mid \hat{T} + \hat{V}_{KS} \mid \Psi^{\bar{\lambda}} \rangle - \langle \Psi^{\lambda=0} \mid \hat{T} + \hat{V}_{KS} \mid \Psi^{\lambda=0} \rangle] + \\ &\quad \int_0^\lambda d\bar{\lambda} \int (n^{\bar{\lambda}} - n)(v_{ext} - v_{KS}) d^3 r \\ &= -\lambda \int_0^\lambda \frac{d\bar{\lambda}}{\bar{\lambda}^2} [\langle \Psi^{\bar{\lambda}} \mid \hat{T} + \hat{V}_{KS} \mid \Psi^{\bar{\lambda}} \rangle - \langle \Psi^{\lambda=0} \mid \hat{T} + \hat{V}_{KS} \mid \Psi^{\lambda=0} \rangle]. \end{aligned} \tag{12}$$

$\tilde{E}_c^\lambda$ IN THE He- AND THE Be-SERIES

Reliable densities of iso-electronic ions were obtained from configuration interaction calculations [13]. Uncontracted even tempered basis sets were used (see Appendix). One-electron systems were treated at Hartree-Fock level. Two- and three-electron systems were described within CI with single and double excitations; in the Be-series, the orbitals were determined in a multi-configuration calculation using the determinants in

Table 1. Correlation energies in mhartree

Atom	exact	CI-HF	DF
H^-	-42.0	-39.7	-41.9
He	-42.0	-41.6	-41.6
Li^+	-43.5	-42.9	-42.9
Be^{2+}	-44.3	-43.6	-43.6
B^{3+}	-44.7	-44.0	-44.0
C^{4+}	-45.1	-44.3	-44.3
N^{5+}	-45.3	-44.5	-44.5
O^{6+}	-45.5	-44.6	-44.9
F^{7+}	-45.6	-44.8	-45.0
Ne^{8+}	-45.7	-44.9	-45.1
Li^-	-72.7	-69.0	-75.8
Be	-94.3	-90.9	-92.8
B^+	-111.4	-108.8	-111.5
C^{2+}	-126.4	-123.8	-127.8
N^{3+}	-140.5	-137.8	-143.4
O^{4+}	-154.1	-151.2	-158.7
F^{5+}	-167.3	-163.9	-173.9
Ne^{6+}	-180.4	-176.8	-189.9

the space including the 2s and 2p orbitals, from which singly and doubly excited determinants were produced. The (quantum chemical) correlation energies obtained are shown in Table 1, where they are compared to estimates of the 'exact' correlation energies. The 'exact' values according to the traditional definition of Wigner and Löwdin [23,24] are taken from Ref. [25], and correspond to the difference between corrected experimental values and Hartree-Fock energies. The 'CI-HF'-values correspond to the difference between our (limited basis set) configuration interaction and Hartree-Fock calculations. The 'DF'-values are those which were obtained with the same basis sets, but according to the definitions given in Eqs. (4) or (10).

Next, Kohn-Sham potentials were generated from these densities. For the He series, only the direct inversion of the Kohn-Sham equation was needed:

$$v_{KS} = \epsilon - \frac{1}{2}\frac{\nabla^2 \varphi}{\varphi} \tag{13}$$

where $\varphi = \sqrt{n/2}$ and ϵ equals minus the ionization potential of the two-electron series. In the Be-series the algorithm is described in Refs. [17,18]: some details are given in the Appendix.

Finally, the matrix elements of the one-body matrix were modified in the program, according to Eq. (1), and the two-electron integrals were multiplied by λ. CI calculations were performed for $\sim$ 30 values of λ between 0 and 1. In order to measure the change of the density we have also calculated the integral:

$$\int \frac{(n^\lambda(\mathbf{r}_1) - n(\mathbf{r}_1))(n^\lambda(\mathbf{r}_2) - n(\mathbf{r}_2))}{r_{12}} d^3r_1 d^3r_2 \tag{14}$$

(cf. Eq. (43) of Ref. [18] and Eq. (7)). The maximum absolute value we get for this integral occurs around $\lambda \approx 0.4$ and is of the order of 10^{-5} hartree, indicating that no important change of the density is present when $0 < \lambda < 1$.

The values of $\tilde{E}_c^\lambda$ are plotted in Fig. 1a. Fig 1b shows the ratio $\tilde{E}_c^\lambda / E_c$. All curves in Fig. 1b are similar. This suggests that a universal behavior of $\tilde{E}_c^\lambda / E_c \approx f(\lambda)$ might

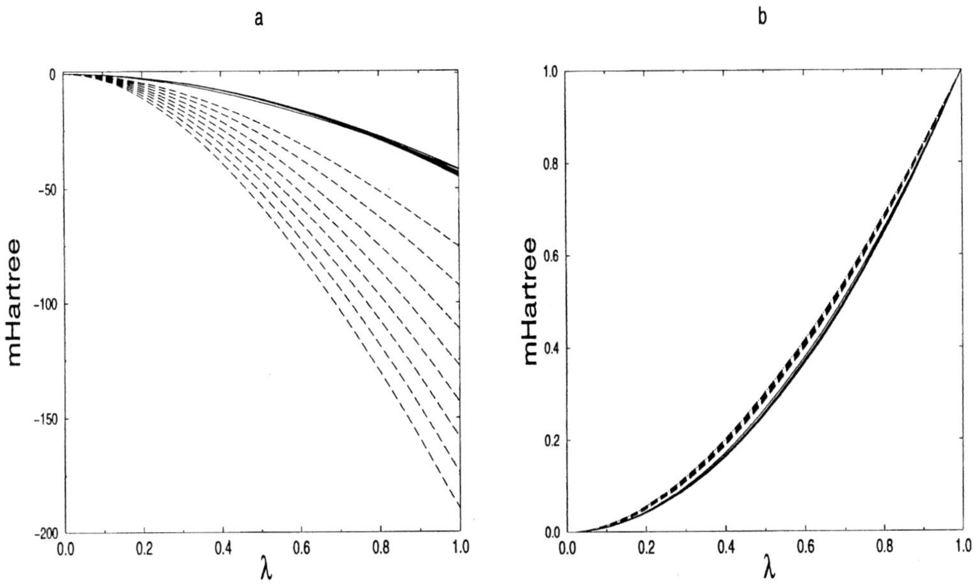

Figure 1. a) Correlation energies $\tilde{E}_c^\lambda$ in the He- and in the Be-series calculated from Eq. (4). All the curves in the He-series (solid lines) appear superimposed at this resolution ; the curves of Li^- to Ne^{6+} (dashed lines) can be easily identified by the monotonic decrease of $\tilde{E}_c^\lambda$ with Z within the series. b) Ratio $\tilde{E}_c^\lambda / E_c$. The H^- curve appears on top of the dashed Be-series curve.

be used in some computational scheme for approximating E_c, if, at a given value $\lambda^\star$ of λ, $E_c^{\lambda^\star}$ can be estimated.

$$E_c = E_c^{\lambda^\star} / f(\lambda)$$

An error in $f(\lambda)$, δf propagates however as $\delta \tilde{E}_c^\lambda = -\tilde{E}_c^{\lambda^\star} \delta f / f^2$ and $f \leq 1$.

Instead of tabulating the values for $\tilde{E}_c^\lambda$, an approximate fit was made with *Mathematica* [19]:

$$\tilde{E}_c^\lambda = E_c \sum_{i=2}^{i_{max}} e_i \lambda^i \tag{15}$$

The values of the coefficients e_i are given in Table 2 (E_c are given in column 'DF' of Table 1).

This is not an expansion in terms of λ. Rather, the coefficients e_i are those of a perturbational series. They also will change with the degree of the polynomial (i_{max}) used for fitting. We have noticed, however, that the coefficients of the lower powers do not change significantly with increasing i_{max}.

Note that the confluence of curves at $\lambda \to 0$ in Fig. 1b comes from the $\approx \lambda^2$ behavior of $\tilde{E}_c^\lambda / \tilde{E}_c$. The e_2 coefficient, is however, system sensitive. For the two-electron systems (with Z $\geq$ 2) it is close to 1. Near-degeneracy increases its value. For a uniform electron gas e_2 has a lnλ divergence.

Figs. 2 and 3 show $\partial \tilde{E}_c^\lambda / \partial \lambda$ and $\partial (\tilde{E}_c^\lambda / \lambda) / \partial \lambda$ vs. λ respectively. Note that, in accordance with Eq. (9), $\partial \tilde{E}_c^\lambda / \partial \lambda$ is monotonically decreasing, and that $\partial (\tilde{E}_c^\lambda / \lambda) / \partial \lambda$ is always negative according to Eq. (11).

Note also that, in accordance with Levy's conjecture [20] the curvature of $\partial \tilde{E}_c^\lambda / \partial \lambda$ is positive and also that all $\partial (\tilde{E}_c^\lambda / \lambda) / \partial \lambda$ increase monotonically with λ.

Table 2. Coefficients of λ^i of the polynomial fit to $\tilde{E}_c^\lambda/E_c$, cf. Eq. (15) (in mhartree)

Atom	e_2	e_3	e_4	e_5
H^-	1.5496	-0.8892	0.4673	-0.1277
He	1.1428	-0.1640	0.0237	-0.0025
Li^+	1.0873	-0.0951	0.0084	-0.0006
Be^{2+}	1.0633	-0.0681	0.0055	-0.0007
B^{3+}	1.0494	-0.0521	0.0029	-0.0002
C^{4+}	1.0407	-0.0426	0.0022	-0.0003
N^{5+}	1.0343	-0.0354	0.0010	0.0000
O^{6+}	1.0300	-0.0314	0.0017	-0.0003
F^{7+}	1.0266	-0.0278	0.0016	-0.0004
Ne^{8+}	1.0234	-0.0234	-0.0003	0.0003
Li^-	1.4412	-0.7618	0.4392	-0.1186
Be	1.3507	-0.5019	0.1863	-0.0351
B^+	1.3332	-0.4580	0.1495	-0.0246
C^{2+}	1.3594	-0.5091	0.1836	-0.0340
N^{3+}	1.3918	-0.5736	0.2282	-0.0464
O^{4+}	1.4405	-0.6759	0.3038	-0.0684
F^{5+}	1.5108	-0.8304	0.4246	-0.1050
Ne^{6+}	1.5858	-1.0058	0.5717	-0.1518

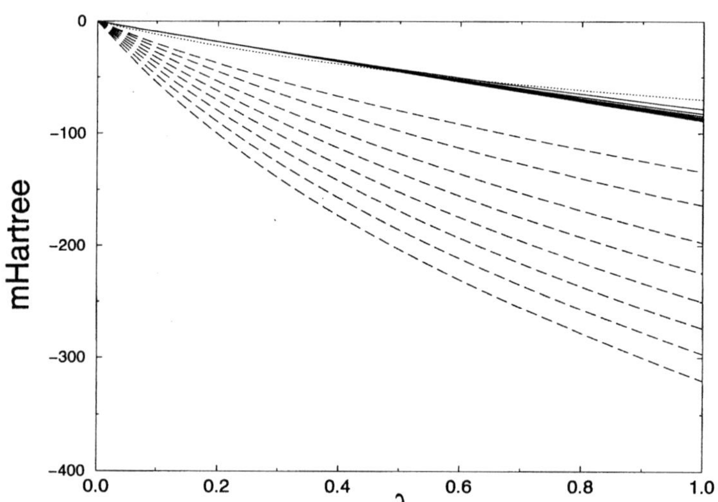

Figure 2. Curves of $\partial\tilde{E}_c^\lambda/\partial\lambda$ calculated from Eq. (8) for the He-series (solid lines, except H^- : dotted line) and for the Be-series (dashed lines); $\partial\tilde{E}_c^\lambda/\partial\lambda$ decreases with Z.

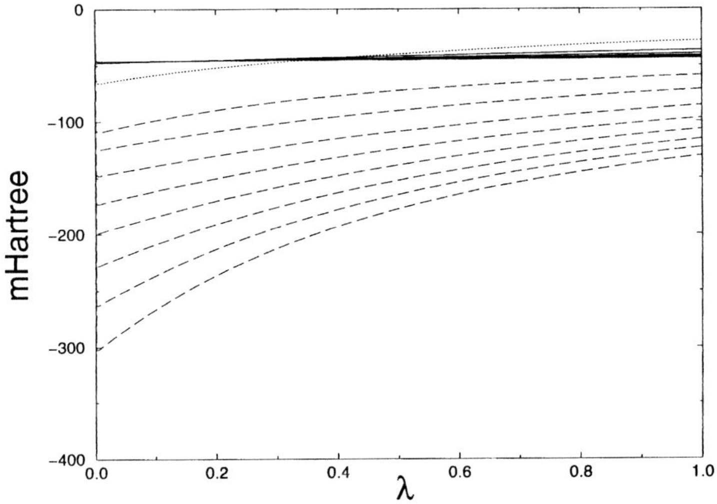

Figure 3. Curves of $\partial(\tilde{E}_c^\lambda/\lambda)/\partial\lambda$ calculated from Eq. (10) for the He-series (solid lines, except H^- : transversal dotted line) and for the Be-series (dashed lines); $\partial(\tilde{E}_c^\lambda/\lambda)/\partial\lambda$ decreases with Z.

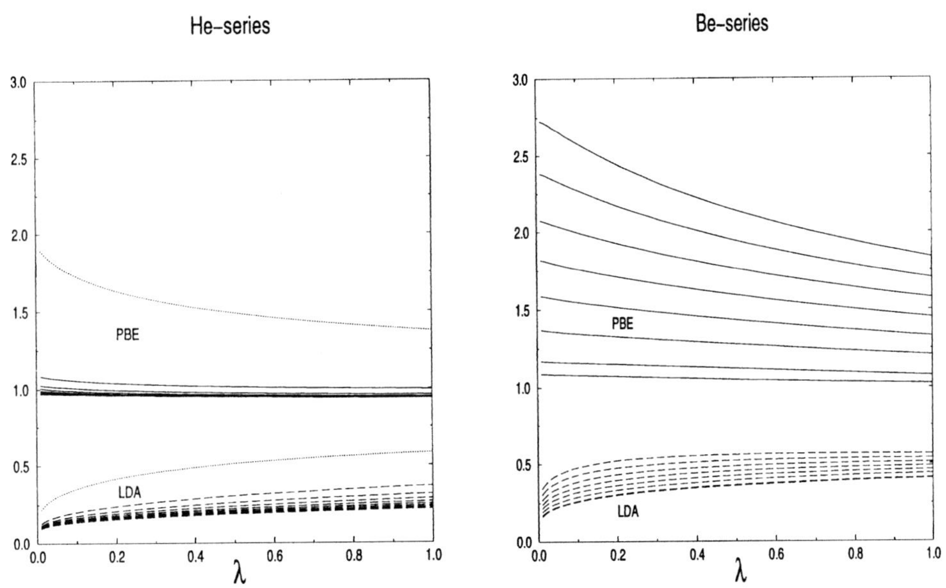

Figure 4. Ratio between exact values of $\tilde{E}_c^\lambda$ calculated from Eq. (4) and those calculated from LDA (dashed lines) or from PBE (solid lines), as a function of λ. All curves appear in the order of their atomic number. Left side: He-series (the two dotted line curves correspond to H^-). Right side: Be-series

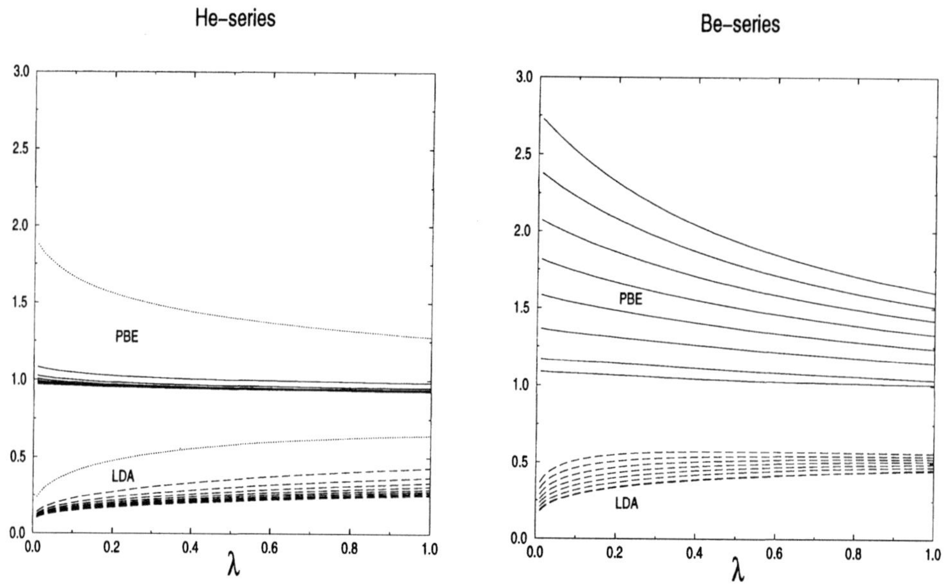

Figure 5. Ratio between exact values of $\partial \tilde{E}_c^\lambda/\partial\lambda$ calculated from Eq. (8) and those calculated from LDA (dashed lines) or from PBE (solid lines), as a function of λ. All curves appear in the order of their atomic number. Left side: He-series (the two dotted line curves correspond to H^-). Right side: Be-series

COMPARISON WITH LDA AND GGA APPROXIMATIONS

Let us now compare our results with approximate ones (LDA or PBE) which were generated via the scaling relationship [21,22]:

$$E_c^\lambda[n] = \frac{1}{\lambda^2} E_c[n_{1/\lambda}] \tag{16}$$

where n_α is the scaled density $n_\alpha(r) = \alpha^3 n(\alpha r)$. We show in Fig. 4 the ratio between $\tilde{E}_c^\lambda$ and the values obtained in density functional approximation. Note the quite good agreement between the exact and PBE values for He and the positive ions of its series. In the Be-series the worsening of the agreement at small λ and large Z is conspicuous. Note also the similarity of the behavior of H^- and the ions of the Be-series. We believe that this is due to the important role of the 2s orbitals in describing the correlation in H^-.

Let us now analyze the quality of the approximate derivative of $\tilde{E}_c^\lambda$. We show in Fig. 5 the ratio between $\partial \tilde{E}_c^\lambda/\partial\lambda$ obtained from our calculations and that obtained from the LDA and the PBE approximations, respectively. While with PBE the ratio stays close to 1 for He and the positive ions of the He-series and at the start of the Be-series, in all the other cases the discrepancy is significant.

It was conjectured that LDA is a better approximation for $\partial \tilde{E}_c^\lambda/\partial\lambda$ when $\lambda \to 1$ [7]. We show in Fig. 6a the ratio between $\partial \tilde{E}_c^\lambda/\partial\lambda$ and the LDA or PBE approximations to it at $\lambda = 1$. Note the similarity of the ratio with that of the exact/approximate correlation energies (Fig. 6b).

Again, a significant improvement of PBE over LDA is manifest in the He-series. In the Be-series a linear increase of the error is, however, also present in PBE.

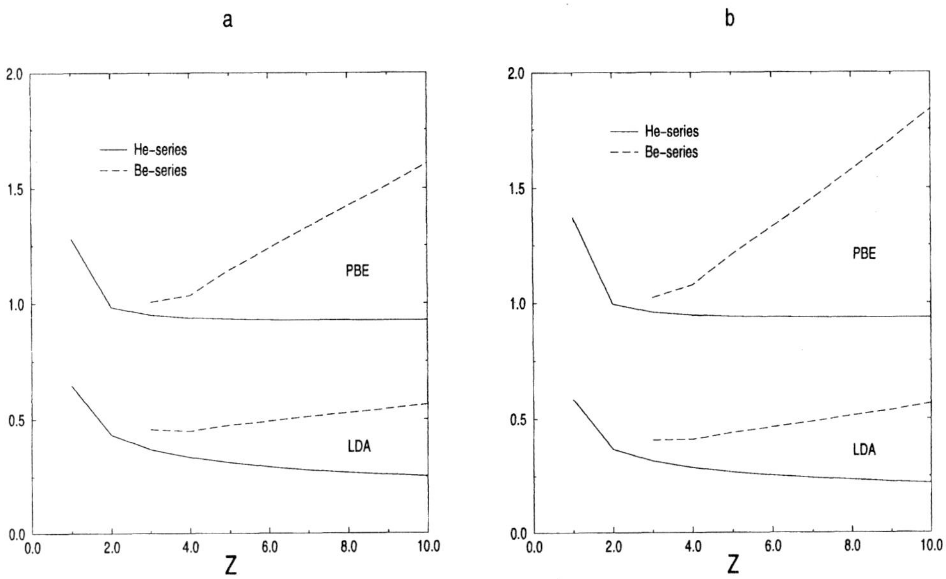

Figure 6. a) Ratio of correlation energies derivatives $\left(\partial\tilde{E}_c^\lambda/\partial\lambda\right)_{\lambda=1}$ calculated from Eq. (8). and those calculated from LDA (lower set of curves) or from PBE (upper set of curves), plotted against the nuclear charge Z. He-series are in solid lines and Be-series in dashed lines, b) the same for correlation energies E_c calculated from Eq. (4)

The simplest approximation for E_c based on the adiabatic scaling is to make a linear approximation for $\partial\tilde{E}_c^\lambda/\partial\lambda$ (cf. Ref. [7] and references therein). This would yield

$$E_c \approx 1/2\left(\partial\tilde{E}_c^\lambda/\partial\lambda\right)_{\lambda=1} \tag{17}$$

(remember that $\left(\partial\tilde{E}_c^\lambda/\partial\lambda\right)_{\lambda=0} = 0$).

In order to show the deviation from this simple approximation we give in Table 3 the ratio:

$$d = 1 - \frac{1/2\left(\partial\tilde{E}_c^\lambda/\partial\lambda\right)_{\lambda=1}}{E_c} \tag{18}$$

Note that the non-linear character (d $\geq$ 0.12) is more pronounced in the Be-series and for H^-, than for He and the positive ions of its series (d $\leq$ 0.06).

CONCLUSIONS

Reliable data have been produced to show the change of the correlation energy with the adiabatic coupling constant. They show the progress made by gradient corrections as well as the room left for further improvement. We hope that the data presented in the paper will be of use in improving the approximate density functionals.

Table 3. Deviation from the linear approximation to E_c (cf. Eq. (17),(18))

Atom	d
He series	
H^-	0.166
He	0.062
Li^+	0.040
Be^{2+}	0.030
B^{3+}	0.023
C^{4+}	0.019
N^{5+}	0.017
O^{6+}	0.017
F^{7+}	0.015
Ne^{8+}	0.014
Be series	
Li^-	0.115
Be	0.116
B^+	0.116
C^{2+}	0.121
N^{3+}	0.128
O^{4+}	0.136
F^{5+}	0.147
Ne^{6+}	0.157

APPENDIX

Basis Sets

Even tempered Gaussian basis sets were used in the calculations. For each angular quantum number, M exponents α_n were produced by the rule $\alpha_n = \alpha_c \delta^{(2n-M+1)/2}$, $n = 1, \ldots$,M. The central exponents α_c, can be found in Table 4, $\delta = 2.1$.

The Kohn-Sham potential and orbitals in the Be-series

For the Be-series we used a numerical procedure to obtain the Kohn-Sham potential and orbitals which has been described in the literature [29,30,31,18,17]. For a given density, n, the orbitals are given by:

$$\begin{aligned} \varphi_1(r) &= \sqrt{n(r)/2}\cos(\theta(r)) \\ \varphi_2(r) &= \sqrt{n(r)/2}\sin(\theta(r)) \end{aligned}$$

where $\theta(r)$ is the solution of a second order differential equation (cf. Ref. [17,18]). Its expression in spherical coordinates is:

$$\frac{d^2\theta(r)}{dr^2} + \frac{d\theta(r)}{dr}\left(\frac{2}{r} + \frac{1}{n(r)}\frac{dn(r)}{dr}\right) = \epsilon \sin(2\theta(r)) \tag{19}$$

subject to the boundary conditions $\theta(\infty) = -\pi/2$, $\theta'(0) = 0$ and $\theta'(\infty) = 0$, the orthogonality of φ_1 and φ_2 being enforced by a Lagrangian multiplier $\epsilon = \epsilon_1 - \epsilon_2$ (see, *e.g.*, Eq. (16) in Ref. [17]), where ϵ_1 and ϵ_2 are the energies associated with $\varphi_1(r)$ and $\varphi_2(r)$, and $n(r)$ is the 'CI' density. In order to obtain $\theta(r)$ the following procedure was followed:

Table 4. Central exponent α_c of the (21s, 7p, 5d, 3f) and (21s, 9p, 5d, 3f) even tempered basis sets in the He and Be-series

Atom	s	p,d,f
	He series	
H^-	8	0.5
He	40	3
Li^+	86	8
Be^{2+}	350	15
B^{3+}	475	23
C^{4+}	663	34
N^{5+}	990	48
O^{6+}	1200	65
F^{7+}	1612	83
Ne^{8+}	2030	100
	Be series	
Li^-	1.8	0.5
Be	4	1
B^+	4.6	3.3
C^{2+}	10	5
N^{3+}	7.5	8
O^{4+}	12	11
F^{5+}	100	13
Ne^{6+}	130	15

1. Choose a value for ϵ and $\theta(r=0)$
2. Solve the differential equation for these values yielding a set of possible φ_i and Kohn-Sham potential, v. ‡ The latter is obtained by inverting the Kohn-Sham equation (cf. Eq. (18) of Ref. [17]) for the higher eigenvalue ϵ_2 (taken from the ionization energy IP as given by the configuration interaction calculation, see Table 5).
3. Calculate the energy of the non-interacting system for the potential just obtained (by using the Hartree-Fock program [13]): $E[v]$.
4. Calculate $E[v] - \int v(r)n(r)d^3r$

According to Ref. [27], Eq. (3.17), this will be maximal when v is the Kohn-Sham potential. We thus repeat the steps 1 to 4 with different values of ϵ and $\theta(0)$ in order to maximize the above expression (by using the simplex procedure [26]). The resulting values of ϵ and $\theta(0)$ are given in Table 5.

ACKNOWLEDGMENTS

We are grateful to H.-J. Werner for providing us the program Molpro and to J.P. Perdew for sending us the PBE subroutines. We would also like to thank O. Bokanowski, M. Levy and C. Umrigar for stimulating discussions, and W. Yang for comments on the manuscript.

‡We have used the routine `odeint` of Ref. [26], because of the stiffness of the differential equation (19), which proved to be difficult to solve with other procedures.

Table 5. The parameters $\epsilon = \epsilon_1 - \epsilon_2$ and $\theta(0)$ calculated from Eq. (19), and $\epsilon_2 = -\text{IP}$

Atom	ϵ	$\theta(0)$	ϵ_2
Li^-	-2.273	0.141	-0.022
Be	-3.882	0.199	-0.342
B^+	-6.722	0.236	-0.924
C^{2+}	-10.366	0.259	-1.759
N^{3+}	-14.749	0.274	-2.845
O^{4+}	-19.938	0.284	-4.182
F^{5+}	-25.950	0.292	-5.770
Ne^{6+}	-32.739	0.299	-7.608

REFERENCES

1. W. Kohn and L. Sham, Phys. Rev. **A140**, 1133 (1965).
2. J. Harris and R.O. Jones, J. Phys. F **4**, 1170 (1974).
3. D.C. Langreth and J.P. Perdew, Solid State Commun. **17**, 1425 (1975).
4. O.Gunnarsson and B.I. Lundqvist, Phys. Rev. B **13**, 4274 (1976).
5. Q. Zhao, M. Levy, and R. G. Parr, Phys. Rev. **A 47**, 918 (1993).
6. A.D. Becke, J. Chem. Phys. **98**, 1372 and 5648 (1993)
7. K. Burke, J.P. Perdew, and M. Ernzerhof, chapter in this volume.
8. R. Hood, M.Y. Chou, A.J. Williamson, G. Rajagopal, R.J. Needs, and W.M.C. Foulkes, Phys. Rev. Lett. **78**, 3359 (1997).
9. J.P. Perdew, E.R. McMullen, and A. Zunger, Phys. Rev. A **23**, 2785 (1981).
10. S.H. Vosko, L. Wilk, H. Nusair, Can. J. Phys. **58**, 1200 (1980).
11. K. Burke, J.P. Perdew, and M. Ernzerhof, Phys. Rev. Lett. **77**, 3865 (1996).
12. P. Hohenberg and W. Kohn, Phys. Rev. B **136**, 864 (1964)
13. MOLPRO is a package of *ab initio* programs written by H.-J. Werner and P. J. Knowles, with contributions from J. Almlöf, R. D. Amos, M. J. O. Deegan, S. T. Elbert, C. Hampel, W. Meyer, K. Peterson, R. Pitzer, A. J. Stone, and P. R. Taylor.
14. M. Levy, Phys. Rev. A **43**, 4637 (1991).
15. M. Levy and J.P. Perdew, Phys. Rev. A **32**, 2010 (1995)
16. A. Savin, Phys. Rev. A **52**, R1805 (1995)
17. F. Aryasetiawan and M. J. Stott, Phys. Rev. B **34**, 4401 (1986).
18. Q. Zhao and R. G. Parr, Phys. Rev. A **46**, 2337 (1992).
19. Stephen Wolfram, *The Mathematica Book*, 3rd edition (Wolfram Media/Cambridge University Press, 1996).
20. M. Levy, in *Recent Developments and Applications of Modern Density Functional Theory*, edited by J. Seminario (Elsevier, Amsterdam, 1996).
21. M. Levy and J.P. Perdew, in *The Single-Particle Density in Physics and Chemistry*, edited by N.H. March and B.M. Deb (Academic, London, 1987).
22. W. Yang, in *Density Matrices and Density Functionals*, edited by R. Erdahl and V.H. Smith, Jr. (Reidel, Boston, 1987).
23. E. Wigner, Phys. Rev. **46**, 1002 (1934).
24. P.-O. Lowdin, Phys. Rev. **97**, 1474 (1955).
25. S.J. Chakravorty, S.R. Gwaltney, E.R. Davidson, F.A. Parpia, and C. Froese Fischer, Phys. Rev. A **47**, 3649 (1993).
26. W.H. Press, S.A. Teukolsky, W.T. Vetterling, and B.P. Flannery, *Numerical Recipes* (Cambridge Univ. Press, 1992).
27. E.H.Lieb Int. J. Quantum. Chem., 24, 243 (1983).
28. K.A. Dawson and N.H. March, J. Chem. Phys. A **81**, 5850 (1984).
29. N.H. March and R.F. Nalewajski, Phys. Rev. A **35**, 525 (1987).
30. A. Nagy and N.H. March, Phys. Rev. A **39**, 5512 (1989).
31. A. Holas and N.H. March, Phys. Rev. A **44**, 5521 (1991).
32. Q. Zhao and R. G. Parr, J. Chem. Phys. **98**, 543 (1993).

DERIVATION OF A GENERALIZED GRADIENT APPROXIMATION: THE PW91 DENSITY FUNCTIONAL

Kieron Burke,* John P. Perdew, and Yue Wang†

Department of Physics and Quantum Theory Group, Tulane University
New Orleans, LA 70118

OUTLINE

Real-space analysis decomposes the exchange-correlation energy of a many-electron system into contributions from all possible interelectronic separations **u**. The density-gradient expansion of the exchange-correlation hole surrounding an electron has a characteristic structure. Its zeroth-order term, the local spin density (LSD) approximation, is a good approximation to both the hole and its cusp at $u = 0$. When the electron density varies slowly over space, addition of each successive term of higher order in ∇ improves the description of the hole at small u, but worsens it at large u. Starting with the second-order gradient expansion, we cut off the spurious large-u contributions in a way that restores the negativity and normalization constraints on the exchange hole, and the normalization constraint on the correlation hole. This procedure defines numerical generalized gradient approximations (GGA's) for the exchange and correlation energies, using no empirical input. We report the results of this construction in detail. This numerical GGA satisfies the most important exact conditions respected by LSD, plus several more (but not all) exact conditions currently known. The PW91 functional is an analytic fit to this functional, designed to respect several further exact conditions including the Lieb-Oxford bound.

HISTORY

Density functional theory is widely used to calculate the ground-state properties of atoms, molecules, solids, and other many-electron systems[1-4]. In principle, this theory would yield the exact ground-state energy and electron spin densities of any system, if the exact dependence of the exchange-correlation energy on the spin densities were

*Permanent address: Department of Chemistry, Rutgers University-Camden, 315 Penn Street, Camden, NJ 08102

†Current address: Department of Biostatistics, School of Public Health University of North Carolina Chapel Hill, NC 27599-7400

Electronic Density Functional Theory: *Recent Progress and New Directions*
Edited by Dobson *et al.*, Plenum Press, New York, 1998

known. This functional is often written as an integral over the system:

$$E_{\rm XC}[n_\uparrow, n_\downarrow] = \int d^3r\ n(\mathbf{r})\ \epsilon_{\rm XC}([n_\uparrow, n_\downarrow], \mathbf{r}), \tag{1}$$

where $\epsilon_{\rm XC}([n_\uparrow, n_\downarrow], \mathbf{r})$ is the exact exchange-correlation energy per particle, and $n(\mathbf{r}) = n_\uparrow(\mathbf{r}) + n_\downarrow(\mathbf{r})$ is the total electron density. (Square brackets indicate a functional dependence.) In the seminal work which established the basic equations of density functional theory, Kohn and Sham [1] also proposed the local spin density (LSD) approximation, which may be written as

$$E_{\rm XC}^{\rm LSD}[n_\uparrow, n_\downarrow] = \int d^3r\ n(\mathbf{r})\ \epsilon_{\rm XC}^{\rm LSD}(n_\uparrow(\mathbf{r}), n_\downarrow(\mathbf{r})), \tag{2}$$

where $\epsilon_{\rm XC}^{\rm LSD}(n_\uparrow, n_\downarrow)$ is the exchange-correlation energy per particle of a uniform electron gas (jellium). This function is now well-known from quantum Monte Carlo data [5-7], and has been accurately fitted to analytic representations [8-10]; we shall use the parametrization of Ref. [10]. LSD is thus a first-principles approximation, in the sense that its parameters are not fitted empirically to calculated or experimental results for any system other than the one in which its form is exact. By construction, LSD is exact for a uniform system, and a good approximation for slowly-varying systems. However, LSD also provides moderate accuracy for real systems where the density varies rapidly over space, which are beyond its obvious range of validity. Three decades after its proposal, LSD remains a popular approximation for realistic solid-state calculations, although it seriously overestimates the atomization energies of molecules and solids.

However, despite its limited accuracy, LSD is a remarkably *reliable* approximation: It reproduces chemical trends, and provides useful information even about systems unlike any previously studied. We attribute this reliability to the first-principles character of LSD, and to LSD's respect for powerful exact constraints (especially the hole constraints of Eqs. (20)-(22) below) which permit a controlled extrapolation from a system of slowly-varying electron density to any real electronic system. Our aim is to retain those good features of LSD in any generalization thereof. The many length scales present in the electron density make free guesswork a futile way to construct $E_{\rm XC}[n_\uparrow, n_\downarrow]$. To be consistently right in practice, a density functional should be right for compelling reasons.

There have been many attempts to improve upon the accuracy of LSD. The most direct is the gradient expansion approximation (GEA), which was already suggested in the original work of Kohn and Sham [1]. GEA is found by considering LSD as the zeroth-order term in a Taylor series for $E_{\rm XC}[n_\uparrow, n_\downarrow]$ about the uniform density, and adding corrections to the next (second) order in the density gradients. Such an expansion can be rigorously performed. The first corrections to the LSD approximation are (in principle) straightforward to calculate, and the addition of these leading corrections to the exchange-correlation energy functional produces the GEA energy functional:

$$E_{\rm XC}^{\rm GEA}[n_\uparrow, n_\downarrow] = E_{\rm XC}^{\rm LSD}[n_\uparrow, n_\downarrow] + \sum_{\sigma,\sigma'} \int d^3r\ C_{\sigma\sigma'}(n_\uparrow, n_\downarrow) \frac{\nabla n_\sigma}{n_\sigma^{2/3}} \cdot \frac{\nabla n_{\sigma'}}{n_{\sigma'}^{2/3}}, \tag{3}$$

where the coefficients $C_{\sigma\sigma'}(n_\uparrow, n_\downarrow)$, which depend weakly on the density, have been calculated by Rasolt and others [11-15]. For systems of slowly-varying density, GEA should improve on LSD. Unfortunately, for real systems, GEA is often *worse* than LSD, providing the wrong sign for the correction to $E_{\rm XC}^{\rm LSD}$ and for $E_{\rm C}$ itself in atoms, molecules, and solids [14,16,17]. The reasons for this failure are discussed in detail below. In a nutshell, GEA violates Eqs. (20)-(22), which are respected by LSD. This violation is known [18,19] to arise from the unphysical long-range behaviors of the GEA exchange and correlation holes, as discussed in our section on the cutoff procedure.

Ma and Brueckner [14] were among the first to recognize and attempt to cure these problems with GEA, but the pioneering fundamental work was performed later by Langreth and co-workers [18,20,21]. The remedies, called [16] generalized gradient approximations (GGA's), assume the following form for the exchange-correlation energy:

$$E_{\rm XC}^{\rm GGA}[n_\uparrow, n_\downarrow] = \int d^3r\ f(n_\uparrow, n_\downarrow, \nabla n_\uparrow, \nabla n_\downarrow), \tag{4}$$

where the function f is chosen by some set of criteria. Many functions f have been proposed in the literature, but consensus is beginning to develop around several which are qualitatively similar for systems of physical interest. These GGA's have been particularly successful in chemical applications, where they tend to reduce the LSD overestimation of molecular binding energies by about a factor of five. In atomic applications, GGA's greatly improve upon the LSD total energy, but improve the first ionization energy and electron affinity only marginally. In the solid state, the (expanded) GGA lattice constants are sometimes more and sometimes less accurate than those of LSD, although in several cases it still remains to separate the error of GGA from the error of computational approximations such as muffin-tinned potentials or pseudopotentials. For a review, see Refs. [22] and [23]; for more recent results, see Refs. [24-37].

Of course, the semi-local form of Eq. (4) cannot encompass the full nonlocality of the exact $E_{\rm XC}[n_\uparrow, n_\downarrow]$. The exact functional is a "square peg" (literally, in view of its derivative discontinuities [38,39] under changes of particle number) which we are trying to force into the "round hole" of a continuum approximation for simplicity and computational convenience. The accuracy of GGA (and LSD) for $E_{\rm XC}$ depends in part upon system-averaging [40], and so may not carry over to the energy density or the potential [41-43], $v_{\rm XC}^\sigma(\mathbf{r}) = \delta E_{\rm XC}/\delta n_\sigma(\mathbf{r})$. Moreover, the energy density is only defined modulo an additive divergence of a vector field, and the potential modulo an additive constant.

A decade ago, Perdew [19] recognized that a simple non-empirical prescription (the real-space cutoff procedure) could be employed to cure the worst problems of GEA, by cutting off the spurious long-range parts of the GEA exchange and correlation holes in order to restore the most important exact conditions satisfied by LSD. The resulting GGA exchange hole for the neon atom has been displayed in Ref. [44]. Since the only input to this prescription is the GEA exchange-correlation hole, the resulting form for f is completely determined by the cutoff procedure and the properties of the slowly-varying electron gas, i.e., this GGA (and so far no other) is a first-principles approximation.

Applying this procedure for exchange, and analytically fitting the resulting numerically-defined f, produces the Perdew-Wang 1986 (PW86) functional for exchange [19,16]. Subsequently, Becke [45] proposed an alternative GGA for the exchange energy, B88, which reproduces the exact asymptotic ($r \to \infty$) behavior of an exchange energy density, and depends on a single parameter which was adjusted to minimize the error in the exchange energies of the rare-gas atoms; this functional has been widely adopted in quantum chemistry. Still later, Perdew and Wang [46] proposed the PW91 functional for exchange, which slightly modifies Becke's form, in order to restore the GEA for slowly-varying densities, and to satisfy the Lieb-Oxford bound [46,47] and various non-uniform scaling conditions [48-50]. The differences [51] between these functionals for systems of physical interest are probably smaller than the error made in using the GGA form, Eq. (4). Thus the real-space cutoff procedure [19,16] may be considered as a justification for *all three* of these exchange functionals.

Until 1991, however, GGA's for correlation [20,17,52] continued to depend upon at least one empirical parameter, and to differ [51] significantly one from another. Indeed, the non-empirical construction of f from the real-space cutoff procedure proved more difficult for correlation than for exchange. A key input to the real-space construction, an accurate analytic representation of the correlation hole in the uniform electron gas with spin densities $n_\uparrow$ and $n_\downarrow$, has only recently been found [53]. Furthermore, the GEA corrections to the correlation hole are imperfectly known. Using a crude model for short-range correlation, Perdew and Wang [46] performed the real-space cutoff of the hole. They fitted the resulting functional to an analytic form and also added several further analytic terms to satisfy other exact conditions, yielding the PW91 correlation functional. As we shall show here, the original PW91 parametrization holds up when a more refined correlation hole is employed.

Our real-space cutoff procedure provides a *unified* derivation of generalized gradient approximations for the exchange and correlation energies, which brings not only conceptual benefits, but also practical ones: The local form of LSD (Eq. (2)) and the semi-local form of GGA (Eq. (4)) are both better suited to exchange and correlation, treated together, than to either one separately [40]. The PW91 GGA displays a strong cancellation [23] between the nonlocalities of exchange and correlation in the range of valence electron densities. As the "most local" [51] of the GGA's, PW91 is the one least likely to overcorrect the subtle errors of LSD for solids.

Is the PW91 GGA more appropriate for solids or for atoms and molecules? Its GEA hole input is better suited to the former, but its sharp radial real-space cutoffs of the hole density are more realistic for the latter. In a solid, the exact holes can have long tails [18,23], which cannot arise in atoms and other systems with compact electron densities. These tails probably limit the accuracy of GGA in extended systems with strongly inhomogeneous densities, e.g., in the calculation of the metal surface energy [23].

This article presents a detailed derivation of the real-space cutoff procedure and its results. We begin by defining the real-space decomposition [19,54] of the exchange-correlation energy, and explaining its importance. Then we discuss the real-space cutoff procedure, and the numerically-defined exchange-correlation hole and energy which it produces. We test the numerical GGA against a number of exact conditions. We next discuss how the numerical results are fitted to analytic forms. This produces the Perdew-Wang 1991 (PW91) generalized gradient approximation. An outline of a more primitive derivation was published earlier [46]. We use atomic units throughout this paper, in which $e^2 = \hbar = m_e = 1$, so that all energies are in hartrees (27.2116 eV) and all distances in bohrs (0.5292 Å), unless explicitly stated otherwise.

Although the basic idea behind the PW91 is simple, its derivation and form are complicated, and its parameters are not seamlessly meshed [55]. Perdew, Burke, and Ernzerhof [56] have recently constructed a simplified derivation of a simplified and formally-improved version of this functional, in which all parameters (other than those in LSD) are fundamental constants. This "PBE" functional, which we recommend for future numerical calculations and formal advances, has also been used to improve [57] the PW91 model for the exchange hole; the correlation hole for PBE is the PW91 correlation hole presented here. Thus the present article, which provides the only full and detailed derivation of the widely-used PW91 functional, also serves as "deep background" for PBE.

REAL-SPACE DECOMPOSITION AND PROBLEMS WITH GEA

An alternative to the point-wise decomposition of Eq. (1) is to write E_{XC} as an integral over interelectronic separations $\mathbf{u}$:

$$E_{XC} = \frac{N}{2} \int d^3u \, \langle n_{XC}(\mathbf{u}) \rangle \frac{1}{u}, \tag{5}$$

where $N = \int d^3r \, n(\mathbf{r})$ is the total number of electrons, and $\langle n_{XC}(\mathbf{u}) \rangle$ is the system-averaged exchange-correlation hole:

$$\langle n_{XC}(\mathbf{u}) \rangle = \frac{1}{N} \int d^3r \, n(\mathbf{r}) \, n_{XC}(\mathbf{r}, \mathbf{r}+\mathbf{u}). \tag{6}$$

Here $n_{XC}(\mathbf{r}, \mathbf{r}+\mathbf{u})$ is the density at $\mathbf{r}+\mathbf{u}$ of the exchange-correlation hole about an electron at $\mathbf{r}$, defined as

$$\begin{aligned} n_{XC}(\mathbf{r}, \mathbf{r}+\mathbf{u}) &= \int_0^1 d\lambda \, n_{XC,\lambda}(\mathbf{r}, \mathbf{r}+\mathbf{u}) \\ &= \int_0^1 d\lambda \, \Big[\langle \Psi_\lambda | \delta\hat{n}(\mathbf{r}) \, \delta\hat{n}(\mathbf{r}+\mathbf{u}) | \Psi_\lambda \rangle / n(\mathbf{r}) - \delta(\mathbf{u}) \Big], \end{aligned} \tag{7}$$

where $\delta\hat{n}(\mathbf{r}) = \hat{n}(\mathbf{r}) - n(\mathbf{r})$ is the density fluctuation operator. The integral over λ is a coupling-constant integration, in which Ψ_λ is the ground state of the Hamiltonian with electron-electron repulsion λ/u, and with external potential $v_\lambda(\mathbf{r})$ adjusted to keep $\langle \Psi_\lambda | \hat{n}(\mathbf{r}) | \Psi_\lambda \rangle = n(\mathbf{r})$ fixed [59] at the physical or $\lambda = 1$ density.

Splitting the hole into separate exchange and correlation contributions (where exchange arises from the lower limit $\lambda = 0$),

$$n_{XC}(\mathbf{r}, \mathbf{r}+\mathbf{u}) = n_X(\mathbf{r}, \mathbf{r}+\mathbf{u}) + n_C(\mathbf{r}, \mathbf{r}+\mathbf{u}), \tag{8}$$

one may easily show from their definitions [4] that these holes satisfy the following exact conditions [19,60,61]:

$$n_X(\mathbf{r}, \mathbf{r}+\mathbf{u}) \leq 0, \tag{9}$$

$$\int d^3u \, n_X(\mathbf{r}, \mathbf{r}+\mathbf{u}) = -1, \tag{10}$$

and

$$\int d^3u \, n_C(\mathbf{r}, \mathbf{r}+\mathbf{u}) = 0. \tag{11}$$

Clearly the system-averaged holes $\langle n_X(\mathbf{u}) \rangle$ and $\langle n_C(\mathbf{u}) \rangle$ also respect Eqs. (9)-(11). In fact, since the Coulomb repulsion is spherically symmetric, only the spherical average [61] of the hole contributes to E_{XC}, i.e.,

$$E_{XC} = 2\pi N \int_0^\infty du \, u \, \langle n_{XC}(u) \rangle_{\text{sph.av.}}, \tag{12}$$

where

$$\langle n_{XC}(u) \rangle_{\text{sph.av.}} = \frac{1}{4\pi} \int d\Omega_{\mathbf{u}} \, \langle n_{XC}(\mathbf{u}) \rangle. \tag{13}$$

This real-space analysis [19,54] is complemented by the wave vector decomposition [59,60,18,20]. We define the Fourier transform of the system-averaged hole as

$$\langle n_{XC}(\mathbf{k})\rangle = \int d^3u \; e^{(-i\mathbf{k}\cdot\mathbf{u})} \; \langle n_{XC}(\mathbf{u})\rangle, \tag{14}$$

so that

$$\langle n_{XC}(\mathbf{u})\rangle = \int \frac{d^3k}{(2\pi)^3} \; e^{(i\mathbf{k}\cdot\mathbf{u})} \; \langle n_{XC}(\mathbf{k})\rangle. \tag{15}$$

Applying this to Eq. (5) yields

$$E_{XC} = \frac{N}{2}\int \frac{d^3k}{(2\pi)^3} \; \langle n_{XC}(\mathbf{k})\rangle \frac{4\pi}{k^2}. \tag{16}$$

Furthermore, from Eq. (7), one can easily show that the Fourier transform of the hole is related to the coupling constant averaged static structure factor $S(\mathbf{k})$ by [18]

$$\langle n_{XC}(\mathbf{k})\rangle = \int_0^1 d\lambda \; S_\lambda(\mathbf{k}) - 1. \tag{17}$$

Provided that $\langle n_{XC}(\mathbf{u})\rangle$ is normalized and sufficiently localized, the long wavelength (i.e., small wave vector) limit of the hole satisfies the perfect screening sum rule:

$$\lim_{k\to 0}\Big[\langle n_{XC}(\mathbf{k})\rangle + 1\Big] = 0. \tag{18}$$

For example, if the non-oscillatory part of $\langle n_{XC}(\mathbf{u})\rangle$ falls off as u^{-5} or faster when $u \to \infty$, then $\big[\langle n_{XC}(\mathbf{k})\rangle + 1\big]$ is proportional to k^2 when $k \to 0$, as in the uniform electron gas [18] or metallic hydrogen [62].

Now consider the gradient expansion of the exchange-correlation hole itself. The zeroth-order approximation is:

$$n_{XC}^{LSD}(\mathbf{r}, \mathbf{r}+\mathbf{u}) = n_{XC}^{unif}(n_\uparrow(\mathbf{r}), n_\downarrow(\mathbf{r}); u), \tag{19}$$

where $n_{XC}^{unif}(n_\uparrow, n_\downarrow; u)$ is the exchange-correlation hole of the uniform electron gas (jellium) with spin densities $n_\uparrow$ and $n_\downarrow$ at separation u from the electron. Insertion of this hole into Eqs. (5) and (6) reproduces Eq. (2). Because the exact hole is approximated by the hole of another physical system, i.e., that of jellium, $n_{XC}^{LSD}(\mathbf{r}, \mathbf{r}+\mathbf{u})$ and $\langle n_{XC}^{LSD}(\mathbf{u})\rangle$ satisfy the constraints of Eqs. (9)-(11). Even satisfying these constraints, LSD often does *not* approximate the exact hole very well, as the LSD hole cannot include any deviation from spherical symmetry [61]. Nonetheless, LSD yields a good approximation to the spherically- and system-averaged hole, $\langle n_{XC}(u)\rangle_{\text{sph.av.}}$, and satisfaction of the constraints of Eqs. (9)-(11) implies satisfaction of the spherically- and system-averaged constraints:

$$\langle n_X(u)\rangle_{\text{sph.av.}} \le 0, \tag{20}$$

$$4\pi \int_0^\infty du \; u^2 \; \langle n_X(u)\rangle_{\text{sph.av.}} = -1, \tag{21}$$

and

$$4\pi \int_0^\infty du \; u^2 \; \langle n_C(u)\rangle_{\text{sph.av.}} = 0. \tag{22}$$

These conditions constrain [19,61] the integral of Eq. (12), so that LSD yields a good approximation to E_{XC}. Any systematic improvement on LSD should continue to respect Eqs. (20)-(22), and should yield an even better approximation to the system- and spherically-averaged hole, $\langle n_{XC}(u)\rangle_{\text{sph.av.}}$.

Eqs. (20)-(22) alone do not completely explain the accuracy of LSD (or GGA) energies in applications to real systems with rapidly-varying densities. The rest of the story lies in the good accuracy of these approximations in self-consistent calculations of the on-top hole density $\langle n_{XC}(u=0)\rangle$, for reasons that are at least partially understood [63,64,65,40]. This "nearly-exact" condition links the hole to the local density, even when the density is not slowly-varying over space.

The gradient expansion approximation (GEA) to second order in ∇ can be derived as follows: Start with a uniform electron gas, and apply an external potential $v(\mathbf{r})$ which is both weak and slowly-varying over space. Evaluate the exchange-correlation hole and the density $n(\mathbf{r})$ to second order in $v(\mathbf{r})$, then eliminate $v(\mathbf{r})$ to express the hole in terms of $n(\mathbf{r})$ and its low-order derivatives. We apply this prescription to Eq. (6), to find the system-averaged GEA hole density

$$\langle n_{XC}^{\text{GEA}}(\mathbf{u})\rangle = \frac{1}{N}\int d^3r\; n(\mathbf{r})\; n_{XC}^{\text{GEA}}(\mathbf{r},\mathbf{r}+\mathbf{u}). \tag{23}$$

The integrand of Eq. (23) contains a term proportional to $|\nabla n|^2$, and another proportional to $\nabla^2 n$. The latter may be integrated by parts [66] and so included in the former. In the subsequent expressions in this paper, we use the symbol $\tilde{n}_{XC}^{\text{GEA}}(\mathbf{r},\mathbf{r}+\mathbf{u})$ to represent the result of this reduction in order of derivatives, but note that $\tilde{n}_{XC}^{\text{GEA}}$ has no direct physical significance even for a slowly-varying density, and is simply an intermediate quantity for the construction of the system-averaged hole. This leads to a considerable simplification in the form of the GGA functional constructed from the real-space cutoff of the GEA hole, while comparison of the numerical results of Ref. [19] (based upon n_X^{GEA}) with those of Ref. [16] (based upon $\tilde{n}_X^{\text{GEA}}$) shows that GGA exchange energies are little affected by this simplification.

Real-space analysis shows why GEA is typically not an improvement over LSD. Since $n_{XC}^{\text{GEA}}(\mathbf{r},\mathbf{r}+\mathbf{u})$ is an expansion of a hole to finite order, but *not* the hole of any physical system, $\langle n_{XC}^{\text{GEA}}(\mathbf{u})\rangle$ can (and does) violate the negativity and normalization constraints of Eqs. (9)-(11), and hence also of Eqs. (20)-(22), and so is less realistic than LSD.

CUTOFF PROCEDURE

A simple cure for this problem with GEA, suggested by Perdew [19], is to modify the GEA hole in real space so as to restore Eqs. (20)-(22). To avoid any bias in the procedure, this is done in the most straightforward and brutal fashion possible. The negativity condition on the exchange hole is restored by cutting out those pieces where $\tilde{n}_X^{\text{GEA}}(\mathbf{r},\mathbf{r}+\mathbf{u})$ becomes positive, while each normalization condition is restored by introducing a sharp cutoff radius u, outside which the hole is set to zero. Most of the resulting discontinuities are smoothed in the system average (Eq. (6)) of the GGA hole. In the subsections below, we review the details of this construction and its results for both exchange and correlation.

Of course, where the reduced density gradients (s and t as defined below, $|\nabla^2 n|/n^{5/3}$, etc.) are too large, none of these approximations (LSD, GEA, or GGA) should be trusted.

Exchange

In this subsection, we review the real-space cutoff procedure for exchange, whose results were first given by Perdew and Wang [16]. The gradient expansion of the exchange-hole density $n_{\rm X}(\mathbf{r},\mathbf{r}+\mathbf{u})$ is known to second-[66] and even third-order [67] in ∇, and the structure [67] of this expansion is clear : The zeroth-order or LSD term gives the exact hole density at $u=0$, while the first-order term gives the exact contribution of order $\mathbf{u}$. When the density $n(\mathbf{r})$ varies slowly over space, addition of each higher-order term improves the description of the hole at small u, where the m-th order term varies as u^m for $m \leq 3$, but worsens the description for large u, where the m-th order term has a non-oscillatory component proportional to u^{m-4} and a sinusoidally-oscillating component proportional to u^{2m-4}.

By Eq. (5), only a long-range interaction (such as $1/u$) can sample the spurious large-u contribution to $n_{\rm X}^{\rm GEA}$ or $\tilde{n}_{\rm X}^{\rm GEA}$ (or to their correlation counterparts). For a sufficiently short-ranged interaction, the ungeneralized GEA for the exchange-correlation energy would require no real-space cutoff correction and would probably make small relative errors like those of the GEA for the non-interacting kinetic energy (which is essentially its own GGA [68]).

The exchange hole obeys the spin-scaling relation

$$n_{\rm X}\Big([n_\uparrow,n_\downarrow];\mathbf{r},\mathbf{r}'\Big)=\sum_\sigma \frac{n_\sigma(\mathbf{r})}{n(\mathbf{r})}n_{\rm X}\Big([2n_\sigma];\mathbf{r},\mathbf{r}'\Big), \tag{24}$$

where $n_{\rm X}\Big([n];\mathbf{r},\mathbf{r}'\Big)$ is the exchange hole as a functional of the density for a spin-unpolarized system ($n_\uparrow=n_\downarrow=n/2$). This leads to the spin-scaling relation for the exchange energy [69]

$$E_{\rm X}[n_\uparrow,n_\downarrow]=\frac{1}{2}E_{\rm X}[2n_\uparrow]+\frac{1}{2}E_{\rm X}[2n_\downarrow]. \tag{25}$$

Thus we need only the GEA for a spin-unpolarized system, which we write as [16,66]

$$\tilde{n}_{\rm X}^{\rm GEA}(\mathbf{r},\mathbf{r}+\mathbf{u})=-n(\mathbf{r})\ \tilde{y}(\mathbf{r},\mathbf{u})/2, \tag{26}$$

where

$$\tilde{y}(\mathbf{r},\mathbf{u})=J(z)+4L(z)\hat{\mathbf{u}}\cdot\mathbf{s}/3-16M(z)\big(\hat{\mathbf{u}}\cdot\mathbf{s}\big)^2/27-16N(z)s^2/3 \tag{27}$$

and $\tilde{y}(\mathbf{r},0)=1$. Here $\hat{\mathbf{u}}=\mathbf{u}/u$ is a unit vector,

$$\mathbf{s}(\mathbf{r})=\nabla n(\mathbf{r})/\big(2k_F(\mathbf{r})n(\mathbf{r})\big) \tag{28}$$

is the local reduced gradient (which diverges in the exponential tail of the density),

$$k_F(\mathbf{r})=\big(3\pi^2 n(\mathbf{r})\big)^{1/3} \tag{29}$$

is the local Fermi wave vector, and

$$z(\mathbf{r},u)=2k_F(\mathbf{r})\ u \tag{30}$$

is a local reduced separation, measured on the scale of the local Fermi wavelength $2\pi/k_F(\mathbf{r})$. The functions $J(z), L(z), M(z)$, and $N(z)$ are known, oscillating functions of z, given in Eqs. (8)-(11) of Ref. [16], respectively. The function $J(z)$ alone yields the LSD hole. To first order in u, $n_{\rm X}^{\rm GEA} = \tilde{n}_{\rm X}^{\rm GEA} = -n(\mathbf{r}+\mathbf{u})/2$, so that the GEA hole is deeper on the high-density side of the electron.

In our generalized gradient approximation (GGA), the hole is represented as

$$\tilde{n}_{\rm X}^{\rm GGA}(\mathbf{r},\mathbf{r}+\mathbf{u}) = -\frac{1}{2}n(\mathbf{r})\ \tilde{y}(\mathbf{r},\mathbf{u})\ \theta\big(\tilde{y}(\mathbf{r},\mathbf{u})\big)\ \theta\big(u_{\rm X}(\mathbf{r})-u\big), \tag{31}$$

where $\theta(x)$ equals 1 for $x > 0$, and vanishes otherwise. The first step function on the right enforces the negativity condition on $\langle n_{\rm X}(\mathbf{u})\rangle$ of Eq. (20), while the second involves a cutoff separation $u_{\rm X}$ which is chosen to enforce the normalization condition, Eq. (21). Inserting this hole into the real-space decomposition, Eqs. (5) and (6), produces the GGA for exchange, which may be conveniently written as

$$E_{\rm X}^{\rm GGA}[n] = \int d^3r\ n(\mathbf{r})\ \epsilon_{\rm X}^{\rm LSD}\big(n(\mathbf{r})\big)\ F_{\rm X}\big(s(\mathbf{r})\big), \tag{32}$$

where

$$\epsilon_{\rm X}^{\rm LSD}(n) = -3k_F/(4\pi) \tag{33}$$

is the exchange energy per particle for a uniform gas of density n. $F_{\rm X}(s)$ is the enhancement factor over local exchange, which is given by

$$F_{\rm X}(s) = \frac{1}{9}\int_0^{z_{\rm X}} dz\ z\ \tilde{y}_{\rm sph.av.}(z,s), \tag{34}$$

where

$$\tilde{y}_{\rm sph.av.}(z,s) = \frac{1}{4\pi}\int d\Omega_{\mathbf{u}}\ \tilde{y}(z,\mathbf{s},\hat{\mathbf{u}})\ \theta\big(\tilde{y}(z,\mathbf{s},\hat{\mathbf{u}})\big), \tag{35}$$

and where $z_{\rm X} = 2k_F u_{\rm X}$ is the reduced cutoff separation. $z_{\rm X}$ is a function of s determined by the normalization condition, Eq. (21), which becomes

$$-\frac{1}{12\pi}\int_0^{z_{\rm X}} dz\ z^2\ \tilde{y}_{\rm sph.av.}(z,s) = -1. \tag{36}$$

The angular integration over $\Omega_{\mathbf{u}}$ in Eq. (35) is performed analytically (Appendix A) and the z integrations in Eqs. (34) and (36) are performed numerically.

Figure 1 is a plot of $-\tilde{y}_{\rm sph.av.}(z,s)\ z^2/(12\pi)$, the integrand of the normalization integral Eq. (36), as a function of z for $s = 1.0$, for the LSD, GEA, and GGA holes. (For a plot of the GGA hole before spherical averaging, see Fig. 3 of Ref. [70].) The LSD hole falls off in magnitude as $z \to \infty$, and is correctly normalized. The GEA hole, on the other hand, oscillates wildly for large z, and its integral is undefined (except with the help of a convergence factor). The GGA hole is cut off sharply at about $u = 6$, ensuring that it has the correct normalization. Note that it is not equal to the GEA hole for $u < u_{\rm X}$, because of the step function in Eq. (35), which produces derivative discontinuities as a function of z, e.g. at $z \approx 1$ and $z \approx 5$. These sharp cutoffs are of course unphysical, in the sense that the exact hole is smooth, but they are smoothed in $\langle n_{\rm X}^{\rm GGA}(\mathbf{u})\rangle$ by the integration over $\mathbf{r}$ that appears in the system-average of Eq. (6), since $z_{\rm X}$ is a function of $\mathbf{r}$. Any long-range oscillations of the hole are also to some extent averaged away. We stress that it is the GGA results for $\langle n_{\rm X}^{\rm GGA}(\mathbf{u})\rangle$ (and $\langle n_{\rm C}^{\rm GGA}(\mathbf{u})\rangle$) (or better their spherical averages) that should be compared with exact results (where available); see Figs. 4 and 5 of Ref. [71] for the He atom. Just as in

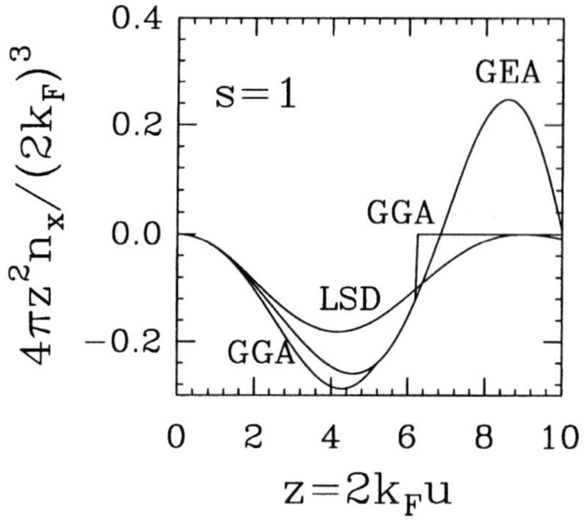

Figure 1. Spherically-averaged exchange hole density $\tilde{n}_X$ for $s = 1$.

GEA, $\tilde{n}_X^{GGA}(\mathbf{r}, \mathbf{r}+\mathbf{u})$ (and $\tilde{n}_C^{GGA}(\mathbf{r}, \mathbf{r}+\mathbf{u})$) are only intermediate quantities, without direct physical significance.

Figure 2 shows numerical results for the reduced cutoff radius ($z_X = 2k_F u_X$) as a function of the reduced density gradient s of Eq. (28). As $s \to 0$, the cutoff radius moves out to ∞, and the GGA exchange hole properly reduces to the GEA hole. As $s \to \infty$, the cutoff radius slowly approaches zero, and the hole becomes highly localized around the electron. In between, the steps in $z_X(s)$ occur when the normalization cutoff passes through a negativity cutoff.

Figure 3 shows numerical results for the enhancement factor $F_X(s)$ over local exchange ($F_X = 1$) as a function of s, as defined by Eq. (34). As $s \to 0$, $F_X \to 1$, reducing to LSD, while F_X grows indefinitely as $s \to \infty$. From Fig. 3, it is clear that the GGA enhancement factor $F_X(s)$ is different from its GEA counterpart [16], being stronger at small s and weaker at large s.

Correlation

Wave vector space cutoffs [18,20,21] provide no GGA correction to the GEA for exchange. In earlier work (P86) [17], the GGA for correlation was constructed from a Langreth-Mehl [20] wave vector space cutoff: The LSD piece of $n_C(\mathbf{r}, \mathbf{k})$ vanishes as k^2 when $k \to 0$, but the gradient piece tends to a positive constant in this limit. To satisfy the normalization constraint of Eq. (11), the gradient part was replaced by zero for $k = |\mathbf{k}|$ less than a cutoff wave vector $k_C = \tilde{f}|\nabla n|/n$. The parameter $\tilde{f}$ was fixed by fitting the correlation energy of the neon atom.

The motivation of Ref. [46] and the present work is to eliminate the need for this empirically-fitted parameter, and to treat exchange and correlation in a *unified* way which can take better advantage of the opposition [23] between the nonlocalities of exchange and correlation for valence electrons. This opposition arises because the exchange-correlation hole is deeper and more localized than the exchange hole [40]. In the uniform electron gas [53,72], and less completely in finite systems [73], the most long-ranged non-oscillatory components of the exchange and correlation holes cancel. It has been found [23] that the present GGA properly accounts for most of the nonlocalities

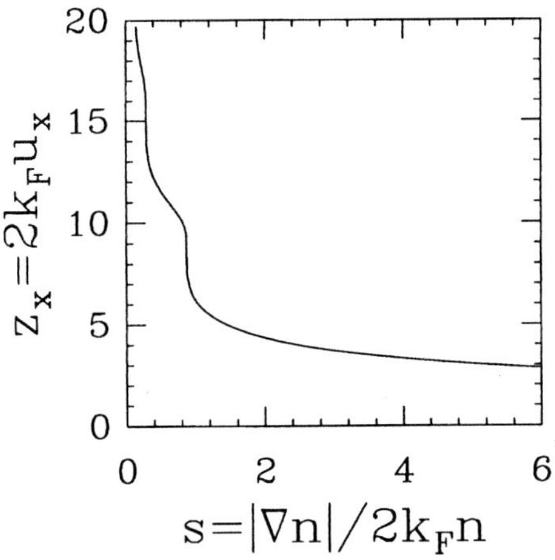

Figure 2. Reduced cutoff separation for the GGA exchange hole as a function of reduced density gradient. For r_s defined in Eq. (40), we find $u_X \lesssim r_s$ when $s \gtrsim 3$.

of $E_X[n_\uparrow, n_\downarrow]$ and of $E_C[n_\uparrow, n_\downarrow]$, and that these GGA nonlocalities tend to cancel in the range of valence electron densities. The residue of this cancellation may not always be a useful correction to LSD, although it often is.

For exchange, the GEA hole is known exactly and analytically, as is the spin-scaling relation, Eq. (24). Moreover, the density n defines only one relevant length scale, the Fermi wavelength $2\pi/k_F$, and hence only one reduced density gradient s. As a result, for a given real-space cutoff procedure, there is neither ambiguity nor uncertainty in the resulting numerical GGA for exchange.

For correlation, the cutoff procedure is more straightforward than for exchange, as there is no negativity constraint like Eq. (9). Moreover, the uncertain large-s behavior is less important, since the GGA correlation hole is cut down to zero in that limit. However, the GEA correlation hole (especially its nonlocal part) is known only imperfectly. In fact, the PW91 correlation functional was originally fitted [46] to numerical results from the cutoff of a relatively crude form for the GEA hole. The present real-space cutoff construction is based on the recent development of an accurate analytic representation [53] for the correlation hole in a uniform electron gas, and of an approximate form for the real-space decomposition of the gradient correction to that hole (see Appendix B), incorporating much of what is currently known. The long-range ($u \to \infty$) oscillations [18,53] of the LSD and GEA correlation holes, which are not well-known, have been neglected.

Correlation introduces a second length scale, the screening length $1/k_s$, where

$$k_s = (4k_F/\pi)^{1/2} \tag{37}$$

is the Thomas-Fermi screening wave vector. Furthermore, unlike the exchange hole of Eq. (24), the correlation hole does not obey a simple scaling relation with ζ, where

$$\zeta(\mathbf{r}) = \left[n_\uparrow(\mathbf{r}) - n_\downarrow(\mathbf{r})\right]/n(\mathbf{r}), \tag{38}$$

is the local relative spin polarization. Even neglecting the small corrections that arise from terms containing $\nabla\zeta$ [74,23,75], the resulting numerical GGA defines a function

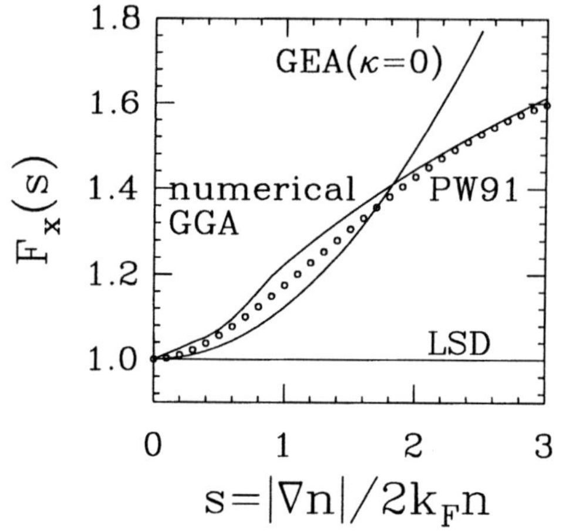

Figure 3. Enhancement factor (Eq. (34)) over local exchange ($F_X = 1$), as a function of reduced density gradient. The PW91 parametrization is indicated by open circles.

of *three* variables:

$$E_C^{GGA}[n_\uparrow, n_\downarrow] = \int d^3r \; n(\mathbf{r}) \; \epsilon_C^{GGA}\big(r_s(\mathbf{r}), \zeta(\mathbf{r}), s(\mathbf{r})\big), \tag{39}$$

where

$$r_s(\mathbf{r}) = \Big[3/\big(4\pi n(\mathbf{r})\big)\Big]^{1/3} \tag{40}$$

is the local Seitz radius. Fortunately, correlation simplifies [46,74-76] in the high-density ($r_s \to 0$) or long-range ($u \to \infty$) limits, where the random phase approximation is valid. Thus the long-range non-oscillatory part of the correlation hole, where the GGA cutoffs are performed, can be accurately constructed. In fact, in Ref. [75], a GGA for correlation was constructed for the high-density limit. Here and in Ref. [46], we extend that construction beyond the high-density limit, but using that limit in many places as a guide.

Since the correlation hole is *not* required to satisfy a negativity constraint like Eq. (9), we do not need the non-spherical component of its GEA density. Thus, we write the spherically-averaged GEA hole as

$$\tilde{n}_C^{GEA}(\mathbf{r}, u) = n_C^{LSD}\big(r_s(\mathbf{r}), \zeta(\mathbf{r}), v(\mathbf{r}, u)\big) + t^2(\mathbf{r}) \; \delta\tilde{n}_C\big(r_s(\mathbf{r}), \zeta(\mathbf{r}), v(\mathbf{r}, u)\big), \tag{41}$$

where

$$t(\mathbf{r}) = |\nabla n(\mathbf{r})|/\big(2k_\zeta(\mathbf{r})n(\mathbf{r})\big) = cs/r_s^{1/2}\phi \tag{42}$$

(with $c = (\pi/4)^{1/2}(9\pi/4)^{1/6} = 1.228$) is another reduced density gradient, which scales with the spin-scaled Thomas-Fermi screening wave vector

$$k_\zeta(\mathbf{r}) = \phi(\zeta(\mathbf{r})) \; k_s(\mathbf{r}). \tag{43}$$

The spin-scaling factor [74,76] is

$$\phi(\zeta) = \big[(1+\zeta)^{2/3} + (1-\zeta)^{2/3}\big]/2, \tag{44}$$

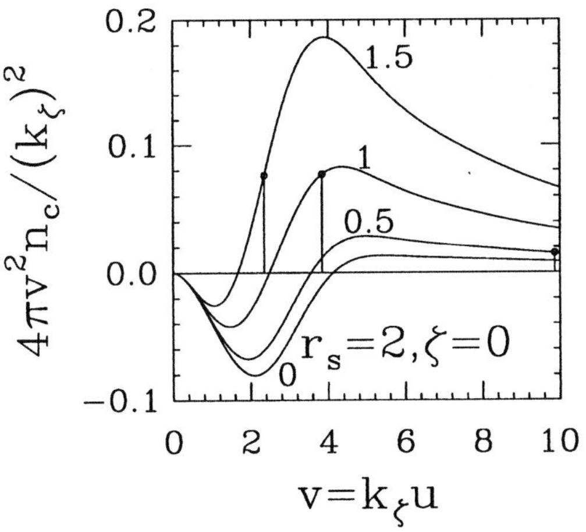

Figure 4. Spherically-averaged correlation hole density $\tilde{n}_C$ for $r_s = 2$ and $\zeta = 0$. GEA holes are shown for four values of the reduced density gradient, $t = |\nabla n|/(2k_\zeta n)$. The vertical lines indicate where the numerical GGA cuts off the GEA hole to make $\int_0^{v_C} dv\ 4\pi\ v^2\ \tilde{n}_C(v) = 0$.

while $v(\mathbf{r}, u)$ is another reduced separation on the Thomas-Fermi length scale:

$$v(\mathbf{r}, u) = k_\zeta(\mathbf{r})u. \tag{45}$$

The LSD correlation hole function, $n_C^{\rm LSD}(r_s, \zeta, v)$, is accurately given by the analytic representation of Perdew and Wang [53], which has been confirmed by recent quantum Monte Carlo calculations [6]. We write this hole in the form

$$n_C^{\rm LSD}(r_s, \zeta, v) = \phi^3\ k_\zeta^2\ A_C(r_s, \zeta, v). \tag{46}$$

In the high-density ($r_s \to 0$) or long-range ($v \to \infty$) limits, A_C depends on v alone, and may be evaluated exactly using the random phase approximation [74-76]. For larger r_s, A_C is written as the sum of long- and short-ranged contributions [53].

The gradient contribution to $\tilde{n}_C^{\rm GEA}$ of Eq. (41) is modeled in a similar fashion, by writing

$$\delta\tilde{n}_C(r_s, \zeta, v) = \phi^3\ k_\zeta^2\ B_C(r_s, \zeta, v), \tag{47}$$

where B_C is also the sum of long- and short-ranged contributions, which are less precisely known than their LSD counterparts. These contributions are described in Appendix B, and are constructed to recover the GEA energy of Eq. (3). A key fact is that Eq. (47) and its normalization integral are positive. While the LSD correlation hole properly integrates to zero, the GEA correlation hole does not.

With the GEA correlation hole fully defined, we construct the spherically-averaged GGA hole

$$\tilde{n}_C^{\rm GGA}(r_s, \zeta, t, v) = \phi^3\ k_\zeta^2\ [A_C(r_s, \zeta, v) + t^2\ B_C(r_s, \zeta, v)]\theta(v_C - v), \tag{48}$$

where $v_C(r_s, \zeta, t)$ is the largest root satisfying the normalization condition

$$\int_0^{v_C} dv\ 4\pi v^2\ [A_C(r_s, \zeta, v) + t^2\ B_C(r_s, \zeta, v)] = 0. \tag{49}$$

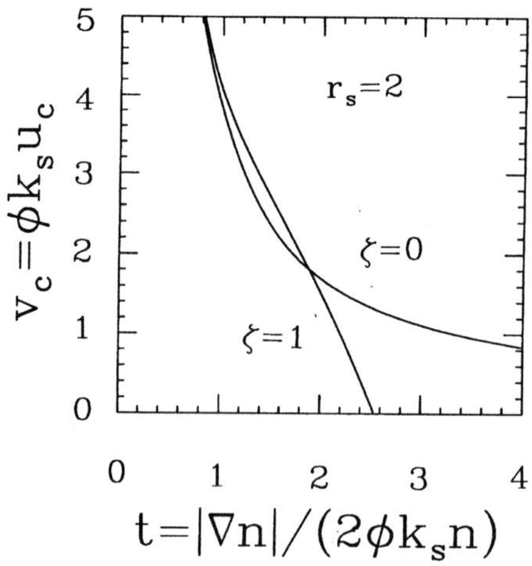

Figure 5. Reduced cutoff separation for the GGA correlation hole, for $r_s = 2$ and $\zeta = 0$ or $\zeta = 1$. For $\zeta = 1$, $v_C = 0$ beyond $t \approx 2.5$. For the case $r_s = 2$ and $\zeta = 0$, $u_C \lesssim r_s$ means $t \gtrsim 1.4$ and $s \gtrsim 1.6$.

Figure 4 is a plot of the spherically-averaged GGA correlation hole for $r_s = 2$ and $\zeta = 0$, for several different values of t. We see that, for a small value of t ($t = 0.5$), the gradient correction to LSD is small and so v_C is large (≈ 10.0), tending to its GEA value (∞) as $t \to 0$. On the other hand, for $t = 1.5$, the gradient correction is nine times larger, causing the cutoff to occur at a much smaller value of v_C ($v_C = 2.3$). In the limit $t \to \infty$, $v_C \to 0$ (Fig. 5), turning off the correlation contribution altogether. In all cases, the GGA correlation hole is more localized than either the LSD or GEA holes.

In Fig. 5, we follow v_C as a function of t for $r_s = 2$, for both the spin-unpolarized ($\zeta = 0$) and the fully spin-polarized ($\zeta = 1$) cases. Note that, for small t, v_C is large, and the curves merge because the long-range contribution to the hole is independent of ζ. The same qualitative behavior occurs for all densities, although for higher densities ($r_s \to 0$) the short-range contribution becomes negligible for all t and the two curves become everywhere identical.

The GGA correlation energy per particle of Eq. (39) is now

$$\begin{aligned}\epsilon_C^{GGA}(r_s,\zeta,s) &= \phi^3 \int_0^{v_C} dv\, \frac{4\pi v^2}{2v}\left[A_C(r_s,\zeta,v) + t^2\, B_C(r_s,\zeta,v)\right] \\ &= \epsilon_C^{LSD}(r_s,\zeta) + H(r_s,\zeta,t). \end{aligned} \tag{50}$$

In Fig. 6, we plot the difference between the GGA and LSD correlation energies for different values of r_s, as a function of t for the spin-unpolarized case ($\zeta = 0$). For $t \to 0$, this figure recovers the t^2 behavior of GEA. Since the GGA correlation energy vanishes at large gradients, the limit as $t \to \infty$ in this figure is precisely $-\epsilon_C^{LSD}$. Fig. 7 shows the same for $\zeta = 1$. Comparison of Figs. 6 and 7 demonstrates the approximate spin-scaling relationship for correlation (when $r_s \lesssim 6$):

$$\epsilon_C^{GGA}(r_s,\zeta;t) \approx \phi^3(\zeta)\, \epsilon_C^{GGA}(r_s,0;t/\phi(\zeta)) \tag{51}$$

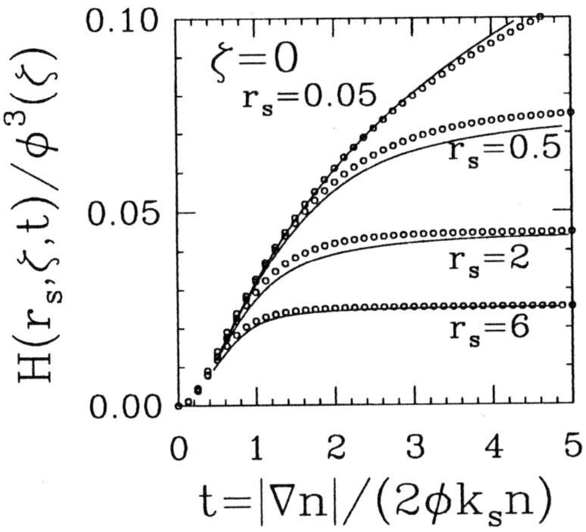

Figure 6. The function $H = \epsilon_C^{GGA} - \epsilon_C^{LSD}$ for several values of r_s for the spin-unpolarized case ($\zeta = 0$). The solid lines are the numerical result of the real-space cutoff procedure, while the open circles are from the PW91 parametrization.

EXACT CONDITIONS

In this section, we describe several exact conditions for all inhomogeneous electronic systems. We show which of these are satisfied by LSD, which are satisfied by the numerical GGA, and which are not.

Some of these conditions apply to the total-density functional $E_{XC}[n]$, for which we write the GGA as

$$E_{XC}^{GGA}[n] = \int d^3r\, n(\mathbf{r})\, \epsilon_X^{LSD}(n(\mathbf{r}))\, F_{XC}(r_s(\mathbf{r}), s(\mathbf{r})). \tag{52}$$

For plots of the enhancement factor $F_{XC}(r_s, s) = F_X(s) + F_C(r_s, s)$, see Refs. [23] and [51]. For futher discussions of exact conditions, see Refs. [22], [54] and [77].

Conditions on the hole

We begin with exact conditions on the (system-averaged) hole of Eqs. (5) and (6). These constraints are among the most important for an approximate functional to satisfy. LSD respects both the negativity constraint and the integral constraints of Eqs. (20)-(22), because the LSD hole is the hole of a possible physical system. By construction, the numerical GGA also respects these constraints.

Another condition satisfied by the exact $n_{XC}(\mathbf{r}, \mathbf{r}+\mathbf{u})$ is the electron-electron cusp condition [78,79]. When $u \to 0$, the Coulomb interaction dominates, causing a cusp at $u = 0$. The relation between the cusp and the value at $u = 0$ is universal. This cusp condition is most easily stated in terms of the pair distribution function for coupling constant λ,

$$g_\lambda(\mathbf{r}, \mathbf{r}') = 1 + n_{XC,\lambda}(\mathbf{r}, \mathbf{r}')/n(\mathbf{r}'), \tag{53}$$

where $n_{XC,\lambda}$ is the λ-dependent hole (the integrand of Eq. (7)). The cusp condition is then

$$g'_\lambda(\mathbf{r}, \mathbf{r}) = \lambda g_\lambda(\mathbf{r}, \mathbf{r}), \tag{54}$$

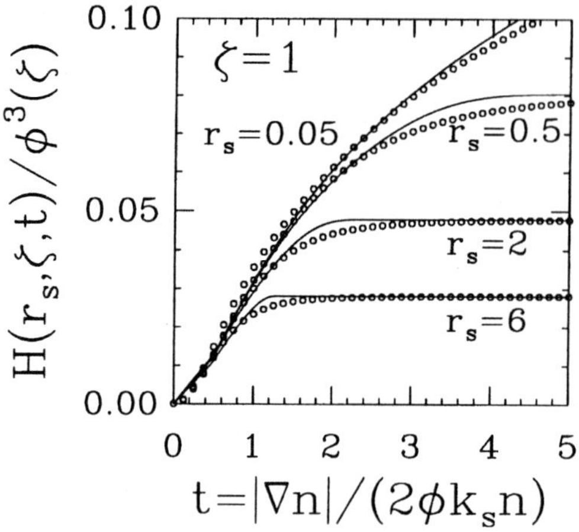

Figure 7. The function $H = \epsilon_C^{GGA} - \epsilon_C^{LSD}$ for several values of r_s for the fully spin-polarized case ($\zeta = 1$). The solid lines are the numerical result of the real-space cutoff procedure, while the open circles are from the PW91 parametrization.

where

$$g'_\lambda(\mathbf{r},\mathbf{r}) = \partial/\partial u|_{u=0} \int d\Omega_{\mathbf{u}}\ g_\lambda(\mathbf{r},\mathbf{r}+\mathbf{u})/(4\pi). \tag{55}$$

This cusp condition can be further decomposed into its exchange and correlation contributions, and also by spin [22,54]. LSD respects all these conditions, because its hole is that of jellium. Furthermore, LSD typically is a very good approximation [40,80] (although not exact [63]) for the hole density at $u = 0$. As discussed in Appendix B, the numerical GGA retains the LSD value of both the hole and its cusp at zero separation, so that GGA also satisfies all these conditions.

Another exact point-wise condition is that the pair distribution function be symmetric under interchange of $\mathbf{r}$ and $\mathbf{r}'$. While no local or semilocal approximation can be symmetric, we have recently shown [22,54] that the system average of this condition, even if spin-decomposed, is trivially satisfied by both LSD and our numerical GGA.

Slowly varying densities

By definition, LSD is exact for the uniform electron gas. As seen in Figs. 3, 6, and 7 the numerical GGA recovers this limit, unlike [51] the Langreth-Mehl [20,21] correlation energy functional (which reduces to the random phase approximation) and the Lee-Yang-Parr [52] functional (which underestimates the magnitude of the uniform-gas correlation energy by about a factor of two [40]). Next we consider the limit of slowly-varying densities, where GEA is exact. Starting from the GEA exchange hole of Eq. (26), we can get a finite gradient coefficient for the energy if we replace the Coulomb interaction $1/u$ by the Yukawa potential $e^{-\kappa u}/u$, and then take the limit $\kappa \to 0$. This yields

$$F_X(s) = 1 + \mu s^2, \tag{56}$$

where $\mu = 7/81$. However, Kleinman and Lee [15] have shown that, starting from the infinite-range (i.e., $\kappa = 0$) Coulomb interaction $1/u$, the correct coefficient for Eq. (56) is not the Sham [13] coefficient 7/81, but rather (10/7) thereof, i.e., $\mu = 10/81$.

We might expect the numerical GGA for exchange to recover Eq. (56) with Sham's coefficient $\mu = 7/81$ as $s \to 0$, but instead we find a linear term:

$$F_X(s) = 1 + 0.1241s + O(s^2)\ (s \to 0), \tag{57}$$

as can be seen clearly in Fig. 3. In the small-s regime, the right-hand side of Eq. (27) is dominated by its first two terms. In the GEA, the linear term vanishes when the hole is spherically-averaged. The numerical GGA, however, enforces the negativity of the exchange hole everywhere, which leaves a linear contribution after spherical averaging. Of course, the correct physical behavior is quadratic in s, but the long-range nature of the Coulomb repulsion, combined with the real-space cutoff procedure, leads to this anomalous result. Thus the numerical cutoff procedure produces incorrect results for very small values of s, where $F_X \approx 1$. We regard the numerical procedure as yielding a good approximation to F_X, rather than to $F_X - 1$.

If desired, the linear term in Eq. (57) can be removed by damping the gradient terms of the GEA exchange hole by the factor $[1 + (bz)^q]^{-1}$, where $z = 2k_F u$. This factor does not alter the hole's behavior at small u. We have not followed this procedure here, but mention that the values $q = 2.5, b = 1/(2\pi)$ yield excellent GGA energies, and generate our most realistic model for the GGA system-averaged hole [57,58].

The numerical GGA correlation energy has no negativity constraint and suffers no such problem, as shown in Appendix B. As $t \to 0$, the GGA correlation energy reduces to the expected GEA (see Appendices B and C).

Uniform scaling relations

Other important constraints arise from fundamental theorems on uniform scaling [81]. We define a uniform scaling of the density by

$$n_\gamma(\mathbf{r}) = \gamma^3 n(\gamma\mathbf{r}), \tag{58}$$

so that the total number of electrons N remains fixed. The reduced density gradients of Eqs. (28) and (42) scale as $s_\gamma(\mathbf{r}) = s(\gamma\mathbf{r})$ and $t_\gamma(\mathbf{r}) = \gamma^{1/2} t(\gamma\mathbf{r})$.

The exact exchange energy scales homogeneously [81]:

$$E_X[n_\gamma] = \gamma\ E_X[n], \tag{59}$$

as do the LSD and GGA exchange energies, as Eq. (32) shows. For a non-degenerate ground-state density, the exact correlation energy has the high-density limit [48,49,80]

$$\lim_{\gamma\to\infty} E_C[n_\gamma] > -\infty. \tag{60}$$

Eq. (60) is violated by LSD but respected (Appendix C) by our numerical GGA. The characteristic $\ln\gamma$ divergence of $E_C^{\mathrm{LSD}}[n_\gamma]$ comes from [76] the long-range tail of the uniform-gas correlation hole, which GGA cuts off.

A fundamental inequality [81] is

$$E_{XC}[n_\gamma] > \gamma\ E_{XC}[n]\ (\gamma > 1). \tag{61}$$

LSD satisfies this condition, because the condition applies to the uniform electron gas. Numerical GGA also obeys this condition, because its curves for $F_{XC}(r_s, s)$ versus s do not cross one another [77]. Furthermore, in the low-density limit,

$$\lim_{\gamma\to 0} \frac{1}{\gamma} E_{XC}[n_\gamma] \equiv B[n] < E_{XC}[n], \tag{62}$$

and $B[n] > -\infty$ [49]. Again this relation is obeyed by both LSD and the numerical GGA [77].

Large density gradients: Non-uniform scaling relations and the Lieb-Oxford bound

As discussed in Appendix D, it is possible to construct densities in which the reduced density gradient $s(\mathbf{r})$ is arbitrarily large almost everywhere. As s increases, the GEA provides a less satisfactory description of the hole even for small separations u, where GEA ignores contributions of order u^2 arising from the fourth- and higher-order gradient corrections to the exchange hole. Moreover, the GGA $F_X(s)$ of Eq. (32) becomes more sensitive to the difference between sharp and diffuse radial cutoffs on u. Our numerical GGA seems trustworthy at best for $0 \lesssim s \lesssim 3$, the range of greatest physical interest [82]. More generally, the LSD form of Eq. (2) and the GGA form of Eq. (4) are not valid when the reduced density gradients (s, t, etc.) are large.

Consequently, exact conditions can lead to inconsistent predictions for the $s \to \infty$ behavior of a GGA. For example, the large-s behavior of the Becke 1988 [45] form for $F_X(s)$ ($\sim s/\ln s$) yields a correct exchange energy density ($\epsilon_X \to -1/2r$) far from a finite system, while a different behavior ($\sim s$) generates [83] the correct limit for the exchange potential ($v_X \to -1/r$). While we cannot get all point-wise conditions right with a GGA, we might still hope to get system-averaged conditions right. We have already seen that our numerical GGA satisfies Eq. (60), a scaling limit in which $t \to \infty$. Several other exact conditions of this sort, which suggest that $F_X(s)$ should vanish or at least remain bounded as $s \to \infty$, are described below.

Consider the one- and two-dimensional density scalings [48-50], respectively

$$n_\gamma^x(x,y,z) = \gamma n(\gamma x, y, z), \tag{63}$$

and

$$n_{\gamma\gamma}^{xy}(x,y,z) = \gamma^2 n(\gamma x, \gamma y, z), \tag{64}$$

which keep the electron number N constant. Both the high-density limit of one-dimensional scaling and the low-density limit of two-dimensional scaling are equivalent to the limit of rapidly-varying densities. In the former case,

$$\lim_{\gamma\to\infty} E_{XC}[n_\gamma^x] > -\infty, \tag{65}$$

and

$$\lim_{\gamma\to\infty} E_C[n_\gamma^x] = 0, \tag{66}$$

while in the latter,

$$\lim_{\gamma\to 0} \frac{1}{\gamma} E_{XC}[n_{\gamma\gamma}^{xy}] > -\infty, \tag{67}$$

and

$$\lim_{\gamma\to 0} \frac{1}{\gamma} E_C[n_{\gamma\gamma}^{xy}] = 0. \tag{68}$$

These constraints [48-50] will all be satisfied by a GGA if [77]

$$\lim_{s\to\infty} s^{1/2}\, F_X(r_s, s) < \infty, \tag{69}$$

and

$$\lim_{s\to\infty} s^{1/2}\, F_C(r_s, s) < \infty. \tag{70}$$

Another constraint affecting the large-s behavior may be derived as follows: Lieb and Oxford [47] proved that

$$E_{XC,\lambda=1}[n_\uparrow, n_\downarrow] \geq C \int d^3r\, n^{4/3}(\mathbf{r}), \tag{71}$$

where, via Eqs. (5) and (7),

$$E_{XC} = \int_0^1 d\lambda\, E_{XC,\lambda}. \tag{72}$$

They also showed that the optimum bounding constant C lies between -1.23 and -1.68. The low-density limit of the uniform gas [10] narrows [46] the range to $-1.43 \geq C \geq -1.68$. Since $E_{XC,\lambda}$ is a monotonically decreasing function of λ, we find [46,77]

$$E_{XC}[n_\uparrow, n_\downarrow] \geq C \int d^3r\, n^{4/3}(\mathbf{r}). \tag{73}$$

Numerical GGA will satisfy Eq. (73) if

$$F_{XC}(r_s, \zeta, s) \leq 2.27, \tag{74}$$

where $F_{XC}(r_s, \zeta, s)$ is the ζ-dependent generalization of $F_{XC}(r_s, s)$ in Eq. (52). (To simplify the present discussion, the small $\nabla\zeta$ contributions to E_X^{GGA} have been suppressed in our notation.) Because there are unphysical systems for which $s(\mathbf{r})$ is arbitrarily large almost everywhere (Appendix D), $F_{XC}(r_s, \zeta, s \to \infty) \leq 2.27$ is a necessary condition for a GGA to satisfy the Lieb-Oxford bound for all possible densities.

Clearly, LSD violates Eqs. (69) and (70), because of its lack of any gradient dependence, but LSD satisfies Eq. (74), because the magnitude of the correlation energy never becomes larger than that of the exchange energy for the uniform gas.

Next we find the behavior of the numerical GGA in the large-s limit. For exchange in this limit, $z_X \to 0$, and one can expand the coefficients in Eq. (27) for small z. The resulting hole is dominated by the quadratic contributions which, when spherically averaged, yield

$$z_X \approx 5.908\, s^{-2/5}, \tag{75}$$

and

$$F_X(s) \approx 0.8862\, s^{2/5}. \tag{76}$$

Thus, although z_X vanishes in the large-s limit, it does not vanish fast enough to make F_X vanish. Numerical GGA violates Eqs. (69) and (74).

Similarly, we can deduce the large-t behavior of the correlation energy. In this case v_C becomes very small, and the correlation energy gets completely annihilated by the cutoff procedure, as shown in Appendix C. Thus numerical GGA does satisfy Eq. (70). GGA correlation "turns off" ($\epsilon_{XC}^{GGA}/\epsilon_X^{GGA} \to 1$) in the limit of large reduced density gradients. This behavior is correct in the tail of the electron density. In contrast, LSD correlation "turns on" ($\epsilon_{XC}^{LSD}/\epsilon_X^{LSD} \to 1.96$) in any low-density limit, and thus also in the tail.

Table 1. Exchange and correlation energies of spherical atoms and ions,[a] in hartrees (1 hartree = 27.2116 eV)

	E_x				E_c			
Atom	LSD	GEA[b]	PW91[c]	exact	LSD	GEA[b]	PW91[d]	exact[e]
H	-0.268	-0.305	-0.307	-0.313	-0.022	0.044	-0.007	-0.000
He	-0.884	-1.007	-1.017	-1.026	-0.113	0.125	-0.046	-0.042
Li^+	-1.421	-1.618	-1.631	-1.652	-0.135	0.260	-0.051	-0.043
Be^{2+}	-1.957	-2.229	-2.245	-2.277	-0.150	0.401	-0.054	-0.044
Li	-1.538	-1.734	-1.763	-1.781	-0.151	0.222	-0.058	-0.045
Be^+	-2.168	-2.441	-2.481	-2.507	-0.173	0.345	-0.062	-0.047
Be	-2.312	-2.581	-2.645	-2.667	-0.224	0.314	-0.094	-0.094
Ne^{6+}	-6.634	-7.370	-7.545	-7.594	-0.334	1.185	-0.123	-0.180
N	-5.893	-6.395	-6.569	-6.596	-0.427	0.567	-0.199	-0.188
Ne	-11.033	-11.775	-12.115	-12.108	-0.743	0.780	-0.382	-0.390
Ar	-27.863	-29.293	-30.123	-30.189	-1.424	1.534	-0.771	-0.722
Zn^{12+}	-54.433	-57.018	-58.437	-58.475	-1.800	3.576	-0.924	–
Zn	-65.642	-68.105	-69.830	-69.640	-2.655	2.467	-1.525	–
Kr	-88.624	-91.651	-93.831	-93.893	-3.269	3.024	-1.914	–
Xe	-170.562	-175.300	-178.986	-179.170	-5.177	4.685	-3.149	–

[a]Hartree-Fock densities from Ref. [86].
[b]The GEA is that for $\kappa = 0$, i.e., Eqs. (25), (32), and (56) with $\mu = 10/81$.
[c]Eqs. (25), (32) and (78).
[d]Eqs. (50) and (83)-(87).
[e]"Exact" correlation energies from Ref. [89].

ANALYTIC APPROXIMATION TO NUMERICAL RESULTS: PW91

As is the case [8-10] for LSD in Eq. (2), the usefulness of the GGA of Eq. (4) is enhanced by an analytic parametrization of the function $f(n_\uparrow, n_\downarrow, \nabla n_\uparrow, \nabla n_\downarrow)$. The real-space cutoff procedure cures some of the worst problems of GEA, but its numerically-defined functional violates several known exact conditions, including one (Eq. (74)) which LSD satisfies. In constructing an analytic fit to the numerical GGA, we restore some of those exact conditions.

Exchange

The GGA for the exchange energy is given by Eqs. (25) and (32), and requires an analytic fit to the numerically-defined $F_x(s)$. The first such fit was that of Perdew and Wang (PW86) [16]:

$$F_x^{PW86}(s) = (1 + 1.296s^2 + 14s^4 + 0.2s^6)^{1/15}. \tag{77}$$

This form was designed to fit the numerical results for most s, but to restore the Sham GEA result [13] for small s. Its large-s behavior is the same as that of the numerical GGA, i.e. $F_x \sim s^{2/5}$, and the coefficient of this term agrees with Eq. (76) to within 1%. Equation (77) clearly violates both Eq. (69) and the Lieb-Oxford bound, Eq. (74).

More recently, Perdew and Wang [46] (PW91) proposed the following more elaborate expression:

$$F_x^{PW91}(s) = \frac{1 + 0.19645s \sinh^{-1}(7.7956s) + \left[0.2743 - 0.1508 \exp(-100s^2)\right]s^2}{1 + 0.19645s \sinh^{-1}(7.7956s) + 0.004s^4}, \tag{78}$$

where

$$\sinh^{-1}(x) = \ln\left[x + \sqrt{1 + x^2}\right], \tag{79}$$

found by modifying the Becke functional form [45]. The Becke form was chosen as it was known to reproduce the exchange energies of atoms and molecules a little better than PW86. In this sense, the PW91 form contains a slight empirical bias. However, for physical values of s, the numerical GGA and Becke forms are very similar. The modifications made appear in Eq. (78) as two additional terms: the Gaussian exponential, and the s^4 term in the denominator. The former is chosen to restore the correct Kleinman [15] GEA for small s. Its Gaussian exponent is somewhat arbitrary, but large enough to recover the original Becke result for $0.2 < s < 3$, the range of reduced density gradients that dominates the exchange energies of atoms [82]. The latter modification changes the large-s behavior, so that both the Lieb-Oxford bound of Eq. (74) and the non-uniform scaling condition of Eq. (69) are now satisfied. Figure 3 shows how close Eq. (78) is to the numerical GGA, which can therefore be considered a first-principles justification of the Becke form.

Although we use Eq. (78), Eq. (77) emerges more directly from the real-space cutoff and seems to provide a better account of the exchange interaction between weakly-overlapped rare-gas atoms [84], and a better description of the hyperfine contact field at the nucleus [85].

In summary, the GGA for exchange may be constructed directly from the result of the real-space cutoff, as in earlier work [16], or from minor modifications which incorporate empirical information and global constraints, as in the present work [46]. Table 1 compares LSD, GEA, GGA, and exact exchange energies for spherical atoms and ions, using near-Hartree-Fock densities [86]. (The exact exchange energy was found by subtracting from the near-Hartree-Fock total energy its kinetic and electrostatic components.)

Correlation

Next we consider correlation, where the GGA is given by Eqs. (39) and (50). We are faced with the practical problem of finding an analytic representation for a numerically-defined function (H) of *three* variables. We first consider correlation in the simpler high-density limit. In this limit, we need only find an analytic fit to a single curve, the $r_s \to 0$ limit of Fig. 6, because we know the simple ζ dependence of H in this limit from the RPA [74]. We fit this curve with the following analytic form:

$$H_o(r_s = 0, \zeta, t) = \phi^3(\zeta)\frac{\beta^2}{2\alpha}\ln\left[1 + \frac{2\alpha}{\beta}t^2\right]. \tag{80}$$

where α and β are constants. For small t, $H_o(r_s = 0, \zeta, t)$ becomes

$$H_o(r_s = 0, \zeta, t) = \phi^3(\zeta)\left[\beta t^2 - \alpha t^4\right], \tag{81}$$

so $\beta = 16(3/\pi)^{1/3}\ C_C(r_s = 0) = 0.06673$, to recover the $\kappa \to 0$ GEA value, which is built into the numerical GGA, while $\alpha = 0.09$, which approximates the quartic behavior of the numerical cutoff result (0.11) given by Eq. (118). For large t we find

$$H_o(r_s = 0, \zeta, t) \approx \phi^3\frac{\beta^2}{\alpha}\ln t. \tag{82}$$

The coefficient $\beta^2/\alpha = 0.05$ in Eq. (82) is smaller than the numerical-GGA value of 0.06218 (see Eq. (113)), and this discrepancy makes this analytic GGA violate Eq. (60), which the numerical GGA respects.

Next, we alter this function to allow for the r_s dependence seen in Fig. 6. Clearly, the most important physical feature is to change the large-t limit, to make $H_o(r_s, \zeta, t)$

level off as a function of t at exactly $-\epsilon_C^{LSD}(r_s,\zeta)$, so as to ensure cancellation of the correlation energy for large gradients. In an earlier attempt [75], this was achieved by a simple step function, but the figures indicate that the transition is not so abrupt (see Appendix C). Thus we generalize the high-density fit to

$$H_o(r_s,\zeta,t) = \phi^3(\zeta)\frac{\beta^2}{2\alpha}\ln\left[1+\frac{2\alpha}{\beta}t^2\left\{\frac{1+At^2}{1+At^2+A^2t^4}\right\}\right]. \tag{83}$$

The functional form of the modification is designed to retain the small-t behavior, while our choice of A is determined by the requirement that ϵ_C^{GGA} of Eq. (50) vanish as $t \to \infty$, yielding

$$A = \frac{2\alpha}{\beta}\left[\exp\left(-2\alpha\epsilon_C^{LSD}(r_s,\zeta)/\phi^3\beta^2\right)-1\right]^{-1}. \tag{84}$$

This analytic fit is close to the numerical GGA as shown in Figs. 6 and 7.

The PW91 functional adds a little further "window-dressing" to the analytic fit. As discussed in the section on slowly varying densities, the gradient coefficient for the GEA exchange energy is different for the Coulomb interaction $1/u$ (Kleinman's [15] $\mu = 10/81$) than for the Yukawa $e^{-\kappa u}/u$ in the limit $\kappa \to 0$ (Sham's [13] $\mu = 7/81$). Plausible arguments [15,87,88] suggest that the GEA for exchange and correlation *together* (E_{XC}) is the same for $\kappa = 0$ and $\kappa \to 0$. While our analytic GGA for exchange, Eq. (78), is designed to recover the correct $\kappa = 0$ result in the limit $s \to 0$, our numerical GGA for correlation and its analytic representation of Eq. (83) recover the different $\kappa \to 0$ result for correlation. To remove this inconsistency, we add a small term H_1, which is negligible unless $s \ll 1$:

$$\epsilon_C^{PW91}(r_s,\zeta,s) = \epsilon_C^{LSD}(r_s,\zeta) + H^{PW91}(r_s,\zeta,t), \tag{85}$$

$$H^{PW91} = H_o + H_1, \tag{86}$$

where

$$H_1(r_s,\zeta,t) = 16\left(\frac{3}{\pi}\right)^{1/3}\left[C_C(r_s)-C_C(0)-3C_X/7\right]\phi^3(\zeta)t^2 e^{-100\phi^4 k_s^2 t^2/k_F^2}. \tag{87}$$

The $\kappa \to 0$ gradient coefficients $C_X = -0.001667$, $C_C(0) = 0.004235$, and $C_C(r_s)$ are taken from Rasolt and Geldart [12]. The Gaussian exponential in Eq. (87) is similar to that of Eq. (78). As pointed out in Ref. [41], this H_1 term improperly makes the left-hand side of Eq. (60) diverge to $+\infty$, but this happens only for unphysically high densities.

Table 1 compares LSD, GEA ($\kappa = 0$), GGA, and exact correlation energies for spherical atoms and ions, using near-Hartree-Fock densities [86].

The functional derivatives $\delta E_X^{GGA}/\delta n_\sigma(\mathbf{r})$ and $\delta E_C^{GGA}/\delta n_\sigma(\mathbf{r})$, which serve as exchange and correlation potentials in the Kohn-Sham self-consistent one-electron Schrödinger equations, are summarized in Eqs. (24) and (26) of Ref. [16], and in Appendix E. Subroutines which evaluate the energy density and potential from the electron spin densities and their derivatives are available by either electronic mail

perdew@mailhost.tcs.tulane.edu

or the world-wide web

http://camchem.rutgers.edu/~kieron/dft_pubs.html.

CONCLUDING REMARKS

The local spin density (LSD) approximation of Eq. (2) transfers information about ground-state many-body effects from the uniform electron gas to real atoms, molecules, and solids, in a way that respects the important hole constraints of Eqs. (20)-(22) and contains accurate information about the on-top hole. The gradient expansion approximation (GEA) of Eq. (3) carries additional information about exchange and correlation in non-uniform systems, but violates these constraints. Through real-space cutoffs of the spurious long-range parts of the GEA exchange and correlation holes, we have restored these constraints. The result is a generalized gradient approximation or GGA (Eq. (4)) for the energy as a functional of the spin densities. (Closely-related GGA's have also been constructed for the correlation contribution to the kinetic energy [68,90,91], for the correlation energy in the random phase approximation [91], and for the anti-parallel-spin correlation energy [91]. GGA nonlocality is easily visualized [25,51] with the help of an enhancement factor F_{XC} like that of Eq. (52). (Like LSD, albeit to a lesser extent, GGA makes a self-interaction error which could be subtracted out on an orbital-by-orbital basis [9,92].

Although our construction is non-empirical and "first-principles" in the sense described in the history section, it is neither unique nor perfect. However, the result has already proved its utility in quantum chemistry and solid-state physics, and seems to change little under refinements (such as our use of an improved uniform-gas correlation hole which was not available when Ref. [46] was written). We intend to continue refining, simplifying [56], and testing (e.g., through a study of system-averaged exchange-correlation holes [58,71]), but we expect [93] that the next major improvement in density functional approximations must go beyond the semi-local form of Eq. (4).

ACKNOWLEGMENTS

This work has been supported by NSF grants DMR92-13755 and DMR95-21353. We thank Matthias Ernzerhof, Jingsong He, Mel Levy, Per Söderlind, and Cyrus Umrigar for useful discussions. K.B. also acknowledges the generous hospitality of both the Physics and Mathematics departments of Trinity College, Dublin, where part of this work was performed.

APPENDIX A: ANGULAR INTEGRAL IN THE CONSTRUCTION OF THE GGA EXCHANGE ENERGY

Let μ be the cosine of the angle between $\mathbf{u}$ and $\mathbf{s}$. Then the integral of Eq. (35) is

$$I = \int \frac{d\Omega_{\mathbf{u}}}{4\pi}\, \tilde{y}\, \theta(\tilde{y}) = \frac{1}{2}\int_{-1}^{1} d\mu\, (A\mu^2 + B\mu + C)\theta(A\mu^2 + B\mu + C), \tag{88}$$

where A, B, and C are independent of μ. The indefinite integral

$$F(\mu) = \big(A\mu^3/3 + B\mu^2/2 + C\mu\big)/2 \tag{89}$$

is simple, but the step function modifies the upper and lower limits. Define the roots $\mu_\pm$ of $A\mu^2 + B\mu + C = 0$ by

$$\mu_\pm = \frac{1}{2}\Big[-\frac{B}{A} \pm \frac{\sqrt{T}}{|A|}\Big], \tag{90}$$

where $T = B^2 - 4AC$. The value of I can be stated for several different cases: If $T < 0$, or $|\mu_\pm| > 1$, then

$$I = \big(F(1) - F(-1)\big)\theta(V), \tag{91}$$

where $V = A - B + C$; if $|\mu_+| > 1$ but $|\mu_-| < 1$, then

$$I = \big(F(\mu_-) - F(-1)\big)\theta(V) + \big(F(1) - F(\mu_-)\big)\theta(-V); \tag{92}$$

if $|\mu_-| > 1$ but $|\mu_+| < 1$, then

$$I = \big(F(\mu_+) - F(-1)\big)\theta(V) + \big(F(1) - F(\mu_+)\big)\theta(-V); \tag{93}$$

finally, if $|\mu_\pm| < 1$, then

$$I = \big(F(\mu_-) - F(-1) + F(1) - F(\mu_+)\big)\theta(V) + \big(F(\mu_+) - F(\mu_-)\big)\theta(-V). \tag{94}$$

APPENDIX B: GRADIENT EXPANSION OF THE CORRELATION HOLE

In real-space construction, the functions $A_C(r_s, \zeta, v)$ and $B_C(r_s, \zeta, v)$ are used to define the LSD and GEA holes, via Eqs. (46) and (47), respectively. Our model for A_C is taken directly from Ref. [53]. The gradient correction to the correlation hole is less precisely known, and we model it in a fashion similar to the LSD contribution.

We first review the model for A_C, which may be written as

$$4\pi v^2 A_C(r_s, \zeta, v) = f(r_s, \zeta, v) = f_1(v) + f_2(r_s, \zeta, v), \tag{95}$$

where $f_1(v)$ is the (non-oscillatory) long-range part, which is known from the high-density limit [76], and f_2 is the short-range, density- and polarization-dependent contribution. The numerical results in the high-density limit have an analytic fit [46,53] of the form

$$f_1(v) = \frac{a_1 + a_2 v + a_3 v^2}{1 + b_1 v + b_2 v^2 + b_3 v^3 + b_4 v^4} \tag{96}$$

where the coefficients $\{a_i, b_i\}$ are constants (with $a_1 = -0.12436, b_4 = 0.0020$), and are given in Eq. (22) of Ref. [53]. (Correction to Ref. [53]: $a_3 = 0.0024317$.) The short-range contribution is modelled as [53]

$$f_2(v) = \big[-a_1 - (a_2 - a_1 b_1)v + c_1 v^2 + c_2 v^3 + c_3 v^4 + c_4 v^5\big] \exp[-p(r_s, \zeta)v^2], \tag{97}$$

where the coefficients $\{c_i\}$ are functions of r_s and ζ, given by Eqs. (38), (39), (43), and (44) of Ref. [53]. Note that the first two terms in Eq. (97) were chosen to cancel the small-v behavior of $f_1(v)$, since $f_1(v) + f_2(v)$ must vanish as v^2 for small v, from Eqs. (95) and (46). The next two coefficients, c_1 and c_2, were chosen to reproduce the correct value of the jellium pair distribution function at zero separation, and its accompanying cusp, respectively. The last two coefficients, c_3 and c_4, were chosen to simultaneously respect the normalization integral

$$\int_0^\infty dv\, f(r_s, \zeta, v) = 0, \tag{98}$$

and reproduce the LSD correlation energy, as fitted by Perdew and Wang [10]:

$$\int_0^\infty dv\ f(r_s,\zeta,v)/(2v) = \epsilon_{\rm C}^{\rm LSD}(r_s,\zeta)/\phi^3. \tag{99}$$

Lastly,

$$p(r_s,\zeta) = \pi k_F d(\zeta)/(4\phi^4) \tag{100}$$

sets the length-scale of the short-range contribution to the LSD correlation hole, where $d(\zeta) = 0.305 - 0.136\zeta^2$ was fitted to quantum Monte Carlo data.

We apply the same techniques to construct a reasonable GEA correlation hole, writing

$$B_{\rm C}(r_s,\zeta,v) = B_{\rm C}^{\rm LM}(v)\ [1-\exp(-pv^2)] + \beta(r_s,\zeta)\ v^2\ \exp(-pv^2), \tag{101}$$

where $B_{\rm C}^{\rm LM}(v)$ is a (non-oscillatory) long-range contribution, and $\beta(r_s,\zeta)\ v^2\ \exp(-pv^2)$ is a short-range, density- and polarization-dependent contribution, with the same length-scale as the short-range contribution to the LSD correlation hole. The form of hole given in Eq. (101) has several desirable features: For any finite value of r_s, as $v \to 0$, the short-range contribution dominates, while as $v \to \infty$, the long-range contribution dominates. Furthermore, for high densities ($r_s \ll 1$) , $p \gg 1$, so that only for $v \ll 1$ is there any deviation from the long-range contribution, while for low densities ($r_s \gg 1$), only for $v \gg 1$ (which is energetically irrelevant) does the long-range piece become significant.

The long-range part is found from the high-density limit, by Fourier transforming the Langreth-Mehl (LM) exponential approximation [20] for the wave vector decomposition,

$$\delta\tilde{n}_{\rm C}^{\rm GEA,LM}(k) = \frac{4\sqrt{3}}{3\pi k_s}\exp\left[-\frac{2\sqrt{3}}{k_s}k\right] \tag{102}$$

for $\zeta = 0$, yielding via Eq. (15) the result [46]

$$B_{\rm C}^{\rm LM}(v) = \left[18\pi^3(1+v^2/12)^2\right]^{-1}. \tag{103}$$

The short-range contribution contains no constant or linear term in v, so that it does not alter the highly accurate LSD on-top hole or its cusp. The coefficient $\beta(r_s,\zeta)$ is determined by the known GEA correction to the energy, given in Eq. (3). Considering just the correlation contribution, and performing the spin sum, we find

$$\begin{aligned}\Delta E_{\rm C}^{\rm GEA} &= \int d^3r\ C_{\rm C}(r_s,\zeta)\ |\nabla n|^2/n^{4/3} \\ &= 16(3/\pi)^{1/3}\int d^3r\ n(\mathbf{r})C_{\rm C}(r_s(\mathbf{r}),\zeta(\mathbf{r}))\ \phi^2(\zeta(\mathbf{r}))\ t^2(\mathbf{r}).\end{aligned} \tag{104}$$

We approximate the polarization dependence of $C_{\rm C}$ by

$$C_{\rm C}(r_s,\zeta) = \phi(\zeta)\ C_{\rm C}(r_s), \tag{105}$$

where $C_{\rm C}(r_s)$ is the unpolarized value. This approximation has been shown to be very accurate in the high-density limit [74]. From Eqs. (104), (1), and (50),

$$C_{\rm C}(r_s) = \frac{\pi}{8}\left(\frac{\pi}{3}\right)^{1/3}\int_0^\infty dv\ v\ B_{\rm C}(r_s,\zeta,v), \tag{106}$$

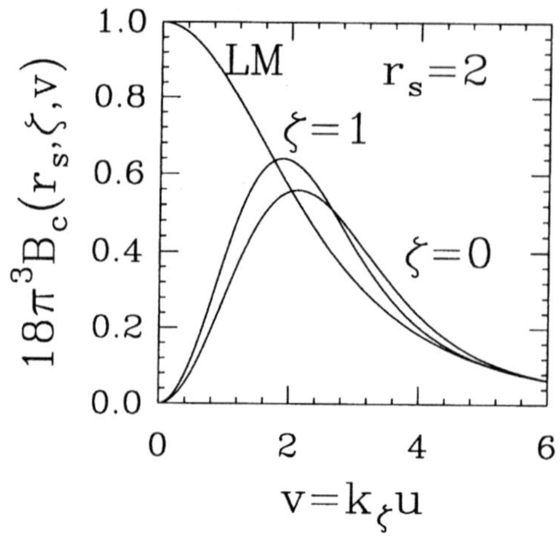

Figure 8. The function $B_C(r_s,\zeta,v)$ which defines the shape of the gradient correction to the correlation hole via Eq. (47), for $r_s = 2$ with $\zeta = 0$ and $\zeta = 1$. Also shown is the entire Langreth-Mehl ($r_s = \zeta = 0$) curve.

which is used to fix $\beta(r_s,\zeta)$:

$$\beta(r_s,\zeta) = \frac{2p^2}{3\pi^3}\left[\frac{C_C(r_s)}{C_C^{LM}} - \tilde{E}_1(12p)\right] \tag{107}$$

where $\tilde{E}_1(x) = x\,\exp(x)\int_x^\infty dt\,\exp(-t)/t$, and $C_C^{LM} = (\pi/3)^{1/3}/(24\pi^2)$ is the GEA correlation coefficient of the Langreth-Mehl hole. While Rasolt and Geldart [12] have given a parametrization of the coefficient $C_C(r_s)$ as a weak function of r_s, we choose here to use just the high-density limit, as derived from the Langreth-Mehl hole, i.e., we set $C_C(r_s)/C_C^{LM} = 1$ in Eq. (107). Fig. 8 shows both the Langreth-Mehl curve and how the short-range contribution changes it for $r_s = 2$ and $\zeta = 0, 1$. The short-range contribution is designed not to change the LSD on-top hole, while preserving the LM gradient contribution to the energy. Note that the v-length scale of the short-range piece is $\approx 1/\sqrt{p}$.

APPENDIX C: LIMITING BEHAVIOR OF NUMERICAL GGA FOR CORRELATION

High-density limit

We consider first the limit in which $r_s \to 0$, keeping t fixed. In this case, the short-range contributions to the hole are only significant in an infinitesimal range of v around $v = 0$, and make no contribution to the normalization or energy integrals. Eq. (49) becomes

$$t = \left[\frac{\int_{v_C}^\infty dv\; f_1(v)}{\int_0^{v_C} dv\; 4\pi v^2\; B_C^{LM}(v)}\right]^{1/2}, \tag{108}$$

which implicitly defines v_C, the cutoff as a function of t at $r_s = 0$, and Eq. (50) becomes

$$H/\phi^3 = -\int_{v_C}^\infty \frac{dv}{2v} f_1(v) + t^2 \int_0^{v_C} dv\; 2\pi v\; B_C^{LM}(v), \tag{109}$$

which yields H, the correction to the LSD energy for $r_s = 0$. Note that both $v_C(t)$ and $H(t)/\phi^3$ are independent of ζ. This is very much like the curve marked $r_s = 0.05$ in Figs. 6 and 7. When $t \ll 1$, we find $v_C \gg 1$, and Eq. (108) becomes

$$v_C = \frac{a_3\pi\sqrt{3}}{4b_4t^2} + O(t^0) \ (t \ll 1), \tag{110}$$

yielding

$$H/\phi^3 = \frac{2t^2}{3\pi^2} - \frac{4b_4t^4}{3\pi^2a_3} + O(t^6) \ (t \ll 1). \tag{111}$$

The first term is the Langreth-Mehl gradient correction. For $t \gg 1$, $v_C \ll 1$ and

$$v_C = \sqrt{\frac{27|a_1|}{2}\frac{\pi}{t}} + O(t^{-2}) \ (t \gg 1). \tag{112}$$

The singular behavior of the first integrand as $v \to 0$ then dominates the resulting energy integral of Eq. (109), producing

$$H/\phi^3 = -\frac{a_1}{2}\ln(t) + O(t^0) \ (t \gg 1), \tag{113}$$

similar to the logarithmic divergence in the LSD correlation energy at high density, i.e., $\epsilon_C^{LSD}/\phi^3 \to -a_1\ln(r_s)/4$ as $r_s \to 0$.

To finish this section, we note that, under uniform scaling, $t(\mathbf{r})$ scales as $r_s^{-1/2}$, and becomes very large as $r_s \to 0$. Thus, under uniform scaling to the high-density limit, we find, from Eq. (113),

$$H/\phi^3 = \frac{a_1}{4}\ln(r_s) \ (\text{uniform scaling}), \tag{114}$$

exactly canceling the logarithmic singularity in the LSD contribution. Thus the numerical GGA correlation energy correctly scales to a constant, as in Eq. (60).

Small gradients

We now take limits of t, keeping r_s fixed. For small values of t, v_C becomes very large. We can expand the normalization integral of Eq. (49) in powers of $1/v_C$, to find

$$v_C = \frac{a_3}{b_4\gamma(r_s,\zeta)t^2} \ (t \to 0). \tag{115}$$

The GEA contribution to the normalization integral is γt^2, where

$$\gamma = \int_0^\infty dv\ 4\pi v^2\ B_C(r_s,\zeta,v) = \frac{4}{\pi\sqrt{3}}\tilde{F}(12p) + \frac{3\pi^{3/2}}{2p^{5/2}}\beta, \tag{116}$$

and

$$\tilde{F}(x) = 1 - \frac{2}{\pi}\int_1^\infty dy\ \sqrt{y-1}\ e^{-x(y-1)}/y^2. \tag{117}$$

Performing the same expansion on the energy integral of Eq. (50), we find

$$\epsilon_C^{GGA} = \epsilon_C^{GEA} - \frac{\phi^3 b_4\gamma^2}{4a_3}t^4 \ (t \to 0). \tag{118}$$

In the high-density ($r_s = 0$) limit, $\gamma = 4/\pi\sqrt{3}$ and $b_4\gamma^2/(4a_3) = 0.11$, as in Eq. (111).

Large Gradients

As $t \to \infty$, the real-space cutoff procedure makes the GGA correlation energy vanish, by making $v_C \to 0$. For small values of v, the correlation hole functions may be expanded about $v = 0$. For the LSD correlation hole, inversion of Eq. (46) yields

$$A_C(v) = \frac{n_C^{LSD}(0)}{\phi^3 k_\zeta^2} + O(v) \ (v \to 0), \tag{119}$$

where the (negative) LSD on-top hole density is given in Ref. [53], while expansion of Eq. (101) yields

$$B_C(v) = (\beta + \frac{p}{18\pi^3}) \ v^2 + O(v^4) \ (v \to 0). \tag{120}$$

Insertion of these results into Eq. (49) gives the asymptotic form of v_C for large t:

$$v_C = \nu(r_s, \zeta)/t \ (t \to \infty), \tag{121}$$

where $\nu = \left(-5n_C^{LSD}(0)/[\phi^3 k_\zeta^2(3\beta + p/(6\pi^3))]\right)^{1/2}$. Then Eq. (50) yields

$$\epsilon_C^{GGA} = \frac{\pi n_C^{LSD}(0)\nu^2}{6k_\zeta^2}\frac{1}{t^2} \ (t \to \infty). \tag{122}$$

While these results apply for all values of ζ other than 1, at exactly $\zeta = 1$ the on-top LSD correlation hole vanishes. Then A_C becomes quadratic in v, because the linear term also vanishes, from Eq. (54). Both the LSD and GEA correlation holes are then quadratic about $v = 0$, and Eq. (49) produces a finite value of t beyond which v_C vanishes (see Fig. 5).

APPENDIX D: DENSITIES FOR WHICH $s(\mathbf{r})$ IS ARBITRARILY LARGE ALMOST EVERYWHERE

Start with any physical density $n_1(\mathbf{r})$ representing N_1 electrons, and replace it by the two-electron density $n(\mathbf{r}) = 2n_1(\mathbf{r})/N_1$. The reduced gradient $s(\mathbf{r})$ of Eq. (28) becomes $(N_1/2)^{1/3} s_1(\mathbf{r})$, and tends to infinity almost everywhere as $N_1 \to \infty$.

Other examples are provided by the non-uniform scaling limits of Eqs. (65) and (67).

APPENDIX E: GGA CORRELATION POTENTIAL

The functional derivative of Eq. (4) is

$$\frac{\delta E_{XC}^{GGA}}{\delta n_\sigma} = \frac{\partial f}{\partial n_\sigma} - \nabla \cdot \left(\frac{\partial f}{\partial \nabla n_\sigma}\right). \tag{123}$$

For ϵ_C of Eq. (1) equal to the last line of Eq. (50) (with r_s, ζ, and t defined by Eqs. (40), (38), and (42), respectively, and for any function H), the correlation potential

$v_C^\sigma([n_\uparrow, n_\downarrow]; \mathbf{r}) = \delta E_C / \delta n_\sigma(\mathbf{r})$ is

$$\begin{aligned}
&\epsilon_C^{\mathrm{LSD}} - \frac{r_s}{3}\frac{\partial \epsilon_C^{\mathrm{LSD}}}{\partial r_s} - (\zeta - \mathrm{sgn}[\sigma])\frac{\partial \epsilon_C^{\mathrm{LSD}}}{\partial \zeta} + H - \frac{r_s}{3}\frac{\partial H}{\partial r_s} + \frac{r_s}{3} t \frac{\partial^2 H}{\partial r_s \partial t} \\
&- (\zeta - \mathrm{sgn}[\sigma])\left[\frac{\partial H}{\partial \zeta} - \frac{\phi'}{\phi} t \frac{\partial H}{\partial t}\right] + \frac{t}{6}\frac{\partial H}{\partial t} + \frac{7t^3}{6}\frac{\partial}{\partial t}\left[t^{-1}\frac{\partial H}{\partial t}\right] \\
&- \frac{\nabla n \cdot \nabla \zeta}{(2k_\zeta)^2 n}\left[\left[t^{-1}\frac{\partial^2 H}{\partial \zeta \partial t}\right] - \frac{\phi'}{\phi}\left\{2\left[t^{-1}\frac{\partial H}{\partial t}\right] + t\frac{\partial}{\partial t}\left[t^{-1}\frac{\partial H}{\partial t}\right]\right\}\right] \\
&- \frac{\nabla n \cdot \nabla |\nabla n|}{(2k_\zeta)^3 n^2}\frac{\partial}{\partial t}\left[t^{-1}\frac{\partial H}{\partial t}\right] - \frac{\nabla^2 n}{(2k_\zeta)^2 n}\left[t^{-1}\frac{\partial H}{\partial t}\right],
\end{aligned} \tag{124}$$

where $\mathrm{sgn}[\sigma]$ is $+1$ for $\sigma = \uparrow$ and -1 for $\sigma = \downarrow$, and $\phi(\zeta)$ is defined in Eq. (44), with $\phi' = d\phi/d\zeta$. We thank Per Söderlind [94] for pointing out that the $\partial^2 H/\partial r_s \partial t$ term was omitted in Eq. (33) of Ref. [46]; this term was present in all our other work, and in our GGA subroutines.

At a nucleus, $\nabla^2 n$ diverges like $(2/r)dn/dr$, and so do the GGA exchange and correlation potentials.

REFERENCES

1. W. Kohn and L.J. Sham, Phys. Rev. **140**, A 1133 (1965).
2. R.O. Jones and O. Gunnarsson, Rev. Mod. Phys. **61**, 689 (1989).
3. R.G. Parr and W. Yang, *Density Functional Theory of Atoms and Molecules* (Oxford, New York, 1989).
4. R.M. Dreizler and E.K.U. Gross, *Density Functional Theory* (Springer-Verlag, Berlin, 1990).
5. D. M. Ceperley and B. J. Alder, Phys. Rev. Lett. **45**, 566 (1980).
6. W. E. Pickett and J. Q. Broughton, Phys. Rev. B **48**, 14859 (1993).
7. G. Ortiz and P. Ballone, Phys. Rev. B **50**, 1391 (1994).
8. S. H. Vosko, L. Wilk, and M. Nusair, Can. J. Phys. **58**, 1200 (1980).
9. J. P. Perdew and A. Zunger, Phys. Rev. B **23**, 5048 (1981).
10. J. P. Perdew and Y. Wang, Phys. Rev. B **45**, 13244 (1992).
11. M. Rasolt and H.L. Davis, Phys. Lett. A **86**, 45 (1981).
12. M. Rasolt and D.J.W. Geldart, Phys. Rev. B **34**, 1325 (1986).
13. L. J. Sham, in *Computational Methods in Band Theory*, edited by P. M. Marcus, J. F. Janak, and A. R. Williams (Plenum, New York, 1971), p. 458.
14. S.-K. Ma and K.A. Brueckner, Phys. Rev. **165**, 18 (1968).
15. L. Kleinman and S. Lee, Phys. Rev. B **37**, 4634 (1988).
16. J.P. Perdew and Y. Wang, Phys. Rev. B **33**, 8800 (1986); **40**, 3399 (1989) (E).
17. J.P. Perdew, Phys. Rev. B **33**, 8822 (1986); **34**, 7406 (1986) (E).
18. D.C. Langreth and J.P. Perdew, Phys. Rev. B **21**, 5469 (1980).
19. J.P. Perdew, Phys. Rev. Lett. **55**, 1665 (1985); **55**, 2370 (1985) (E).
20. D.C. Langreth and M.J. Mehl, Phys. Rev. B **28**, 1809 (1983).
21. C.D. Hu and D.C. Langreth, Phys. Scr. **32**, 391 (1985).
22. K. Burke, J. P. Perdew, and M. Levy, in *Modern Density Functional Theory: A Tool for Chemistry*, edited by J. M. Seminario and P. Politzer (Elsevier, Amsterdam, 1995).
23. J. P. Perdew, J. A. Chevary, S. H. Vosko, K. A. Jackson, M. R. Pederson, D.J. Singh, and C. Fiolhais, Phys. Rev. B **46**, 6671 (1992); **48**, 4978 (1993) (E).
24. N. A. W. Holzwarth and Y. Zeng, Phys. Rev. B **49**, 2351 (1994).
25. L. Stixrude, R. E. Cohen, and D. J. Singh, Phys. Rev. B **50**, 6442 (1994).
26. P. Dufek, P. Blaha, and K. Schwarz, Phys. Rev. B **50** 7279 (1994).
27. P. Söderlind, O. Eriksson, J.M. Willis, and B. Johannson, Phys. Rev. B **50**, 7291 (1994).
28. G. Kresse, J. Furthmüller, and J. Hafner, Phys. Rev. B **50**, 13181 (1994).
29. M. Causá and A. Zupan, Chem. Phys. Letters **220**, 145 (1994).
30. A. Khein, D.J. Singh, and C.J. Umrigar, Phys. Rev. B **51**, 4105 (1995).
31. D. Porezag and M.R. Pederson, J. Chem. Phys. **102**, 9345 (1995).

32. J.C. Grossman, L. Mitas, and K. Raghavachari, Phys. Rev. Lett. **75**, 3870 (1995); **76**, 1006 (1996) (E).
33. L. Stixrude and R.E. Cohen, Science, **267**, 1972 (1995).
34. B. Hammer and J.K. Nørskov, Nature **376**, 238 (1995).
35. A. Gross, B. Hammer, M. Scheffler, and W. Brenig, Phys. Rev. Lett. **73**, 3121 (1994).
36. N. Moll, M. Bockstedte, M. Fuchs, E. Pehlke, and M. Scheffler, Phys. Rev. B **52**, 2550 (1995).
37. D.R. Hamann, Phys. Rev. Lett. **76**, 660 (1996).
38. J.P. Perdew, R.G. Parr, M. Levy, and J.L. Balduz, Jr., Phys. Rev. Lett. **49**, 1691 (1982).
39. J. P. Perdew, in *Density Functional Methods in Physics*, edited by R.M. Dreizler and J. da Providencia (Plenum, NY, 1985), p. 265.
40. K. Burke, M. Ernzerhof, and J.P. Perdew, *Why semi-local functionals work: Accuracy of the on-top hole density*, work in progress.
41. C. J. Umrigar and X. Gonze, in *High Performance Computing and its Application to the Physical Sciences*, Proceedings of the Mardi Gras 1993 Conference, edited by D. A. Browne *et al* (World Scientific, Singapore, 1993).
42. C. Filippi, C. J. Umrigar, and M. Taut, J. Chem. Phys. **100**, 1290 (1994).
43. C. J. Umrigar and X. Gonze, Phys. Rev. A **50**, 3827 (1994).
44. M. Slamet and V. Sahni, Phys. Rev. B **44**, 10921 (1991).
45. A.D. Becke, Phys. Rev. A **38**, 3098 (1988).
46. J.P. Perdew, in *Electronic Structure of Solids '91*, edited by P. Ziesche and H. Eschrig (Akademie Verlag, Berlin, 1991), page 11.
47. E. H. Lieb and S. Oxford, Int. J. Quantum Chem. **19**, 427 (1981).
48. M. Levy, Int. J. Quantum Chem. **S23**, 617 (1989).
49. M. Levy, Phys. Rev. A **43**, 4637 (1991).
50. A. Görling and M. Levy, Phys. Rev. A **45**, 1509 (1992).
51. J.P. Perdew and K. Burke, Int. J. Quantum Chem. **57**, 309 (1996).
52. C. Lee, W. Yang, and R.G. Parr, Phys. Rev. B **37**, 785 (1988).
53. J. P. Perdew and Y. Wang, Phys. Rev. B **46**, 12947 (1992): **56**, 7018 (1997) (E).
54. K. Burke and J. P. Perdew, Int. J. Quantum Chem. **56**, 199 (1995).
55. E. Engel and S.H. Vosko, Phys. Rev. B **47**, 13164 (1993). See Fig. 2.
56. J.P. Perdew, K. Burke, and M. Ernzerhof, Phys. Rev. Lett. **77**, 3865 (1996); **78**, 1396 (1997) (E).
57. J.P. Perdew, K. Burke, and Y. Wang, Phys. Rev. B **54**, 16533 (1996).
58. K. Burke, J.P. Perdew, and M. Ernzerhof, *System-averaged exchange-correlation holes*, work in progress.
59. D.C. Langreth and J.P. Perdew, Solid State Commun. **17**, 1425 (1975).
60. D.C. Langreth and J.P. Perdew, Phys. Rev. B **15**, 2884 (1977).
61. O. Gunnarsson and B.I. Lundqvist, Phys. Rev. B **13**, 4274 (1976).
62. D.M. Ceperley and B.J. Alder, Phys. Rev. B **36**, 2092 (1987).
63. K. Burke, J. P. Perdew, and D. C. Langreth, Phys. Rev. Lett. **73**, 1283 (1994).
64. J.P. Perdew, A. Savin, and K. Burke, Phys. Rev. A **51**, 4531 (1995).
65. K. Burke and J.P. Perdew, Mod. Phys. Lett. B **9**, 829 (1995).
66. E.K.U. Gross and R.M. Dreizler, Z. Phys. A **302**, 103 (1981).
67. Y. Wang, J. P. Perdew, J. A. Chevary, L. D. MacDonald, and S. H. Vosko, Phys. Rev. A **41**, 78 (1990).
68. J.P. Perdew, Phys. Lett. A **165**, 79 (1992).
69. G.L. Oliver and J.P. Perdew, Phys. Rev. A **20**, 397 (1979).
70. J. P. Perdew, in *Condensed Matter Theories, Vol. 2*, edited by P. Vashishta, R. K. Kalia, and R. F. Bishop (Plenum, NY, 1987), p. 89.
71. M. Ernzerhof, J.P. Perdew, and K. Burke, in *Density Functional Theory*, edited by R. Nalewajski (Spinger-Verlag, Berlin, 1996).
72. D.C. Langreth and J.P. Perdew, Phys. Lett. A **92**, 451 (1982).
73. M. Ernzerhof, K. Burke, and J.P. Perdew, J. Chem. Phys. 105, 2798 (1996).
74. Y. Wang and J. P. Perdew, Phys. Rev. B **43**, 8911 (1991).
75. J.P. Perdew, Physica B **172**, 1 (1991).
76. Y. Wang and J. P. Perdew, Phys. Rev. B **44**, 13298 (1991).
77. M. Levy and J. P. Perdew, Phys. Rev. B **48**, 11638 (1993).
78. J. C. Kimball, Phys. Rev. A **7**, 1648 (1973).
79. E. R. Davidson, *Reduced Density Matrices in Quantum Chemistry* (Academic Press, New York, 1976).

80. J.P. Perdew, M. Ernzerhof, K. Burke, and A. Savin, Int. J. Quantum Chem. **61**, 197 (1997).
81. M. Levy and J.P. Perdew, Phys. Rev. A **32**, 2010 (1985).
82. A. Zupan, J.P. Perdew, K. Burke, and M. Causá, Int. J. Quantum Chem. 61, 835 (1997).
83. E. Engel, J.A. Chevary, L.D. MacDonald, and S.H. Vosko, Z. Phys. D **23**, 7 (1992).
84. D.J. Lacks and R.G. Gordon, Phys. Rev. A **47**, 4681 (1993).
85. L. A. Eriksson, O. L. Malkina, V. G. Malkin, and D. R. Salahub, J. Chem. Phys. **100**, 5066 (1994).
86. E. Clementi and C. Roetti, At. Data Nucl. Data Tables **14**, 177 (1974).
87. D. C. Langreth and S. H. Vosko, Adv. in Quantum Chem. **21**, 175 (1990).
88. D.J.W. Geldart, E. Dunlap, M.L. Glasser, and M.R.A. Shegelski, Solid State Commun. **88**, 81 (1993).
89. S.J. Chakravorty, S.R. Gwaltney, E.R. Davidson, F.A. Parpia, and C. Froese Fischer, Phys. Rev. A **47**, 3649 (1993).
90. A. Görling, M. Levy, and J. P. Perdew, Phys. Rev. B **47**, 1167 (1993).
91. J.P. Perdew, Int. J. Quantum Chem. S **27**, 93 (1993).
92. Y. Li and J.B. Krieger, Phys. Rev. A **41**, 1701 (1990).
93. K. Burke, J.P. Perdew, and M. Ernzerhof, Int. J. Quantum Chem. **61**, 287 (1997).
94. P. Söderlind, doctoral dissertation, Uppsala University (1994).

CORRELATION ENERGY IN A HIGH-DENSITY LIMIT FROM ADIABATIC CONNECTION PERTURBATION THEORY

Mel Levy [a], Stanislav Ivanov [a], and Andeas Görling [a,b]

[a]Chemistry Department and Quantum Theory Group
Tulane University, New Orleans, Louisiana 70118
[b]Lehrstuhl für Theoretische Chemie
Technische Universität München
85747 Garching, Germany

INTRODUCTION

Here the derivation of the Görling-Levy adiabatic perturbation theory is reviewed for the correlation energy, $E_c[n]$, in a high-density limit for finite systems, with particular emphasis on the resultant expression in terms of Kohn-Sham orbitals, for the second-order contribution, $E_c^{(2)}[n]$. This contribution occurs through an asymptotic uniform scaling of $n(\mathbf{r})$. The quantity $2E_c^{(2)}[n]$ is especially important because it is the initial slope in the adiabatic connection method (coupling-constant formula) for $E_c[n]$, even where n does not belong to a high-density system. The recent bounds and equalities of Ivanov *et al*, involving $E_c^{(2)}[n]$ and its functional derivative, are compared against the numerical results from functionals of Perdew-Burke-Ernzerhof and Lee-Yang-Parr.

Density functional theory [1-5] provides, arguably, the most widely used and powerful computational formalism for large molecules and solids. The theory enables one to replace the complicated traditional interacting wavefunction approach by the simpler electron density and the noninteracting wavefunction which yields that density.

Correlation energies in density-functional theory (DFT) must be approximated. For this purpose, knowledge of exact properties is necessary. With this in mind, it is now known that the exact correlation energy for use as part of a full DFT calculation, $E_c[n]$ [1-8], satisfies [8-11] the expansion

$$E_c[n_\lambda] = E_c^{(2)}[n] + \lambda^{-1}E_c^{(3)}[n] + \lambda^{-2}E_c^{(4)}[n] + \ldots \tag{1}$$

for high-enough λ, where

$$n_\lambda(x, y, z) = \lambda^3 n(\lambda x, \lambda y, \lambda z). \tag{2}$$

Hence, $E_c[n_\lambda]$ is bounded, and is equal to $E_c^{(2)}[n]$, as $\lambda \to \infty$.

Electronic Density Functional Theory: *Recent Progress and New Directions*
Edited by Dobson *et al.*, Plenum Press, New York, 1998

It is our purpose to briefly review expansion (1) through the adiabatic perturbation theory of Görling and Levy [9], which arrives at the formal expression for the second-order energy, $E_c^{(2)}[n]$, in terms of Kohn-Sham orbitals. Further, the practical variation-perturbation expression for $E_c^{(2)}[n]$ shall also be presented. As we shall discuss, the quantity $2E_c^{(2)}[n]$ is particularly valuable because it is the initial slope in the adiabatic connection method (coupling-constant formula) for $E_c[n]$.

Numerical bounds for $E_c^{(2)}[n]$, as recently generated by Ivanov and Levy [12], shall be given for hydrogenic densities, corresponding to two, three, nine, ten, and eleven electrons. For two electrons, where the bounds become an equality, we also review the recently derived closed-form equality of Ivanov, Lopez-Boada, Görling, and Levy [13], that connects $E_c^{(2)}[n]$ to an integral involving its functional derivative.

Relations shall also be presented involving the correlation energy [9], ${}^{HF}E_c[n]$, whose functional derivative is meant to be added to the ordinary Fock potential to produce, upon self-consistency, the exact ground-state energy and density. Finally, comparisons shall be made, for the mentioned relations involving $E_c[n]$ and ${}^{HF}E_c[n]$, with the correlation energy functionals of Perdew-Burke-Ernzerhof [14] and Lee-Yang-Parr [15].

In atomic units (a.u. are to be used throughout the paper), let's define the Hamiltonian operator for the system of interest as

$$\hat{H} = \hat{T} + \hat{V}_{ee} + \sum_{i=1}^{N} v(\mathbf{r}_i), \tag{3}$$

where $\hat{T}$ is the kinetic energy operator,

$$\hat{T} = -\sum_{i=1}^{N} \frac{1}{2}\nabla_i^2 \tag{4}$$

where $\hat{V}_{ee}$ is the operator for the electron-electron repulsion,

$$\hat{V}_{ee} = \sum_{j=1}^{N}\sum_{i>j}^{N} \frac{1}{|\mathbf{r}_i - \mathbf{r}_j|}, \tag{5}$$

and where $v(r)$ is the local spin-independent multiplicative potential of interest.

The ground-state energy of the system under investigation is obtained [1-5] by

$$E^{GS} = \min_{n(\mathbf{r})\to N}\{F[n] + \int v(\mathbf{r})n(\mathbf{r})d\mathbf{r}\}, \tag{6}$$

where $F[n]$ is the universal Hohenberg-Kohn functional of the density. The constrained-search definition [16] of $F[n]$ is

$$F[n] = \min_{\Psi\to n(\mathbf{r})} < \Psi \left|\hat{T} + \hat{V}_{ee}\right| \Psi > \equiv < \Psi[n] \left|\hat{T} + \hat{V}_{ee}\right| \Psi[n] > . \tag{7}$$

In other words, $\Psi[n]$ is that antisymmetric wavefunction that minimizes $< \hat{T} + \hat{V}_{ee} >$ and yields the density $n(\mathbf{r})$.

For computational purposes $F[n]$ must be approximated and it is commonly partitioned [3-6] as

$$F[n] = T_s[n] + U[n] + E_x[n] + E_c[n]. \tag{8}$$

In Eq. (8), $T_s[n]$ is the non-interacting kinetic energy

$$T_s[n] = < \Psi_0[n] \left| \hat{T} \right| \Psi_0[n] >, \tag{9}$$

where $\Psi_0[n]$ is the Kohn-Sham (KS) wavefunction. That is, by definition, $\Psi_0[n]$ is the wavefunction that minimizes just $< \hat{T} >$ and yields the same density as $\Psi[n]$. Except for certain degenerate cases, $\Psi_0[n]$ is a single determinant [17]. The Hartree electron-electron repulsion energy $U[n]$ is given by

$$U[n] = \frac{1}{2} \int \int \frac{n(\mathbf{r}_1) n(\mathbf{r}_2)}{|\mathbf{r}_1 - \mathbf{r}_2|} d\mathbf{r}_1 d\mathbf{r}_2. \tag{10}$$

The exchange energy, $E_x[n]$, defined in terms of the KS single determinant, is

$$E_x[n] = < \Psi_0[n] \left| \hat{V}_{ee} \right| \Psi_0[n] > - U[n], \tag{11}$$

and the correlation energy, $E_c[n]$, is defined as

$$E_c[n] = < \Psi[n] \left| \hat{T} + \hat{V}_{ee} \right| \Psi[n] > - < \Psi_0[n] \left| \hat{T} + \hat{V}_{ee} \right| \Psi_0[n] > . \tag{12}$$

With these definitions, Eqs. (9)-(12), the following simple scaling occurs with $E_x[n]$, [6]:

$$E_x[n_\lambda] = \lambda E_x[n], \tag{13}$$

Upon uniform scaling of the electron density, Eq. (2), $E_c[n]$ does not scale homogeneously. Instead, $E_c[n]$ satisfies the following high-density relationship [8-11]:

$$\lim_{\lambda \to \infty} E_c[n_\lambda] = E_c^{(2)}[n]. \tag{14}$$

In other words, $E_c[n_\lambda]$ is bounded, as $\lambda \to \infty$, and is equal to a second-order energy. It is important to stress here that, in contrast to Eq. (14), the LDA correlation energy diverges as $-\ln(\lambda)$, as $\lambda \to \infty$, because of the Gell-Mann and Brueckner contribution to the correlation energy per particle of a uniform electron gas [18].

AN ADIABATIC CONNECTION PERTURBATION THEORY

The link between uniform scaling of the electron density and scaling of the electron-electron interaction has been developed by Görling and Levy in a series of papers [9,19]. The effective potential and the electron-electron interaction along the coupling constant path, which connects a noninteracting and a fully interacting system with the same electron density, have been used by them for a DFT perturbation theory. (For simplicity, non-degenerate situations shall be assumed here.)

To derive Eq. (14), we start with [6,20,21]

$$E_c^\alpha[n] = \alpha^2 E_c[n_\lambda]; \quad \alpha = \lambda^{-1} , \tag{15}$$

where

$$E_c^\alpha[n] = < \Psi_\alpha[n] \left| \hat{T} + \alpha \hat{V}_{ee} \right| \Psi_\alpha[n] > - < \Psi_0[n] \left| \hat{T} + \alpha \hat{V}_{ee} \right| \Psi_0[n] >, \tag{16}$$

or

$$E_c^\alpha[n] =< \Psi_\alpha[n] \left| \hat{H}_\alpha[n] \right| \Psi_\alpha[n] > - < \Psi_0[n] \left| \hat{H}_\alpha[n] \right| \Psi_0[n] >, \tag{17}$$

According to the constrained-search orientation of DFT, the wavefunction $\Psi_\alpha[n]$ is the one that minimizes the expectation value $< \hat{T} + \alpha\hat{V}_{ee} >$and yields the density $n(\mathbf{r})$. In Eq. (17), the Hamiltonian $\hat{H}_\alpha[n]$ keeps the density fixed at $n(\mathbf{r})$ as α changes [9]. $\hat{H}_\alpha[n]$ has the form

$$\hat{H}_\alpha[n] = \hat{T} + \alpha\hat{V}_{ee} + \sum_{i=1}^{N} v_\alpha([n]; \mathbf{r}_i) \tag{18}$$

Without concerning ourselves, at this time, with the exact form of $v_\alpha([n]; \mathbf{r})$, we review the perturbation expansion for $E_c^\alpha[n]$ for small enough α. Start with

$$E_c^\alpha[n] = E_c^{(0)}[n] + \alpha E_c^{(1)}[n] + \alpha^2 E_c^{(2)}[n] + \alpha^3 E_c^{(3)}[n] + \dots . \tag{19}$$

The first term on the right-hand side of Eq. (19) is equal to zero. This follows directly from the definition of $E_c^\alpha[n]$, i.e.

$$E_c^{\alpha=0}[n] = 0 = E_c^{(0)} \tag{20}$$

Next, apply the Hellmann-Feynman theorem to Eq. (17) to obtain

$$E_c^{(1)}[n] = \left. \frac{\partial E_c^\alpha[n]}{\partial \alpha} \right|_{\alpha=0} = 0 \tag{21}$$

Consequently, from Eqs. (20) and (21), it follows that the perturbation expansion for $E_c^\alpha[n]$ begins with the quadratic term in α. That is

$$E_c^\alpha[n] = \alpha^2 E_c^{(2)}[n] + \alpha^3 E_c^{(3)}[n] + \dots \tag{22}$$

The combination of Eqs. (15) and (22) leads to Eq. (1), and therefore

$$\lim_{\lambda \to 0} E_c[n_\lambda] = E_c^{(2)}[n]. \tag{23a}$$

Also, note that for a high-enough density $n(\mathbf{r})$, Eq. (1) implies [9,22,23]

$$E_c^{(j)}[n_\lambda] = \lambda^{2-j} E_c^{(j)}[n]. \tag{23b}$$

For numerical evaluation purposes, it remains for us to identify $E_c^{(2)}[n]$ through traditional perturbation theory. With this in mind, first we will investigate the form of the Hamiltonian $\hat{H}_\alpha[n]$ (more specifically the form of $v_\alpha([n]; \mathbf{r})$) that keeps the density fixed along the adiabatic path. To express $v_\alpha([n]; \mathbf{r})$, note that the ground-state energy and density are obtained from

$$E_\alpha = \min_{\rho(\mathbf{r}) \to N} \{F_\alpha[\rho] + \int v_\alpha([n]; \mathbf{r})\rho(\mathbf{r})d\mathbf{r}\} \tag{24}$$

or

$$E_\alpha = F_\alpha[n] + \int v_\alpha([n]; \mathbf{r})d\mathbf{r} \tag{25}$$

so that

$$v_\alpha([n];\mathbf{r}) = -\left.\frac{\delta F_\alpha[\rho]}{\delta\rho}\right|_{\rho=n} \tag{25a}$$

where

$$F_\alpha[\rho] = \left< \Psi_\alpha[\rho] \left| \hat{T} + \alpha\hat{V}_{ee} \right| \Psi_\alpha[\rho] \right> \tag{26}$$

Here $\rho(\mathbf{r})$ is a density and $n(\mathbf{r})$ is the ground-state density of $\hat{H}_\alpha[n]$. [Note that $n(\mathbf{r})$, in turn, is a trial density for the variational problem in Eq. (6).]

Partition $F_\alpha[\rho]$ as

$$F_\alpha[\rho] = \left< \Psi_0[\rho] \left| \hat{T} + \alpha\hat{V}_{ee} \right| \Psi_0[\rho] \right> + E_c^\alpha[\rho] \tag{27}$$

or

$$F_\alpha[\rho] = F_0[\rho] + \alpha U[\rho] + \alpha E_x[\rho] + E_c^\alpha[\rho]. \tag{28}$$

Then by Eqs. (25a) and (28), we have

$$v_\alpha([n];\mathbf{r}) = v_0([n];\mathbf{r}) - \alpha u([n];\mathbf{r}) - \alpha v_x([n];\mathbf{r}) - v_c^\alpha([n];\mathbf{r}) \tag{29}$$

where

$$v_0([n];\mathbf{r}) = -\left.\frac{\delta F_0[\rho]}{\delta\rho}\right|_{\rho=n}, \tag{30}$$

$$u([n];\mathbf{r}) = \left.\frac{\delta U[\rho]}{\delta\rho}\right|_{\rho=n}, \tag{31}$$

$$v_x([n];\mathbf{r}) = \left.\frac{\delta E_x[\rho]}{\delta\rho}\right|_{\rho=n}, \tag{32}$$

and

$$v_c^\alpha([n];\mathbf{r}) = \left.\frac{\delta E_c^\alpha[\rho]}{\delta\rho}\right|_{\rho=n}. \tag{33}$$

But, from Eq. (22), for small enough α, we have

$$v_c^\alpha([n];\mathbf{r}) = \alpha^2 \left.\frac{\delta E_c^{(2)}[\rho]}{\delta\rho}\right|_{\rho=n} + \alpha^3 \left.\frac{\delta E_c^{(3)}[\rho]}{\delta\rho}\right|_{\rho=n} + \ldots . \tag{34}$$

With

$$v_c^{(j)}([n];\mathbf{r}) = \left.\frac{\delta E_c^{(j)}[\rho]}{\delta\rho}\right|_{\rho=n} \tag{35}$$

the combination of Eqs. (29) and (34) leads to

$$v_\alpha([n],\mathbf{r}) = v_0([n],\mathbf{r}) - \alpha u([n],\mathbf{r}) - \alpha v_x([n],\mathbf{r}) - \alpha^2 v_c^{(2)}([n],\mathbf{r}) - \alpha^3 v_c^{(3)}([n],\mathbf{r}) - \ldots . \tag{36}$$

This enables us to express $\hat{H}_\alpha[n]$ in a form convenient for a perturbation expansion for small enough α. Namely

$$\hat{H}_\alpha[n] = \hat{H}_0[n] + \alpha\hat{H}^{(1)}[n] + \alpha^2\hat{H}^{(2)}[n] + \alpha^3\hat{H}^{(3)}[n] + \ldots \ , \tag{37}$$

where

$$\hat{H}^{(1)}[n] = \hat{V}_{ee} - \sum_{i=1}^{N}\{u([n],\mathbf{r}_i) + v_x([n],\mathbf{r}_i)\} \equiv \hat{V}_{ee} - \sum_{i=1}^{N} v_{xu}([n],\mathbf{r}_i) \tag{37a}$$

and

$$\hat{H}^{(j)}[n] = -\sum_{i=1}^{N} v_c^{(j)}([n],\mathbf{r}_i); \quad j \geq 2 \ . \tag{37b}$$

Now take the derivative of $E_c^\alpha[n]$ with respect to α (see Eqs. (22) and (17)) and use the Hellmann-Feynman theorem, to obtain

$$2\alpha E_c^{(2)}[n] + 3\alpha^2 E_c^{(3)}[n] + \ldots = < \Psi_\alpha[n] \left| \frac{\partial \hat{H}_\alpha[n]}{\partial\alpha} \right| \Psi_\alpha[n] > - < \Psi_0[n] \left| \frac{\partial \hat{H}_\alpha[n]}{\partial\alpha} \right| \Psi_0[n] > . \tag{38}$$

By making use of Eq. (37), we obtain from Eq. (38)

$$\begin{aligned} 2\alpha E_c^{(2)}[n] + 3\alpha^2 E_c^{(3)}[n] + \ldots = &< \Psi_\alpha[n] \left| \hat{H}^{(1)}[n] + 2\alpha\hat{H}^{(2)}[n] + \ldots \right| \Psi_\alpha[n] > \\ &- < \Psi_0[n] \left| \hat{H}^{(1)}[n] + 2\alpha\hat{H}^{(2)}[n] + \ldots \right| \Psi_0[n] > . \end{aligned} \tag{39}$$

Next, utilize the fact that $\Psi_\alpha[n]$ is the ground-state solution to

$$\hat{H}_\alpha[n]\Psi_\alpha[n] = E_\alpha\Psi_\alpha[n]. \tag{40}$$

and take into account the form of $\hat{H}_\alpha[n]$ for small enough α. The following perturbation series for $\Psi_\alpha[n]$ and E_α result:

$$\Psi_\alpha[n] = \Psi_0[n] + \alpha\Psi^{(1)}[n] + \alpha^2\Psi^{(2)}[n] + \ldots \tag{41}$$

and

$$E_\alpha = E_0 + \alpha E^{(1)} + \alpha^2 E^{(2)} + \ldots \tag{42}$$

By substituting the expression for $\Psi_\alpha[n]$, Eq. (41), into Eq. (39) and equating all terms of order α, we arrive at

$$E_c^{(2)}[n] = \frac{1}{2}\left\{< \Psi_0[n] \left| \hat{H}^{(1)} \right| \Psi^{(1)}[n] > + < \Psi^{(1)}[n] \left| \hat{H}^{(1)} \right| \Psi_0[n] >\right\} \tag{43}$$

The form of $\Psi^{(1)}[n]$ is easily found by following standard non-degenerate perturbation theory. Replace $\Psi_\alpha[n]$ and E_α by their respective series expansions given by Eqs. (41) and (42) in formula (40), and equate all terms of order one in α, to obtain the familiar

$$\hat{H}_0[n]\Psi^{(1)}[n] + H^{(1)}[n]\Psi_0[n] = E_0\Psi^{(1)}[n] + E^{(1)}\Psi_0[n] \ . \tag{44}$$

Next, we proceed in the usual fashion which should be quite familiar to the reader. Expand $\Psi^{(1)}[n]$ in terms of the eigenfunctions of $\hat{H}_0[n]$, and utilize Eq. (44) to obtain

$$\Psi^{(1)}[n] = \sum_{k=1}^{\infty} \frac{< \Phi_k \left| H^{(1)}[n] \right| \Psi_0 >}{E_0 - E_0^k} \Phi_k. \tag{45}$$

In Eq. (45), Φ_k is the k-th eigenfunction of $\hat{H}_0[n]$ and E_0^k is its corresponding eigenvalue. (Observe that $\Phi_0 = \Psi_0$.)

By substituting Eq. (45) in Eq. (43) and making use of the fact that $\hat{H}^{(1)}[n]$ is an Hermitian operator, we arrive at our desired result

$$E_c^{(2)}[n] = \sum_{k=1}^{\infty} \frac{\left| < \Phi_k \left| \hat{H}^{(1)}[n] \right| \Psi_0 > \right|^2}{E_0 - E_0^k} \tag{46}$$

or

$$\lim_{\lambda \to \infty} E_c[n_\lambda] = E_c^{(2)}[n] = \sum_{k=1}^{\infty} \frac{\left| < \Phi_k \left| \hat{V}_{ee} - \sum_{i=1}^N \{u([n]; \mathbf{r}_i) + v_x([n]; \mathbf{r}_i)\} \right| \Psi_0 > \right|^2}{E_0 - E_0^k} \tag{47}$$

NEW RELATIONS INVOLVING HIGH-DENSITY LIMIT OF E$_c$[n] AND ITS FUNCTIONAL DERIVATIVE

Görling and Levy, by means of their DFT perturbation theory, have derived [19] the following expression connecting $E_c^{(2)}[n]$ and $v_c^{(2)}([n^k]; \mathbf{r})$:

$$E_c^{(2)}[n] - \sum_{k=2}^{N} \int d\mathbf{r}\, v_c^{(2)}[n_k]; \mathbf{r})\, \phi_k^*(\mathbf{r})\phi_k(\mathbf{r}) =$$
$$\sum_{k=2}^{N} \sum_{i=1}^{k-1} \sum_{m=k}^{\infty} \frac{\left| < \phi_m(\mathbf{r}) \left| u([n^k]; \mathbf{r}_i) + v_x([n^k]; \mathbf{r}_i) - u([n^{k-1}]; \mathbf{r}_i) - v_x([n^{k-1}]; \mathbf{r}_i) \right| \phi_i(\mathbf{r}) > \right|^2}{\epsilon_i - \epsilon_m} \tag{48}$$

In the above formula, $v_c^{(2)}([n^k]; \mathbf{r})$, $u([n^k]; \mathbf{r})$ and $v_x([n^k]; \mathbf{r})$ are the functional derivatives, respectively, of $E_c^{(2)}[n^k]$, $U[n^k]$, and $E_x[n^k]$, with respect to the density. The density $n^k(\mathbf{r})$ is obtained through

$$n^k(\mathbf{r}) = \sum_{m=1}^{k} \phi_m^*([n]; \mathbf{r}) \phi_m([n]; \mathbf{r}) \tag{49}$$

In Eq. (48), ϵ_m are the orbital energies corresponding to the KS orbitals $\phi_m(\mathbf{r})$ of Ψ_0. (Also observe that $n^N(\mathbf{r}) = n(\mathbf{r})$). Since the derivation of Eq. (48) is based upon the fact that the energy of the highest occupied Kohn-Sham orbital corresponds to the ionization energy [24,25], formula (48) holds for systems whose potentials approach zero as $|\mathbf{r}|$ goes to infinity.

Ivanov *et al* [13] have very recently presented the right-hand side of Eq. (48) in a closed form. For a two-electron density, they have expressed the difference $E_c^{(2)}[n] - \frac{1}{2} \int d\mathbf{r} v_c^{(2)}([n]; \mathbf{r}) n(\mathbf{r})$ as a functional of the density $n(\mathbf{r})$ only, and for any spherically symmetric diamagnetic two-electron density, they have obtained an analytical expression. As an example, they have considered an exponential two-electron density,

Table 1. Comparison of second-order energy quantities for a hydrogenic density $n(\mathbf{r}) = (2a^3/\pi)\exp(-2ar)$

	$E_c^{(2)}[n]$	$\int d\mathbf{r}\, v_c^{(2)}([n];\mathbf{r})\, n(\mathbf{r})$
Exact Value	-0.046663	0.017677
$E_c^{PBE}[n]$	-0.047896	-0.073013
$E_c^{LYP}[n]$	-0.056481	-0.108922

$n(\mathbf{r}) = (2a^3/\pi)\exp(-2ar)$ with $a > 0$. By using a known result for the second-order energy for He, when electron-electron repulsion is treated as the perturbation, along with their closed form expression, they have identified each component in the difference $E_c^{(2)}[n] - \frac{1}{2}\int d\mathbf{r} v_c^{(2)}([n];\mathbf{r})n(\mathbf{r})$ as

$$E_c^{(2)}[n] = -0.046663 \tag{50}$$

and

$$\int d\mathbf{r} v_c^{(2)}([n],\mathbf{r})n(\mathbf{r}) = 0.017677 \tag{51}$$

We compare, in Table 1, the values of $E_c^{(2)}[n]$ and $\int d\mathbf{r}\, v_c^{(2)}([n];\mathbf{r})\, n(\mathbf{r})$ with those of $E_c^{(2)APP}[n]$ and $\int d\mathbf{r}\, v_c^{(2)APP}([n];\mathbf{r})\, n(\mathbf{r})$, corresponding to two approximate (APP) correlation energy functionals with gradients, the recently derived GGA of Perdew, Burke, and Ernzerhof (PBE) [14], and the one of Lee, Yang, and Parr (LYP) [15].

As far as $E_c^{(2)}[n]$ is concerned, both functionals give numbers close to the theoretically-calculated value. Unfortunately, neither of the two functionals has the correct functional derivative as required by condition (51); each approximation for $\int d\mathbf{r}\, v_c^{(2)}([n];\mathbf{r})\, n(\mathbf{r})$ has the wrong sign. This result is somewhat expected because the present approximations have not been designed with this condition in mind.

HIGH-DENSITY SCALING LIMIT OF ${}^{HF}E_c[n]$ AND ITS CONNECTION TO $E_c[n]$ AND $E_c^{QC,(2)}[n]$

Another type of correlation energy functional, ${}^{HF}E_c[n]$, is defined [8,9] by

$${}^{HF}E_c[n] =< \Psi_n \left| \hat{T} + \hat{V}_{ee} \right| \Psi_{n}> - < \Phi_n^{HF} \left| \hat{T} + \hat{V}_{ee} \right| \Phi_n^{HF} > , \tag{52}$$

where Φ_n^{HF} is the single determinant which minimizes the expectation value $< \hat{T}+V_{ee} >$ and yields the density $n(\mathbf{r})$. The functional derivative of ${}^{HF}E_c[n]$ is meant to be added to the HF non-local potential in order to produce HF-like equations, but whose solution upon self-consistency, in principle, leads to the exact density and to the exact ground-state energy. (Without this added functional derivative, the familiar HF density and energy are obtained upon self-consistency.)

For the high-density scaling limit of ${}^{HF}E_c[n]$, Görling and Levy [9], using their adiabatic perturbation theory, arrived at

$${}^{HF}E_c[n_\lambda] = {}^{HF}E_c^{(2)}[n] + \lambda^{-1}\, {}^{HF}E_c^{(3)}[n] + \lambda^{-2}\, {}^{HF}E_c^{(4)}[n] + \ldots , \tag{53}$$

Table 2. Comparison of $E_c^{QC(2)}$ with $E_c^{APP}[n_\lambda]$, as $\lambda \to \infty$, where n is a hydrogenic ground-state density of $\hat{H}_0$

DENSITY	$E_c^{QC(2)}$	$\lim_{\lambda\to\infty} E_c^{PBE}[n_\lambda]$	$\lim_{\lambda\to\infty} E_c^{LYP}[n_\lambda]$
2-electron	-0.0467	-0.0479	-0.0565
3-electron	-0.0537	-0.0584	-0.0991
9-electron	-0.3694	-0.3856	-0.4648
10-electron	-0.4278	-0.4577	-0.5275
11-electron	-0.4534	-0.4753	-0.5868

with

$$\lim_{\lambda\to\infty} {}^{HF}E_c[n_\lambda] = {}^{HF}E_c^{(2)}[n]$$

$$= \sum_{k=1}^{\infty} \frac{\left| < \Phi_k \left| \hat{V}_{ee} - \sum_{i=1}^{N} \{ {}^{HF}\hat{v}_x([\Psi_0]; \mathbf{r}_i) + u([n]; \mathbf{r}_i) \} \right| \Psi_0 > \right|^2}{E_0 - E_k^0}. \quad (54)$$

In the above formula, ${}^{HF}\hat{v}_x([\Psi_0]; \mathbf{r})$ is the familiar non-local Fock exchange operator, but here built from the one-particle Kohn-Sham orbitals of Ψ_0 instead of from the Hartree-Fock orbitals.

Since there are very few known exact values for use for comparisons with approximations to $E_c[n]$ and ${}^{HF}E_c[n]$, Ivanov and Levy [12] have very recently established the link between the high-density scaling limit of the *unknown* correlation energies $\lim_{\lambda\to\infty} E_c[n_\lambda]$ and $\lim_{\lambda\to\infty} {}^{HF}E_c[n_\lambda]$, and *known* results for the second-order Z^{-1} quantum chemistry correlation energy, $E_c^{QC,(2)}$ [26-28]. They have shown that

$$\lim_{\lambda\to\infty} E_c[n_\lambda] \leq E_c^{QC(2)} \quad (55a)$$

with an equality for only the two-electron case. Moreover, Ivanov and Levy [12] have also derived that, except for the existence of certain degeneracies, the correlation energy functional ${}^{HF}E_c[n]$ actually satisfies the *equality*

$$\lim_{\lambda\to\infty} {}^{HF}E_c[n_\lambda] = E_c^{QC(2)}. \quad (55b)$$

To test approximations to $E_c[n_\lambda]$ and ${}^{HF}E_c[n_\lambda]$, for hydrogenic densities $n(\mathbf{r})$, as $\lambda \to \infty$, they have taken [12] the exact values [27,28] for the second-order energy $E_c^{QC(2)}$ for five different hydrogenic densities. The two- and ten-electron densities are generated from the non-degenerate ground-state wavefunctions of the non-interacting hydrogenic $\hat{H}_0$ in Eq. (18), where V_0 is set equal to $-Z/r$. The ground state of this $\hat{H}_0$ in Eq. (18) is degenerate for three, nine and eleven electrons. However, Eqs. (55a) and (55b) still hold because of symmetry [26]. The three-, nine- and eleven-electron densities were obtained from those wavefunctions $\Psi_0[n]$ that correspond to the configurations $1s^2 2s$, $1s^2 2s^2 2p^5$, $1s^2 2s^2 2p^6 3s$.

We compare, in Table 2, the values of $E_c^{QC(2)}$ with those of $E_c^{APP}[n_\lambda]$, as $\lambda \to \infty$, where $E_c^{APP}[n_\lambda]$ refers in the table to either the functional of Lee, Yang, and Parr (LYP) or that of Perdew, Burke, and Ernzerhof (PBE).

The second column in the table is $\lim_{\lambda\to\infty} {}^{HF}E_c[n_\lambda] = E_c^{QC(2)}$, where $n(\mathbf{r})$, as described in the preceding paragraph, is the hydrogenic groundstate density of $\hat{H}_0$ in Eq. (18). For $E_c[n]$, as required by Eq. (55a) for more than two electrons, both tested approximations give values more negative than the respective values for $E_c^{QC(2)}$. For a two-electron density, the *equality* $\lim_{\lambda\to\infty} E_c[n_\lambda] = \lim_{\lambda\to\infty} {}^{HF}E_c[n_\lambda] = E_c^{QC(2)}$ is desired.

CLOSING REMARKS

Approximations to $E_c^{(2)}[n]$, to the desired accuracy, may be obtained through the variation-perturbation method of Hylleraas [29]. It can be shown that according to this approach, as applied to our present problem, we have

$$E_c^{(2)}[n] = \min_g \frac{\left|< g \left| E^{(1)}[n] - \hat{H}^{(1)}[n] \right| \Psi_0[n] >\right|^2}{< g \left| E_0[n] - \hat{H}_0[n] \right| g >} \tag{56}$$

where $\hat{H}^{(1)}[n]$ is given in Eq. (37a), and where

$$E^{(1)}[n] =< \Psi_0[n] \left| \hat{H}^{(1)}[n] \right| \Psi_0[n] > . \tag{57}$$

In Eq. (56), a minimizing g is $\Psi^{(1)}[n]$. Eq. (56) is quite suitable for use when g is expanded in some finite basis set.

The quantitiy $2E_c^{(2)}[n]$ is especially important because, as the potential energy contribution [9] to $E_c^{(2)}[n]$, it is the initial slope [9] (slope at zero coupling constant) in the adiabatic connection formula for the correlation energy [8,30-34]. In fact, quite good use of the evaluation of $2E_c^{(2)}[n]$ has recently been made by Ernzerhof [35] and by Perdew, Burke, and Ernzerhof [36-39] in their modeling of the adiabatic connection integrand, not only near zero but all the way to unity, to produce very encouraging results in their hybrid schemes.

Finally, formal extensions of the ${}^{HF}E_c[n]$ idea, to include generalized Kohn-Sham and hybrid formulations where, say, only a part of the exact exchange energy (expressed in terms of orbitals) and corresponding Fock potentials are employed, have recently been studied in depth [40,41]. Bounds and expressions analogous to Eq. (54) apply in these generalized hybrid formulations [41]. Further work is now in progress along these lines.

In a simple manner, we have reviewed the fact that, upon uniform scaling of the density, $E_c[n]$ is bounded, i.e. $\lim_{\lambda\to\infty} E_c[n_\lambda] = E_c^{(2)}[n] > -\infty$, and we have reviewed the identification of the exact form of $E_c^{(2)}[n]$ by means of adiabatic perturbation theory. We have also provided numerical results for $E_c^{(2)}[n]$ and $\int d\mathbf{r}\, v_c^{(2)}([n];\mathbf{r})\, n(\mathbf{r})$, where $v_c^{(2)}([n];\mathbf{r})$ is the functional derivative of $E_c^{(2)}[n]$, for the simple two-electron density $n(\mathbf{r}) = (2a^3/\pi)\exp(-2ar)$.

Since there are very few known values of $E_c[n]$ for use for comparisons with approximations to $E_c[n]$, we have introduced relationships that connect *known* results for the second-order quantum chemistry correlation energy, $E_c^{QC(2)}$, in the well-studied $1/Z$ expansion, as $Z \to \infty$, and the unknown $E_c^{(2)}[n]$. Moreover, we have reviewed the fact that, except in certain degenerate cases, $\lim_{\lambda\to\infty} {}^{HF}E_c[n_\lambda]$ *equals* $E^{QC(2)}$, where ${}^{HF}E_c[n]$ is defined slightly differently from $E_c[n]$. The functional derivative of ${}^{HF}E_c[n]$ is meant to be added to the Hartree-Fock (HF) non-local potential, leading to Hartree-Fock-like equations, but with the generation of the exact ground-state energy and density. Numerical results obtained from two of the most widely used approximations have been compared to the exact values.

REFERENCES

1. P. Hohenberg and W. Kohn, Phys. Rev. **136**, B864 (1964).
2. W. Kohn and L. J. Sham, Phys. Rev. **140**, A1133 (1965).
3. R. G. Parr and W. Yang, *Density Functional Theory of Atoms and Molecules* (Oxford University Press, New York, 1989) and references therein.
4. R. M. Dreizler and E. K. U. Gross, *Density Functional Theory*, (Springer-Verlag, Berlin-New York, 1990) and references therein.
5. R.O. Jones and O. Gunnarsson, Rev. Mod. Phys. **61**, 689 (1989).
6. M. Levy and J. P. Perdew, Phys. Rev. A **32**, 2010 (1985).
7. M. Levy, W. Yang, and R.G. Parr, J. Chem. Phys. **83**, 2334 (1985).
8. M. Levy, Phys. Rev. A **43**, 4637 (1991), and references therein.
9. A. Görling and M. Levy, Phys. Rev. B **47**, 13105 (1993).
10. M. Levy, Int. J. Quantum Chem. S **23**, 617 (1989).
11. A. Görling and M. Levy, Phys. Rev. A **45**, 1509 (1992).
12. S. Ivanov and M. Levy, unpublished.
13. S. Ivanov, R. Lopez-Boada, A. Görling, and M. Levy, unpublished.
14. J. P. Perdew, K. Burke, and M. Ernzerhof, Phys. Rev. Lett. **77**, 3865 (1996).
15. C. Lee, W. Yang, and R. Parr, Phys. Rev. B **37**, 785 (1988).
16. M. Levy, Proc. Natl. Acad. Sci (USA) **76**, 6062 (1979): M Levy, Bull. Am. Phys. Soc. **24**, 626 (1979).
17. M. Levy, Phys. Rev. A **26**, 1200 (1982).
18. J. P. Perdew E.R. McMullen, and A. Zunger, Phys. Rev. A **23**, 2785 (1981).
19. A. Görling and M. Levy, Int. J. Quantum Chem. S **29**, 93 (1995); M. Levy and A. Görling, Phys. Rev. A **51**, 2851 (1995).
20. M. Levy and J. P. Perdew, in *Single-Particle Density in Physics and Chemistry*, edited by N.H. March and B.D. Deb (Academic, London, 1987).
21. W. Yang, in *Density Matrices and Density Functionals*, edited by R. Erdahl and V.H. Smith, Jr. (Reidel, Boston, 1987).
22. A. Görling and M. Levy, Phys. Rev. A **52**, 4493 (1995).
23. S. Liu and R.G. Parr, Phys. Rev. A **53**, 2211 (1996).
24. J.P. Perdew, R.G. Parr, M. Levy, and J.L. Balduz, Jr., Phys. Rev. Lett. **49**, 1691 (1982); M. Levy, J.P. Perdew, and V. Sahni, Phys. Rev. A **30**, 2745 (1984).
25. C.O. Almbladh and U. Von Barth, Phys. Rev. B **31**, 3231 (1985).
26. J. Linderberg and H. Shull, J. Mol. Spectroscopy **5**, 1 (1960); and references therein.
27. E.P. Ivanova and U.I. Safronova, J. Phys.B: Atom. Molec. Phys. **8**, 1591 (1975); and references therein.
28. S.J. Chakravorty and E.R. Davidson, J. Phys. Chem. **100**, 6167 (1996); S.J. Chakravorty, S.R. Gwaltney, E.R. Davidson, F.A. Parpia and C.F. Fisher, Phys. Rev. A **47**, 3649 (1993); E.R. Davidson, S.A Hagstrom, S.J. Chakravorty, V.M. Umar, and C.F. Fisher, Phys. Rev. A **44**, 7071 (1991).
29. E.A. Hylleraas, Z. Phys. **65**, 209 (1930).
30. D.C. Langreth and J.P. Perdew, Solid State Commun. **17**, 1425 (1975).
31. O. Gunnarsson and B.I. Lundqvist, Phys. Rev. B **13**, 4274 (1976).
32. A.D. Becke, J. Chem. Phys. **98**, 1372 (1993).
33. A.D. Becke, J. Chem. Phys. **98**, 5648 (1993).
34. M. Levy, N. A. March, and N.C. Handy, J.Chem. Phys. **104**, 1989 (1996).
35. M. Ernzerhof, Chem. Phys. Lett. **263**, 499 (1996).
36. J. P. Perdew, M. Ernzerhof, and K. Burke, J. Chem. Phys. **105**, 9982 (1996).
37. M. Ernzerhof, K. Burke, and J. P. Perdew, in *Recent Developments in Density Functional Theory*, edited by J.M. Seminario, *Theoretical and Computational Chemistry* series (Elsevier, Amsterdam, 1996).
38. K. Burke, M. Ernzerhof, and J. P. Perdew, Chem. Phys. Lett. **265**,115 (1997).
39. M. Ernzerhof, J. P. Perdew, and K. Burke, Int. J. Quantum Chem. **64**, 285 (1997).
40. A. Seidl, A. Görling, P. Vogl, J.A. Majewski, and M. Levy, Phys. Rev. B **53**, 3764 (1996).
41. A. Görling and M. Levy, J. Chem. Phys. **106**, 2675 (1997).

CALCULATING ELECTRONIC ENERGIES FROM KOHN-SHAM EFFECTIVE POTENTIALS

Robert C. Morrison[1] and Robert G. Parr[2]

[1] Department of Chemistry, East Carolina University
Greenville, North Carolina 27858
[2] Department of Chemistry, University of North Carolina at Chapel Hill
Chapel Hill, North Carolina 27599

INTRODUCTION

The problem of finding the exact energy of a system given the exact density is well known, though not solved. Here, approximate formulas for calculating the energy from the density are examined using density scaling relations and a functional expansion of the energy. The term that is first order in the functional expansion can be evaluated exactly from the Kohn-Sham effective potential. The term that is second order is evaluated using the Fermi-Amaldi-like correction for the Coulomb energy. An analysis of the approximation of Parr and Ghosh [Phys. Rev. A 51, 3564 (1995)] and of Morrison and Parr [Phys. Rev. A 53, R2918 (1996)] is made using the functional expansion.

SUMMARY OF EXISTING METHODS OF OBTAINING TOTAL ENERGY

The Hohenberg-Kohn [1] functional of the density can be expressed as

$$E_v[\rho] = \int v_0(\mathbf{r})\rho(\mathbf{r})d^3r + F[\rho] \tag{1}$$

The potential $v_0(\mathbf{r})$ is the external potential of interest and $F[\rho]$ comprises the kinetic energy and electron repulsion energy contributions to the total energy. The usual implementation of the Hohenberg-Kohn functional in practical applications is through the Kohn-Sham [2] equations

$$[-\frac{1}{2}\nabla^2 + v_{ks}(\mathbf{r})]\,\phi_i(\mathbf{r}) = \epsilon_i\phi_i(\mathbf{r}) \tag{2}$$

where $v_{ks}(\mathbf{r})$ is the Kohn-Sham effective potential. The density is given by

Electronic Density Functional Theory: *Recent Progress and New Directions*
Edited by Dobson *et al.*, Plenum Press, New York, 1998

$$\rho(r) = \sum_i \phi_i(\mathbf{r})^2 \tag{3}$$

The noninteracting kinetic energy is defined by

$$T_s =< D| - \frac{1}{2}\sum_i \nabla_i^2 |D > \tag{4}$$

where D is the Slater determinant of the Kohn-Sham orbitals i. The Hohenberg-Kohn functional is

$$E_v[\rho] = \int v_0(\mathbf{r})\rho(\mathbf{r})d^3r + T_s[\rho] + J[\rho] + E_{xc}[\rho] \tag{5}$$

where $J[\rho]$ is the Coulomb energy and $E_{xc}[\rho]$ is the exchange-correlation energy.

A method for obtaining the Kohn-Sham effective potential from the density has been developed and implemented [3-6] using the constrained search [7,8] formulation. Other methods for determining the Kohn-Sham effective potential from the density have also been developed [9-12]. Although the exact density functional for the energy is unknown, the Kohn-Sham effective potential is completely determined from the density and is easily computed.

Parr and Ghosh (PG) [13] have addressed the problem of obtaining the total energy directly from the density. PG suggest partitioning the energy functional as

$$E_v = T_s[\rho] + \int v_0(\mathbf{r})\rho(\mathbf{r})d^3r + J[\rho](1 - \frac{1}{N}) + G_{xc}[\rho] \tag{6}$$

and assume that G_{xc} is a homogeneous functional of degree one in ρ. This leads to

$$G_{xc}[\rho] = \int \frac{\delta G_{xc}[\rho]}{\delta\rho(\mathbf{r})}\rho(\mathbf{r})d^3r \equiv \int v_G(\mathbf{r})\rho(\mathbf{r})d^3r \tag{7}$$

and gives the result

$$\begin{aligned} E &= \sum_{i(occ)} \epsilon_i - J(1 - \frac{1}{N}) - \left[\int \nu_G(r)\rho(r)d^3r - G_{xc}\right] \\ &= \sum_{i(occ)} \epsilon_i - J\left(1 - \frac{1}{N}\right) \end{aligned} \tag{8}$$

The parts of the energy that are homogeneous of degree one in the density are contained in the sum over the Kohn-Sham occupied orbital eigenvalues. Energy components which are homogeneous of degree higher than one, such as J, are over accounted for in the orbital eigenvalue sum and must be subtracted, as they are in Eq. (8).

We (MP) [14] have proposed approximating $G_{xc} - T_c$ by

$$G_{xc} - T_c = \int v_G(\mathbf{r})\rho(\mathbf{r})d^3r. \tag{9}$$

The quantity G_{xc} is obtained by adding $1/2[G_{xc} + T_c]$ to $1/2[G_{xc} - T_c]$. The quantity $G_{xc} + T_c$ can be determined exactly from the virial theorem if the exact density is used. Burke *et al* have proposed calculating E_{xc} from $1/2[E_{xc} + T_c]$ and $1/2[E_{xc} - T_c]$ [15].

ENERGY FORMULAS FROM DENSITY SCALING

We now use relations from the scaling of the density to derive an exact formula for obtaining the energy from the density. In the next section we analyze an approximation to the exact formula.

The scaled reference density $\rho_0^\gamma(\mathbf{r}) = \gamma^3\rho_0(\gamma\mathbf{r})$ is the exact density for the Hamiltonian [16]

$$\hat{H}^\gamma = \hat{T} + \gamma\hat{V}_{ee} + \sum_i \gamma^2 v_0(\gamma\hat{\mathbf{r}}_i) \tag{10}$$

where the potential $v_0(\gamma\mathbf{r})$ is the scaled external potential, assumed to be a Coulomb potential.

It is convenient to define the following modified scaled quantities:

$$v^\gamma(\mathbf{r}) = \gamma^2 v_0(\gamma\mathbf{r})\ ;$$

$$T_s^\gamma = \sum_i < \gamma^{3/2}\phi_i(\gamma\mathbf{r})| - \frac{1}{2}\nabla^2|\gamma^{3/2}\phi_i(\gamma\mathbf{r}) >$$

where $\gamma^{3/2}\phi_i(\gamma\mathbf{r})$ are the Kohn-Sham orbitals that produce $\rho^\gamma(\mathbf{r})$;

$$V_{ext}^\gamma = \int \rho^\gamma(\mathbf{r})v^\gamma(\mathbf{r})d^3r\ ;$$

and

$$J^\gamma = \int\int \frac{\gamma\rho^\gamma(\mathbf{r})\rho^\gamma(\mathbf{r}')d^3rd^3r'}{|\mathbf{r}-\mathbf{r}'|}\ .$$

Then the total energy in Eq. (6) can be expressed as a sum of the Kohn-Sham kinetic energy part plus a single term representing the remainder of the total energy:

$$E^\gamma = T_s^\gamma + \mathcal{E}^\gamma \tag{11}$$

where

$$\mathcal{E}^\gamma = V_{ext}^\gamma + (1 - \frac{1}{N})J^\gamma + G_{xc}^\gamma \tag{12}$$

The components of the energy have the scaling property

$$A^\gamma = \gamma^2 A \tag{13}$$

where A is T_s, V_{ext}, J, E_{xc}, G_{xc}, or E.

The derivative of $\mathcal{E}^\gamma$ with respect to γ evaluated at $\gamma = 1$ is twice $\mathcal{E}$:

$$\left(\frac{d\mathcal{E}^\gamma}{d\gamma}\right)_{\gamma=1} = 2\mathcal{E}\ . \tag{14}$$

The chain rule can be used to express the derivative as

$$\begin{aligned}\left(\frac{d\mathcal{E}^\gamma}{d\gamma}\right)_{\gamma=1} &= \int \left(\frac{\delta\mathcal{E}}{\delta\rho(\mathbf{r})}\right)_{v,u} \left(\frac{d\rho^\gamma(\mathbf{r})}{d\gamma}\right)_{\gamma=1} d^3r + \int \left(\frac{\delta\mathcal{E}}{\delta v_0(\mathbf{r})}\right)_{\rho,u} \left(\frac{dv_0^\gamma(\mathbf{r})}{d\gamma}\right)_{\gamma=1} d^3r \\ &+ \int\int \left(\frac{\delta\mathcal{E}}{\delta u(\mathbf{r},\mathbf{r}')}\right)_{\rho,v} \left(\frac{du^\gamma(\mathbf{r},\mathbf{r}')}{d\gamma}\right)_{\gamma=1} d^3rd^3r'\end{aligned} \tag{15}$$

The two-electron repulsion depends on γ,

$$u^{\gamma}(\mathbf{r},\mathbf{r}') = \frac{\gamma}{|\mathbf{r}-\mathbf{r}'|}, \tag{16}$$

as do the density and the external potential. The first term in Eq. (15) is $V_{ext} + J(1 - 1/N) + G_{xc} + T_c$ from the virial theorem [17,18]. The second term is V_{ext}. In order for the derivative to be equal to $2\mathcal{E}$, the third term must be $J(1 - 1/N) + G_{xc} - T_c$. For $\mathcal{E}$ we now have the exact result

$$\begin{aligned}\mathcal{E} = & \frac{1}{2}\int v_{ks}(r)\left(\frac{d\rho^{\gamma}(r)}{d\gamma}\right)_{\gamma=1} d^3r + \frac{1}{2}\int \rho_0(r)\left(\frac{dv^{\gamma}(r)}{d\gamma}\right)_{\gamma=1} d^3r \\ & +\frac{1}{2}\int\int\left(\frac{\delta\mathcal{E}}{\delta u(\mathbf{r},\mathbf{r}')}\right)_{\rho,v}\left(\frac{du^{\gamma}(\mathbf{r},\mathbf{r}')}{d\gamma}\right)_{\gamma=1} d^3r d^3r' \end{aligned} \tag{17}$$

The first two terms can be evaluated quite easily from the potentials and the scaled density. From the last term we can obtain the potential energy part of G_{xc}:

$$\int\int\left(\frac{\delta G_{xc}}{\delta u(\mathbf{r},\mathbf{r}')}\right)\left(\frac{du^{\gamma}(\mathbf{r},\mathbf{r}')}{d\gamma}\right)_{\gamma=1} d^3r d^3r' = G_{xc} - T_c \tag{18}$$

In a previous communication MP have made an approximation for $G_{xc} - T_c$ which produces the energy formula

$$\begin{aligned} E = & T_s[\rho] + \int v(r)\rho(r)d^3r + J[\rho](1 - \frac{1}{N}) \\ & +\frac{1}{2}\int v_G(r)\left(\frac{d\rho^{\gamma}(r)}{d\gamma}\right)_{\gamma=1} d^3r + \frac{1}{2}\int v_G(r)\rho(r)d^3r. \end{aligned} \tag{19}$$

The only approximation is the last term. The virial theorem can be used to show that the next to the last term is exactly $(1/2)(G_{xc} + T_c)$, and it is easily evaluated. The last term is approximately $(1/2)(G_{xc} - T_c)$. All other terms in the Eq. (19) are known exactly, if the exact density is used.

A FUNCTIONAL EXPANSION OF $\mathcal{E}$

Suppose we chose to approximate the total energy less the Kohn-Sham kinetic energy — the quantity $\mathcal{E}$ of Eq. (17). We begin by considering a functional expansion of $\mathcal{E}[\rho]$ [19,20]:

$$\mathcal{E}[\rho] = \mathcal{E}^{(0)} + \mathcal{E}^{(1)} + \mathcal{E}^{(2)} + \ldots \tag{20}$$

$$\begin{aligned} = & \ \mathcal{E}^{(0)} + \int \frac{\delta\mathcal{E}}{\delta\rho(\mathbf{r})}\rho(\mathbf{r})d^3r - \frac{1}{2}\int\int \frac{\delta^2\mathcal{E}}{\delta\rho(\mathbf{r})\delta\rho(\mathbf{r}')}\rho(\mathbf{r})\rho(\mathbf{r}')d^3rd^3r' \\ & +\frac{1}{3!}\int\int\int \frac{\delta^3\mathcal{E}}{\delta\rho(\mathbf{r})\delta\rho(\mathbf{r}')\delta\rho(\mathbf{r}'')}\rho(\mathbf{r})\rho(\mathbf{r}')\rho(\mathbf{r}'')d^3rd^3r'd^3r'' - \ldots. \end{aligned} \tag{21}$$

By carrying out the expansion for $\mathcal{E}$ instead of E, the question of the homogeneity of T_s in the density [21] is avoided. The term $\mathcal{E}^{(1)}$ can be evaluated from the Kohn-Sham effective potential

Table 1. Calculated values of $\mathcal{E}$ and the first two terms in the functional expansion of $\mathcal{E}$. $\mathcal{E}^{(1)}$ is calculated from Eq. (22), $\mathcal{E}^{(2)}$ is calculated from Eq. (23). Δ_{PG} is the error in the approximation $\mathcal{E}^{(1)} + \mathcal{E}^{(2)}$. Δ_{MP} is the error in the approximation $\mathcal{E}^{(1)} + \mathcal{E}^{(2)} + T_c$. Numerical values are based on Table II Ref. [6].

	$E - T_s$ [a]	$\mathcal{E}^{(1)}$	$\mathcal{E}^{(2)}$	$\mathcal{E}^{(1)} + \mathcal{E}^{(2)}$	$\mathcal{E}^{(1)} + \mathcal{E}^{(2)} + T_c$	Δ_{PG}	Δ_{MP}
He	-5.77	-4.67	1.03	-5.70	-5.66	-0.07 [b]	0.11
Li	-14.92	-11.81	2.71	-14.52	-14.48	-0.40	0.44
Be	-29.26	-23.70	5.42	-29.11	-29.04	-0.14	0.22
B	-49.21	-39.71	9.32	-49.03	-48.94	-0.18	0.27
C	-75.562	-60.67	14.88	-75.55	-75.43	-0.01	0.13
N	-108.91	-86.73	22.44	-109.17	-109.02	0.26	-0.11
O	-149.91	-117.87	32.06	-149.93	-149.74	0.02	0.18
F	-199.19	-154.98	44.29	-199.27	-199.03	0.08	0.16
Ne	-257.56	-198.37	59.48	-257.85	-257.55	0.29	0.01
Na	-324.16	-249.78	72.76	-322.54	-322.23	-1.62	1.93
Mg	-399.71	-310.22	87.91	-398.13	-397.79	-1.58	1.92
Al	-484.24	-378.85	104.26	-483.11	-482.76	-1.13	1.48
Si	-578.20	-455.97	123.83	-579.80	-579.45	1.61 [b]	-1.25
P	-681.93	-538.68	143.44	-682.12	-681.72	0.19	0.22
S	-795.59	-629.38	166.16	-795.54	795.15	-0.05	0.44
Cl	-919.59	-728.54	191.37	-919.91	-919.50	0.33	0.09
Ar	-1054.56	-835.25	219.31	-1054.56	-1054.35	0.002 [b]	0.21

[a]Values obtained from Ref [6]
[b]The values for He, Si, and Ar differ slightly from PG [13] which are based on revised numerical values. Data reported here are based on Table II of Ref [6].

$$\int \frac{\delta \mathcal{E}}{\delta \rho(\mathbf{r})} \rho(\mathbf{r}) d^3r = \int v_{ks}(\mathbf{r}) \rho(\mathbf{r}) d^3r = V_{ext} + 2(1 - \frac{1}{N}) J + \int v_G(\mathbf{r}) \rho(\mathbf{r}) d^3r$$
$$= \sum_i \epsilon_i - T_s \tag{22}$$

The approximation we may make, embodying the Fermi-Amaldi idea [22], is for $\mathcal{E}^{(2)}$:

$$\mathcal{E}^{(2)} = -(1 - 1/N) J. \tag{23}$$

In Table 1 are values of $\mathcal{E} = E_{total} - T_s$, $\mathcal{E}^{(1)}$, $\mathcal{E}^{(2)}$, $\mathcal{E}^{(1)} + \mathcal{E}^{(2)}$, and $\mathcal{E}(1) + \mathcal{E}(2) + T_c$. The values of E_{total} are the total energies calculated from the wave functions that produced the electron densities used in the calculations. The sum $T_s + \mathcal{E}^{(1)} + \mathcal{E}^{(2)}$ is the PG approximation for E_{total}. In the PG model $\mathcal{E}^{(1)}$ contains T_c in the form of Eq. (7), but in the MP model it is not present in $\mathcal{E}^{(1)}$ because the approximation of Eq. (19) is used. To obtain $\mathcal{E}$ in the expansion of Eq. (20) using the MP model, T_c must be added to the sum. The values of Δ_{PG} are the errors in approximating $\mathcal{E}$ with $\mathcal{E}^{(1)}+ \mathcal{E}^{(2)}$, and Δ_{MP} are errors in approximating $\mathcal{E}$ with $\mathcal{E}^{(1)}+ \mathcal{E}^{(2)} + T_c$. The average magnitude of the error for the PG approximation is 0.47 compared to 0.27 when using Eq. (19) to calculate E. The average magnitude of Δ_{MP} is 0.54, just twice the average error magnitude using Eq. (19). The MP approximation in Eq. (19) does not represent a better model for the *exchange-correlation* energy than the PG model. More accurate total energies are produced using the MP model simply because the term in the energy that is approximated appears with a factor of 1/2, reducing the error by almost 1/2.

DISCUSSION

The exact formula for determining the total energy from the density and the external potential remains unknown, although models for its approximation are becoming increasingly more accurate. The total energy for a given $\rho_0(\mathbf{r})$ and $v_0(\mathbf{r})$ could be determined either from the first-order reduced density matrix $\gamma_1(\mathbf{r}_1 \mid \mathbf{r}_{1'})$ or the pair density $\gamma_2(\mathbf{r}_1, \mathbf{r}_2)$. The total kinetic energy could be determined from γ_1, or the total potential energy could be determined from γ_2, and the total energy then determined from virial relations.

Since the quantity $G_{xc} - T_c$ has no kinetic energy contribution, one might expect to obtain good approximations from accurate models of the pair density. Such models would avoid questions concerning the kinetic energy contribution to the correlation energy.

We continue to hope and expect that exact formulas for obtaining the total energy from the density will be found in the near future. To support our optimism, note that one could easily get the total energy if one just knew the true densities of a species S and all of its positive ions: Just add the highest orbital energies for S, S^+, S^{++}, ..., determined by the present method [3-6]. (It is also true, incidentally, that the ground-state electron density of S is the sum of the Fukui functions for $S, S^+, S^{++}, \ldots$.).

ACKNOWLEDGMENTS

We are grateful to the National Science Foundation for grants to the University of North Carolina at Chapel Hill and to East Carolina University. We acknowledge helpful discussions with John P. Perdew, Mel Levy, and Shubin Liu.

REFERENCES

1. P. Hohenberg and W. Kohn, Phys. Rev. **136**, B864 (1964).
2. W. Kohn and L. J. Sham, Phys. Rev. **140**, A1133 (1965).
3. Q. Zhao and R. G. Parr, Phys. Rev. A **46**, 2337 (1992).
4. Q. Zhao and R. G. Parr, J. Chem. Phys. **98**, 543 (1993).
5. Q. Zhao, R. C. Morrison, and R. G. Parr, Phys. Rev. A **50**, 2138 (1994).
6. R. C. Morrison and Q. Zhao, Phys. Rev. A **51**, 1980 (1995).
7. M. Levy, Proc. Natl. Acad. Sci. USA **76**, 2062 (1979).
8. M. Levy and J. P. Perdew, pp. 11-30 in *Density Functional Methods in Physics*, edited by R. M. Dreizler and J. de Providencia (Plenum, New York, 1985).
9. F. Aryasetiawan and M. J. Stott, Phys. Rev. B **38**, 2974 (1988).
10. A. Görling, Phys. Rev. A **46**, 3753 (1992).
11. Y. Wang and R. G. Parr, Phys. Rev. A **47**, R1591 (1993).
12. R. van Leeuwen and E. J. Baerends, Phys. Rev. A **49**, 2421 (1994).
13. R. G. Parr and S. K. Ghosh, Phys. Rev. A **51**, 3564 (1995).
14. R. C. Morrison and R. G. Parr, Phys. Rev. A **53**, R2918 (1996).
15. K. Burke, J. P. Perdew, and M. Levy, Phys. Rev. A **53**, R2915 (1996).
16. M. Levy, W. Yang, and R. G. Parr, J. Chem. Phys. **83**, 2334 (1985).
17. F. Averill and G. S. Painter, Phys. Rev. B **24**, 6795 (1981).
18. M. Levy and J. P. Perdew, Phys. Rev. A **32**, 2010 (1985).
19. S. Liu and R. G. Parr, Phys. Rev. A (in press).

20. R. G. Parr, S. Liu, A. A. Kugler, and Á. Nagy, Phys. Rev. A **52**, 969 (1995).
21. Á. Nagy, Phys. Rev. A **47**, 2715 (1993).
22. E. Fermi and E. Amaldi, Mem. R. Accad. Italia **6**, 117 (1934).

A CORRELATION-ENERGY FUNCTIONAL FOR ADDITION TO THE HARTREE-FOCK ENERGY

Leslie C. Wilson [1,2] and Stanislav Ivanov [2,3]

[1] Chemistry Department, Loyola University
New Orleans, Louisiana 70118

[2]Quantum Theory Group, Tulane University
New Orleans, Louisiana 70118

[3]Chemistry Department, Tulane University
New Orleans, Louisiana 70118

INTRODUCTION AND DEFINITIONS

This article concerns the Wilson-Levy (WL) correlation-energy functional, Wigner-like with gradients, which was developed in 1990. Its numerical success in density functional theory (DFT) has been limited when used as part of a full DFT calculation. It has nonetheless been shown to be surprisingly accurate when used with Hartree-Fock densities as a "tack-on" functional to the completed Hartree-Fock energy. Fuentealba and Savin in 1994 pointed out that the WL yields better ionization potentials and dissociation energies for a number of atoms and molecules than other DFT correlation-energy functionals when used in that way. Here possible reasons for this will be discussed.

In the present Section, the topic will be introduced and some definitions will be given. In the following Section, selected theoretical requirements for the exact correlation-energy functional in DFT will be reviewed and recently-derived constraints will be presented. In the next Section, results obtained from the WL functional for the newly-derived conditions will be shown to be promising compared with quantum chemistry values. Various numerical results obtained by means of the WL and other DFT correlation-energy functionals will be compared with traditional quantum chemistry values. In the final Section some conclusions will be drawn.

Electronic calculations using density functionals are significantly more manageable than conventional wavefunction methods due to the replacement of the complicated many-body wavefunction method with the simpler approach of electronic density [1-6].

Considerable attention has been paid to the construction and improvement of accurate approximations to the exact correlation energy in density functional theory (DFT). Among these are the approximations by Perdew and Wang stemming from corrections to the local density approximation (LDA) or the local spin density (LSD) approximation [3,4,6]: PW86 [7], PW91 [8], and a very recently derived functional by Perdew, Burke, and Ernzerhof (PBE) [9]. Other recent correlation-energy functionals

Electronic Density Functional Theory: *Recent Progress and New Directions*
Edited by Dobson *et al.*, Plenum Press, New York, 1998

have been constructed by Becke (B) [10], Lee, Yang and Parr (LYP) [11], and Wilson and Levy (WL) [12].

The PW and LYP functionals have been very popular worldwide, especially in conjunction with Becke's exchange functional [13]. The WL, similar in its Wigner-like structure [17] to the LYP, has not been competitive with these approximations, but has given promising results when used as a "tack-on" energy to the completed Hartree-Fock (HF) calculations to produce a very good approximation to the ground-state energy.

The use of functionals enables one not only to replace traditional time-consuming methods for computing the correlation energy by much simpler density functional calculations, but also to obtain respectable values [3,4, 15,16]. Fuentealba and Savin [17] have illustrated that the WL proved to be particularly effective when used as a "tack-on" functional to the completed HF energies. Their calculations demonstrated that the WL would improve HF results for ionization potentials and dissociation energies for numerous atoms and molecules, frequently surpassing the results from other DFT correlation- energy functionals when used as "tack-on". (See two Sections on.)

In atomic units (a.u. shall be used throughout the paper unless otherwise indicated) the Hamiltonian operator of the system under investigation is

$$\hat{H} = \hat{T} + \hat{V}_{ee} + \sum_{i=1}^{N} v_0(\mathbf{r}_i) \tag{1}$$

with $\hat{T}$ being the kinetic energy operator

$$\hat{T} = -\sum_{i=1}^{N} \frac{1}{2} \nabla_i^2, \tag{2}$$

$\hat{V}_{ee}$ being the operator of the electron-electron repulsion,

$$\hat{V}_{ee} = \sum_{j=1}^{N} \sum_{i=j+1}^{N} \frac{1}{|\mathbf{r}_i - \mathbf{r}_j|}, \tag{3}$$

and $v_0(\mathbf{r})$ being a local spin-independent multiplicative operator. The Hohenberg-Kohn functional, $F[n]$, [1] is a universal functional of the density, formally identified [5] as:

$$F[n] = < \Psi_n^{\min} \left| \hat{T} + \hat{V}_{ee} \right| \Psi_n^{\min} > \tag{4}$$

where $\Psi_n^{\min}$ is that antisymmetric wavefunction that minimizes $< \hat{T} + \hat{V}_{ee} >$ and yields the ground-state density $n(\mathbf{r})$. In Kohn-Sham formalism of DFT [2], $F[n]$ is most commonly partitioned in the following way:

$$F[n] = T_s[n] + U[n] + E_{xc}[n]. \tag{5}$$

In Eq.(5), $T_s[n]$ is the non-interacting kinetic energy and is defined [5,18] as

$$T_s[n] = < \Phi_n^{\min} \left| \hat{T} \right| \Phi_n^{\min} > , \tag{6}$$

with $\Phi_n^{\min}$ being the Kohn-Sham (KS) wavefunction, usually a single determinant, that minimizes just $< \hat{T} >$ and yields the same ground-state density $n(\mathbf{r})$. The Hartree electron-electron repulsion energy $U[n]$ is given by

$$U[n] = \frac{1}{2} \int \int \frac{n(\mathbf{r}_1) n(\mathbf{r}_2)}{|\mathbf{r}_1 - \mathbf{r}_2|} d\mathbf{r}_1 d\mathbf{r}_2. \tag{7}$$

The exchange-correlation energy, $E_{xc}[n]$, may be further partitioned [19] into the exchange component $E_x[n]$, known exactly from the KS single determinant, and the unknown correlation energy $E_c[n]$. $E_x[n]$ is given [18] by

$$E_x[n] = < \Phi_n^{\min} \left| \hat{V}_{ee} \right| \Phi_n^{\min} > -U[n] \ , \tag{8}$$

and $E_c[n]$ is defined [18-22] as

$$E_c[n] = < \Psi_n^{\min} \left| \hat{T} + \hat{V}_{ee} \right| \Psi_n^{\min} > - < \Phi_n^{\min} \left| \hat{T} + \hat{V}_{ee} \right| \Phi_n^{\min} > \ , \tag{9}$$

or,

$$E_c[n] = T_c[n] + V_c[n] \ , \tag{10}$$

with

$$T_c[n] = < \Psi_n^{\min} \left| \hat{T} \right| \Psi_n^{\min} > - < \Phi_n^{\min} \left| \hat{T} \right| \Phi_n^{\min} > \tag{11}$$

and

$$V_c[n] = < \Psi_n^{\min} \left| \hat{V}_{ee} \right| \Psi_n^{\min} > - < \Phi_n^{\min} \left| \hat{V}_{ee} \right| \Phi_n^{\min} > \ . \tag{12}$$

Next, for the Hamiltonian in Eq.(1), we shall define the quantum chemistry correlation energy as a functional of the HF density $E_c^{QC}[n^{HF}]$, [23,24]. Namely

$$E_c^{QC}[n^{HF}] \equiv E^{GS} - E^{HF} = < \Psi^{GS} \left| \hat{H} \right| \Psi^{GS} > - < \Phi^{HF} \left| \hat{H} \right| \Phi^{HF} > \tag{13}$$

In Eq.(13), Ψ^{GS} is the ground state of $\hat{H}$ and yields the ground-state density $n(\mathbf{r})$, and Φ^{HF} is the single determinant that minimizes $< \hat{H} >$. Note that the density obtained from Φ^{HF} is *different* from $n(\mathbf{r})$ [25].

Similarly to Eqs.(11)-(13), we can partition $E_c^{QC}[n^{HF}]$ as

$$E_c^{QC}[n^{HF}] = T_c^{QC}[n^{HF}] + V_c^{QC}[n^{HF}] \ , \tag{14}$$

with

$$T_c^{QC}[n^{HF}] = < \Psi^{GS} \left| \hat{T} \right| \Psi^{GS} > - < \Phi^{HF} \left| \hat{T} \right| \Phi^{HF} > \tag{15}$$

and

$$V_c^{QC}[n^{HF}] = < \Psi^{GS} \left| \hat{V}_{ee} \right| \Psi^{GS} > - < \Phi^{HF} \left| \hat{V}_{ee} \right| \Phi^{HF} > . \tag{16}$$

For the purpose of arriving at the very best approximations to $E_c^{QC}[n^{HF}]$, one needs to know as many exact properties of $E_c^{QC}[n^{HF}]$ as possible. Unfortunately, there are only a few known constraints on $E_c^{QC}[n^{HF}]$. Levy [26] has shown that

$$\lim_{\lambda \to \infty} E_c^{QC}[n_\lambda^{HF}] > -\infty \ , \tag{17}$$

where

$$n_\lambda^{HF}(x, y, z) = \lambda^3 n_\lambda^{HF}(\lambda x, \lambda y, \lambda z) \ . \tag{18}$$

Ivanov and Levy [27] have very recently derived the following identity

$$\lim_{\lambda \to \infty} E_c[n_\lambda^{HF}] \leq \lim_{\lambda \to \infty} E_c^{QC}[n_\lambda^{HF}] \tag{19}$$

with an equality for only two-electron densities.

Since there appear to be some important similarities and connections between $E_c[n]$ and $E_c^{QC}[n^{HF}]$, we summarize a significant number of theoretical conditions for $E_c[n]$. We believe that these constraints for $E_c[n]$ might inspire the reader to explore the exact properties of $E_c^{QC}[n^{HF}]$.

Directly from the definition of $E_c[n]$, Eq.(9) [19,28], it follows that

$$E_c[n] \leq 0 \tag{I}$$

An important technique for deriving the exact properties of the unknown density functionals is the scaling of the electron density $n(\mathbf{r})$. The behavior of energy functionals as density of the system is scaled uniformly (equally along all axes) or scaled non-uniformly (unequally) has been investigated extensively by Levy, Perdew, Görling and Ou-Yang [8,19,20,29-32]. Uniform and nonuniform scaling, when used with other constraints, turns out to be very powerful for constructing new approximations or improving upon the properties of already existing ones.

Let us define three types of scaled densities [19,29]:

$$n_\lambda(x,y,z) = \lambda^3 n(\lambda x, \lambda y, \lambda z) \ , \tag{20}$$

$$n_\lambda^x(x,y,z) = \lambda n(\lambda x, y, z) \ , \tag{21}$$

$$n_{\lambda\lambda}^{xy}(x,y,z) = \lambda^2 n(\lambda x, \lambda y, z) \ . \tag{22}$$

Equation (20) represents a uniformly scaled density, and Eqs. (21) and (22) non-uniformly scaled densities.

In general, we use the following notation

$$n_{\alpha\beta\gamma}^{xyz}(x,y,z) = \alpha\beta\gamma n(\alpha x, \beta y, \gamma z) \ . \tag{23}$$

For instance, Eq.(21) is the result for $\alpha = \beta = \gamma = \lambda$, and Eq.(22) follows from Eq.(23) for $\alpha = \beta = \lambda$ and $\gamma = 1$.

The correlation-energy functional $E_c[n]$ as defined in Eq.(9), incorporates a complicated dimensionality upon uniform scaling of the density. In contrast to $T_s[n]$, $U[n]$ and $E_x[n]$ which scale homogeneously, i.e. $T_s[n_\lambda] = \lambda^2 T_s[n]$, $U[n_\lambda] = \lambda U[n]$, and $E_x[n_\lambda] = \lambda E_x[n]$, $E_c[n]$ exhibits more complex behavior [19] when the density is scaled uniformly, i.e.

$$E_c[n_\lambda] > \lambda E_c[n] \ , \ for \ \lambda > 1 \ . \tag{II}$$

An identity developed by Levy and Perdew [19], known as the virial theorem in DFT which connects $T_c[n]$ to $E_c[n]$ is

$$T_c[n] = -E_c[n] + \left.\frac{\partial E_c[n_\lambda]}{\partial \lambda}\right|_{\lambda=1} \ . \tag{III}$$

Some more recent constraints [33] involving derivatives of $E_c[n_\lambda]$ with respect to the scaling parameter are:

$$E_c[n_\lambda] \leq \lambda E_c[n] + \lambda(\lambda - 1)\{\left.\frac{\partial E_c[n_\gamma]}{\partial \gamma}\right|_{\gamma=1} - E_c[n]\} \ , \tag{IV}$$

$$2E_c[n_\lambda] - 2\lambda \frac{\partial E_c[n_\lambda]}{\partial \lambda} + \lambda^2 \frac{\partial^2 E_c[n_\lambda]}{\partial \lambda^2} \leq 0 \ , \tag{V}$$

$$\lim_{\lambda \to 0} \lambda^{-1} E_c[n_\lambda] - E_c[n] \leq E_c[n] - \left. \frac{\partial E_c[n_\lambda]}{\partial \lambda} \right|_{\lambda=1} \leq 0 \ . \tag{VI}$$

Further, it has been shown [25,30] that:

$$\lim_{\lambda \to \infty} E_c[n_\lambda] > -\infty \ , \tag{VII}$$

$$\lim_{\lambda \to 0} (\lambda^{-1} E_c[n_\lambda]) = -b[n] \ , \tag{VIII}$$

where $b[n]$ is a positive constant such that $b[n] \leq U[n] + Ex[n] < U[n]$ [25,30].

Another bound involving $\lim_{\lambda \to 0}(\lambda^{-1} E_c[n_\lambda])$ derived by Oxford and Lieb [34] is

$$\lim_{\lambda \to 0} (\lambda^{-1} E_c[n_\lambda]) \geq -1.68 \int n^{4/3}(\mathbf{r}) d\mathbf{r} - E_x[n] \ . \tag{IX}$$

Upon non-uniform coordinate scaling given by Eq.(21) the exact correlation functional satisfies [29,35]:

$$\lim_{\lambda \to \infty} E_c[n_\lambda^x] = 0 \ , \tag{X}$$

$$\lim_{\lambda \to 0} E_c[n_\lambda^x] = 0 \ . \tag{XI}$$

When the assumption of a Taylor series in is made, Görling and Levy [31] have shown that the following versions of constraints X and XI are also true:

$$\lim_{\lambda \to \infty} (\lambda E_c[n_\lambda^x]) = const. \ , \tag{XII}$$

$$\lim_{\lambda \to 0} (\lambda^{-1} E_c[n_\lambda^x]) = 0 \ . \tag{XIII}$$

Similarly, when the density is scaled non-uniformly as in Eq.(22), they have shown that

$$\lim_{\lambda \to \infty} E_c[n_{\lambda\lambda}^{xy}] = 0 \ , \tag{XIV}$$

$$\lim_{\lambda \to 0} E_c[n_{\lambda\lambda}^{xy}] = 0 \ , \tag{XV}$$

$$\lim_{\lambda \to 0} (\lambda^{-1} E_c[n_{\lambda\lambda}^{xy}]) = 0 \ , \tag{XVI}$$

$$\lim_{\lambda \to 0} (\lambda^{-2} E_c[n_{\lambda\lambda}^{xy}]) = const. \ . \tag{XVII}$$

Constraints XIV, XVI and XVII are valid under the assumption of a Taylor series expansion . Conditions XII, XIII and XVII are much more demanding constraints than X, XI and XV, respectively, i.e. substantially more difficult to satisfy. For different combinations of α, β and γ, Görling and Levy [31] have derived:

$$\lim_{\lambda \to \infty} E_c[n_{\lambda\lambda^{-1}}^{xy}] = 0 \ , \tag{XVIII}$$

$$\lim_{\lambda\to\infty}(\lambda E_c[n^{xy}_{\lambda\lambda^{-1}}]) = 0 \ , \tag{XIX}$$

$$\lim_{\lambda\to 0} E_c[n^{xy}_{\lambda\lambda^{-1}}] = 0 \ , \tag{XX}$$

$$\lim_{\lambda\to 0}(\lambda^{-1} E_c[n^{xy}_{\lambda\lambda^{-1}}]) = 0 \ , \tag{XXI}$$

$$\lim_{\lambda\to\infty} E_c[n^{xyz}_{\lambda\lambda\lambda^{-1}}] = 0 \ , \tag{XXII}$$

$$\lim_{\lambda\to 0}(\lambda^{-1} E_c[n^{xyz}_{\lambda\lambda\lambda^{-1}}]) = 0 \ , \tag{XXIII}$$

$$\lim_{\lambda\to 0}(\lambda^{-2} E_c[n^{xyz}_{\lambda\lambda\lambda^{-1}}]) = const. \ . \tag{XXIV}$$

Levy and Perdew [36] have also shown two more demanding scaling constraints based on the definition of $A[n]$ resembling constraint IX. If $A[n] = \lim_{\lambda\to 0}(\lambda^{-1}E_c[n_\lambda])$ then the following, they maintain, must hold true:

$$\lim_{\lambda\to\infty} A[n^x_\lambda] > -\infty \ , \tag{XXV}$$

$$\lim_{\lambda\to 0}(\lambda^{-1} A[n^x_\lambda]) > -\infty \ . \tag{XXVI}$$

Additional constraints from coordinate scaling in the coupling-constant integration formula for Ec[n] [22] are:

$$E_c[n] = \int_0^1 V_c^\alpha[n] d\alpha \ , \tag{XVII}$$

where V_c^α is $< \hat{V}_{ee} > -(U[n] + E_x[n])$, and where the expectation value $< \hat{V}_{ee} >$ is with that wavefunction which minimizes $< \hat{T} + \alpha \hat{V}_{ee} >$ and yields the same density $n(\mathbf{r})$ at every α. It follows that

$$\frac{\partial}{\partial\alpha} V_c^\alpha[n] \leq 0; \ \ all \ \alpha \geq 0 \ . \tag{XXVIII}$$

Moreover, the fundamental scaling property of $V_c^\alpha[n]$ is

$$V_c^\alpha[n] = \alpha V_c[n_\lambda]; \ \ \lambda = \alpha^{-1} \ , \tag{XXIX}$$

leading to

$$\lim_{\alpha\to 0} V_c^\alpha[n] = 0 \ , \tag{XXX}$$

$$\lim_{\alpha\to\infty} V_c^\alpha[n] > -\infty \ . \tag{XXXI}$$

The analog of condition XXIX for the whole $E_c^\alpha[n]$ reads

$$E_c^\alpha[n] = \alpha^2 E_c[n_\lambda]; \ \ \alpha = \lambda^{-1}. \tag{XXXII}$$

Görling and Levy [37] have recently introduced an adiabatic DFT perturbation theory. The effective potential and the electron-electron interaction along the coupling constant path, which connects a non-interacting and a fully interacting system with the *same* electron density, have been used for this DFT perturbation theory. By using their perturbation theory, they have derived the following high-density scaling expansion for $E_c[n]$:

$$E_c[n_\lambda] = E_c^{(2)}[n] + \lambda^{-1}E_c^{(3)}[n] + \lambda^{-2}E_c^{(4)}[n] + \ldots \tag{XXXIII}$$

and they have identified the second-order energy $E_c^{(2)}[n]$ as:

$$\lim_{\lambda\to\infty} E_c[n_\lambda] = E_c^{(2)}[n] = \sum_{k=1}^{\infty} \frac{\left|< \Phi_k \left|\Delta\hat{V}\right| \Phi_0 >\right|^2}{E_0 - E_k} \quad , \tag{XXXIV}$$

with

$$\Delta\hat{V} = \hat{V}_{ee} - \sum_{i=1}^{N}\{u([n];\mathbf{r}_i) + v_x([n];\mathbf{r}_i)\}, \tag{24}$$

and where

$$u([n];\mathbf{r}_i) = \int \frac{n(\mathbf{r})}{|\mathbf{r}_i - \mathbf{r}|} d\mathbf{r} \quad . \tag{25}$$

In Eq.(24), the Hartree potential $u([n];\mathbf{r})$ and the exchange potential $v_x([n];\mathbf{r})$ are local multiplicative potentials obtained by taking functional derivatives of $U[n]$ and $E_x[n]$ with respect to the density $n(\mathbf{r})$. The density $n(\mathbf{r})$ is obtained from the non-interacting Φ_0, $\Phi_0 \equiv \Phi_n^{\min}$.

Ivanov *et al* [38] have very recently expressed the difference $E_c^{(2)}[n] - \frac{1}{2}\int d\mathbf{r}\, v_c^{(2)}([n];\mathbf{r})\, n(\mathbf{r})$, where $v_c^{(2)}([n];\mathbf{r})$ is the functional derivative of $E_c^{(2)}[n]$ with respect to the density, for a two-electron system as a functional of the density $n(\mathbf{r})$ only. For any spherically symmetric two-electron density, the analytical expression has been obtained and for an exponential two-electron density, $n(\mathbf{r}) = (2a^3/\pi)\exp(-2ar)$ with $a > 0$, they have found that

$$E_c^{(2)}[n] = -0.046663 \ , \tag{XXXV}$$

and

$$\int d\mathbf{r}\, v_c^{(2)}([n];\mathbf{r})\, n(\mathbf{r}) = 0.017677 \ . \tag{XXXVI}$$

In order to generate more numerical data for $E_c^{(2)}[n]$ and $v_c^{(2)}([n];\mathbf{r})$, Ivanov and Burke [39] have applied the results from Ref. [38] to a Gaussian density $n(\mathbf{r}) = 2(2a^3/\pi)^{3/2}\exp(-2ar^2)$ with $a > 0$, and they have arrived at

$$E_c^{(2)}[n] = -0.049702 \ , \tag{XXXVII}$$

and

$$\int d\mathbf{r}\, v_c^{(2)}([n];\mathbf{r})\, n(\mathbf{r}) = -0.071216 \ . \tag{XXXVIII}$$

C.J. Huang and C.J. Umrigar [40] have also obtained the same numerical result for condition XXXVII, i.e. $E_c^{(2)}[n] = -0.049702$, for $n(\mathbf{r}) = 2(2a^3/\pi)^{3/2}\exp(-2ar^2)$ with $a > 0$.

It is also accepted that the exact value for the correlation energy of Helium [41] is

$$E_c[n^{HF}] = -0.0420 \ . \qquad \text{(XXXIX)}$$

Last but not least, the only system for which the correlation-energy functional is known is the uniform electron gas [3,4,6]. Even though the density of this model system is far away from the varying densities of real molecules, the uniform electron gas limit imposes a stringent constraint that must be satisfied by the very best approximations to $E_c[n]$.

These constraints obviously do not exhaust every possible property of the exact correlation-energy functional, but they do represent some of the most demanding ones.

COMPARISON OF SELECTED DFT FUNCTIONALS: REASONS FOR RELATIVE SUCCESS OF WL FOR APPROXIMATING $E_c^{QC}[n^{HF}]$

The WL was originally constructed to be a Wigner-like functional with gradients that obeyed certain coordinate scaling constraints when the density is scaled uniformly, conditions VII and VIII. The impressively succinct Wigner formula of 1938 for the homogeneous electron gas [14] not only satisfies both uniform scaling requirements, but also has the simplest form possible for a local functional to obey VII and VIII. The Wigner functional reads

$$E_c^W[n] = -\int \frac{0.44n(\mathbf{r})}{7.8 + r_s} d\mathbf{r} \ , \qquad (26)$$

where $r_s = (4\pi n(\mathbf{r})/3)^{-1/3}$. (In this Section, E_c^{APP} signifies an approximation to $E_c[n]$ and " APP" is replaced by the abbreviation corresponding to the specific functional under consideration.)

$E_c^W[n]$ gives fairly good values for correlation energies [42], consistently better than the VWN [41]. In 1976, McWeeny suggested an expression comparable to the Wigner formula as an improvement for calculating correlation energies, which was a single simple formula for both low and high density limits [44]. Two years later, Brual and Rothstein [45] presented a Wigner-like formula parametrized to agree with the exact value for the Helium atom. Given its simplicity, the Brual-Rothstein formula (BR) gives surprisingly improved values for the correlation energies for some atoms and ions. Similar to the Wigner, McWeeny and BR formulas, the WL gives the correct limits upon uniform scaling, but does not obey most non-uniform scaling constraints which were unknown at the time of its construction.

Interestingly, constraints XXVIII-XXXI will always be obeyed under three stipulations [42]:

1. the form of the functional is Wigner-like;
2. constraints VII and VIII for $E_c[n]$ are obeyed;
3. coefficients for the functional are constrained never to give a positive answer for the correlation energy, condition I.

This last condition is relatively easy to achieve for a Wigner-like functional if the functional can be parametrized such that the numerator coefficients are required to have only negative values, and the denominator coefficients only positive. The Wilson-Levy functional [12] does not always obey constraints XXVIII-XXXI because stipulation 3) was not built into the construction of the functional.

The relative success of the WL, when used as "tack-on" to the HF energy, might be related to the fact that $\lim_{\lambda\to\infty} E_c[n_\lambda^{HF}]$ is bounded and to assumptions used in the optimization of WL. Parametrization of the WL was mainly achieved by making use of constraint XXXIX and the assumption that the value of $(\partial E_c[n_\lambda^{HF}]/\partial\lambda)_{\lambda=1}$ is relatively small.

The virial expression in traditional QC for an atom with Hamiltonian $\hat{H}$ is

$$< \Psi^{GS} \left| \hat{H} \right| \Psi^{GS} >= - < \Psi^{GS} \left| \hat{T} \right| \Psi^{GS} > \quad . \tag{27}$$

It can be shown that for atoms,

$$E_c^{QC}[n^{HF}] = -T^{QC}[n^{HF}] \tag{28}$$

where $E_c^{QC}[n^{HF}]$ and $T^{QC}[n^{HF}]$ are given by Eqs. (13) and (15).

When HF densities are used in III, an assumption was made for the parametrization of WL that the term $(\partial E_c[n_\lambda^{HF}]/\partial\lambda)_{\lambda=1}$ is relatively small, which at least appears to be the case for He [46,47]. Given that $E_c^{QC}[n^{HF}] \approx E_c[n]$, comparison of condition III and Eq.(28) would imply that $(\partial E_c[n_\lambda^{HF}]/\partial\lambda)_{\lambda=1}$ is indeed relatively small in magnitude. It is believed that the use of this assumption partially explains the success of the WL within traditional QC calculations.

Furthermore, experimental values for the kinetic energy contribution, $T_c[n]$, to correlation energy tend to support the idea that $T_c[n] < |E_c[n]|$. This implies that $(\partial E_c[n_\lambda^{HF}]/\partial\lambda)_{\lambda=1} \leq 0$, another potentially important and informative constraint, if proven [48].

The Wigner-like approximations together with Becke (B) [10], and Lee, Yang and Parr (LYP) [11] formulas clearly do not give the correct limit for uniform densities. However, they do satisfy conditions VII and VIII. Becke's formula has its source in the concepts of properties of the exchange-correlation hole in many-electron systems while the LYP functional is a restatement of the correlation-energy formula developed by Colle and Salvetti [49]. Their results are competitive with the correlation-energy functionals based upon the LDA, as those of Langreth and Mehl, and Perdew and co-workers. Most functionals based on the LDA, or LSD, do not satisfy conditions VII and X. However, the Perdew-Wang functional of 1991 (PW91) and all later versions have the positive feature that, in the process of derivation, most coordinate scaling constraints are automatically obeyed.

The Wilson-Levy functional is Wigner-like in form, but features two gradient terms, one in the numerator and one in the denominator. While it is similar to the functionals of Becke and Lee, Yang and Parr in that it does not give correct results in the limit of the uniform electron gas, it nonetheless has one of the simplest semi-local forms. The spin-polarized form of the WL functional correctly gives zero for all one-electron species, and is given by

$$E_c^{WL}[n_\alpha, n_\beta] = -\int \frac{n(\mathbf{r})(a + b\frac{|\nabla n(\mathbf{r})|}{n^{4/3}(\mathbf{r})})(1-\zeta^2)^{1/2}}{c + d(\frac{|\nabla n_\alpha(\mathbf{r})|}{n_\alpha^{4/3}(\mathbf{r})} + \frac{|\nabla n_\beta(\mathbf{r})|}{n_\beta^{4/3}(\mathbf{r})}) + r_s} d\mathbf{r} \tag{29}$$

where $n(\mathbf{r}) = n_\alpha(\mathbf{r}) + n_\beta(\mathbf{r})$, $r_s = (4\pi n(\mathbf{r})/3)^{-1/3}$, $\zeta = (n_\alpha - n_\beta)/(n_\alpha + n_\beta)$ (α signifies spin-up and β signifies spin-down densities), and a= -0.74860, b = 0.06001, c = 3.60073

Table 1. Correlation Energies for Closed-Shell Species

Species	$E_c^{PW91}[n]$[a]	$E_c^{B}[n]$[b]	$E_c^{LYP}[n]$[c]	$E_c^{WL}[n]$[d]	Exact [e]
He	-0.046	-0.042	-0.0437	-0.0420	-0.042
Li^+	-0.051	-0.050[f]	-0.0475	-0.0442	-0.043
Be^{2+}	-0.054	-0.055[d]	-0.0490	-0.0452	-0.044
Li^-		-0.070[f]	-0.073[f]	-0.0805	-0.0725[g]
Be	-0.094	-0.092	-0.095	-0.095	-0.094
B^+		-0.107[f]	-0.107	-0.100	-0.111[f]
Ne^{6+}	-0.123			-0.108	-0.180
F^-		-0.364[d]	-0.361[f]	-0.364	-0.397[g]
Ne	-0.382	-0.391	-0.383	-0.383	-0.390
Mg		-0.466	-0.459[d]	-0.444	-0.444
Ar	-0.771	-0.785	-0.751	-0.787	-0.722
Kr	-1.914		-1.748	-1.899	
Xe	-3.149		-2.743	-3.151	

[a]From Ref. [8] [b]From Ref. [10] [c]From Ref. [11] [d]From Ref. [12]
[e]From Ref. [50] [f]From Ref. [51] [g]See Ref. [52]

Table 2. Correlation Energies for Open-Shell Species

Species	$E_c^{PW91}[n]$[a]	$E_c^{B}[n]$[b]	$E_c^{LYP}[n]$[c]	$E_c^{WL}[n]$[d]	Exact[e]
H	-0.007	0.0000	0.0000	0.0000	0.0000
He^+		0.0000	0.0000	0.0000	0.0000
Li^{2+}		0.0000	0.0000	0.0000	0.0000
Li	-0.058	-0.055	-0.053	-0.046	-0.045
B		-0.129	-0.128	-0.129	-0.125
C		-0.166	-0.161	-0.160	-0.157[b]
N	-0.199	-0.202	-0.193	-0.188	-0.188
C^-		-0.179[f]	-0.171[f]	-0.177	-0.183
Na		-0.419	-0.408[d]	-0.399	-0.398[c]
P		-0.590	-0.566[d]	-0.554	-0.553[b]

[a]From Ref. [8] [b]From Ref. [10] [c]From Ref. [11] [d]From Ref. [12]
[e]From Ref. [50] [f]From Ref. [51]

and d = 0.90000. Numerical results for atoms and ions calculated by $E_c^{WL}[n]$, Eq. (29), compare favorably with results obtained from the PW91, Becke's functional (B), and the LYP as demonstrated in Tables 1 and 2.

The original conjecture by Wilson and Levy that the magnitude of $(\partial E_c[n_\lambda]/\partial\lambda)_{\lambda=1}$ be small for atoms and molecules was the primary identity for parametrization for the WL and has gained credibility in DFT with recent calculations and comparisons of values from certain functionals: see Table 3.

When compared to traditional QC values, correlation-energy functionals tend to over-estimate the magnitude of the correlation energy. This effect is offset by a typical underestimation of the exchange energy obtained from exchange-energy functionals. This is probably why the PW91 and the LYP fare better in DFT than the WL does and, conversely, why the WL operates better to produce "tack-on" energies. The use of HF densities in DFT correlation-energy functionals and the subsequent comparison of energy values obtained from DFT functionals to their traditional QC counterparts are common practices. It could be argued, however, if the results from DFT functionals should be used in QC because DFT correlation energy is different from QC correlation energy by definition, even though they are expected to be numerically close.

For a finite system, according to constraint XXXIII [33], $E_c[n]$ is relatively insensitive to uniform scaling as $\lambda \to \infty$, i.e. $\lim_{\lambda\to\infty} E_c[n_\lambda] \approx E_c[n]$.

Table 3. Values of $(\partial E_c[n_\lambda]/\partial\lambda)_{\lambda=1}$ for Correlation Energy Approximations

Species	$E_c^{LDA}[n]^a$	$E_c^{W}[n]^b$	$E_c^{PW}[n]^a$	$E_c^{LYP}[n]^a$	$E_c^{WL}[n]^b$	Exact[c]
He	-0.045	-0.013	-0.008	-0.009	-0.0023	-0.0055
Be	-0.0861	-0.030	-0.021	-0.036	-0.0096	
Ne	-0.247	-0.042	-0.069		-0.0175	
Ar	-0.451	-0.072	-0.140		-0.0401	
Xe	-1.439	-0.142	-0.483			

[a]From Ref. [33] [b]From Refs. [42] and [48] [c]From Refs. [46], [47]

Table 4. Correlation energies and $(\partial E_c[n_\lambda]/\partial\lambda)_{\lambda=1}$ from Wilson-Levy and Becke functionals

Species	$(\partial E_c^B[n_\lambda]/\partial\lambda)_{\lambda=1}$	$E_c^B[n]^a$	Exact	$E_c^{WL}[n]$	$(\partial E_c^{WL}[n_\lambda]/\partial\lambda)_{\lambda=1}$
He	-0.0165	-0.0419	-0.0420	-0.0420	-0.0023
Li^+	-0.0150	-0.0498	-0.0435	-0.0442	-0.0015
Be^{2+}	-0.443	-0.0546	-0.0443	-0.0452	-0.0012

[a]From Ref. [42]

As can be seen from Table 3, the values for $(\partial E_c[n_\lambda]/\partial\lambda)_{\lambda=1}$ for the LDA are far too negative. This is explained by Görling and Levy [31] to be a result of the correlation energy being too sensitive to coordinate scaling of the density and, they suggest, might explain the "well-known overbinding in the LDA in molecules and solids." There also appears to be a relationship between how small this term is and numerical accuracy of a functional. The WL functional, for instance, gives better correlation energy values for the 2- electron series, He, Li^+, and Be^{2+}, than Becke's functional does. When the term $(\partial E_c[n_\lambda]/\partial\lambda)_{\lambda=1}$was obtained for these two functionals, the values for the WL were much smaller than those for Becke's functional.

Similarly, investigation into this term for the LDA and the original Wigner uncovered the same tendency that the smaller this value, the closer the correlation energy was to exact values for a particular functional [33].

Pragmatically speaking, the WL seems to be much more effective in traditional QC than in DFT [17], although exact values of DFT correlation energies are known only in a few cases and it is difficult to assess the WL performance in DFT. It is the purpose of this paper to alert the reader to the potential value of the WL for improving accuracy in traditional QC calculations.

Ivanov and Levy [54] have very recently established a link between the high-density scaling limit of the correlation energy, $E_c^{(2)}[n]$, and known results for the second-order energy $E_c^{QC(2)}$ from the high-Z asymptotic expansion for the traditional QC correlation energy. They have shown that, for a non-degenerate ground state, $E_c^{(2)}[n] \leq E_c^{QC(2)}$, with an equality only for two-electron densities. The density $n(\mathbf{r})$ is obtained from hydrogenic orbitals. It can be seen from Table 5 that the WL obeys the strict inequality for more than two electrons as required for any viable approximation to $E_c[n]$.

In Table 6, we present the results of applying conditions XXXV and XXXVI to some of the most widely used approximations to $E_c[n]$. As far as $E_c^{(2)}[n]$ is concerned, all tested functionals give numbers close to the theoretically-calculated value. Unfortunately, all three functionals do not have the correct functional derivatives as required by condition XXXVI; $\int d\mathbf{r}\, v_c^{(2)}([n];\mathbf{r})\, n(\mathbf{r})$ has the wrong sign for $n(\mathbf{r}) = (2a^3/\pi)\exp(-2ar)$ with $a > 0$.

For Gaussian densities $n(\mathbf{r}) = 2(2a/\pi)^{3/2}\exp(-2ar^2)$ with $a > 0$, on the other hand, all tested approximations give negative numbers for both $E_c^{(2)}[n]$ and

Table 5. Comparison of $E_c^{QC(2)}$ with $E_c^{(2),WL}[n]$

Density	$E_c^{QC(2)a}$	$E_c^{(2),WL}[n]$
2-electron	-0.0467	-0.0480
3-electron	-0.0537	-0.0568
9-electron	-0.3694	-0.3826
10-electron	-0.4278	-0.4504
11-electron	-0.4534	-0.4718

[a]From Ref. [54]

Table 6. [a]Second-Order Results with $n(\mathbf{r}) = (2a^3/\pi)\exp(-2ar)$

	$E_c^{(2)}[n]$	$\int d\mathbf{r}\, v_c^{(2)}([n];\mathbf{r})\, n(\mathbf{r})$
Exact Value	-0.046663	0.017677
$E_c^{PBE}[n]$	-0.047896	-0.073013
$E_c^{LYP}[n]$	-0.056481	-0.108922
$E_c^{WL}[n]$	-0.048030	-0.073415

[a]From Ref. [38]

$\int d\mathbf{r}\, v_c^{(2)}([n];\mathbf{r})\, n(\mathbf{r})$, as required by constraints XXXVII and XXXVIII: see Table 7.

The most impressive collection of calculations to date that attest to the relative effectiveness of the WL for approximating $E_c^{QC(2)}[n^{HF}]$ was published by Fuentealba and Savin [17]. Some of the more accurate ionization potentials obtained by the WL are given in Table 8 and dissociation energies are listed in Table 9.

CONCLUSIONS

While the use of density functionals seems to have a place in traditional QC calculations, the idea of using an approximation to DFT correlation energy as an approximation to $E_c^{QC}[n^{HF}]$ should be handled cautiously. The common practice of using traditional quantum chemistry energy values to measure the performance of approximate functionals in DFT should be also subject to questioning. There is a subtle difference in the definitions of the correlation energy in DFT and the QC correlation energy, even if the same reference density is used. Even though the numerical values for $E_c[n]$ and $E_c^{QC}[n^{HF}]$ are expected to be close, almost the same for small systems, it has been shown that $E_c[n_\lambda] \leq \lim_{\lambda\to\infty} E_c^{QC}[n_\lambda^{HF}]$, with an equality only for the two-electron case. It is possible that $E_c^{QC}[n^{HF}]$ possesses similar asymptotic behavior upon uniform scaling as $\lambda \to 0$, and non-uniform scaling of the HF density. Deriving more properties of $E_c^{QC}[n^{HF}]$ is essential for arriving at very good approximations.

Table 7. [a] Second-Order Results with $n(\mathbf{r}) = 2(2a/\pi)^{3/2}\exp(-2ar^2)$

	$E_c^{(2)}[n]$	$\int d\mathbf{r}\, v_c^{(2)}([n];\mathbf{r})\, n(\mathbf{r})$
Exact Value	-0.049702 [b]	-0.071216
$E_c^{PBE}[n]$	-0.081347	-0.109562
$E_c^{LYP}[n]$	-0.035371	-0.101855
$E_c^{WL}[n]$	-0.078170	-0.106126

[a]From Ref. [39] [b]Also from Ref. [40]

Table 8. [a]Ionization Potentials (in eV)

System	HF	HF+PW91	HF+WL	Expt.
Li	5.34	5.53	5.39	5.39
Be	8.04	8.91	9.31	9.32
N	13.96	14.65	14.57	14.53
F	15.72	16.96	17.22	17.42
Be_2	6.19	6.96	7.33	7.46
B_2	7.34	8.34	8.72	8.99
BeH	8.13	8.47	8.29	8.20
BH	8.45	9.42	9.75	9.76

[a]From Ref. [17]

Table 9. [a]Dissociation Energies (in eV)

System	HF	HF+PW91	HF+B	HF+LYP	HF+WL	Expt.
Li_2	0.16	0.76	0.73	0.89	1.37	1.06
Be_2	-0.33	-0.04	-0.03	-0.19	0.22	0.11
B_2	0.87	1.54	1.58	1.36	1.99	3.06
C_2	0.68	2.39	2.39	2.34	4.01	6.32
N_2	5.03	7.51	7.45	7.67	9.30	9.91
O_2	1.15	3.13	2.84	2.67	3.38	5.21
F_2	-1.37	-0.69	-0.36	-0.53	-0.41	1.66

[a]From Ref. [17]

The presented tables by no means assure that the WL will give reliable results for all systems, but the functional does show promise. As a point of interest, it should be noted that "exact" values differ among publications. In recent papers, Davidson and co-workers [55,56] have refined values for non-relativistic correlation energies and some of these values vary from previously accepted ones, particularly in the case of Argon.

The exact correlation-energy functional $E_c[n]$, as part of $F[n]$, clearly obeys all theoretical constraints, but some of these constraints may have a bigger effect on accuracy while others may be inconsequential for particular systems.

Realistically, known constraints for $E_c[n]$ should be studied and prioritized since obedience to all of them, even just the requirements of Section II, is likely impossible. Scaling requirements seem to affect numerical efficacy [42], and might serve as important guideposts for construction, including the relatively small magnitude of $(\partial E_c[n_\lambda]/\partial\lambda)_{\lambda=1}$ which might have more significance than previously realized.

Although not a derived functional when compared to the functionals by Perdew and co-workers, and by Becke, the WL produces very good results in the framework of traditional QC calculations, usually surpassing the results obtained from theoretically better justified and more complex DFT functionals. Whether this is a result of built-in properties of the exact $E_c^{QC}[n^{HF}]$, or it is purely coincidental, is unknown.

Insisting on simple and manageable functionals ensures a larger range of application in the future. Since working with the WL has been computationally very simple in both DFT and traditional QC, different versions of it incorporating more constraints could be tried.

We would like to thank Professor Mel Levy for his time and very important suggestions in revisions of this paper. This research has been supported in part by Research Corporation, NSF and the Center for Photoinduced Processes.

REFERENCES

1. P. Hohenberg and W. Kohn, Phys. Rev. **136**, B864 (1964).
2. W. Kohn and L. J. Sham, Phys. Rev. **140**, A1133 (1965).
3. R. G. Parr and W. Yang, *Density Functional Theory of Atoms and Molecules* (Oxford University Press, New York, 1989).
4. R. M. Dreizler and E. K. U. Dross, *Density Functional Theory* (Springer-Verlag, Berlin-New York, 1990).
5. M. Levy, Proc. Natl. Acad. Sci (USA) **76**, 6062 (1979).
6. R. O. Jones and O. Gunnarsson, Rev. Mod. Phys. **61**,689 (1989).
7. J. P. Perdew, Phys. Rev. B **33**,8822 (1986); **34**, 7406 (E) (1986).
8. J. P. Perdew in *Electronic Structure of Solids '91*, edited by P. Ziesche and H. Eschrig (Akademie Verlag, Berlin, 1991); J. P. Perdew and Y. Wang, Phys. Rev. B **45**, 13244 (1992); J. P. Perdew, K. Burke, and Y. Wang, Phys. Rev. B **54**, 16538 (1996): see also chapter by Perdew *et al* in present Volume .
9. J. P. Perdew, K. Burke, and M. Ernzerhof, Phys. Rev. Lett. **77,** 3865 (1996).
10. A. D. Becke, J. Chem. Phys. **88**, 1053 (1988).
11. C. Lee, W. Yang, and R. Parr, Phys. Rev. B **37**, 785 (1988).
12. L. C. Wilson and M. Levy, Phys. Rev. B **41**, 12930 (1990).
13. A. D. Becke, J. Chem. Phys. **33**, 3098 (1988).
14. E. P. Wigner, Trans Faraday Soc. **34**, 678 (1938).
15. E. Kryachko and E. Ludeña, *Density Functional Theory of Many-electron Systems* (Kluwer, Dordrecht, 1990).
16. N. H. March, *Electron Density Theory of Atoms and Molecules* (Academic Press, New York, 1992).
17. P. Fuentealba and A. Savin, Chem. Phys. Lett. **217**, 566 (1994).
18. V. Sahni, J. Gruenebaum, and J.P. Perdew, Phys. Rev. B **26**, 4371 (1982).
19. M. Levy and J. P. Perdew, Phys. Rev. A **32**, 2010 (1985).
20. V. Sahni and M. Levy, Phys. Rev. B **33** (1986) 3869.
21. J. Harris and R. O. Jones, J. Phys. F **4**, 1170 (1974).
22. D. C. Langreth and M. J. Mehl, Phys. Rev. B **28**, 1030 (1983).
23. R. A. Harris and L. R. Pratt, J. Chem. Phys. **83**, 4024 (1985).
24. M. Levy, in *Density Matrices and Density Functionals*, edited by R. Erdahl and V.H. Smith, Jr. (Reidel, Boston, 1987), pp.479-498. This book constitutes the proceedings of the A. J. Coleman Symposium, Kingston, Canada, 1985.
25. M. Levy, Phys. Rev. A **43**, 4637 (1991).
26. M. Levy, Int. J. Quantum.Chem. **61**, 281 (1997).
27. S. Ivanov and M. Levy, unpublished.
28. C. Filippi, C. J. Umrigar, and M. Taut, J. Chem. Phys, **100**, 1290 (1994).
29. H. Ou-Yang and M. Levy, Phys. Rev. A **42**, 155 (1990).
30. M. Levy, Int. J. Quantum Chem. S **23**, 617 (1989).
31. A. Görling and M. Levy, Phys. Rev. A **45**, 1509 (1992).
32. M. Levy, Phys. Rev. A **26**, 1200 (1982).
33. M. Levy in *Density Functional Theory*, edited by E. K. U. Gross and R. M. Dreizler (Springer-Verlag, Berlin-New York, 1990).
34. E. H. Lieb and S. Oxford, Int. J. Quantum Chem. **19**, 427 (1981).
35. M. Levy and H. Ou-Yang, Phys. Rev. A **42**, 651 (1990).
36. M. Levy and J. P. Perdew, Phys. Rev. B **48**, 11639 (1993).
37. A. Görling and M. Levy, Phys. Rev. B **47**, 13105(1993).
38. S. Ivanov, R. Lopez-Boada, A. Görling, and M. Levy, unpublished
39. S. Ivanov, K. Burke, and M. Levy, unpublished.
40. C. J. Huang and C. J. Umrigar, unpublished.
41. A. J. Thakkar and V. H. Smith, Jr., Phys. Rev. A **15**, 1 (1977)
42. L. C. Wilson, Chem. Phys. **181**, 337 (1994).
43. S. J. Vosko, L. Wilk and M. Nusair, Can. J. Phys. Rev. **58**, 12100 (1980).
44. R. McWeeny, in: *The New World of Quantum Chemistry*, edited by B. Pullman and R. Parr (D. Reidel Publishing Company, Boston, 1976), 3.
45. G. Brual, Jr. and S. M. Rothstein, J. Chem Phys. **69**, 1177 (1978).
46. C. J. Umrigar and X. Gonze, in *High performance Computing and its Application to the Physical Sciences*, proceedings of the Mardi Gras '93 Conference, edited by D. A. Browne *et al* (World

Scientific, 1993).
47. C. J. Umrigar and X. Gonze, Phys. Rev. A **50**, 3827 (1994).
48. L. C. Wilson, unpublished.
49. R. Colle and D. Salvetti, Theor. Chem. Acta **37**, 329 (1975).
50. S. J. Chakravorty, S. R. Gwaltney, E. R. Davidson, F. A. Parpia, and C. F. Fischer, Phys. Rev. A **47**, 3649 (1993).
51. B. Miehlich, A. Savin, H. Stoll, and H. Preuss, Chem. Phys. Lett. **157**, 200 (1989).
52. J. B. Lagrowski and S.H. Vosko, J. Phys. B: At . Mol. Opt. Phys. **21**, 203 (1988).
53. C. Filippi, X. Gonze, and C. J. Umrigar, in *Recent Developments and Applications of Modern Density Functional Theory*, Theoretical and Computational Chemistry, Vol. 4, edited by J. M. Seminario (Elsevier, Amsterdam, 1996).
54. S. Ivanov and M. Levy, unpublished.
55. E. R. Davidson, S. A. Hagstrom, S. J. Chakravorty, V. M. Umar, and C. F. Fischer, Phys. Rev. A **44**, 7071 (1991).
56. S. J. Chakravorty and E. R. Davidson, J. Phys. Chem. **100**, 6167 (1996).

RELATIVISTIC CORRECTIONS TO THE EXCHANGE-CORRELATION ENERGY FUNCTIONAL

E. Engel, S. Keller, and R. M. Dreizler

Institut für Theoretische Physik, Universität Frankfurt
Robert-Mayer-Str. 8-10, D-60054 Frankfurt/Main, Germany

INTRODUCTION

During recent years the development of fully relativistic methods for electronic structure calculations has been a major trend both in atomic physics and computational quantum chemistry (see e.g.[1-5]). Looking at Gold as the prime candidate for relativistic effects this trend is easy to understand: When going from a nonrelativistic to a relativistic treatment the ionization potential of Au increases by 30% [6], the bond length of AuH is contracted by 13% and its dissociation energy increases by about 50% [6]. The differences are even more dramatic for the electron affinity: For Au a nonrelativistic CI-calculation [7] gives a value of 1.02 eV compared with 2.28 eV obtained with a fully relativistic Coupled-Cluster calculation [4]. Moreover, the fact that the relativistic Hartree-Fock (RHF) approach leads to 7.48 eV [4] for the electron affinity of Au while the nonrelativistic HF value is 0.10 eV, clearly demonstrates that relativistic and exchange-correlation (xc) effects are nonadditive.

Since the introduction of Generalized Gradient Approximations (GGAs) [8-10] density functional theory (DFT) has gained considerable attraction in the quantum chemistry community. In fact, it has been shown by studying a variety of small molecules with light constituents [11,12] that for binding geometries and atomization energies DFT methods can successfully compete with the traditional *ab initio* approaches. As soon as heavy elements are involved, however, the situation is not as clear. All applications of relativistic DFT (RDFT) [13-16] have so far essentially been restricted to a relativistic treatment of the DFT kinetic energy, thus ignoring the nonadditivity of relativistic and xc-effects. Although the relativistic Local Density Approximation (RLDA) has been evaluated rather early [14,15,17] its application is hampered by both the well-known deficiencies of the nonrelativistic LDA and its inaccurate description of relativistic corrections [18,19]. This is e.g. obvious from the 50% overestimation of the Au_2 dissociation energy, found in a fully relativistic LDA-KS-calculation [20]. Only very recently has a relativistic extension of the GGA (RGGA) for the exchange-only (x-only) energy been put forward [21], applications to molecules in fully relativistic Kohn-Sham (KS) calculations are not yet available.

In this contribution we analyze the RGGA on the basis of exact relativistic x-only

Electronic Density Functional Theory: *Recent Progress and New Directions*
Edited by Dobson *et al.*, Plenum Press, New York, 1998

results obtained with the relativistic Optimized-Potential-Method (ROPM) [22-24,19] (after a brief review of RDFT, which serves to define the various concepts involved, and an outline of our RGGA approach). In particular, we present first selfconsistent ROPM calculations including the complete transverse interaction (usually only taken into account on the weakly relativistic Breit level or evaluated perturbatively*) and use the resulting x-only potentials to investigate the accuracy of the RGGA on a microscopic level. In response to the deficiencies of the RLDA for the correlation energy we extend the RGGA to include relativistic correlation effects. Moreover, by applying the relativistic extension scheme to different nonrelativistic GGAs its general validity is demonstrated.

RELATIVISTIC KOHN-SHAM EQUATIONS

The starting point for the derivation of relativistic KS-equations is the standard QED Hamiltonian $\hat{H}_e$ augmented by an external potential term, representing the nuclei and, if present, applied static electric and/or magnetic fields (we use $\hbar = c = 1$ throughout),

$$\hat{H} = \hat{H}_e + \int d^3r \; v^{\mu}_{ext} \; \hat{j}_{\mu} \qquad ; \qquad \hat{j}_{\mu} = \frac{1}{2}[\hat{\bar{\psi}}, \gamma_{\mu}\hat{\psi}] \; , \tag{1}$$

where the current operator $\hat{j}_{\mu}$ has been written in a charge conjugation 'invariant' form. As usual in relativistic quantum field theory both $\hat{H}$ and $\hat{j}_{\mu}$ require renormalization. The energy of the negative continuum states can e.g. be eliminated via

$$\hat{H}_R = \hat{H} - <0|\hat{H}_e|0> \; ,$$

where $|0>$ represents the vacuum corresponding to $\hat{H}_e$. Additional counterterms are necessary to deal with the UV-divergencies. In the subsequent discussion we will always assume that the quantities involved are renormalized without explicitly displaying the relevant counterterms.

On the basis of the assumption that there exists a minimum principle† for the renormalized ground state energy E_{tot} one can then directly establish a Hohenberg-Kohn theorem [13-15]: E_{tot} is a unique functional of the (renormalized) ground state four current j_{μ} which can be written as

$$E_{tot}[j_{\mu}] = F[j_{\mu}] + \int d^3r \; v^{\mu}_{ext} \; j_{\mu} \; , \tag{2}$$

with $F[j_{\mu}]$ being a universal functional of j_{μ}. Note that the gauge dependence of the corresponding QED ground state is not transferred into $E_{tot}[j_{\mu}]$ as the ground state energy and four current are gauge invariant quantities (up to trivial energy shifts‡).

One can then represent the exact current j_{μ} of the interacting system via auxiliary

*Inclusion of the complete orbital-dependent transverse RHF exchange potential would lead to gauge dependent results.

†Note that its existence is the guiding principle in the construction of the standard renormalization scheme.

‡Only static gauge transformations are relevant in this context.

single particle four spinors,

$$j^\nu(\mathbf{x}) = j_V^\nu(\mathbf{x}) + j_D^\nu(\mathbf{x}) \tag{3}$$

$$j_V^\nu(\mathbf{x}) = \frac{1}{2}\Big\{ \sum_{\epsilon_k \le -m} \overline{\varphi}_k(\mathbf{x})\gamma^\nu\varphi_k(\mathbf{x}) - \sum_{-m<\epsilon_k} \overline{\varphi}_k(\mathbf{x})\gamma^\nu\varphi_k(\mathbf{x}) \Big\} \tag{4}$$

$$j_D^\nu(\mathbf{x}) = \sum_{-m<\epsilon_k\le\epsilon_F} \overline{\varphi}_k(\mathbf{x})\gamma^\nu\varphi_k(\mathbf{x}) \ , \tag{5}$$

and define the corresponding noninteracting kinetic energy

$$T_s[j^\nu] = T_{s,V}[j^\nu] + T_{s,D}[j^\nu] \tag{6}$$

$$T_{s,V}[j^\nu] = \frac{1}{2}\int d^3x \Big\{ \sum_{\epsilon_k\le -m} \overline{\varphi}_k(\mathbf{x})\Big[-i\gamma\cdot\nabla + m\Big]\varphi_k(\mathbf{x}) - \sum_{-m<\epsilon_k} \overline{\varphi}_k(\mathbf{x})\Big[-i\gamma\cdot\nabla + m\Big]\varphi_k(\mathbf{x}) \Big\} \tag{7}$$

$$T_{s,D}[j^\nu] = \int d^3x \sum_{-m<\epsilon_k\le\epsilon_F} \overline{\varphi}_k(\mathbf{x})\Big[-i\gamma\cdot\nabla + m\Big]\varphi_k(\mathbf{x}) \ . \tag{8}$$

Here j_V^ν represents the KS-equivalent of the usual vacuum polarization current and $T_{s,V}$ is the kinetic contribution to the KS Casimir energy. j_D^ν and $T_{s,D}$, on the other hand, result from the discrete bound 'electrons'. Minimizing E_{tot} with respect to the φ_k one finds the relativistic KS-equations in their most general form,

$$\gamma^0\Big\{ -i\gamma\cdot\nabla + m + \gamma_\nu\big[v_{ext}^\nu(\mathbf{x}) + v_H^\nu(\mathbf{x}) + v_{xc}^\nu(\mathbf{x})\big]\Big\}\varphi_k(\mathbf{x}) = \epsilon_k\varphi_k(\mathbf{x}) \ , \tag{9}$$

where the relativistic Hartree and xc-potentials are given by[§]

$$v_H^\nu(\mathbf{x}) = e^2\int d^3y \, \frac{j^\nu(\mathbf{y})}{|\mathbf{x}-\mathbf{y}|} \quad ; \quad v_{xc}^\nu(\mathbf{x}) = \frac{\delta E_{xc}[j^\mu]}{\delta j_\nu(\mathbf{x})} \tag{10}$$

$$E_{xc}[j^\mu] = F[j^\mu] - T_s[j^\mu] - \frac{e^2}{2}\int d^3x \int d^3y \, \frac{j^\nu(\mathbf{x})\, j_\nu(\mathbf{y})}{|\mathbf{x}-\mathbf{y}|} \ . \tag{11}$$

This set of equations has to be solved selfconsistently, leading to the exact ground state four current and energy, in principle.

Due to the four component structure of the relativistic KS potential and in particular the selfconsistent treatment of radiative corrections Eqs.(3-10) are not very useful in practice. However, having in mind electronic structure calculations for atoms, molecules and solids with heavy constituents, two simplifications offer themselves:

(i) If no external magnetic fields are present the four current version of RDFT (presented so far) can be reduced to a density-only variant: The complete proof of the existence theorem can be gone through using just the ground state density $j^0 = n$ as basic RDFT-variable [19], leading to the statement that E_{tot} for such systems can be written as a unique functional of n only,

$$E_{tot}[j^\mu] \xrightarrow{B\to 0} \tilde{E}_{tot}[n] \ .$$

[§]Note that in the relativistic Hartree term (last term of Eq.(11)) the retarded electron-electron interaction via the photon propagator of QED reduces to the static Coulomb interaction due to the time-independence of $\mathbf{j}$ (compare [16]).

This does not imply that the spatial components $\mathbf{j}(\mathbf{r})$ of the ground state four current vanish, but rather that they have to be interpreted as functionals of the density, $\mathbf{j}([n];\mathbf{r})$. Correspondingly, the functional $\tilde{E}_{tot}[n]$ is not identical with $E_{tot}[j^0 = n, \mathbf{j} = \mathbf{0}]$, indicated by $\tilde{}$. For brevity, however, the $\tilde{}$ will be dropped from now on, as the density-only variant of RDFT will be used exclusively in the following.

(ii) In addition, the selfconsistent treatment of vacuum corrections is not necessary in standard applications of RDFT: These contributions only lead to shifts in absolute ground state energies, but do not affect the physically relevant quantities like ionization potentials or dissociation energies. One can thus safely neglect j_V^μ, $T_{s,V}$ and the vacuum contributions to E_{xc}, at least on the selfconsistent level, thus enormously reducing the complexity of the KS-equations (3-10).

The subsequent discussion will be based on these two simplifications. The resulting electrostatic, no-sea KS-equations read

$$\Big\{ -i\alpha\cdot\nabla + \beta m + v_{ext}(\mathbf{r}) + v_H(\mathbf{r}) + v_{xc}(\mathbf{r})\Big\}\varphi_k(\mathbf{r}) = \epsilon_k\varphi_k(\mathbf{r}), \tag{12}$$

where the Hartree and xc-potentials are now given by

$$v_H(\mathbf{r}) = e^2\int d^3r' \,\frac{n(\mathbf{r}')}{|\mathbf{r}-\mathbf{r}'|} + \frac{\delta E_H^T[n]}{\delta n(\mathbf{r})} \quad ; \quad v_{xc}(\mathbf{r}) = \frac{\delta E_{xc}[n]}{\delta n(\mathbf{r})} \tag{13}$$

$$E_H^T = -\frac{e^2}{2}\int d^3r \int d^3r' \,\frac{\mathbf{j}(\mathbf{r})\cdot\mathbf{j}(\mathbf{r}')}{|\mathbf{r}-\mathbf{r}'|} \tag{14}$$

$$n(\mathbf{r}) = \sum_{-m<\epsilon_k\le\epsilon_F} \varphi_k^+(\mathbf{r})\,\varphi_k(\mathbf{r}) \quad ; \quad \mathbf{j}(\mathbf{r}) = \sum_{-m<\epsilon_k\le\epsilon_F} \varphi_k^+(\mathbf{r})\,\alpha\,\varphi_k(\mathbf{r})\,. \tag{15}$$

Note that the KS-current (15) of density-only RDFT in general is no longer identical with the exact current, leading to a corresponding redefinition of the transverse Hartree energy E_H^T.

In a final step one can isolate the two sources for relativistic corrections, i.e. the relativistic kinematics of the electrons and the photons, by decomposing the electron-electron interaction into a longitudinal (Coulomb) and a transverse (Breit + ...) component,

$$D^0_{\mu\nu}(x-y) = g_{\mu 0}g_{\nu 0}\delta(x^0-y^0)\frac{e^2}{|\mathbf{x}-\mathbf{y}|} + D^T_{\mu\nu}(x-y)\,. \tag{16}$$

Here $D^0_{\mu\nu}$ represents the free photon propagator of QED, which mediates the electron-electron interaction, and $D^T_{\mu\nu}$ is its transverse part. Complete neglect of $D^T_{\mu\nu}$ leads to the longitudinal limit of RDFT, in which only the electrons are dealt with on the relativistic level. This limit of RDFT corresponds to the (no-pair) Dirac-Coulomb Hamiltonian in standard quantum chemistry. All photon retardation effects are contained in $D^T_{\mu\nu}$, whose inclusion leads to the RDFT equivalent of the Dirac-Coulomb-Breit (DCB) Hamiltonian (ignoring the fact that $D^T_{\mu\nu}$ agrees with the Breit interaction only to lowest order in $1/c^2$).

RELATIVISTIC OPTIMIZED-POTENTIAL-METHOD

In analogy to the nonrelativistic case [26-28] the exchange-only energy of RDFT is defined via the relativistic Fock term [29-31], using the KS-spinors [19,16],

$$E_x[n] = -\frac{e^2}{2}\int d^3r \int d^3r' \sum_{-m<\epsilon_k,\epsilon_l\leq\epsilon_F} \frac{\cos(\omega_{kl}|\mathbf{r}-\mathbf{r}'|)}{|\mathbf{r}-\mathbf{r}'|}\,\overline{\varphi}_k(\mathbf{r})\gamma_\mu\varphi_l(\mathbf{r})\overline{\varphi}_l(\mathbf{r}')\gamma^\mu\varphi_k(\mathbf{r}')\ , \tag{17}$$

where $\omega_{kl} = |\epsilon_k - \epsilon_l|$ and the φ_k have to be interpreted as functionals of n. In (17) we have used Feynman gauge for $D^0_{\mu\nu}$. The correlation energy is then simply given by

$$E_c[n] = E_{xc}[n] - E_x[n]\ , \tag{18}$$

with E_{xc} as in (11), taking into account the redefinition of $\mathbf{j}$, Eq.(15). In the longitudinal limit (17) reduces to the more familiar form [24,19]

$$E_x^L[n] = -\frac{e^2}{2}\int d^3r \int d^3r' \sum_{-m<\epsilon_k,\epsilon_l\leq\epsilon_F} \frac{\varphi_k^+(\mathbf{r})\varphi_l(\mathbf{r})\varphi_l^+(\mathbf{r}')\varphi_k(\mathbf{r}')}{|\mathbf{r}-\mathbf{r}'|}\ . \tag{19}$$

The difference between E_x and E_x^L is the transverse exchange energy, $E_x^T = E_x - E_x^L$.

The x-only potential corresponding to either (17) or (19) can not be evaluated by directly taking the functional derivative (13). To calculate this potential one uses the fact that there is a one-to-one correspondence between the ground state density and the total KS-potential v^{KS}. Consequently the minimization of $E_{tot}[n]$ with respect to $n(\mathbf{r})$ is equivalent to

$$\frac{\delta E_{tot}[n]}{\delta v^{KS}(\mathbf{r})} = 0 \tag{20}$$

together with a simultaneous solution of the KS-equations (12-15) (in order to establish the functional dependence between v^{KS} and n). Eq.(20) leads to the ROPM integral equation for the x-only potential v_x,

$$\int d^3r'\ K(\mathbf{r},\mathbf{r}')\ v_x(\mathbf{r}') \ = \ Q(\mathbf{r}) \tag{21}$$

with

$$K(\mathbf{r},\mathbf{r}') = \sum_{-m<\epsilon_k\leq\epsilon_F} \varphi_k^+(\mathbf{r})\ G_k(\mathbf{r},\mathbf{r}')\ \varphi_k(\mathbf{r}') + c.c. \tag{22}$$

$$Q(\mathbf{r}) = \sum_{-m<\epsilon_k\leq\epsilon_F} \int d^3r'\ \varphi_k^+(\mathbf{r})\ G_k(\mathbf{r},\mathbf{r}')\ \frac{\delta E_x[n]}{\delta\varphi_k^+(\mathbf{r})} + c.c. \tag{23}$$

$$G_k(\mathbf{r},\mathbf{r}') = \sum_{l\neq k} \frac{\varphi_l(\mathbf{r})\ \varphi_l^+(\mathbf{r}')}{\epsilon_l - \epsilon_k}\ , \tag{24}$$

which has to be solved selfconsistently together with the KS-equations. Some remarks are appropriate:

- The ROPM integral equation determines v_x only up to a global constant, which, for finite systems, is fixed via $v_x(\mathbf{r}) \underset{|\mathbf{r}|\to\infty}{\longrightarrow} 0$, in analogy to the nonrelativistic case.

Table 1. Longitudinal x-only ground state energies: Selfconsistent ROPM, RHF, RLDA and RGGA results for neutral atoms with closed subshells (in *hartree* [34]).

Atom	$-E_{tot}^{L}$	$E_{tot}^{L} - E_{tot}^{L,ROPM}$		
	ROPM	RHF	RLDA	RPW91
He	2.862	0.000	0.138	0.006
Be	14.575	-0.001	0.350	0.018
Ne	128.690	-0.002	1.062	-0.024
Mg	199.932	-0.003	1.376	-0.001
Ar	528.678	-0.005	2.341	0.041
Ca	679.704	-0.006	2.656	0.026
Zn	1794.598	-0.014	4.140	-0.262
Kr	2788.848	-0.013	5.565	-0.021
Sr	3178.067	-0.013	5.996	-0.008
Pd	5044.384	-0.016	7.707	-0.067
Cd	5593.299	-0.020	8.213	-0.033
Xe	7446.876	-0.019	9.800	0.085
Ba	8135.625	-0.019	10.289	0.059
Yb	14067.621	-0.048	13.272	-0.893
Hg	19648.826	-0.039	17.204	-0.250
Rn	23601.969	-0.035	19.677	0.004
Ra	25028.027	-0.034	20.460	-0.006

- The ROPM allows one to include the transverse exchange selfconsistently without introducing a gauge dependence. In contrast to the transverse RHF-potential the local, non-orbital-dependent x-only potential (21) does not produce gauge dependent spinors [25]. Thus all gauges for the photon propagator give the same exchange energy and potential, justifying the particular choice of Feynman gauge in (17).

- The ROPM procedure is by no means restricted to x-only functionals, as is obvious from the general form of its ingredients K and Q. It can also be used to deal with orbital-dependent correlation energy functionals such as the Colle-Salvetti functional [32,33].

- As the current $\mathbf{j}$ is a trivial functional of the φ_k, the previous statement particularly applies to $\mathbf{j}$-dependent functionals like E_H^T, thus establishing a transition from the four current to the density-only variant of (R)DFT.

While the (R)OPM definition of exchange within (R)DFT has a number of conceptual advantages [35], its legitimacy finally relies on its success in practice. In Table 1 we list the x-only ground state energies E_{tot}^{L} obtained from selfconsistent ROPM calculations utilizing only the longitudinal Fock energy (19) for neutral atoms with closed subshells. Also given is the difference between these energies and the corresponding RHF values [36]. These differences explicitly demonstrate the somewhat more restricted variational freedom of the ROPM as compared to the RHF scheme with its nonlocal exchange potential. The absolute magnitude of the differences, however, is marginal. The ROPM thus explicitly demonstrates the ability of local (i.e. multiplicative) xc-potentials to capture the complete physics (at least on the x-only level).

Table 2 gives the corresponding comparison including the transverse exchange. However, while the ROPM data have been obtained from a selfconsistent calculation

Table 2. Total relativistic x-only ground state energies: Selfconsistent ROPM, RLDA and (R)GGA results for neutral atoms with closed subshells in comparison with perturbative RHF data (in *hartree* [34]).

Atom	$-E_{tot}^{L+T}$	$E_{tot}^{L+T} - E_{tot}^{L+T,ROPM}$			
	ROPM	RHF(p)	RLDA	RPW91	PW91
He	2.862	0.000	0.138	0.006	0.006
Be	14.575	-0.001	0.351	0.018	0.017
Ne	128.674	-0.002	1.080	-0.024	-0.043
Mg	199.900	-0.003	1.408	-0.001	-0.037
Ar	528.546	-0.005	2.458	0.041	-0.111
Ca	679.513	-0.006	2.818	0.026	-0.195
Zn	1793.840	-0.014	4.702	-0.263	-1.146
Kr	2787.429	-0.012	6.543	-0.022	-1.683
Sr	3176.359	-0.011	7.149	-0.010	-2.014
Pd	5041.098	-0.013	9.765	-0.069	-3.953
Cd	5589.496	-0.016	10.556	-0.035	-4.537
Xe	7441.173	-0.011	13.161	0.083	-6.705
Ba	8129.161	-0.009	14.051	0.058	-7.652
Yb	14053.750	-0.021	20.888	-0.894	-17.660
Hg	19626.705	0.008	29.161	-0.257	-27.253
Rn	23573.354	0.029	35.207	-0.009	-35.145
Ra	24996.946	0.038	37.395	-0.022	-38.267

with transverse contributions, the RHF energies result from perturbatively adding the transverse exchange energies to the longitudinal RHF ground state energy. As a consequence the RHF energies are somewhat lower than the ROPM values for low Z where E_x^T is rather irrelevant. For heavy atoms, on the other hand, the variational treatment of E_x^T leads to a lowering of the energy which dominates over the energy differences resulting from the reduced variational freedom of the ROPM. From comparing selfconsistent and perturbative ROPM energies one extracts a selfconsistency effect of roughly $1eV$, which is also the size of many higher order QED contributions [37]. On the level of accuracy one tries to achieve in the spectroscopy of high-Z ions a selfconsistent treatment of the transverse exchange thus seems useful.

The numerical solution of the ROPM integral equation is rather involved, so that for applications to complex systems more efficient, but still accurate approximations are required. On the level of orbital-dependent potentials the KLI-approximation [38] can be extended to the relativistic domain [39]. On the other hand, the density functionals most often used at present are GGAs, whose relativistic extension will be discussed in the following Section.

RELATIVISTIC GENERALIZED GRADIENT APPROXIMATION

The starting point for the derivation of RGGAs is the RLDA [14,15,19],

$$E_{xc}^{RLDA}[n] = \int d^3r \; e_{xc}^{RHEG}(n) = \int d^3r \; e_{xc}^{HEG}(n) \; \Phi_{xc,0}(\beta) \; . \tag{25}$$

In (25) the xc-energy density e_{xc}^{RHEG} of the relativistic homogeneous electron gas (HEG) has been decomposed into a product of its nonrelativistic limit e_{xc}^{HEG} and a relativistic correction factor $\Phi_{xc,0}$, whose density dependence is most conveniently written in terms of

$$\beta = \frac{(3\pi^2 n)^{1/3}}{m} \ . \tag{26}$$

In the case of the x-only RLDA $\Phi_{xc,0}$ e.g. reads

$$\Phi_{x,0}(\beta) = 1 - \frac{3}{2}\left[\frac{\sqrt{1+\beta^2}}{\beta} - \frac{\mathrm{Arsh}(\beta)}{\beta^2}\right] \overset{\beta<<1}{=} 1 - \frac{2}{3}\beta^2 + \ \ldots \ , \tag{27}$$

which can be further decomposed into a longitudinal and a transverse part using (16) (for details see [18,16]).

The RLDA forms the basis for RGGAs, which take into account the inhomogeneity of the system via the most simple resummed form of the straightforward gradient expansion (GEA). In the case of the x-only energy the 'general' form¶ of the RGGA is

$$E_x^{RGGA}[n] = \int d^3r \ e_x^{HEG}(n) \ \Big[\Phi_{x,0}(\beta) + g(\xi)\Phi_{x,2}(\beta)\Big] \ , \tag{28}$$

where $\xi = [\nabla n/(2(3\pi^2 n)^{1/3} n)]^2$ is the dimensionless characteristic density gradient and $g(\xi)$ represents a nonrelativistic gradient correction in the form of a GGA. The β-dependence of $\Phi_{x,2}$ can, in principle, be obtained from the weakly inhomogeneous limit, in which the RGGA should approach the relativistic GEA,

$$E_x^{RGGA}[n] \overset{\xi<<1}{=} E_x^{RGEA}[n] = \int d^3r \ e_x^{HEG}(n) \ \Big[\Phi_{x,0}(\beta) + \frac{10}{81}\xi\Phi_{x,2}(\beta) + \ \ldots\Big] \ . \tag{29}$$

The ingredients of the RGEA, on the other hand, are determined by the long-wavelength expansion of the first order static irreducible current-current response function of the RHEG, $\Pi^{(1)}_{\mu\nu}(q^0 = 0, |\mathbf{q}|)$, which, however, is not available in the literature. Moreover, at present it is not clear whether the RGEA (29) exists at all, as a calculation of $\Phi_{x,2}$ using the Kirznits technique‖ leads to an infrared divergent $\Phi_{x,2}$ [40]. While the Kirznits gradient expansion gives an incorrect (but finite) gradient coefficient for the nonrelativistic $E_x[n]$ [41] and thus does not allow a definitive conclusion concerning the RGEA, the form of the infrared divergence nevertheless indicates a logarithmic $|\mathbf{q}|$-dependence of $\Pi^{(1)}_{00}$,

$$\Pi^{(1)}_{00}(q^0 = 0, |\mathbf{q}|) \overset{|\mathbf{q}|<<2k_F}{=} a + b\,\mathbf{q}^2 + c\,\mathbf{q}^2\ln(|\mathbf{q}|/k_F) + \ \ldots \ . \tag{30}$$

Unfortunately, the complete $\mathbf{q}$-dependence of $\Pi^{(1)}_{00}$ is rather difficult to evaluate (compare [42-45]), prohibiting an immediate verification of (30).

In this situation it seems legitimate to take a pragmatic point of view and construct a semiempirical RGGA [21], following the approach of Becke [8]. In the present context this means

(i) choose a suitable $g(\xi)$ and keep it fixed in the following. (Both the B88 [8] and the PW91 [10] GGA have been considered: explicit results are given for the latter)

¶(28) is not the most general form of the RGGA, in that only one single relativistic correction factor is used for the complete gradient contribution, involving arbitrary powers of ξ. However, in the relativistic high-density regime $g(\xi)$ is usually dominated by the lowest order in ξ (apart from the physically irrelevant immediate vicinity of the nucleus).

‖The Kirznits commutator expansion is a real space version of the GEA which, in the case of $E_x[n]$, requires a regularization scheme in order to deal with the long range of the Coulomb interaction.

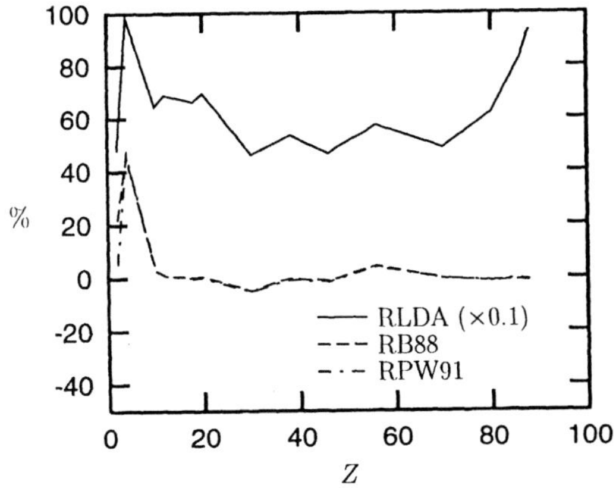

Figure 1. Relativistic contribution ΔE_x to the x-only energy: Percentage deviation of selfconsistent RLDA and RGGA results from ROPM-data for neutral atoms. The RB88 and RPW91 results are virtually indistinguishable on this scale.

(ii) make a sufficiently flexible ansatz for $\Phi_{x,2}(\beta)$, using the fact that due to time reversal invariance $\Phi_{x,2}$ must be an even function of β (here a [2/2]-Padé approximant has been used),

$$\Phi_{x,2}(\beta) = \frac{1 + a_1\beta^2 + a_2\beta^4}{1 + b_1\beta^2 + b_2\beta^4} \; . \tag{31}$$

Note that the ultrarelativistic limit of very large β is not relevant in practice as even for the heaviest atoms $\beta \leq 10$.

(iii) fit the parameters in the ansatz to the exact relativistic correction

$$\Delta E_x = E_x^R - E_x^{NR} \tag{32}$$

in the x-only energies of closed subshell atoms, obtained from ROPM calculations (in the case of the PW91-GGA this leads to $a_1 = 2.2156$, $a_2 = 0.66967$, $b_1 = 1.3267$, $b_2 = 0.79420$ for the longitudinal part of the functional and $a_1 = 3.5122$, $a_2 = 0.62525$, $b_1 = 1.3313$, $b_2 = 0.10066$ for its transverse part).

The longitudinal and complete relativistic x-only ground state energies obtained with this RGGA are listed in Tables 1,2 in comparison with the ROPM and RLDA values. In both cases the RGGA reduces the error of the RLDA with respect to the ROPM standard by roughly two orders of magnitude. While the improvement is smaller for more subtle quantities than total ground state energies [21], it always remains substantial. Also given in Table 2 are the x-only energies obtained with the nonrelativistic PW91-GGA used in the relativistic KS-equations, demonstrating the necessity to introduce relativistic correction factors: In fact, in the relativistic regime of heavy atoms the nonrelativistic PW91-GGA overbinds as much as the RLDA underbinds.

The errors in the x-only ground state energies are directly related to the accuracy of the RGGA exchange energies. In Fig. 1 we plot the percentage deviation** of the

**Neutral Mercury may serve as an example for the absolute size of ΔE_x: Its exact nonrelativistic E_x^{NR} is -345.24 *hartree* compared with a ΔE_x of $+2.61$ *hartree* [34]. Note, however, that the latter value results from a subtle cancellation of the relativistic contribution to E_x^L and E_x^T (for Hg $E_x^L - E_x^{NR} = -19.44$ *hartree* and $E_x^T = 22.05$ *hartree*). As these two components have a different physical origin their size is more characteristic of the overall importance of relativistic corrections.

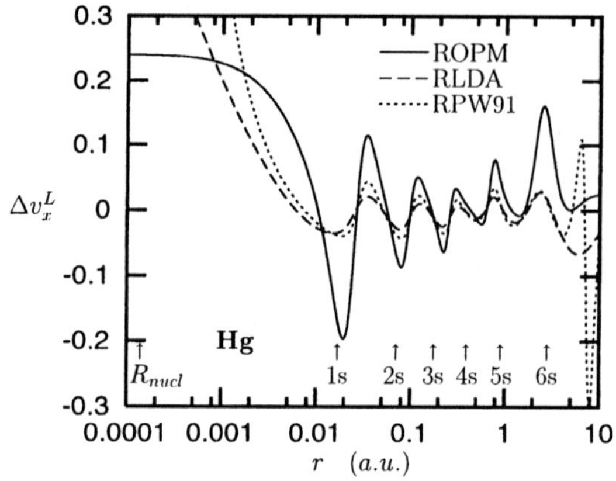

Figure 2. Relativistic contribution to the longitudinal x-only potential: Δv_x^L from selfconsistent ROPM, RLDA and RGGA calculations for neutral Hg.

RLDA and two RGGAs for the total relativistic correction ΔE_x, Eq.(32). As is obvious from Fig. 1 the RGGA reproduces ΔE_x as accurately as the total exchange energies, while the RLDA misrepresents ΔE_x by a factor of 5-10 (note the reduced scale for the RLDA). Fig. 1 also demonstrates that the RGGA concept is independent of its nonrelativistic basis, as the two GGAs considered give essentially identical values for ΔE_x.

The availability of the exact x-only potential via ROPM calculations also allows one to analyze the RGGA on a more microscopic level. In Fig. 2 we show the percentage relativistic contribution to the longitudinal x-only potential,

$$\Delta v_x^L(r) = \left[v_x^L(r) - v_x^{NR}(r)\right]/v_x^{NROPM}(r) \ , \tag{33}$$

for both the RLDA and the PW91-RGGA in comparison with the OPM standard, using neutral Hg as a prototype for all atoms. A number of features are obvious from Fig. 2:

- Δv_x^L reflects the contraction of s- and p-shells when going from a nonrelativistic to a relativistic treatment.

- Δv_x^L is of the order of 5-10% even for the outermost shells. This is the microscopic analogue of the shift of the $6s$-eigenvalue from the nonrelativistic OPM value of -0.262 *hartree* to the ROPM result of -0.329 *hartree*.

- Both approximate functionals are 'in phase' with the ROPM oscillations, but the amplitude of the RGGA oscillations is larger. However, the improvement that the RGGA offers for Δv_x^L is by far not as impressive as that for ΔE_x.

- The RLDA and the RGGA both deviate substantially from the ROPM for very small r. However, the immediate vicinity of the nucleus is physically irrelevant as both approximations give x-only potentials which in this regime are small compared with the nuclear potential.

In Fig. 3 we have split Δv_x^L into the contribution resulting from the difference between the selfconsistent relativistic density n^R and its nonrelativistic counterpart n^{NR} for fixed (nonrelativistic) functional, $\Delta v_x^L(dens) = (v_x^{NR}[n^R] - v_x^{NR}[n^{NR}])/v_x^{NROPM}$, and the contribution resulting from the relativistic modification of the functional for fixed density, $\Delta v_x^L(fctl) = (v_x^L[n^R] - v_x^{NR}[n^R])/v_x^{NROPM}$. The latter quantity directly

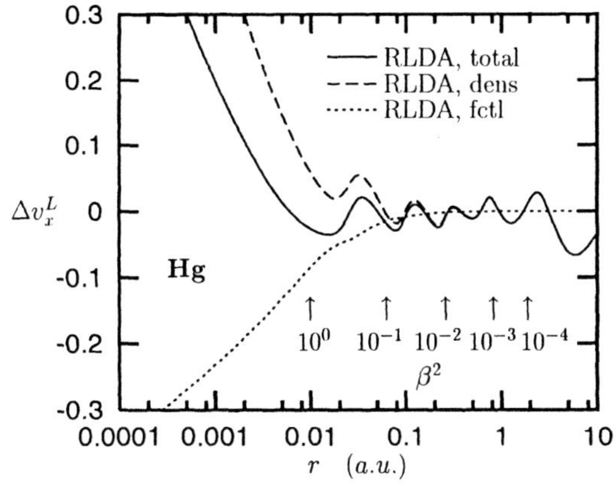

Figure 3. Decomposition of Δv_x^L into density (dens) and functional (fctl) contribution: RLDA-results for neutral Hg.

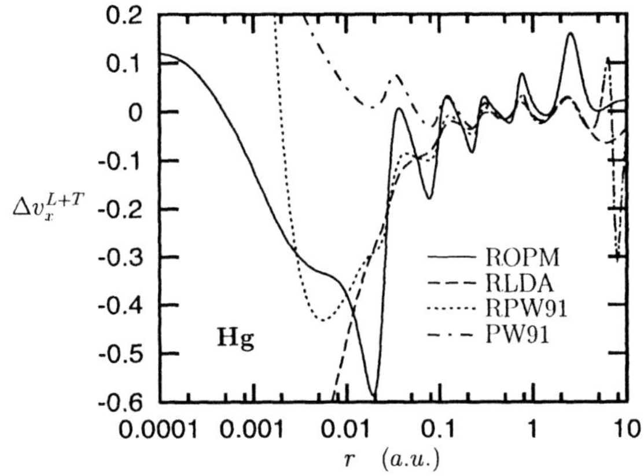

Figure 4. Complete relativistic contribution to the x-only potential: Δv_x^{L+T} from selfconsistent ROPM, RLDA, RGGA and GGA calculations for neutral Hg.

illustrates the r-regime in which the relativistic correction factors deviate from 1. $\Delta v_x^L(dens)$, on the other hand, shows that the oscillatory structure in Δv_x^L is also present if a nonrelativistic xc-functional is used in the relativistic KS-equations, i.e. that the oscillations are completely due to the relativistic treatment of $T_s[n]$, which dominates the difference between n^R and n^{NR}.

Finally, Fig. 4 displays the complete relativistic contribution to the x-only potential, $\Delta v_x^{L+T} = (v_x^L + v_x^T - v_x^{NR})/v_x^{NROPM}$. As is obvious from a comparison with Fig. 2, v_x^T contributes substantially for the innermost shells. In fact, v_x^T and the relativistic contribution to v_x^L do not cancel to the extent observed for the x-only energy, with v_x^T dominating over $v_x^L - v_x^{NR}$. Moreover, while the RGGA roughly mimics the ROPM in the relevant regime, the nonrelativistic GGA fails completely. The same picture emerges from the corresponding eigenvalues listed in Table 3. However, more important than the eigenvalues of the innermost shells are those for the outermost orbitals. In this respect one finds that the relativistic corrections to $E_x[n]$ shift e.g. the $5p$-eigenvalues of Hg by 12 *mhartree* compared with 38 *mhartree* resulting from the inclusion of the nonrelativistic correlation potential. Only the $5d$- and $6s$-orbitals remain unaffected. Moreover, in contrast to the correlation potential the relativistic corrections to $E_x[n]$

Table 3. Single particle energies ($-\epsilon_{nlj}$) for neutral Mercury from ROPM and RLDA calculations in comparison with PW91 results, based on its relativistic x-only (R-x) as well as its nonrelativistic x-only (NR-x) and xc (NR-(x+c)) form (in *hartree* [34]).

Level	OPM	LDA	PW91		
	R-x	R-x	R-x	NR-x	NR-(x+c)
1S1/2	3036.871	3027.914	3036.466	3050.287	3050.287
2S1/2	538.444	536.824	538.052	540.181	540.209
2P1/2	516.198	515.615	516.110	518.473	518.498
2P3/2	445.422	445.025	445.285	446.880	446.907
3S1/2	127.956	127.397	127.704	128.137	128.180
3P1/2	117.994	117.736	117.861	118.315	118.357
3P3/2	102.302	102.084	102.155	102.459	102.500
3D3/2	86.069	85.934	85.962	86.123	86.163
3D5/2	82.692	82.559	82.585	82.725	82.765
4S1/2	28.361	27.931	28.037	28.130	28.179
4P1/2	24.090	23.770	23.820	23.912	23.959
4P3/2	20.321	19.980	20.025	20.082	20.129
4D3/2	13.397	13.133	13.152	13.170	13.215
4D5/2	12.689	12.423	12.442	12.456	12.501
4F5/2	3.766	3.571	3.574	3.561	3.604
4F7/2	3.613	3.417	3.420	3.407	3.450
5S1/2	4.394	4.267	4.281	4.295	4.333
5P1/2	3.004	2.885	2.889	2.901	2.939
5P3/2	2.360	2.213	2.221	2.227	2.267
5D3/2	0.507	0.364	0.369	0.369	0.406
5D5/2	0.440	0.297	0.302	0.301	0.339
6S1/2	0.330	0.220	0.224	0.225	0.253

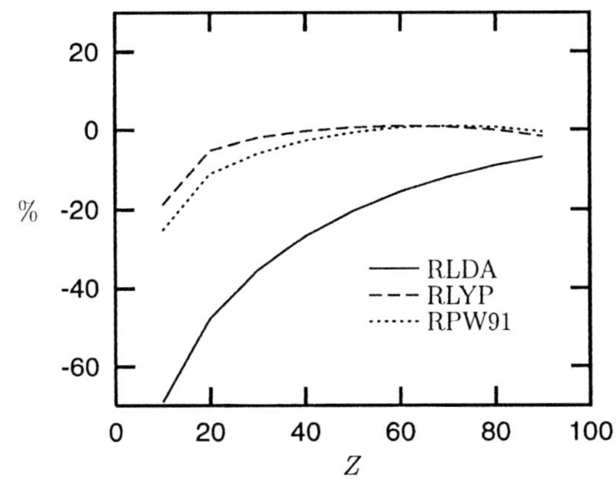

Figure 5. Relativistic correction ΔE_c to the correlation energy: Percentage deviation of selfconsistent RLDA- and RGGA-results from MBPT2-data for Ne isoelectronic series.

do not shift all eigenvalues into the same direction: While f-orbitals are stabilized, s-, p- and most d-eigenvalues become less attractive.

Given the semiempirical approach to the x-only RGGA and the involved structure of the relevant diagrammatic contributions to $\Pi_{\mu\nu}$, a first principles derivation of an RGGA for the correlation energy seems out of question. In fact, even the RLDA for $E_c[n]$ is not completely known: In contrast to the RPA diagrams [17], the second order exchange graphs, a contribution to E_c required for the relativistic regime [16], have not been evaluated beyond the ultrarelativistic limit [46]. In addition, some nonrelativistic GGAs for $E_c[n]$ [9] are not even based on the LDA. Fortunately, the relativistic corrections to E_c are quantitatively much smaller than their x-only counterparts [19], so that a more crude treatment might be sufficient. We have thus introduced only one overall correction factor for the complete correlation GGA,

$$E_c^{RGGA}[n] = \int d^3r \; e_c^{GGA}(n, (\nabla n)^2, \dots) \; \Phi_c(\beta) \; , \tag{34}$$

using a fixed nonrelativistic form for e_c^{GGA} (both the LYP [9] and the PW91 [10] GGA have been considered). The parameters of the [2/2]-Padé ansatz for $\Phi_c(\beta)$ (analogous to (31)) have been fitted to the DCB second order perturbation theory (MBPT2) correlation energies for the Ne isoelectronic series [5], which seems to be the most systematic set of quantum chemical E_c available to date[††] (Note that on the present level of sophistication the small difference between the exact RDFT correlation energy and the corresponding quantum chemical energy, originating from the KS treatment of the kinetic energy, is irrelevant). In addition, $\Phi_c(\beta)$ has been fitted to the relativistic correction to E_c,

$$\Delta E_c = E_c^R - E_c^{NR} \; , \tag{35}$$

rather than the total E_c^R, in order to suppress errors in the individual values of E_c^R and E_c^{NR} as much as possible (for PW91 the resulting parameters are $a_1 = 1.9407$, $a_2 = 0.14435$, $b_1 = 0.28142$ and $b_2 = 0.004723$ — the longitudinal and the transverse part of the functional have not been fitted individually).

[††] Only selected systems have been considered within the framework of relativistic CI calculations (see e.g. [4]). For a fit of $\Phi_c(\beta)$, however, a series of E_c covering the complete Z-range is required.

The success of this very simple approach can be seen from Fig. 5, where ΔE_c is plotted for the Ne isoelectronic series[‡‡]. Again both RGGAs give very similar results for ΔE_c, in spite of their rather different nonrelativistic form. Compared with the RLDA the RGGAs improve the accuracy of ΔE_c by more than one order of magnitude. The situation is not as clear for the few quantum chemical results for ΔE_c, which have not been included in the fitting procedure. E.g. for neutral Xe the RGGA gives a ΔE_c of 145 *mhartree*, compared with the RLDA result of 80 *mhartree*, the DCB-MBPT2 energy of 146 *mhartree* [5] and the DCB-Coupled-Cluster value of 103 *mhartree* [47]. While the RGGA agrees very well with the quantum chemical E_c on the MBPT2 level, to which it has been fitted, the discrepancy between the MBPT2 and the Coupled-Cluster E_c is quite substantial, illustrating the uncertainty in the reference data used to construct Φ_c. Clearly, as soon as more accurate systematic data for ΔE_c become available, the fit should be adjusted.

ACKNOWLEDGMENTS

Financial support by the Deutsche Forschungsgemeinschaft (project Dr 113/20-1,2) is gratefully acknowledged.

REFERENCES

1. *The Effects of Relativity in Atoms, Molecules and the Solid State*, edited by S. Wilson, I. P. Grant and B. L. Gyorffy (Plenum, New York, 1991).
2. S. A. Blundell, W. R. Johnson, Z. W. Liu, and J. Sapirstein, Phys. Rev. A **37**, 3768 (1989).
3. E. Ilyabaev and U. Kaldor, Chem. Phys. Lett. **194**, 95 (1992).
4. E. Eliav, U. Kaldor, and Y. Ishikawa, Phys. Rev. A **49**, 1724 (1994).
5. Y. Ishikawa and K. Koc, Phys. Rev. A **50**, 4733 (1994).
6. U. Kaldor and B. A. Hess, Chem. Phys. Lett. **230**, 1 (1994).
7. A. Pizlo, G. Jansen, B. A. Hess, and W. von Niessen, J. Chem. Phys. **98**, 3945 (1993).
8. A. D. Becke, Phys. Rev. A **38**, 3098 (1988).
9. C. Lee, W. Yang, and R. G. Parr, Phys. Rev. B **37**, 785 (1988).
10. J. P. Perdew, in *Electronic Structure of Solids 1991*, edited by P. Ziesche and H. Eschrig (Akademie Verlag, Berlin, 1991).
11. A. D. Becke, J. Chem. Phys. **96**, 2155 (1992).
12. B. G. Johnson, P. M. W. Gill, and J. A. Pople, J. Chem. Phys. **98**, 5612 (1993).
13. A. K. Rajagopal and J. Callaway, Phys. Rev. B **7**, 1912 (1973).
14. A. K. Rajagopal, J. Phys. C **11**, L943 (1978).
15. A. H. MacDonald and S. H. Vosko, J. Phys. C **12**, 2977 (1979).
16. E. Engel and R. M. Dreizler, in: *Density Functional Theory II*, edited by R. F. Nalewajski, *Topics in Current Chemistry*, Vol. 181 (Springer, Berlin, 1996), p.1.
17. M. V. Ramana and A. K. Rajagopal, Phys. Rev. A **24**, 1689 (1981).
18. M. V. Ramana, A. K. Rajagopal, and W. R. Johnson, Phys. Rev. A **25**, 96 (1982).
19. E. Engel, S. Keller, A. Facco Bonetti, H. Müller, and R. M. Dreizler, Phys. Rev. A **52**, 2750 (1995).
20. S. Varga, W.-D. Sepp, B. Fricke, E. Engel and R. M. Dreizler, unpublished.
21. E. Engel, S. Keller, and R. M. Dreizler, Phys. Rev. A **53**, 1367 (1996).
22. R. T. Sharp and G. K. Horton, Phys. Rev. **90**, 317 (1953).
23. J. D. Talman and W. F. Shadwick, Phys. Rev. A **14**, 36 (1976).
24. B. A. Shadwick, J. D. Talman, and M. R. Norman, Comp. Phys. Commun. **54**, 95 (1989).
25. E. Engel, S. Keller, and R. M. Dreizler, unpublished.
26. V. Sahni, J. Gruenebaum, and J. P. Perdew, Phys. Rev. B **26**, 4371 (1982).
27. D. C. Langreth and M. J. Mehl, Phys. Rev. B **28**, 1809 (1983).

[‡‡]The nonrelativistic E_c^{NR} of -5.086 *hartree* (MBPT2) for neutral Mercury and the corresponding ΔE_c of -0.486 *hartree* may serve as an indication of the absolute size of the relativistic corrections.

28. L. J. Sham, Phys. Rev. B **32**, 3876 (1985).
29. I. P. Grant, Proc. Roy. Soc. (London) **A262**, 555 (1961).
30. Y.-K. Kim, Phys. Rev. **154**, 17 (1967).
31. J. B. Mann and W. R. Johnson, Phys. Rev. A **4**, 41 (1971).
32. R. Colle and O. Salvetti, Theoret. Chim. Acta (Berl.) **37**, 329 (1975).
33. T. Grabo and E. K. U. Gross, Chem. Phys. Lett. **240**, 141 (1995).
34. In all our calculations the nuclei were represented by uniformly charged spheres with nuclear radii given by $R_{nucl} = 1.0793\ A^{1/3} + 0.73587\ fm$, A being the atomic mass (weighted by isotopic abundances) taken from Table III.7 of K. Hisaka *et al* (Particle Data Group), Phys. Rev. D **45**, Number 11, Part II (1992), unless explicitly stated otherwise. The speed of light had been set to $c = 137.0359895$.
35. M. Levy and J. P. Perdew, Phys. Rev. A **32**, 2010 (1985).
36. K. G. Dyall, I. P. Grant, C. T. Johnson, F. A. Parpia, and E. P. Plummer, Comp. Phys. Commun. **55**, 425 (1989).
37. I. Lindgren, H. Persson, S. Salomonson, P. Sunnergren, to be published in *Proceedings of the 15th International Conference on Atomic Physics, Amsterdam, 5-9 August 1996* (World Scientific, Singapore).
38. J. B. Krieger, Y. Li and G. J. Iafrate, Phys. Lett. **146** A, 256 (1990).
39. T. Kreibich, E. K. U. Gross, and E. Engel, Phys. Rev. A **56**, in press.
40. E. Engel, diploma thesis (1983).
41. E. K. U. Gross and R. M. Dreizler, Z. Phys. A **302**, 103 (1981).
42. P. R. Antoniewicz and L. Kleinman, Phys. Rev. B **31**, 6779 (1985).
43. L. Kleinman and S. Lee, Phys. Rev. B **37**, 4634 (1988).
44. J. A. Chevary and S. H. Vosko, Phys. Rev. B **42**, 5320 (1990).
45. E. Engel and S. H. Vosko, Phys. Rev. B **42**, 4940 (1990); Phys. Rev. B **44**, 1446(E) (1991).
46. B. A. Freedman and L. D. McLerran, Phys. Rev. D **16**, 1130, 1147, 1169 (1977).
47. E. Eliav, U. Kaldor, and Y. Ishikawa, Int. J. Quantum Chem. **28**S, 205 (1994).

Part III

INVITED CHAPTERS ON EXCITED STATES, TIME-DEPENDENT DFT AND THE VAN DER WAALS INTERACTION

ARE UNOCCUPIED KOHN-SHAM EIGENVALUES RELATED TO EXCITATION ENERGIES?

C. J. Umrigar,[1] A. Savin,[2] and Xavier Gonze[3]

[1]Cornell Theory Center and Laboratory of Atomic and Solid State Physics
Cornell University, Ithaca, New York 14853

[2]Laboratoire de Chimie Théorique (CNRS)
Université de Paris

[3] Unité P.C.P.M.
Université Catholique de Louvain,
B-1348 Louvain-la-Neuve, Belgium

INTRODUCTION

In the Kohn-Sham density functional method [1,2,3,4], the true interacting-electron system is replaced by a system of non-interacting electrons in an effective potential, $v_{\rm eff}$, defined by the requirement that the density of the non-interacting electrons equals the true density. The single particle orbitals and their eigenenergies were originally introduced as a mathematical artifact in order to achieve a good approximation to the kinetic energy, leaving only a relatively small term, the exchange-correlation energy $E_{\rm xc}$, to be approximated in practical implementations of the theory. It was later shown [5] that the energy of the highest occupied orbital is in fact the negative of the ionization energy. However, most approximate functionals (such as the commonly used local density approximation [2,3,4]) yield poor approximations to it. The energies of the other occupied orbitals and of the unoccupied orbitals do not have a rigorous correspondence to excitation energies. Nevertheless, it is common practice to compare eigenvalue differences to optical spectra of molecules and solids. Since these comparisons are made using Kohn-Sham eigenvalues obtained from approximate exchange-correlation functionals, it is not clear how much of the discrepancy between theory and experiment would persist if the true Kohn-Sham eigenvalues were to be used. In this paper we show that there is a surprising degree of agreement between the exact ground-state Kohn-Sham eigenvalue differences and excitation energies, for excitations from the highest occupied orbital to the unoccupied orbitals.

Electronic Density Functional Theory: *Recent Progress and New Directions*
Edited by Dobson *et al.*, Plenum Press, New York, 1998

Table 1. Excitation energies of He in Hartree atomic units. The theoretical energies of Drake [12] and the eigenvalue differences are for infinite nuclear mass and neglect relativity. The experimental energies are from Ref. [15].

Transition	Final State	Experiment	Drake	Kohn-Sham eigenvalue differences
$1s \to 2s$	$2\,^3$S	0.72833	0.72850	0.7460
	$2\,^1$S	0.75759	0.75775	
$1s \to 2p$	$1\,^3$P	0.77039	0.77056	0.7772
	$1\,^1$P	0.77972	0.77988	
$1s \to 3s$	$3\,^3$S	0.83486	0.83504	0.8392
	$3\,^1$S	0.84228	0.84245	
$1s \to 3p$	$2\,^3$P	0.84547	0.84564	0.8476
	$2\,^1$P	0.84841	0.84858	
$1s \to 3d$	$1\,^3$D	0.84792	0.84809	0.8481
	$1\,^1$D	0.84793	0.84809	
$1s \to 4s$	$4\,^3$S	0.86704	0.86721	0.8688
	$4\,^1$S	0.86997	0.87014	

METHOD

The first step in the calculation is the evaluation of accurate densities. In the case of the 2-electron systems they were obtained [6] by numerically integrating extremely accurate wave functions [7] that yield energies that are correct to 14 and 16 significant digits respectively for He and Li^+. In the case of Be, the density was obtained [8] by a combination of variational and diffusion quantum Monte Carlo, using a wave function that recovers 99.2% of the correlation energy in variational Monte Carlo and 99.9% of the correlation energy in diffusion Monte Carlo. Efficient implementations [9,10] of the variational and diffusion Monte Carlo methods are employed to rapidly sample the many-dimensional space of the wave functions.

The true Kohn-Sham potential is then obtained from the density. For the two-electron systems, it is obtained directly from the single-particle Kohn-Sham equation for the sole occupied orbital. For systems with more electrons, the potential is expanded in a set of functions and the expansion parameters varied such that the potential yields the true density [8,11]. All the Kohn-Sham eigenvalues can then be obtained by solving the Kohn-Sham equation

$$-\frac{\nabla^2\psi}{2} + v_{\text{eff}}\psi = \epsilon\psi. \tag{1}$$

RESULTS

In Tables 1,2 and 3, we show the excitation energies of He, Li^+ and Be respectively, obtained from differences of Kohn-Sham eigenvalues, and compare them to the corre-

Table 2. Excitation energies of Li^+ in Hartree atomic units. The theoretical energies of Drake [12] and the eigenvalue differences are for infinite nuclear mass and neglect relativity. The experimental energies are from Ref. [15].

Transition	Final State	Experiment	Drake	Kohn-Sham eigenvalue differences
$1s \to 2s$	$2\,{}^3S$	2.1690	2.1692	2.2082
	$2\,{}^1S$	2.2330	2.2390	
$1s \to 2p$	$1\,{}^3P_o$	2.2521	2.2522	2.2733
	$1\,{}^1P_o$	2.2865	2.2866	
$1s \to 3s$	$3\,{}^3S$	2.5277	2.5278	2.5375
	$3\,{}^1S$	2.5460	2.5462	
$1s \to 3p$	$2\,{}^3P_o$	2.5493	2.5495	2.5554
	$2\,{}^1P_o$	2.5595	2.5597	
$1s \to 3d$	$1\,{}^3D$	2.5572	2.5574	2.5576
	$1\,{}^1D$	2.5574	2.5575	
$1s \to 4s$	$4\,{}^3S$	2.6426	2.6428	2.6466
	$4\,{}^1S$	2.6499	2.6501	

Table 3. Excitation energies of Be in Hartree atomic units. The eigenvalue differences are for infinite nuclear mass and neglect relativity. The experimental energies are from Ref. [15].

Transition	Final State	Experiment	Kohn-Sham eigenvalue differences
$2s \to 2p$	$1\,^3$P	0.100153	0.1327
	$1\,^1$P	0.193941	
$2s \to 3s$	$2\,^3$S	0.237304	0.2444
	$2\,^1$S	0.249127	
$2s \to 3p$	$2\,^3$P	0.267877	0.2694
	$2\,^1$P	0.274233	
$2s \to 3d$	$1\,^3$D	0.282744	0.2833
	$1\,^1$D	0.293556	
$2s \to 4s$	$3\,^3$S	0.293921	0.2959
	$3\,^1$S	0.297279	
$2s \to 4p$	$3\,^3$P	0.300487	0.3046
	$3\,^1$P	0.306314	
$2s \to 4d$	$2\,^3$D	0.309577	0.3098
	$2\,^1$D	0.313390	
$2s \to 5s$	$4\,^3$S	0.314429	0.3153
	$4\,^1$S	0.315855	

sponding experimental energies. Of course, the single-electron Kohn-Sham energies do not distinguish between the energies within a multiplet. The energies obtained from the eigenvalue differences, lie between the the experimental singlet and triplet excitation energies with the exception of only the (1s → 3d) $1D$ excitation for He and Li^+. The calculated energies are for an infinite mass nucleus and do not include relativity, whereas the experimental numbers are, of course, for finite mass nuclei and are relativistic. In the case of the two-electron systems it is possible to calculate exceedingly accurate infinite nuclear mass non-relativistic total energies [12,13]. The excitation energies obtained from these calculations are also shown for He in Table 1. We observe that now even the (1s → 3d) $1D$ excitation energy of He obtained from the eigenvalue differences agrees with the true calculated excitation energy to the number of digits shown. To the best of our knowledge, this remarkable agreement has not been noticed before for any real system, though it has been noticed for a model semiconductor [14].

It should be emphasised that the agreement exists only for accurate Kohn-Sham eigenvalues. Those obtained from popular approximate density functionals, such as the local density approximation (LDA) and various generalized gradient approximations (GGAs) are nowhere close to being sufficiently accurate. In fact, it is not even possible to make such a comparison with LDA or GGA orbitals because for neutral atoms the LDA or GGA potentials bind few, if any, unoccupied orbitals. This, of course, is due to the fact that the exchange-correlation potential of these approximate functionals decays considerably faster than $-1/r$.

DISCUSSION

It should come as no surprise that the higher excitations are accurately given by the eigenvalue differences since that follows from the following two well-known facts. First, the highest occupied eigenvalue is exact, *i.e.* it is minus the ionization energy [5]. Second, the higher unoccupied eigenvalues approach a Rydberg series since the exchange-correlation potential approaches $-1/r$ at large distances. Hence, it is to be expected that the higher excitations will be progressively better approximated by the eigenvalue differences. However, this argument does not explain the agreement observed for the lower lying states. Second, the agreement for the higher lying states is better than can be explained by the above argument since the manner in which the eigenvalues approach the Rydberg limit is also in agreement with the excitation energies. The deviation of the eigenvalues from a Rydberg series is given by the quantum defect.

$$\delta_n = n - \sqrt{\frac{-1}{2\epsilon_n}}, \tag{2}$$

where ϵ_n is the nth eigenvalue. In the left panel of Fig. 1 we compare the S and P quantum defects of He from the Kohn-Sham eigenvalues to those obtained for the $^3S,^1S,^3P$ and ^1P states obtained from highly accurate calculations [12]. The right panel of Fig. 1 is a similar plot for Be, except that the comparison is made to experiment [15] rather than to accurate theoretical calculations. It is apparent that both for He and for Be the quantum defects obtained from the Kohn-Sham eigenvalues are in rather good agreement with the average of the true singlet and triplet quantum defects. A similar agreement has previously been observed for Neon [16]. Since the quantum defects describe the deviation from a Rydberg series, this finding cannot be explained in terms of the well-known fact that the true Kohn-Sham potential decays at large distances as $-(Z-N+1)/r$, where Z is the nuclear charge and N is the number of electrons. Clearly some further explanation is needed.

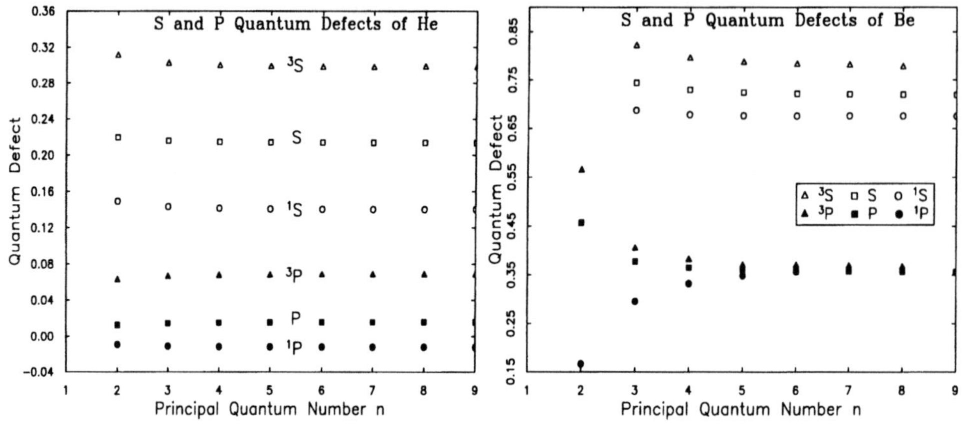

Figure 1. Comparison of the S and P quantum defects of He and Be obtained from the Kohn-Sham eigenvalues with accurate $^3S, ^1S, ^3P$ and ^1P quantum defects. For He, the latter are obtained from accurate theoretical calculations [12] while for Be they are obtained from experiment [15].

We now argue that the explanation lies in the fact that the Kohn-Sham orbitals and quasiparticle amplitudes (defined below) satisfy the same equation not only up to order $(1/r)$ but up to order $(1/r^4)$ inclusive. Consider an N-electron system in its ith state Ψ_i^N, obtained by exciting an electron from the highest occupied level of the ground state Ψ_0^N. If we keep the 1st electron fixed, we have an $(N-1)$-electron function which we can expand in any complete basis set, with the expansion coefficients f being functions of the position of the 1st electron. Choosing the complete set to be the eigenstates of the $N-1$ electron system, we have

$$\Psi_i^N(r_1, r_2, ..., r_N) = \sum_{j=0}^{\infty} f_{ij}(r_1)\Psi_j^{N-1}(r_2, ..., r_N). \tag{3}$$

The expression on the right hand side is antisymmetric because the left hand side is. We will refer to the f_{ij} as quasiparticle amplitudes but we note that they are not the usual Dyson quasiparticle amplitudes.

At large distances, the quasiparticle amplitudes f_{i0} obey the equation

$$\Big(-\frac{\nabla^2}{2} - \frac{(Z-N+1)}{r} + \frac{Q}{r^3} - \frac{\alpha}{2r^4} + \mathcal{O}(\frac{1}{r^5})\Big) f_{i0} = (E_i^N - E_0^{N-1}) f_{i0} \ , \tag{4}$$

where Q and α are the quadrupole moment and the dipole polarizability of the ground state Ψ_0^{N-1} of the $(N-1)$-electron system. Further it can be shown that

$$f_{ij} \sim f_{i0}/r^2 \quad \text{for} \quad j \neq 0. \tag{5}$$

Equations (4) and (5) are generalizations of equations obtained by Almbladh and von Barth [18] for the ground $(i=0)$ state of the N-electron system.

At large distances, the Kohn-Sham orbitals ϕ_i satisfy the equation

$$\Big(-\frac{\nabla^2}{2} - \frac{(Z-N)}{r} + v_{xc}(r) + \mathcal{O}(\exp)\Big)\phi_i = \epsilon_i \phi_i \tag{6}$$

Now consider the densities resulting from the many-body quasiparticle equations and the Kohn-Sham equation:

$$\rho_i^{\rm MB} = \sum_{j=0}^{\infty} f_{ij}^2 \to f_{i0}^2 \left(1 + \mathcal{O}(\frac{1}{r^4})\right) \quad \text{as} \quad r \to \infty, \tag{7}$$

$$\rho_0^{\rm KS} = \sum_{j=1}^{N} \phi_j^2 \to \phi_N^2 \left(1 + \mathcal{O}(\exp)\right) \quad \text{as} \quad r \to \infty, \tag{8}$$

Equating the ground state many-body and Kohn-Sham densities we obtain:

$$\rho_0^{\rm KS} = \rho_0^{\rm MB} \Rightarrow \begin{cases} \epsilon_N = (E_0^N - E_0^{N-1}) \\ \phi_N = f_{00}\left(1 + \mathcal{O}(\frac{1}{r^4})\right) \\ V_{xc} = -\frac{1}{r} + \frac{Q}{r^3} - \frac{\alpha}{2r^4} + \mathcal{O}(\frac{1}{r^5}). \end{cases} \tag{9}$$

The first of these follows from the fact that the exponential decay is governed only by the right hand side of Eqs. (4) and (6) and is the well-known result that the highest occupied KS eigenvalue yields the ionization energy [5]. The second of these follows from Eqs. (7) and (8). The third follows from relating the coefficients of a $1/r$ expansion of the density or equivalently the highest KS orbital to a $1/r$ expansion of $v_{\rm xc}$. If the KS and the quasiparticle orbitals are equal to $\mathcal{O}(\frac{1}{r^n})$, then it is easy to show that the potentials must be equal to $\mathcal{O}(\frac{1}{r^{n+1}})$

Of course all the KS orbitals satisfy the same equation, while the quasiparticle amplitudes all satisfy different equations, but the quasiparticle equations differ from each other and from the KS equation only at order $\mathcal{O}(\frac{1}{r^5})$.

The question now arises: Is the fact the the highest occupied eigenvalue is correct plus the fact that the Kohn-Sham and the quasiparticle potentials agree to order $1/r^4$ sufficient to pin down the unoccupied eigenvalues with sufficient accuracy? We conjecture the answer is yes, at least for sufficiently smooth potentials. In order to provide evidence for our conjecture we constructed two model potentials both of which yield the true occupied Kohn-Sham eigenvalue for Helium. The first of these potentials agrees with the true Kohn-Sham potential to order $1/r$ while the second agrees up to order $1/r^4$. The procedure used to construct the model potentials has similarities to the standard procedure used to generate pseudopotentials but we do not impose the norm-conservation condition.

Motivated by the long-range asymptotic behavior of many-body wave functions [19], we construct these potentials by choosing the Kohn-Sham orbital to be

$$\psi \propto (r+1)^a e^{-\zeta r}\left(1 + \frac{q_1}{(r+1)} + \frac{q_2}{(r+1)^2} + \frac{q_3}{(r+1)^3}\right), \tag{10}$$

where $\zeta = \sqrt{2E_{\rm ion}}$, $a = Z_{\rm eff}/\zeta - 1$, $Z_{\rm eff} = Z - N + 1$, Z is the nuclear charge, N is the number of electrons and $E_{\rm ion}$ is the ionization energy. The Kohn-Sham potential that yields this orbital is obtained simply by solving the single-particle Schrödinger equation for the potential. The first potential, which agrees with the true Kohn-Sham potential to order $1/r$, is obtained by setting $q_1 = q_2 = q_3 = 0$. The second potential, which

Table 4. Comparison of the Kohn-Sham eigenvalues for He with those with those of two model systems that have the same occupied eigenvalue and have potentials that agree with the He potential to order $1/r$ and $1/r^4$.

Orbital	Model		Helium
	$\mathcal{O}(1/r)$	$\mathcal{O}(1/r^4)$	
1s	-0.9037	-0.9037	-0.9037
2s	-0.1748	-0.1588	-0.1577
2p	-0.1594	-0.1284	-0.1266
3s	-0.0696	-0.0648	-0.0645
3p	-0.0653	-0.0566	-0.0561
3d	-0.0621	-0.0557	-0.0556
4s	-0.0370	-0.0350	-0.0349

agrees with the true Kohn-Sham potential to order $1/r^4$, is obtained by setting

$$\begin{aligned} q_1 &= p_1 - a, \\ q_2 &= p_2 - \frac{a(a-1)}{2} + (1-a)q_1, \\ q_3 &= p_3 - \frac{a(a-1)(a-2)}{6} - \frac{a(a-1)}{2}q_1 + a(q_1 - q_2) - q_1 + 2q_2, \end{aligned}$$

where

$$\begin{aligned} p_1 &= \frac{1}{\zeta}\left(\frac{-Z_{\text{eff}}(Z_{\text{eff}}-\zeta)}{2\zeta^2} + \frac{l(l+1)}{2}\right), \\ p_2 &= \frac{1}{2\zeta}\left(Q + p_1\left(-\frac{(Z_{\text{eff}}-\zeta)(Z_{\text{eff}}-2\zeta)}{2\zeta^2}\right) + \frac{l(l+1))}{2}\right), \\ p_3 &= \frac{1}{3\zeta}\left(\frac{-\alpha}{2} + p_1 Q + p_2\left(-\frac{(Z_{\text{eff}}-2\zeta)(Z_{\text{eff}}-3\zeta)}{2\zeta^2}\right) + \frac{l(l+1))}{2}\right), \end{aligned} \tag{11}$$

where Q is the quadrupole moment and α the dipole polarizability of the residual He^+ ion.

In Table 4, we compare the eigenvalues obtained from the two model potentials with the KS eigenvalues for He. It is apparent that the eigenvalues of the second model are closer, by a factor of 20 or more, to the eigenvalues of He than are the eigenvalues of the first model. The differences of the eigenvalues of the second model and those of He are of comparable magnitude to the differences of the true He Kohn-Sham eigenvalue differences from the excitation energies. Hence, it appears that our conjectured explanation is correct. We have performed the same test for Li^+ and find very similar results. In Fig. 2 we compare the two model potentials to the true Kohn-Sham potential. The external potential has been subtracted from each of the potentials. Clearly, the three potentials differ considerably at short distances, hence precluding the possibility that the agreement in the eigenvalues is merely due to a chance agreement of the $\mathcal{O}(1/r^4)$ model potential with the true Kohn-Sham potential.

In conclusion we have discovered that excitations from the highest occupied orbital are in surprising agreement with differences of Kohn-Sham eigenvalues and we present a plausible explanation for this finding. The present results indicate that useful information about excited states could be obtained from ground-state single-particle energies provided that the approximate functional used yields single-particle energies that are sufficiently close to the true Kohn-Sham values. Unfortunately the popular LDA and

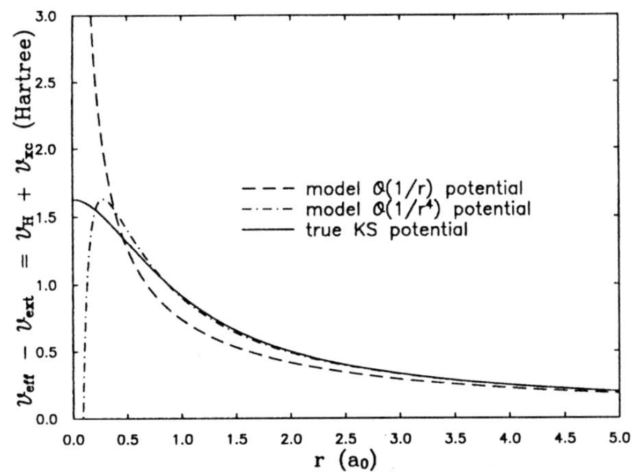

Figure 2. Comparison of true Kohn-Sham effective potential, with two model potentials. The external potential has been subtracted from each of the potentials. All three potentials have the same highest occupied eigenvalue. At large distances, the first model potential agrees with the Kohn-Sham potential to $\mathcal{O}(1/r)$ while the second model potential agrees with the Kohn-Sham potential to $\mathcal{O}(1/r^4)$. It is apparent that the three potentials are very different at short distances.

GGA functionals are nowhere near sufficiently accurate. The exact exchange-only potential, obtained from the optimized effective potential (OEP) method, does better [20] since it decays as $-1/r$ at large distances but it is much inferior to the true Kohn-Sham potential. Hence it is necessary to combine it with an accurate correlation potential but unfortunately all existing correlation functionals, that we have studied, yield potentials that bear little resemblance [11] to the true correlation potential. The Kohn-Sham eigenvalues are also the starting point for two more accurate methods for calculating excitation energies, namely the time-dependent density functional theory [21,22] and density functional perturbation theory [17,20]. Again, the success of these methods (for all excitations except those within the same principal quantum number shell as the highest occupied level) relies on having functionals that yield good approximations to the true Kohn-Sham eigenvalues.

ACKNOWLEDGMENTS

We thank John Morgan and Jonathan Baker for making available their program for calculating very accurate two-electron wave functions and Raffaele Resta, Mel Levy, Hardy Gross and Sharad Patil for useful discussions. The computations were performed on IBM RS-6000 workstations provided by the Cornell Theory Center. This work was supported by the Office of Naval Research and NATO.

REFERENCES

1. P. Hohenberg and W. Kohn, Phys. Rev. **136**, B864 (1964).
2. W. Kohn and L. J. Sham, Phys Rev. **140**, A1133 (1965).
3. R. G. Parr and W. Yang, *Density Functional Theory of Atoms and Molecules* (Oxford University Press, 1989).
4. R. M. Dreizler and E. K. U. Gross, *Density Functional Theory* (Springer-Verlag, Berlin, 1990).
5. J. P. Perdew, R. G. Parr, M. Levy and J. L. Balduz, Phys. Rev. Lett. **49**, 1691 (1982); M. Levy, J. P. Perdew and V. Sahni Phys. Rev. B **30**, 2745 (1994); U. von Barth in *Electronic Structure*

of Complex Systems, edited by P. Phariseau and W. M. Temmerman, NATO ASI Series B, vol. 113 (1984).
6. C. J. Umrigar and X. Gonze, Phys. Rev. A **50**, 3827 (1994).
7. The wave function used is a minor modification of that in D. E. Freund, B. D. Huxtable and J. D. Morgan, Phys. Rev. A **29**, 980 (1984). The precise form of the wave function used is described in Ref. [6].
8. C. J. Umrigar and X. Gonze, in *High Performance Computing and its Application to the Physical Sciences*, proceedings of the Mardi Gras '93 Conference, edited by D.-A. Browne *et al* (World Scientific, Singapore, 1993); and unpublished.
9. C. J. Umrigar, Phys. Rev. Lett. **71**, 408 (1993).
10. C. J. Umrigar, P. Nightingale and K. J. Runge, J. Chem. Phys. **99**, 2865 (1993).
11. C. Filippi, C. J. Umrigar and X. Gonze in *Theoretical and Computational Chemistry: Recent Developments in Density Functional Theory*, edited by J. Seminario (Elsevier, 1996).
12. G. W. F. Drake in *Casimir Forces: Theory and Recent Experiments on Atomic Systems*, edited by F. S. Levin and D. A. Micha, (Plenum, New York 1993); G. W. F. Drake and Zong-Chao Yan, Chem. Phys. Lett. **229**, 486 (1994).
13. G. W. F. Drake in *Handbook of Atomic, Molecular, and Optical Physics*, edited by G. W. F. Drake (AIP Press, New York, 1996), pp. 154-171; E. Riis, A. G. Sinclair, O. Poulsen, G. W. F. Drake, W. R. C. Rowlwy and A. P. Levick, Phys. Rev. A **49**, 207 (1994); private communication.
14. O. Gunnarson and K. Schönhammer, Phys. Rev. Lett. **56**, 1968 (1986).
15. S. Bashkin and J. D. Stoner, *Atomic Energy Levels and Grotrian Diagrams*, vol. 1. (North Holland 1975, addendum 1978).
16. Abdullah Al-Sharif, R. Resta and C. J. Umrigar, unpublished.
17. A. Görling and M. Levy, Int. J. Quant. Chem, Symp. **29**, 93 (1995); Phys. Rev. A **50**, 196 (1994).
18. C. -O. Almbladh and U. von Barth, Phys. Rev. B **31**, 3231 (1985).
19. S. H. Patil, J. Phys. B **23**, 1 (1990); J. Chem. Phys. **95**, 4245 (1991).
20. C. Filippi, C. J. Umrigar and X. Gonze, unpublished.
21. Mark E. Casida in *Theoretical and Computational Chemistry: Recent Developments in Density Functional Theory*, edited by J. Seminario (Elsevier, 1996).
22. M. Petersilka and E. K. U. Gross, Int. J. Quant. Chem. **30**, 1393 (1996); M. Petersilka , U. J. Gossman and E. K. U. Gross, contribution in this volume.

TIME-DEPENDENT OPTIMIZED EFFECTIVE POTENTIAL IN THE LINEAR RESPONSE REGIME

M. Petersilka, U. J. Gossmann, and E. K. U. Gross

Institut für Theoretische Physik
Universität Würzburg
Am Hubland, D-97074 Würzburg, Germany

TIME-DEPENDENT DENSITY FUNCTIONAL THEORY

Time-dependent density functional theory of spin-polarized systems is based on the fact [1,2] that the exact time-dependent spin densities $n_\sigma(\mathbf{r}t) = \langle\Psi(t)|\hat{n}_\sigma(\mathbf{r})|\Psi(t)\rangle$ of an interacting many-particle system subject to time-dependent potentials $v_{\text{ext}\,\sigma}(\mathbf{r}t)$ can be calculated from the orbitals of an auxiliary noninteracting system, i. e.

$$n_\sigma(\mathbf{r}t) = \sum_j^{N_\sigma} |\phi_{j\sigma}(\mathbf{r}t)|^2 \quad . \tag{1}$$

The orbitals $\phi_{j\sigma}(\mathbf{r}t)$ satisfy the time-dependent Kohn-Sham equations (atomic units are used throughout)

$$i\,(\partial/\partial t)\,\phi_{j\sigma}(\mathbf{r}t) = \left(-\nabla^2/2 + v_\sigma(\mathbf{r}t)\right)\phi_{j\sigma}(\mathbf{r}t) \quad , \quad j = 1, \ldots, N_\sigma \tag{2}$$

with suitable initial conditions $\phi_{j\sigma}(\mathbf{r}t_0)$ chosen to represent the spin densities $n_{0\,\sigma}(\mathbf{r}) = \langle\Psi_0|\hat{n}_\sigma(\mathbf{r})|\Psi_0\rangle$ of the initial (many-body) state Ψ_0 at time $t = t_0$. The time-dependent local effective potential in Eq. (2) is given by the sum of the external potential, the Hartree-potential and the time-dependent exchange-correlation (xc) potential

$$v_\sigma(\mathbf{r}t) = v_{\text{ext}\,\sigma}(\mathbf{r}t) + \int d^3r'\,\frac{n(\mathbf{r}'t)}{|\mathbf{r}-\mathbf{r}'|} + v_{\text{xc}\,\sigma}(\mathbf{r}t) \quad , \tag{3}$$

with the total density $n(\mathbf{r}t) = \sum_\sigma n_\sigma(\mathbf{r}t)$. To date, most applications of the time-dependent Kohn-Sham scheme (1) – (3) fall in the linear response regime (see e.g. [3] for a recent review). In this limit, one considers electronic systems subject to external potentials of the form

$$v_{\text{ext}\,\sigma}(\mathbf{r}t) = \begin{cases} v_{0\,\sigma}(\mathbf{r}) & ; \quad t \le t_0 \\ v_{0\,\sigma}(\mathbf{r}) + v_{1\,\sigma}(\mathbf{r}t) & ; \quad t > t_0 \end{cases} \tag{4}$$

where $v_{0\,\sigma}(\mathbf{r})$ denotes the static external potential of the unperturbed system (e. g. a nuclear Coulomb potential plus a static magnetic field that couples to the electronic

Electronic Density Functional Theory: *Recent Progress and New Directions*
Edited by Dobson *et al.*, Plenum Press, New York, 1998

spin only) and $v_{1\sigma}(\mathbf{r}t)$ is a time-dependent perturbation. We assume that at times $t \leq t_0$ the system is in the ground state corresponding to $v_{0\sigma}(\mathbf{r})$. In this case, the time-dependent density is a functional of the external potential alone also at $t = t_0$ and the full 1-1 correspondence

$$\{v_{\text{ext}\,\sigma}(\mathbf{r}t)\} \overset{1-1}{\longleftrightarrow} \{n_\sigma(\mathbf{r}t)\} \tag{5}$$

is guaranteed for the interacting system [1,2]. This 1-1 mapping can be established for any particle-particle interaction, in particular also for vanishing particle-particle interaction. Consequently, the time-dependent density $n(\mathbf{r}t)$ uniquely determines the time-dependent effective potential (3) as well:

$$\{v_\sigma(\mathbf{r}t)\} \overset{1-1}{\longleftrightarrow} \{n_\sigma(\mathbf{r}t)\} \quad . \tag{6}$$

Combining (5) and (6) we see that the time-dependent external potential $v_{\text{ext}\,\sigma}(\mathbf{r}t)$ and the effective potential $v_\sigma(\mathbf{r}t)$ are also in 1-1 correspondence:

$$\{v_\sigma(\mathbf{r}t)\} \overset{1-1}{\longleftrightarrow} \{v_{\text{ext}\,\sigma}(\mathbf{r}t)\} \quad . \tag{7}$$

The above statements can be used to establish a relation between the density-density response function of the noninteracting (Kohn-Sham) system

$$\chi_{\text{s}\,\sigma\sigma'}(\mathbf{r}t, \mathbf{r}'t') = \left.\frac{\delta n_\sigma(\mathbf{r}t)}{\delta v_{\sigma'}(\mathbf{r}'t')}\right|_{v[n_0]} \tag{8}$$

and the density-density response function of the interacting electron system

$$\chi_{\sigma\sigma'}(\mathbf{r},t,\mathbf{r}',t') = \left.\frac{\delta n_\sigma(\mathbf{r},t)}{\delta v_{\text{ext}\,\sigma'}(\mathbf{r}',t')}\right|_{v_0} = \sum_\nu \int d^3x \int d\tau \left.\frac{\delta n_\sigma(\mathbf{r},t)}{\delta v_\nu(\mathbf{x},\tau)} \frac{\delta v_\nu(\mathbf{x},\tau)}{\delta v_{\text{ext}\,\sigma'}(\mathbf{r}',t')}\right|_{n_0} . \tag{9}$$

Obviously, they are connected via the functional derivative

$$\begin{aligned} \left.\frac{\delta v_\sigma(\mathbf{r},t)}{\delta v_{\text{ext}\,\sigma'}(\mathbf{r}',t')}\right|_{n_0} &= \delta_{\sigma\sigma'}\delta(\mathbf{r}-\mathbf{r}')\,\delta(t-t') \\ &+ \sum_\nu \int d^3x \int d\tau \left(\frac{\delta(t-\tau)}{|\mathbf{r}-\mathbf{x}|} + \frac{\delta v_{\text{xc}\,\sigma}(\mathbf{r},t)}{\delta n_\nu(\mathbf{x},\tau)}\right) \frac{\delta n_\nu(\mathbf{x},\tau)}{\delta v_{\text{ext}\,\sigma'}(\mathbf{r}',t')} . \end{aligned} \tag{10}$$

of the map (7) by the Dyson-type relation

$$\begin{aligned} \chi_{\sigma\sigma'}(\mathbf{r}t,\mathbf{r}'t') &= \chi_{\text{s}\,\sigma\sigma'}(\mathbf{r}t,\mathbf{r}'t') + \sum_{\nu\nu'} \int d^3x \int d\tau \int d^3x' \int d\tau'\, \chi_{\text{s}\,\sigma\nu}(\mathbf{r},t,\mathbf{x},\tau) \\ &\times \left(\frac{\delta(\tau-\tau')}{|\mathbf{x}-\mathbf{x}'|} + f_{\text{xc}\,\nu\nu'}(\mathbf{x},\tau,\mathbf{x}',\tau')\right) \chi_{\nu'\sigma'}(\mathbf{x}',\tau',\mathbf{r}',t') , \end{aligned} \tag{11}$$

where the so-called time-dependent exchange-correlation kernel

$$f_{\text{xc}\,\sigma\sigma'}(\mathbf{r},t,\mathbf{r}',t') := \left.\frac{\delta v_{\text{xc}\,\sigma}(\mathbf{r},t)}{\delta n_{\sigma'}(\mathbf{r}',t')}\right|_{n_0} \tag{12}$$

is a functional of the ground-state spin densities $\{n_{0\,\sigma}(\mathbf{r})\}$. Eqs. (11) and (12) constitute an exact representation of the linear density response of the interacting system. A detailed discussion of its formal properties and of applications can be found in [3].

In the following, we are going to describe an approach to the construction of $f_{\mathrm{xc}\,\sigma\sigma'}(\mathbf{r}t, \mathbf{r}', t')$ based on the recently proposed [4] time-dependent version of the so-called optimized potential (OEP). This method leads to $v_{\mathrm{xc}\,\sigma}$ as a function of $(\mathbf{r}t)$ via solution of an integral equation rather than to $v_{\mathrm{xc}\,\sigma}$ as an explicit functional of the spin densities. The definition (12) can therefore not be applied directly. It turns out however that the central equation (11) holds for a quantity $f_{\mathrm{xc}\,\sigma\sigma'}(\mathbf{r}, t, \mathbf{r}', t')$ defined through an integral equation which is quite analogous to - though more complicated than - the integral equation for the OEP.

THE TIME-DEPENDENT OPTIMIZED EFFECTIVE POTENTIAL

For stationary systems, a given expression for the total energy $E[\varphi_{j\sigma}]$ of an N-electron system as a functional of a set of spin orbitals $\{\varphi_{j\sigma}(\mathbf{r})\}$ (e. g. the Hartree-Fock total energy functional in the exchange-only case) is the starting point for the construction of the OEP [5,6]. Then, the variationally best *local* effective potential is determined for each spin orientation such that, when inserted in a stationary single-particle Schrödinger equation, it yields the set of $N = \sum_\sigma N_\sigma$ eigenfunctions (corresponding to the N_σ lowest eigenvalues) that minimize $E[\varphi_{j\sigma}]$. Although built from orbitals, this scheme is in fact a true density functional method, since, by virtue of the theorems of Hohenberg and Kohn and Kohn and Sham, the computed orbitals are functionals of the density.

In practice, the full OEP scheme is computationally considerably more demanding than traditional density functional schemes since it requires the numerical solution of an integral equation for each $v_{\mathrm{xc}\,\sigma}(\mathbf{r})$. There exists, however, an approximate OEP scheme, recently proposed by Krieger, Li and Iafrate (KLI) [7,8], which is numerically as easy to handle as the ordinary Kohn-Sham scheme. This simplified OEP has been applied very successfully to the calculation of atomic properties [9,10,11].

In order to derive a time-dependent generalization of the OEP we consider an N-electron system subject to a potential of the form (4) which, for all times up until t_0, has been in the ground state associated with the static external potential $v_{0\sigma}(\mathbf{r})$. We assume that the corresponding stationary OEP problem has been solved for that system, i. e. a local effective potential for each spin orientation and a set of N spin orbitals $\{\varphi_{j\sigma}\}$ (with energy eigenvalues $\varepsilon_{j\sigma}$) minimizing a given energy functional $E[\varphi_{j\sigma}]$ are assumed to be known. Our goal is to determine the time evolution of the system under the influence of the total external potential $v_{\mathrm{ext}\sigma}(\mathbf{r}t) = v_{0\sigma}(\mathbf{r}) + v_{1\sigma}(\mathbf{r}t)$ from t_0 up until an arbitrary later time t_1. Note that in this section we do not linearize with respect to $v_1(\mathbf{r}t)$ but develop the full time-dependent theory. The construction of an optimized local effective potential starts with the quantum mechanical action

$$A[\phi_{j\sigma}] = \sum_\sigma \sum_j^{N_\sigma} \int_{-\infty}^{t_1} dt \int d^3r\, \phi_{j\sigma}^*(\mathbf{r}t) \left(i\,\partial/\partial t + \nabla^2/2\right) \phi_{j\sigma}(\mathbf{r}t)$$

$$-\sum_\sigma \int_{-\infty}^{t_1} dt \int d^3r\, n_\sigma(\mathbf{r}t) v_{\mathrm{ext}\sigma}(\mathbf{r}t) - \frac{1}{2}\int_{-\infty}^{t_1} dt \int d^3r \int d^3r'\, \frac{n(\mathbf{r}t)n(\mathbf{r}'t)}{|\mathbf{r}-\mathbf{r}'|} - A_{\mathrm{xc}}[\phi_{j\sigma}] \tag{13}$$

written as a functional of $N = \sum_\sigma N_\sigma$ time-dependent spin orbitals $\{\phi_{j\sigma}(\mathbf{r}t)\}$. In the following we will not refer to any specific approximation for the exchange-correlation functional $A_{\mathrm{xc}}[\phi_{j\sigma}]$. We mention however, that in an exchange-only theory A_{xc} would be replaced by the time-dependent Hartree-Fock (TDHF) expression

$$A_x = -(1/2)\sum_{\sigma}\sum_{i,j}^{N_\sigma}\int_{-\infty}^{t_1} dt \int d^3r \int d^3r'\ \phi^*_{i\sigma}(\mathbf{r}'t)\phi_{j\sigma}(\mathbf{r}'t)\phi_{i\sigma}(\mathbf{r}t)\phi^*_{j\sigma}(\mathbf{r}t)\ /\ |\mathbf{r}-\mathbf{r}'|. \tag{14}$$

The orbitals are solutions of the time-dependent Schrödinger equation (2) with the initial condition $\phi_{j\sigma}(\mathbf{r}t) = \varphi_{j\sigma}(\mathbf{r})\exp[-i\varepsilon_{j\sigma}(t-t_0)]$ for $-\infty < t \le t_0$. The local effective potential $v_\sigma(\mathbf{r}t)$ has to be determined in such a way that the $\{\phi_{j\sigma}(\mathbf{r}t)\}$, resulting from Eq. (2), render the total action functional $A[\phi_{j\sigma}]$ stationary. Therefore, we have to solve the following variational problem:

$$\frac{\delta A[\phi_{j\sigma}]}{\delta v_\sigma(\mathbf{r}t)} = \sum_{\sigma'}\int_{-\infty}^{+\infty} dt' \int d^3r' \sum_{j}^{N_{\sigma'}} \left(\frac{\delta A[\phi_{j\sigma}]}{\delta \phi_{j\sigma'}(\mathbf{r}'t')}\frac{\delta \phi_{j\sigma'}(\mathbf{r}'t')}{\delta v_\sigma(\mathbf{r}t)} + \frac{\delta A[\phi_{j\sigma}]}{\delta \phi^*_{j\sigma'}(\mathbf{r}'t')}\frac{\delta \phi^*_{j\sigma'}(\mathbf{r}'t')}{\delta v_\sigma(\mathbf{r}t)} \right) = 0. \tag{15}$$

In order to compute the functional derivative $\delta A/\delta\phi_{j\sigma}$, the first term of Eq. (13) has to be integrated by parts with respect to the time coordinate. We impose the usual boundary condition on $\phi_{j\sigma}(\mathbf{r}t)$ at $t = t_1$, i. e. $\delta\phi_{j\sigma}(\mathbf{r}t_1) = 0$, thus obtaining a zero boundary contribution. The other boundary contribution at $t = -\infty$ vanishes, too, because the action functional (13), in order to be well-defined, is to be calculated by introducing the usual factor $\exp(\eta t)$ in the integrand and taking $\lim_{\eta\to 0^+}$ after the integration. Then, substituting Eq. (3) and making use of the fact that $\phi^*_{j\sigma}$ solves the complex conjugate of the Schrödinger equation (2), we find

$$\frac{\delta A[\phi_{j\sigma}]}{\delta \phi_{j\sigma}(\mathbf{r}'t')} = [v_{\mathrm{xc}\,\sigma}(\mathbf{r}'t') - u_{\mathrm{xc}j\sigma}(\mathbf{r}'t')]\,\phi^*_{j\sigma}(\mathbf{r}'t')\,\theta(t_1-t') \quad , \tag{16}$$

where

$$u_{\mathrm{xc}j\sigma}(\mathbf{r}t) = \frac{1}{\phi^*_{j\sigma}(\mathbf{r}t)}\frac{\delta A_{\mathrm{xc}}[\phi_{j\sigma}]}{\delta\phi_{j\sigma}(\mathbf{r}t)} \tag{17}$$

and $\theta(x)$ denotes the usual step function (1 for $x > 0$, 0 for $x < 0$). An analogous expression is obtained for $\delta A/\delta\phi^*_{j\sigma}$ which, for all reasonable (i. e. real) functionals $A[\phi_{j\sigma}]$, is the complex conjugate of (16).

In order to evaluate $\delta A/\delta v_\sigma$ from Eq. (15), we further need the functional derivatives $\delta\phi_{j\sigma}/\delta v_\sigma$ and $\delta\phi^*_{j\sigma}/\delta v_\sigma$. To this end, we consider the orbitals $\{\phi_{j\sigma}(\mathbf{r}t)\}$ as unperturbed states, remembering that at $t = t_1$ the orbitals are held *fixed* with respect to variations in the total potential. We therefore start from $t = t_1$, subject the system to an *additional* small perturbation $\delta v_\sigma(\mathbf{r}t)$ and let it evolve *backwards* in time. The corresponding perturbed wave functions $\phi'_{j\sigma}(\mathbf{r}t)$ are determined by the backward Schrödinger equation

$$i\,(\partial/\partial t)\,\phi'_{j\sigma}(\mathbf{r}t) = \left(-\nabla^2/2 + v_\sigma(\mathbf{r}t) + \delta v_\sigma(\mathbf{r}t)\right)\phi'_{j\sigma}(\mathbf{r}t) \quad , \quad j = 1, \ldots, N_\sigma \tag{18}$$

with the initial condition $\phi'_{j\sigma}(\mathbf{r}t_1) = \phi_{j\sigma}(\mathbf{r}t_1)$. The resulting first-order correction to the wave function $\phi_{j\sigma}(\mathbf{r}t)$ under the influence of $\delta v_\sigma(\mathbf{r}t)$ is given by

$$\delta\phi_{j\sigma}(\mathbf{r}t) = i\sum_{k=1}^{\infty}\int_t^{t_1} dt' \int d^3r'\ \phi^*_{k\sigma}(\mathbf{r}'t')\delta v_\sigma(\mathbf{r}'t')\phi_{j\sigma}(\mathbf{r}'t')\phi_{k\sigma}(\mathbf{r}t) \ . \tag{19}$$

from which we can read off the desired functional derivative

$$\frac{\delta\phi^{bw}_{j\sigma'}(\mathbf{r}'t')}{\delta v_\sigma(\mathbf{r}t)} = i\delta_{\sigma\sigma'}\sum_{k=1}^{\infty}\phi^*_{k\sigma}(\mathbf{r}t)\phi_{j\sigma}(\mathbf{r}t)\phi_{k\sigma}(\mathbf{r}'t')\,\theta(t_1-t)\,\theta(t-t') \quad . \tag{20}$$

Once again, $\delta\phi^*_{j\sigma'}/\delta v_\sigma$ leads to the complex conjugate expression. We can now insert (16) and (20) in the variational equation (15), and the result is the time-dependent OEP (TDOEP) integral equation for the local exchange-correlation potential $v_{\mathrm{xc}\,\sigma}(\mathbf{r}t)$:

$$\int_{-\infty}^{t_1} dt' \int d^3r' \left(iK_\sigma(\mathbf{r}t,\mathbf{r}'t') \sum_j^{N_\sigma} \phi_{j\sigma}(\mathbf{r}t)\phi^*_{j\sigma}(\mathbf{r}'t')\left[v_{\mathrm{xc}\,\sigma}(\mathbf{r}'t') - u_{\mathrm{xc}j\sigma}(\mathbf{r}'t')\right] + c.c.\right) = 0\,. \tag{21}$$

The kernel $K_\sigma(\mathbf{r}t,\mathbf{r}'t') = \sum_{k=1}^{\infty}\phi^*_{k\sigma}(\mathbf{r}t)\phi_{k\sigma}(\mathbf{r}'t')\theta(t-t')$ can be identified with the Green's function of the system, which satisfies the differential equation

$$\left[i\,\partial/\partial t' - \left(-\nabla'^2/2 + v_\sigma(\mathbf{r}'t')\right)\right]K_\sigma(\mathbf{r}t,\mathbf{r}'t') = -i\delta(\mathbf{r}-\mathbf{r}')\delta(t-t') \tag{22}$$

with the initial condition $K_\sigma(\mathbf{r}t,\mathbf{r}'t') = 0$ for $t' > t$. The TDOEP scheme is now complete: the integral equation (21) has to be solved for $v_{\mathrm{xc}\,\sigma}(\mathbf{r}t)$ in combination with the Schrödinger equation (2) and the differential equation (22) for $K_\sigma(\mathbf{r}t,\mathbf{r}'t')$, both with the appropriate initial conditions. It is easy to show that in the time interval $[-\infty, t_1]$ the exchange-correlation potential $v_{\mathrm{xc}\,\sigma}(\mathbf{r}t)$ is unique up to within an additive, purely time-dependent function $c(t)$ (as expected in view of the Runge-Gross theorem). Also it can be demonstrated [12] that for time-independent external potentials ($v_{1\sigma}(\mathbf{r}t) \equiv 0$) the TDOEP reduces to the stationary OEP.

The implementation of the full TDOEP scheme is an extremely demanding task. It is therefore highly desirable to construct approximations of $v_{\mathrm{xc}\,\sigma}(\mathbf{r}t)$ which are *explicit* functionals of the orbitals $\{\phi_{j\sigma}\}$, thereby avoiding the need to solve the integral equation numerically. With similar manipulations as developed by KLI for the stationary case [8,9] one can derive the so-called TDKLI expression [4]

$$\begin{aligned}\tilde{v}_{\mathrm{xc}\,\sigma}(\mathbf{r}t) &= \frac{1}{n_\sigma(\mathbf{r}t)}\sum_j^{N_\sigma} n_{j\sigma}(\mathbf{r}t)\,\frac{1}{2}\left(u_{\mathrm{xc}j\sigma}(\mathbf{r}t) + u^*_{\mathrm{xc}j\sigma}(\mathbf{r}t)\right)\\ &+ \frac{1}{n_\sigma(\mathbf{r}t)}\sum_j^{N_\sigma} n_{j\sigma}(\mathbf{r}t)\left[\overline{\tilde{v}}_{\mathrm{xc}j\sigma}(t) - \frac{1}{2}\left(\overline{u}_{\mathrm{xc}j\sigma}(t) + \overline{u}^*_{\mathrm{xc}j\sigma}(t)\right)\right]\\ &+ \frac{i}{4n_\sigma(\mathbf{r}t)}\sum_j^{N_\sigma}\nabla^2 n_{j\sigma}(\mathbf{r}t)\int_{-\infty}^{t} dt'\left(\overline{u}_{\mathrm{xc}j\sigma}(t') - \overline{u}^*_{\mathrm{xc}j\sigma}(t')\right)\,.\end{aligned} \tag{23}$$

The tilde indicates that this is only an approximate solution of the TDOEP equation (21) and the overbar is a shorthand notation for the orbital average, i.e. $\overline{\tilde{v}}_{\mathrm{xc}j\sigma}(t) = \int d^3r\, n_{j\sigma}(\mathbf{r}t)\tilde{v}_{\mathrm{xc}j\sigma}\mathbf{r}(t)$. The last term of Eq. (23) vanishes identically for a large class of exchange-correlation functionals A_{xc} including all functionals depending on $\{\phi_{j\sigma}\}$ only through the combinations $\phi_{j\sigma}(\mathbf{r}t)\phi^*_{j\sigma}(\mathbf{r}'t)$ (such as the TDHF functional, Eq. (14)). Eq. (23) is still an integral equation for $\tilde{v}_{\mathrm{xc}\sigma}$. In contrast to the full TDOEP equation it can be solved analytically [7]: multiplying Eq. (23) by $n_{k\sigma}(\mathbf{r}t)$ and integrating over all space yields

$$\overline{\tilde{v}}_{\mathrm{xc}k\sigma}(t) = \overline{w}_{\mathrm{xc}k\sigma}(t) + \sum_j^{N_\sigma} M_{kj\sigma}(t)\overline{\tilde{v}}_{\mathrm{xc}j\sigma}(t) \quad , \tag{24}$$

where

$$w_{\mathrm{xc}\,\sigma}(\mathbf{r}t) = \frac{1}{n_\sigma(\mathbf{r}t)} \sum_j^{N_\sigma} n_{j\sigma}(\mathbf{r}t)\, \frac{1}{2}\left(u_{\mathrm{xc}j\sigma}(\mathbf{r}t) + u^*_{\mathrm{xc}j\sigma}(\mathbf{r}t)\right)$$
$$- \frac{1}{n_\sigma(\mathbf{r}t)} \sum_j^{N_\sigma} n_{j\sigma}(\mathbf{r}t) \left[\frac{1}{2}\left(\overline{u}_{\mathrm{xc}j\sigma}(t) + \overline{u}^*_{\mathrm{xc}j\sigma}(t)\right)\right]$$
$$+ \frac{i}{4n_\sigma(\mathbf{r}t)} \sum_j^{N_\sigma} \nabla^2 n_{j\sigma}(\mathbf{r}t) \int_{-\infty}^{t} dt' \left(\overline{u}_{\mathrm{xc}j\sigma}(t') - \overline{u}^*_{\mathrm{xc}j\sigma}(t')\right) \tag{25}$$

and

$$M_{kj\sigma}(t) = \int d^3r \, \frac{n_{k\sigma}(\mathbf{r}t) n_{j\sigma}(\mathbf{r}t)}{n_\sigma(\mathbf{r}t)} \quad . \tag{26}$$

Solving Eq. (24) for $\overline{\tilde{v}}_{\mathrm{xc}j\sigma}(t)$ requires inversion of the $N_\sigma \times N_\sigma$ matrix $A_{kj\sigma}(t) = \delta_{kj} - M_{kj\sigma}(t)$ and leads to

$$\overline{\tilde{v}}_{\mathrm{xc}j\sigma}(t) = \sum_k^{N_\sigma} \left(A_\sigma^{-1}(t)\right)_{jk} \overline{w}_{\mathrm{xc}k\sigma}(t) \quad . \tag{27}$$

When Eq. (27) is substituted into Eq. (23), one obtains $\tilde{v}_{\mathrm{xc}\sigma}(\mathbf{r}t)$ as an explicit functional of the orbitals $\{\phi_{j\sigma}(\mathbf{r}t)\}$. The TDKLI scheme has been applied successfully to the nonperturbative description of atoms [13-15] and clusters [16] in strong laser pulses.

THE EXCHANGE CORRELATION KERNEL WITHIN THE TDOEP

In the previous section we presented the TDOEP scheme as a method to treat a time-dependent interacting many particle system nonperturbatively. We now discuss the linear density response to the time-dependent part of the external potential in the TDOEP, and show how it is related to the TDDFT linear response formalism. Note that to recover the Dyson-type equation (11) we have to establish an equation of the form

$$\left.\frac{\delta v_{\mathrm{xc}\,\sigma}(\mathbf{r}t)}{\delta v_{\mathrm{ext}\,\sigma'}(\mathbf{r}'t')}\right|_{v^{(0)}} = \sum_\nu \int d^3y \int d\tau \, f_{\mathrm{xc}\,\sigma\nu}(\mathbf{r}t, \mathbf{y}\tau) \chi_{\nu\sigma'}(\mathbf{y}\tau, \mathbf{r}'t') \tag{28}$$

where

$$\chi_{\nu\sigma'}(\mathbf{y}\tau, \mathbf{r}'t') = \frac{\delta n_\sigma(\mathbf{y}\tau)}{\delta v_{\mathrm{ext}\,\sigma'}(\mathbf{r}'t')} \quad . \tag{29}$$

Within the TDOEP scheme the exchange-correlation potential does not depend explicitly on the spin densities n_σ. The density-dependence is only implicit through the orbitals. Therefore the evaluation of the functional derivative $\delta v_{\mathrm{xc}}/\delta v_{\mathrm{ext}'}$ is more complicated than in ordinary TDDFT. The integral equation (21) does not even define $v_{\mathrm{xc}\,\sigma}$ as an explicit functional of only the occupied orbitals $\{\phi_{j\sigma}; j = 1 \ldots N_\sigma\}$ but contains the Greens function $K_\sigma(\mathbf{r}t, \mathbf{r}'t')$. For the moment we therefore imagine $v_{\mathrm{xc}\,\sigma}$ to be a functional of all orbitals $\{\phi_{j\sigma}\}$. By applying the chain rule for functional derivatives on the left hand side of Eq. (28) several times we obtain

$$\frac{\delta v_{\mathrm{xc}\,\sigma}(\mathbf{r}t)}{\delta v_{\mathrm{ext}\,\sigma'}(\mathbf{r}'t')} = \sum_{\nu\nu'\tilde{\nu}} \int d^3y \int d\tau \int d^3y' \int d\tau' \int d^3\tilde{y} \int d\tilde{\tau}$$
$$\times \sum_m \left(\frac{\delta v_{\mathrm{xc}\,\sigma}(\mathbf{r}t)}{\delta \phi_{m\nu}(\mathbf{y}\tau)} \frac{\delta \phi_{m\nu}(\mathbf{y}\tau)}{\delta v_{\nu'}(\mathbf{y}'\tau')} + \frac{\delta v_{\mathrm{xc}\,\sigma}(\mathbf{r}t)}{\delta \phi^*_{m\nu}(\mathbf{y}\tau)} \frac{\delta \phi^*_{m\nu}(\mathbf{y}\tau)}{\delta v_{\nu'}(\mathbf{y}'\tau')} \right) \frac{\delta v_{\nu'}(\mathbf{y}'\tau')}{\delta n_{\tilde{\nu}}(\tilde{\mathbf{y}}\tilde{\tau})} \frac{\delta n_{\tilde{\nu}}(\tilde{\mathbf{y}}\tilde{\tau})}{\delta v_{\mathrm{ext}\,\sigma'}(\mathbf{r}'t')} \quad . \tag{30}$$

Comparison of (30) and (28) yields the following representation for the exchange-correlation kernel

$$f_{\mathrm{xc}\,\sigma\sigma'}(\mathbf{r}t,\mathbf{r}'t') = \Bigg[\sum_{\nu\nu'} \int d^3y \int d\tau \int d^3y' \int d\tau'$$

$$\times \sum_m \left(\frac{\delta v_{\mathrm{xc}\,\sigma}(\mathbf{r}t)}{\delta\phi_{m\nu}(\mathbf{y}\tau)} \frac{\delta\phi_{m\nu}(\mathbf{y}\tau)}{\delta v_{\nu'}(\mathbf{y}'\tau')} + \frac{\delta v_{\mathrm{xc}\,\sigma}(\mathbf{r}t)}{\delta\phi^*_{m\nu}(\mathbf{y}\tau)} \frac{\delta\phi^*_{m\nu}(\mathbf{y}\tau)}{\delta v_{\nu'}(\mathbf{y}'\tau')} \right) \frac{\delta v_{\nu'}(\mathbf{y}'\tau')}{\delta n_{\sigma'}(\mathbf{r}'t')} \Bigg]_{\phi_{j\sigma}(t)=\phi^{(0)}_{j\sigma}(t)} . \quad (31)$$

Here the functional derivatives $\delta\phi_{j\sigma'}/\delta v_\sigma$ and $\delta v_{\sigma'}/\delta n_\sigma$ refer to the mappings

$$v_\sigma \longrightarrow \{\phi_{j\sigma}\} \quad (32)$$

$$v_\sigma \longleftrightarrow \{n_{j\sigma}\} \quad (33)$$

defined by solution of the Schrödinger equation (2) for $t > t_0$ under the initial condition $\phi^{(0)}_{j\sigma}(\mathbf{r}t_0) = \varphi_{j\sigma}(\mathbf{r})$. The correspondence (33) between potentials and spin densities is one-to-one by the Runge-Gross theorem. The left hand side of Eq. (31) is to be evaluated at the stationary orbitals $\phi^{(0)}_{j\sigma}(\mathbf{r}t) = \varphi_{j\sigma}(\mathbf{r})e^{-\epsilon_{j\sigma}(t-t_0)}$ corresponding to $v(\mathbf{r}t) = v^{(0)}(\mathbf{r})$ for all times. For $\delta\phi_{j\sigma'}/\delta v_\sigma$ we obtain from first-order perturbation theory

$$\frac{\delta\phi_{j\sigma}(\mathbf{r}t)}{\delta v_{\sigma'}(\mathbf{r}'t')} = -i\delta_{\sigma\sigma'}\,\phi_{j\sigma}(\mathbf{r}'t')\sum_k \phi_{k\sigma}(\mathbf{r}t)\phi^*_{k\sigma}(\mathbf{r}'t')\theta(t-t') \quad , \quad (34)$$

and for $\delta\phi^*_{j\sigma'}/\delta v_\sigma$ the complex conjugate. The derivative $\delta v_{\sigma'}/\delta n_\sigma$ is the inverse of the noninteracting density-density response function

$$\begin{aligned}\tilde{\chi}_{\mathrm{s}\,\sigma\sigma'}(\mathbf{r}t,\mathbf{r}'t') &= \frac{\delta n_\sigma(\mathbf{r}t)}{\delta v_{\sigma'}(\mathbf{r}'t')} \\ &= -i\theta(t-t')\delta_{\sigma\sigma'}\sum_{j,k}\left(f_{k\sigma} - f_{j\sigma} \right) \phi_{j\sigma}(\mathbf{r}t)\phi^*_{j\sigma}(\mathbf{r}'t')\phi_{k\sigma}(\mathbf{r}'t')\phi^*_{k\sigma}(\mathbf{r}t)\end{aligned} \quad (35)$$

($\tilde{\chi}_\mathrm{s}$ reduces to χ_s of (8) after inserting the stationary orbitals $\phi^{(0)}(t)$; the $f_{j\sigma}$ are occupation factors). Thus multiplying Eq. (31) with $\tilde{\chi}_{\mathrm{s}\sigma'\tilde{\sigma}}(\mathbf{r}'t',\tilde{\mathbf{r}}\tilde{t})$ from the right and integrating over $\mathbf{r}'t'$ yields the following integral equation for the exchange-correlation kernel

$$\int dt' \int d^3r' \, \Bigg\{ -i \sum_m \left(f_{m\sigma'}\, f_{\mathrm{xc}\,\sigma\sigma'}(\mathbf{r}t,\mathbf{r}'t') - g^{(m)}_{\mathrm{xc}\,\sigma\sigma'}(\mathbf{r}t,\mathbf{r}'t') \right)$$

$$\times \left[\phi_{m\sigma'}(\mathbf{y}\tau)\phi^*_{m\sigma'}(\mathbf{r}'t')\sum_k \phi_{k\sigma'}(\mathbf{r}'t')\phi^*_{k\sigma'}(\mathbf{y}\tau)\theta(t'-\tau) \right]_{\phi_{j\sigma}(t)=\phi^{(0)}_{j\sigma}(t)} + \text{c.c.} \Bigg\} = 0 . \quad (36)$$

where we used the abbreviation

$$g^{(m)}_{\mathrm{xc}\,\sigma\sigma'}(\mathbf{r}t,\mathbf{r}'t') = \left[\frac{1}{\phi^*_{m\sigma'}(\mathbf{r}'t')} \frac{\delta v_{\mathrm{xc}\,\sigma}(\mathbf{r}t)}{\delta\phi_{m\sigma'}(\mathbf{r}'t')} \right]_{\phi_{j\sigma}(t)=\phi^{(0)}_{j\sigma}(t)} . \quad (37)$$

We emphasize that Eq.(36) is by no means restricted to the TDOEP but defines the xc-kernel in any scheme involving a representation of a local xc-potential as a functional of the orbitals rather than the spin densities. We observe that Eq.(36) is formally identical with the integral equation for the time-dependent OEP xc-potential (21) with $v_{\mathrm{xc}\,\sigma}(\mathbf{r}'t')$ and $u_{\mathrm{xc}\,j\sigma}(\mathbf{r}'t')$ replaced by $f_{\mathrm{xc}\,\sigma\sigma'}(\mathbf{r}t,\mathbf{r}'t')$ and $g^{(j)}_{\mathrm{xc}\,\sigma\sigma'}(\mathbf{r}t,\mathbf{r}'t')$, respectively. Consequently it can be treated by the same approximation methods once the quantity $g^{(j)}_{\mathrm{xc}}$ has been obtained from a functional $v_{\mathrm{xc}}[\{\phi_{j\sigma}\}]$.

Within the full TDOEP scheme one can still not evaluate the derivatives in (37) directly because one has only an implicit knowledge of the functional form of $v_{\mathrm{xc}}[\{\phi_{j\sigma}\}]$. An exact procedure to overcome this difficulty is presented in the appendix and can be outlined as follows: Acting with $\delta/\delta\phi_{j\sigma}$ on Eq. (21) one generates a system of integral equations for the derivatives $\delta v_{\mathrm{xc}}/\delta\phi_{j\sigma}$ of the xc-potential. These relations can be combined with Eq. (37) to a single integral equation defining f_{xc} in terms of derivatives $\delta A_{\mathrm{xc}}/\delta\phi_{j\sigma}$ and $\delta^2 A_{\mathrm{xc}}/\delta\phi_{j\sigma}\delta\phi_{j'\sigma'}$ of the xc-part of the action functional.

For practical calculations, however, it is desirable to devise a sufficiently simple analytic approximation of f_{xc}. To this end we consider the expression

$$v^{\mathrm{appr}}_{\mathrm{xc}\,\sigma}(\mathbf{r}t) = \sum_{j}^{N_\sigma} \frac{|\psi_{j\sigma}(\mathbf{r}t)|^2}{2n_\sigma(\mathbf{r}t)} \left[u'_{\mathrm{xc}\,j\sigma}(\mathbf{r}t) + u'^{*}_{\mathrm{xc}\,j\sigma}(\mathbf{r}t)\right] \tag{38}$$

as an approximate solution to (21), which means that we keep only the first term on the right hand side of Eq. (23). In the static exchange-only case this procedure leads to the so-called Slater approximation [17]. Now we apply the same approximation to (36), i.e., we set

$$f^{\mathrm{appr}}_{\mathrm{xc}\,\sigma\sigma'}(\mathbf{y}\tau,\mathbf{r}'t') = \sum_{j} \frac{|\varphi_{j\sigma'}(\mathbf{r})|^2}{2n_{0\,\sigma'}(\mathbf{r})} \left[g^{(j)}_{\mathrm{xc}\,\sigma\sigma'}(\mathbf{y}\tau,\mathbf{r}'t') + c.c.\right] \tag{39}$$

and use the explicit analytical form (38) to evaluate (37). In the TD x-only approximation defined by (14) this scheme yields the compact expression

$$f^{\mathrm{appr}}_{\mathrm{x}\,\sigma\sigma'}(\mathbf{r}t,\mathbf{r}'t') = -\,\delta(t-t')\,\delta_{\sigma\sigma'}\,\frac{\left|\sum_k f_{k\sigma}\varphi_{k\sigma}(\mathbf{r})\varphi^{*}_{k\sigma}(\mathbf{r}')\right|^2}{|\mathbf{r}-\mathbf{r}'|\,n_{0\,\sigma}(\mathbf{r})\,n_{0\,\sigma}(\mathbf{r}')}\,. \tag{40}$$

We remark that in general the Fourier transform of the xc-kernel defined by Eq. (36) is frequency-dependent (even in the TD x-only case), a feature which is not accounted for in the present treatment of this equation. However, for the special case of a two-electron system treated within TD x-only theory, Eqs. (40) and (38) are the *exact* solutions of the respective integral equations, as is easily checked.

CALCULATION OF EXCITATION ENERGIES

The traditional Hohenberg-Kohn theorem guarantees that every observable of a stationary physical system can be expressed in terms of its ground-state density. In principle, this is also true for the set of excited-state energies: Knowledge of the external potential, which is a unique functional of the ground-state density, allows the calculation of any excited state and its energy. In practice, however, the description of excited states within stationary density-functional theory is a notoriously difficult subject [18-33]. In particular, the Kohn-Sham eigenvalues, introduced as purely mathematical constructs into density-functional theory, cannot simply be interpreted as excited-state energies.

In this section we shall discuss a different approach to the calculation of excitation energies which is based on time-dependent density functional theory in the linear response regime [34,35]. Similar ideas were recently proposed by Casida and coworkers [36] on the basis of the one-particle density matrix.

Since the response functions (8) and (9) as well as the xc kernel (12) are functionals of the ground-state density only, this approach represents a way of explicitly constructing functionals for excitation energies in terms of ground state properties.

To calculate excitation energies from time-dependent density functional theory we use the fact that the frequency-dependent linear density response of a finite system exhibits discrete poles at the true excitation energies of the unperturbed system. Taking the Dyson-type equation (11) for the interacting density-density response function, the exact linear density response can be written as the linear density response of a noninteracting (Kohn-Sham) system to an effective perturbation. The fundamental idea is to use this formally exact representation of the linear density response to calculate the shifts of the Kohn-Sham orbital eigenvalue differences (which are the poles of the Kohn-Sham response function) toward the true excitation energies (which are the poles of the interacting response function).

Inserting the Fourier transform (with respect to time) of relation (11) into the equation

$$n_{1\sigma}(\mathbf{r},\omega)=\sum_{\nu}\int d^3y\,\chi_{\sigma\nu}(\mathbf{r},\mathbf{y};\omega)v_{1\nu}(\mathbf{y},\omega) \tag{41}$$

for the frequency-dependent linear (spin) density response, one obtains the response equation

$$n_{1\sigma}(\mathbf{r},\omega)=\sum_{\nu}\int d^3y\,\chi_{\mathrm{s}\,\sigma\nu}(\mathbf{r},\mathbf{y};\omega)v_{\mathrm{s}\,1\nu}(\mathbf{y},\omega) \tag{42}$$

of non-interacting particles subject to the effective perturbation

$$v_{\mathrm{s}\,1\nu}(\mathbf{y},\omega)=v_{1\nu}(\mathbf{y},\omega)+\sum_{\nu'}\int d^3y'\left(\frac{1}{|\mathbf{y}-\mathbf{y}'|}+f_{\mathrm{xc}\,\nu\nu'}(\mathbf{y},\mathbf{y}';\omega)\right)n_{1\nu'}(\mathbf{y}',\omega)\,. \tag{43}$$

Equations (42) and (43) have to be solved self-consistently for the frequency-dependent linear spin density response $n_{1\sigma}(\mathbf{r},\omega)$.

The response-function χ_{s} of the Kohn-Sham system can be expressed in terms of the unperturbed static Kohn-Sham spin orbitals $\varphi_{j\sigma}$ as

$$\chi_{\mathrm{s}\,\sigma\sigma'}(\mathbf{r},\mathbf{r}';\omega)=\delta_{\sigma\sigma'}\sum_{j,k}(f_{k\sigma}-f_{j\sigma})\frac{\varphi_{j\sigma}(\mathbf{r})\varphi^*_{k\sigma}(\mathbf{r})\varphi^*_{j\sigma}(\mathbf{r}')\varphi_{k\sigma}(\mathbf{r}')}{\omega-(\epsilon_{j\sigma}-\epsilon_{k\sigma})+i\eta} \tag{44}$$

where $f_{k\sigma}$, $f_{j\sigma}$ are occupation factors (1 or 0). The summations in (44) run over all unperturbed Kohn-Sham orbitals, including the continuum states.

Note that the equations (42) and (43) are valid not only in the conventional formulation but also in the OEP formulation of time-dependent density functional theory, where the same Dyson-type relation (11) holds true as has been demonstrated in the previous section. In particular, one can employ approximate exchange-correlation kernels in Eq. (43) which were derived in the framework of the TDOEP.

In order to calculate the shifts from the KS single-particle energy differences towards the true excitation energies Ω of the interacting system, we rewrite Eq. (42)

together with Eq. (43) as

$$\sum_{\nu'}\int d^3y'\left(\delta_{\sigma\nu'}\delta(\mathbf{r}-\mathbf{y}')-\sum_{\nu}\int d^3y\,\chi_{s\,\sigma\nu}(\mathbf{r},\mathbf{y};\omega)\right.$$
$$\left.\times\left(\frac{1}{|\mathbf{y}-\mathbf{y}'|}+f_{\mathrm{xc}\,\nu\nu'}(\mathbf{y},\mathbf{y}';\omega)\right)\right)n_{1\nu'}(\mathbf{y}',\omega)$$
$$=\sum_{\nu}\int d^3y\,\chi_{s\,\sigma\nu}(\mathbf{r},\mathbf{y};\omega)v_{1\nu}(\mathbf{y},\omega)\,. \tag{45}$$

In general, the true excitation energies Ω are not identical with the Kohn-Sham excitation energies $\epsilon_{j\sigma}-\epsilon_{k\sigma}$. Therefore, the right-hand side of Eq. (45) remains finite for $\omega\to\Omega$. Since, on the other hand, the exact spin-density response $n_{1\sigma}$, has poles at the true excitation energies Ω, the integral operator acting on $n_{1\sigma}$ on the left-hand side of Eq. (45) cannot be invertible for $\omega\to\Omega$ (Assuming the existence of the inverse operator, its action on both sides of Eq. (45) results in a finite right-hand side for $\omega\to\Omega$. This leads to a contradiction since $n_{1\sigma}$, remaining on the left-hand side, has a pole at $\omega=\Omega$).

Consequently, the true excitation energies Ω are characterized as those frequencies where the eigenvalues of the integral operator acting on the spin-density vector in Eq. (45) vanish. Integrating out the delta-function in Eq. (45), the true excitation energies Ω are those frequencies, where the eigenvalues $\lambda(\omega)$ of

$$\sum_{\nu'}\int d^3y'\sum_{\nu}\int d^3y\,\chi_{s\,\sigma\nu}(\mathbf{r},\mathbf{y};\omega)\left(\frac{1}{|\mathbf{y}-\mathbf{y}'|}+f_{\mathrm{xc}\,\nu\nu'}(\mathbf{y},\mathbf{y}';\omega)\right)\gamma_{\nu'}(\mathbf{y}',\omega)=$$
$$\lambda(\omega)\gamma_\sigma(\mathbf{r},\omega) \tag{46}$$

satisfy

$$\lambda(\Omega)=1\,. \tag{47}$$

This condition rigorously determines the true excitation spectrum of the interacting system at hand.

To facilitate the notation, we introduce double indices $q\equiv(j,k)$ so that $\omega_{q\sigma}\equiv\epsilon_{j\sigma}-\epsilon_{k\sigma}$ denotes the excitation energy of the single-particle transition $(k\sigma\to j\sigma)$. Moreover, we define

$$\Phi_{q\sigma}(\mathbf{r}):=\phi_{k\sigma}(\mathbf{r})^*\phi_{j\sigma}(\mathbf{r})\,, \tag{48}$$
$$\alpha_{q\sigma}:=f_{k\sigma}-f_{j\sigma} \tag{49}$$

and set

$$\xi_{q\sigma}(\omega):=\sum_{\nu'}\int d^3y'\int d^3y\,\Phi_{q\sigma}(\mathbf{y})^*\left(\frac{1}{|\mathbf{y}-\mathbf{y}'|}+f_{\mathrm{xc}\,\sigma\nu'}(\mathbf{y},\mathbf{y}';\omega)\right)\gamma_{\nu'}(\mathbf{y}',\omega)\,. \tag{50}$$

Using these definitions, Eq. (46) can be recast into

$$\sum_q\frac{\alpha_{q\sigma}\Phi_{q\sigma}(\mathbf{r})}{\omega-\omega_{q\sigma}+i\eta}\xi_{q\sigma}(\omega)=\lambda(\omega)\gamma_\sigma(\mathbf{r},\omega)\,. \tag{51}$$

Solving this equation for $\gamma_\sigma(\mathbf{r},\omega)$ and reinserting the result on the right-hand side of Eq. (50) we arrive at

$$\sum_{\sigma'}\sum_{q'}\frac{M_{q\sigma\,q'\sigma'}(\omega)}{\omega-\omega_{q'\sigma'}+i\eta}\xi_{q'\sigma'}(\omega)=\lambda(\omega)\xi_{q\sigma}(\omega)\,, \tag{52}$$

where we have introduced the matrix elements

$$M_{q\sigma\, q'\sigma'}(\omega) = \alpha_{q'\sigma'} \int d^3r \int d^3r'\, \Phi^*_{q\sigma}(\mathbf{r}) \left(\frac{1}{|\mathbf{r}-\mathbf{r}'|} + f_{\mathrm{xc}\sigma\sigma'}(\mathbf{r},\mathbf{r}';\omega) \right) \Phi_{q'\sigma'}(\mathbf{r}') \,. \tag{53}$$

Note that the summation in Eq. (52) extends over an infinite number of single-particle transitions $q'\sigma'$ between occupied and unoccupied Kohn-Sham orbitals, so that up to this point, no approximations have been made.

In order to actually calculate the excitation spectrum, the eigenvalue problem (52) has to be truncated in one way or another. One possibility is to expand all quantities in Eq. (52) about one particular KS-orbital energy difference $\omega_{p\tau}$ [34,35]. The true excitation energies Ω are then determined by the solution of

$$\lambda(\Omega) = \frac{A(\omega_{p\tau})}{\Omega - \omega_{p\tau}} + B(\omega_{p\tau}) + \ldots = 1 \tag{54}$$

For non-degenerate single-particle poles $\omega_{p\tau}$, the coefficients in Eq. (54) are given by

$$A(\omega_{p\tau}) = M_{p\tau\, p\tau}(\omega_{p\tau}) \tag{55}$$

and

$$B(\omega_{p\tau}) = \left. \frac{dM_{p\tau p\tau}}{d\omega} \right|_{\omega_{p\tau}} + \frac{1}{M_{p\tau p\tau}(\omega_{p\tau})} \sum_{q'\sigma' \neq p\tau} \frac{M_{p\tau\, q'\sigma'}(\omega_{p\tau}) M_{q'\sigma'\, p\tau}(\omega_{p\tau})}{\omega_{p\tau} - \omega_{q'\sigma'} + i\eta} \,. \tag{56}$$

If the pole $\omega_{p\tau}$ is $\wp$-fold degenerate, $\omega_{p_1\tau_1} = \omega_{p_2\tau_2} = \ldots = \omega_{p_\wp\tau_\wp} \equiv \omega_0$, the lowest-order coefficient A in Eq. (54) is determined by a $\wp$-dimensional matrix equation

$$\sum_{k=1}^{\wp} M_{p_i\tau_i\, p_k\tau_k}(\omega_0) \xi^{(n)}_{p_k\tau_k} = A_n(\omega_0) \xi^{(n)}_{p_i\tau_i} \,, \quad i = 1 \ldots \wp \,, \tag{57}$$

leading to $\wp$ different coefficients $A_1 \ldots A_\wp$, in general. For excitation energies Ω close to ω_0, the lowest-order term of the above Laurent expansion will dominate the series. In this single pole approximation (SPA), Eq. (54) reduces to

$$\lambda_n(\Omega) \approx \frac{A_n(\omega_0)}{\Omega - \omega_0} = 1 \,. \tag{58}$$

The condition (47) and its complex conjugate, $\lambda^*(\Omega) = 1$, finally lead to to a simple (approximate) expression for the exact excitation energies.

$$\Omega_n \approx \omega_0 + \Re A_n(\omega_0) \,. \tag{59}$$

For closed-shell systems, every Kohn-Sham orbital eigenvalue is degenerate with respect to spin, i.e. the spin multiplet structure is absent in the bare Kohn-Sham eigenvalue spectrum. Within the SPA, the dominant terms in the corrections to the Kohn-Sham eigenvalues towards the true multiplet energies naturally emerge from the solution of the (2×2) eigenvalue problem

$$\sum_{\sigma'=\uparrow,\downarrow} M_{p\sigma p\sigma'}(\omega_0) \xi_{p\sigma'}(\omega_0) = A \xi_{p\sigma}(\omega_0) \,. \tag{60}$$

Then, the resulting excitation energies are:

$$\Omega_{1,2} = \omega_0 + \Re \left\{ M_{p\uparrow p\uparrow} \pm M_{p\uparrow p\downarrow} \right\} . \tag{61}$$

Using the explicit form of the matrix elements (53) one finds

$$\Omega_1 = \omega_0 + 2\Re \int d^3r \int d^3r' \, \Phi_p^*(\mathbf{r}) \left(\frac{1}{|\mathbf{r} - \mathbf{r}'|} + f_{\mathrm{xc}}(\mathbf{r}, \mathbf{r}'; \omega_0) \right) \Phi_p(\mathbf{r}') \tag{62}$$

$$\Omega_2 = \omega_0 + 2\Re \int d^3r \int d^3r' \, \Phi_p^*(\mathbf{r}) \mu_0^2 G_{\mathrm{xc}}(\mathbf{r}, \mathbf{r}'; \omega_0) \Phi_p(\mathbf{r}') \tag{63}$$

where, since we are dealing with spin saturated systems, we have dropped the spin-index of $\Phi_{p\sigma}$ for simplicity.The xc-kernels appearing in Eqs. (62) and (63) are given by

$$f_{\mathrm{xc}}(\mathbf{r}, \mathbf{r}'; \omega) = \frac{1}{4} \sum_{\sigma, \sigma' = \pm 1} f_{\mathrm{xc}\sigma\sigma'}(\mathbf{r}, \mathbf{r}'; \omega) \tag{64}$$

$$G_{\mathrm{xc}}(\mathbf{r}, \mathbf{r}'; \omega) = \frac{1}{4\mu_0^2} \sum_{\sigma, \sigma' = \pm 1} (\sigma \cdot \sigma') f_{\mathrm{xc}\sigma\sigma'}(\mathbf{r}, \mathbf{r}'; \omega) . \tag{65}$$

The latter kernel, G_{xc}, is responsible for the exchange and correlation effects in the Kohn-Sham equation for the linear response of the frequency-dependent magnetization density $m(\mathbf{r}, \omega)$ [2]. The fact that the magnetization density response naturally involves spin-flip processes suggests that Ω_2 can be attributed to spin triplet excitation energies of many-electron systems. The corresponding spin singlet excitation energies, on the other hand, are given by Ω_1.

Apart from the explicit calculation of the dominant contributions to the shifts of the Kohn-Sham eigenvalue differences towards the true excitation energies, one can, in principle, solve Eq. (52) directly. At the frequencies $\omega = \Omega$ Eq. (52) can be written as

$$\sum_{q'\sigma'} \left(M_{q\sigma\, q'\sigma'}(\Omega) + \delta_{q\sigma\, q'\sigma'} \omega_{q\sigma} \right) \beta_{q'\sigma'}(\Omega) = \Omega \beta_{q\sigma}(\Omega) , \tag{66}$$

where we have introduced

$$\beta_{q\sigma}(\Omega) := \xi_{q\sigma}(\Omega) / (\Omega - \omega_{q\sigma}) . \tag{67}$$

Hence, the exact excitation energies Ω are the exact solutions of the nonlinear matrix-equation (66). Truncation of the infinite-dimensional matrix in Eq. (66) amounts to the approximation of $\chi^{(0)}$ by a finite sum

$$\chi^{(0)}(\mathbf{r}, \mathbf{r}', \omega) \approx \sum_{\sigma = \uparrow\downarrow} \sum_{q}^{Q} \alpha_q \frac{\Phi_q(\mathbf{r}) \Phi_q(\mathbf{r}')}{\omega - \omega_{q\sigma}} . \tag{68}$$

Thus, instead of expanding about a single pole, we explicitly take into account several poles of the noninteracting response function. Taking the adiabatic approximation for the xc-kernels, the matrix elements $M_{q\sigma\, q'\sigma'}$ become real and frequency independent, and the excitation energies Ω are then given as the eigenvalues of the $(Q \times Q)$ matrix $M_{q\sigma\, q'\sigma'}(\Omega = 0) + \delta_{q\sigma\, q'\sigma'} \omega_{q\sigma}$.

Results for the He Atom

In this section we report numerical results for excitation energies of the He atom using the above formalism of time-dependent density functional theory. The scheme generally involves three different types of approximations: (i) In the calculation of the KS orbital energies one employs some approximation of the static xc potential v_{xc}. (ii) The xc kernel f_{xc} needs to be approximated. (iii) The infinite-dimensional eigenvalue problem (52) (or, equivalently, (66)) must be truncated in one way or another. The aim of the present work is to investigate the effects due to the approximations (ii) and (iii). In order to eliminate the errors (i) associated with the approximation for the ground-state KS potential we employ the *exact* xc potential of the He atom [37]. Table 1 shows the excitation energies of neutral helium, calculated in the single-pole approximation (SPA) (Eqs. (62) and (63)) and obtained from the truncated matrix equation (66) in comparison with results of the nonrelativistic variational calculation of Kono and Hattori [38] (column 8). The matrix equation (66) was solved using $N = 34$ unoccupied Kohn-Sham orbitals of s or p symmetry. For each symmetry class the resulting dimension of the (fully coupled but truncated) matrix in Eq. (66) is $(4N \times 4N)$ (due to the spin-degeneracy of the KS orbitals of Helium and the fact that the frequency dependent Kohn-Sham response function is symmetric in the complex plane with respect to the imaginary axis).

The excitation energies resulting from equation (66) are listed in column 5 for f_{xc}^{ALDA}, the so called adiabatic local density approximation for the exchange-correlation kernel [39], where the parametrization of Vosko, Wilk and Nusair [40] has been used for the correlation energy per particle. In column 7 we list the excitation energies obtained with the TDOEP kernel f_{x-only}^{appr} (Eq. (40)).

Within the accuracy of the figures given in the table, the results are converged with respect to variations of the dimension of the matrix in Eq. (66). Hence, the remaining difference between the results of the variational calculation [38] given in column 8 and the values obtained from our TDDFT formalism (column 5 and 7) is entirely due to the approximations in the xc kernels.

As also noted by Umrigar [41], it is a remarkable fact that for the exact Kohn-Sham potential of Helium, the Kohn-Sham excitation energies $\epsilon_j - \epsilon_k$, denoted $\Delta\omega_{KS}$ (column 3), are already very close to the exact spectrum, with $\Delta\omega_{KS}$ lying always between the singlet and the triplet energies. Hence, our method yields correspondingly small corrections towards singlet and triplet energies.

Comparing the results of the single-pole approximation (SPA) in columns 4 and 6 to the solutions of the "full" matrix equations in columns 5 and 7, respectively, we see that there is very little change (from a few hundredth of a percent to at most one half percent) in the resulting excitation energies. Thus we conclude that in helium the single-pole approximation gives the dominant correction to the Kohn-Sham excitation spectrum.

Figure 1 shows the quantum defects obtained from the calculated excitation energies. Except for the $1P$ series, the results for the quantum defects obtained with the (x-only) TDOEP kernel (40) are closest to the exact values. Since expression (40) is the exact solution of the integral equation (36) for two-electron systems in the x-only limit, the remaining difference between the TDOEP results and the exact values can be attributed to the lack of correlation terms in f_{x-only}^{TDOEP}. Their explicit incorporation (within ALDA) into f_{xc} reduces the singlet-triplet splitting, as can be seen from Table 2.

The TDOEP scheme offers a starting point for the construction of an approximate

Table 1. Comparison of the excitation energies of neutral helium, calculated from the exact xc potential by using approximate xc kernels. All values are in Hartrees.

State	$k \to j$	$\Delta\omega_{KS}$	ALDA (xc) SPA	ALDA (xc) full[a]	TDOEP (x-only) SPA	TDOEP (x-only) full[a]	exact[b]
2^3S	$1s \to 2s$	0.7460	0.7357	0.7351	0.7232	0.7207	0.7285
2^1S			0.7718	0.7678	0.7687	0.7659	0.7578
3^3S	$1s \to 3s$	0.8392	0.8366	0.8368	0.8337	0.8343	0.8350
3^1S			0.8458	0.8461	0.8448	0.8450	0.8425
4^3S	$1s \to 4s$	0.8688	0.8678	0.8679	0.8667	0.8671	0.8672
4^1S			0.8714	0.8719	0.8710	0.8713	0.8701
5^3S	$1s \to 5s$	0.8819	0.8814	0.8815	0.8808	0.8811	0.8811
5^1S			0.8832	0.8835	0.8830	0.8832	0.8825
6^3S	$1s \to 6s$	0.8888	0.8885	0.8885	0.8882	0.8883	0.8883
6^1S			0.8895	0.8898	0.8894	0.8896	0.8892
7^3S	$1s \to 7s$	0.8929	0.8927	0.8927	0.8925	0.8926	0.8926
7^1S			0.8933	0.8935	0.8932	0.8934	0.8931
8^3S	$1s \to 8s$	0.8955	0.8954	0.8954	0.8952	0.8953	0.8953
8^1S			0.8958	0.8959	0.8957	0.8958	0.8956
9^3S	$1s \to 9s$	0.8972	0.8972	0.8972	0.8971	0.8971	0.8971
9^1S			0.8975	0.8976	0.8974	0.8975	0.8974
2^3P	$1s \to 2p$	0.7772	0.7702	0.7698	0.7693	0.7688	0.7706
2^1P			0.7764	0.7764	0.7850	0.7844	0.7799
3^3P	$1s \to 3p$	0.8476	0.8456	0.8457	0.8453	0.8453	0.8456
3^1P			0.8483	0.8483	0.8500	0.8501	0.8486
4^3P	$1s \to 4p$	0.8722	0.8714	0.8715	0.8712	0.8713	0.8714
4^1P			0.8726	0.8726	0.8732	0.8733	0.8727
5^3P	$1s \to 5p$	0.8836	0.8832	0.8832	0.8831	0.8831	0.8832
5^1P			0.8838	0.8838	0.8841	0.8842	0.8838
6^3P	$1s \to 6p$	0.8898	0.8895	0.8895	0.8895	0.8895	0.8895
6^1P			0.8899	0.8899	0.8901	0.8901	0.8899
7^3P	$1s \to 7p$	0.8935	0.8933	0.8933	0.8933	0.8933	0.8933
7^1P			0.8935	0.8936	0.8937	0.8937	0.8936
8^3P	$1s \to 8p$	0.8959	0.8958	0.8958	0.8958	0.8958	0.8958
8^1P			0.8959	0.8959	0.8960	0.8960	0.8959
9^3P	$1s \to 9p$	0.8975	0.8975	0.8975	0.8974	0.8975	0.8975
9^1P			0.8976	0.8976	0.8976	0.8976	0.8976
Mean abs. dev.[c]			0.0011	0.0010	0.0010	0.0010	
Mean percentage error			0.15%	0.13%	0.13%	0.13%	

[a]Using the lowest 34 unoccupied orbitals of s and p symmetry, respectively.
[b]Nonrelativistic variational calculation [38].
[c]Mean value of the absolute deviations from the exact values.

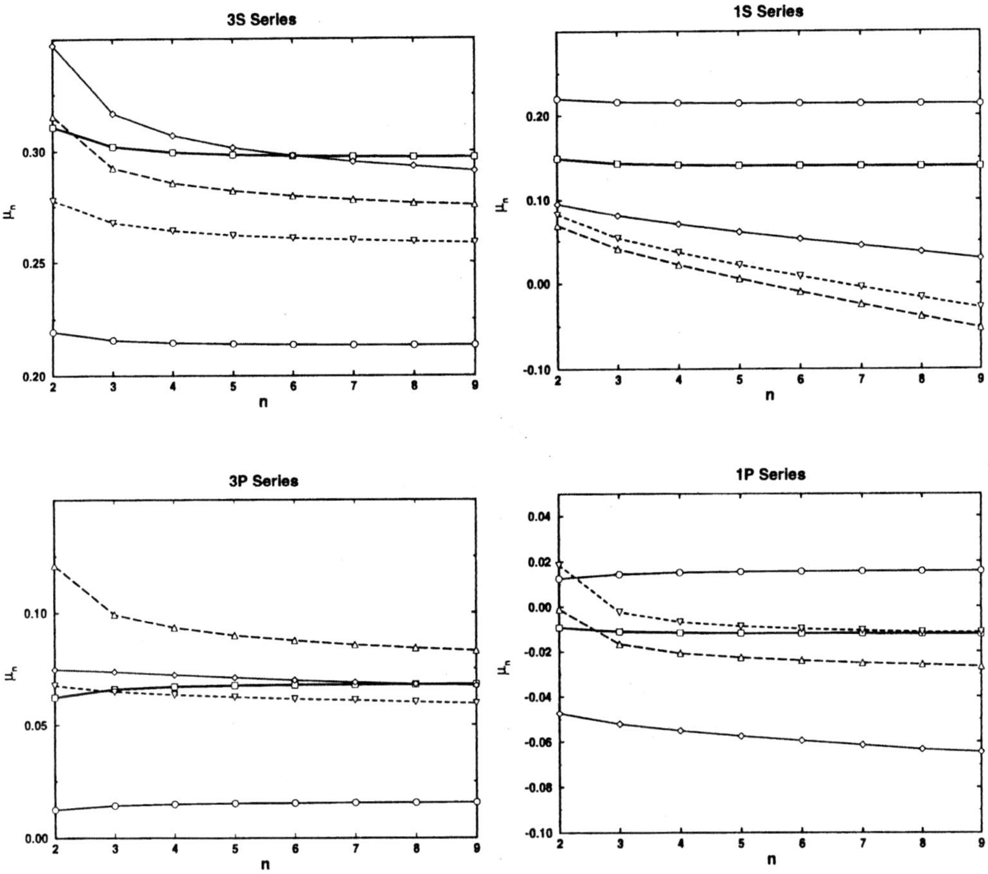

Figure 1. Quantum defects μ_n of the S and P series in Helium, plotted versus the quantum number n, calculated from the exact Kohn-Sham potential of Helium using various approximations for the exchange correlation kernel (triangles up: $f_{\text{x-only}}^{\text{ALDA}}$, triangles down: $f_{\text{xc}}^{\text{ALDA}}$ using the parametrization of Ref. [40], diamonds: $f_{\text{x-only}}^{\text{TDOEP}}$ (Eq. 40)). Also shown are the values from a variational calculation (squares). The circles denote the quantum defects of the uncorrected Kohn-Sham eigenvalues.

Table 2. Singlet-triplet separations in neutral helium calculated from the exact xc potential by using various approximate xc kernels. Calculated from Eq. (0.65), using the lowest 34 unoccupied orbitals of s and p symmetry. All values are in mHartrees.

	ALDA		TDOEP	
State	xc[a]	x-only	x-only	exact[b]
$2S$	32.7	42.2	45.2	29.3
$3S$	9.4	11.1	10.8	7.4
$4S$	4.0	4.7	4.3	2.9
$5S$	2.1	2.4	2.2	1.4
$6S$	1.3	1.4	1.2	0.8
$7S$	0.8	0.9	0.8	0.5
$8S$	0.6	0.6	0.5	0.3
$9S$	0.4	0.5	0.4	0.2
$2P$	6.6	16.7	15.6	9.3
$3P$	2.6	4.5	4.7	2.9
$4P$	1.1	1.8	2.0	1.3
$5P$	0.6	0.9	1.0	0.6
$6P$	0.3	0.5	0.6	0.4
$7P$	0.2	0.3	0.4	0.2
$8P$	0.1	0.2	0.3	0.2
$9P$	0.1	0.2	0.2	0.1
dev.[c]	0.7	2.0	2.0	

[a]Including correlation contributions in the form of Vosko, Wilk and Nusair [40].
[b]Taken from Ref. [38]
[c]Mean absolute deviation from the exact values

xc kernel, which is *not* based on the homogeneous electron gas. Work along the lines of improving the TDOEP kernels by the inclusion of suitable correlation contributions is in progress.

On the other hand, we agree with the authors of Ref. [42] that the inaccuracies introduced by *approximate* ground-state exchange-correlation *potentials* can be substantial. In the LDA and in the popular GGAs for instance, the highest occupied orbital eigenvalue is in error by about a factor of two, due to spurious self-interaction. There may be error cancellations for the lower Kohn-Sham eigenvalue differences, but in general one should not expect to get a reliable (Kohn-Sham) spectrum in LDA and GGAs, because the respective potentials have the wrong behavior for large r. In addition, this causes the number of (unoccupied) bound KS states to be *finite*. It is very unlikely that these defects will be cured by better approximations of $f_{\rm xc}$ alone, since the terms containing $f_{\rm xc}$ only give corrections to the underlying Kohn-Sham eigenvalue spectrum. Hence, the quantitiative calculation of excitation energies heavily depends on the accuracy of the ground-state potential employed.

APPENDIX
The TDOEP integral equation for the exchange-correlation kernel

Within the TDOEP the functional derivatives $\delta v_{\rm xc}/\delta\phi_{j\sigma}$ in (37) have to be extracted from the integral equation for the potential (21). To this end we rewrite the latter as

$$\sum_{\sigma'}\int dt'\int d^3r'\;\Big\{\;\tilde\chi_{{\rm s}\,\sigma\sigma'}(\mathbf{r}t,\mathbf{r}'t')v_{xc\,\sigma'}(\mathbf{r}'t')$$
$$-i\,\delta_{\sigma\sigma'}\sum_{j,l}\Big(u_{xc\,j\sigma'}(\mathbf{r}'t')-u^*_{xc\,l\sigma'}(\mathbf{r}'t')\Big)\phi_{j\sigma'}(\mathbf{r}t)\phi^*_{j\sigma'}(\mathbf{r}'t')\;\phi_{l\sigma'}(\mathbf{r}'t')\phi^*_{l\sigma'}(\mathbf{r}t)\theta(t-t')\;\Big\}=0, \tag{69}$$

where the functions $\tilde\chi_{{\rm s}\,\sigma\sigma'}$ are defined in (35). We now act with $\delta/\delta\phi_{m\nu'}$ on Eq. (69) which yields a system of coupled integral equations

$$\sum_{\sigma'}\int dt'\int d^3r'\;\tilde\chi_{{\rm s}\,\sigma\sigma'}(\mathbf{r}t,\mathbf{r}'t')\frac{\delta v_{xc\,\sigma'}(\mathbf{r}'t')}{\delta\phi_{m\nu'}(\mathbf{y}'\tau')}=$$
$$\sum_{\sigma'}\int dt'\int d^3r'\;\Bigg(\frac{\delta}{\delta\phi_{m\nu'}(\mathbf{y}'\tau')}\Big\{+i\,\delta_{\sigma\sigma'}\sum_{j,l}\Big(\,u_{xc\,j\sigma'}(\mathbf{r}'t')-u^*_{xc\,l\sigma'}(\mathbf{r}'t')\,\Big)$$
$$\times\;\phi_{j\sigma'}(\mathbf{r}t)\phi^*_{j\sigma'}(\mathbf{r}'t')\;\phi_{l\sigma'}(\mathbf{r}'t')\phi^*_{l\sigma'}(\mathbf{r}t)\theta(t-t')\;\Big\}\;-\;\frac{\delta\tilde\chi_{{\rm s}\,\sigma\sigma'}(\mathbf{r}t,\mathbf{r}'t')}{\delta\phi_{m\nu'}(\mathbf{y}'\tau')}\;v_{xc\,\sigma'}(\mathbf{r}'t')\;\Bigg), \tag{70}$$

and similarly for $\delta/\delta\phi^*_{m\nu'}$. To make use of Eq. (70) we transform the integral equation (36) for $f_{\rm xc}$ by acting with $\tilde\chi_{{\rm s}\,\sigma\sigma'}$ from the left

$$\sum_{\sigma'\nu'}\int dt'\int d^3r'\int d\tau'\int d^3y'\;\tilde\chi_{{\rm s}\,\sigma\sigma'}(\mathbf{r}t,\mathbf{r}'t')f^{OEP}_{xc\,\sigma'\nu'}(\mathbf{r}'t',\mathbf{y}'\tau')\tilde\chi_{{\rm s}\,\nu',\nu}(\mathbf{y}'\tau',\mathbf{y}\tau)$$
$$=\sum_{\sigma'\nu'}\int dt'\int d^3r'\int d\tau'\int d^3y'\;\sum_m\tilde\chi_{{\rm s}\,\sigma\sigma'}(\mathbf{r}t,\mathbf{r}'t')$$
$$\times\left[\frac{\delta v_{xc\,\sigma'}(\mathbf{r}'t')}{\delta\phi_{m\nu'}(\mathbf{y}'\tau')}\frac{\delta\phi_{m\nu'}(\mathbf{y}'\tau')}{\delta v_\nu(\mathbf{y}\tau)}+\frac{\delta v_{xc\,\sigma'}(\mathbf{r}'t')}{\delta\phi^*_{m\nu'}(\mathbf{y}'\tau')}\frac{\delta\phi^*_{m\nu'}(\mathbf{y}'\tau')}{\delta v_\nu(\mathbf{y}\tau)}\right]. \tag{71}$$

The functional derivatives $\delta\phi_{j\sigma'}/\delta v_\sigma$ are given by (34). Now we can insert (70) in (71) and obtain

$$\sum_{\sigma'\nu'}\int dt'\int d^3r'\int d\tau'\int d^3y'\; \chi^{(0)}_{\sigma\sigma'}(\mathbf{r}t,\mathbf{r}'t')f^{OEP}_{xc\,\sigma'\nu'}(\mathbf{r}'t',\mathbf{y}'\tau')\chi^{(0)}_{\nu',\nu}(\mathbf{y}'\tau',\mathbf{y}\tau)$$
$$=\sum_{\sigma'\nu'}\int dt'\int d^3r'\int d\tau'\int d^3y'\sum_m\Bigg[$$
$$\Bigg\{\Bigg(+i\delta_{\sigma\sigma'}\frac{\delta}{\delta\phi_{m\nu'}(\mathbf{y}'\tau')}\sum_{j,l}\Big(u_{xc\,j\sigma'}(\mathbf{r}'t')-u^*_{xc\,l\sigma'}(\mathbf{r}'t')\Big)\,\phi_{j\sigma'}(\mathbf{r}t)\phi^*_{j\sigma'}(\mathbf{r}'t')$$
$$\times\,\phi_{l\sigma'}(\mathbf{r}'t')\phi^*_{l\sigma'}(\mathbf{r}t)\theta(t-t')\Bigg)-\frac{\delta\tilde{\chi}_{s\,\sigma\sigma'}(\mathbf{r}t,\mathbf{r}'t')}{\delta\phi_{m\nu'}(\mathbf{y}'\tau')}v_{xc\,\sigma'}(\mathbf{r}'t')\Bigg\}\frac{\delta\phi_{m\nu'}(\mathbf{y}'\tau')}{\delta v_\nu(\mathbf{y}\tau)}$$
$$+\Bigg\{\Bigg(+i\delta_{\sigma\sigma'}\frac{\delta}{\delta\phi^*_{m\nu'}(\mathbf{y}'\tau')}\sum_{j,l}\Big(u_{xc\,j\sigma'}(\mathbf{r}'t')-u^*_{xc\,l\sigma'}(\mathbf{r}'t')\Big)\,\phi_{j\sigma'}(\mathbf{r}t)\phi^*_{j\sigma'}(\mathbf{r}'t')$$
$$\times\,\phi_{l\sigma'}(\mathbf{r}'t')\phi^*_{l\sigma'}(\mathbf{r}t)\theta(t-t')\Bigg)-\frac{\delta\tilde{\chi}_{s\,\sigma\sigma'}(\mathbf{r}t,\mathbf{r}'t')}{\delta\phi^*_{m\nu'}(\mathbf{y}'\tau')}v_{xc\,\sigma'}(\mathbf{r}'t')\Bigg\}\frac{\delta\phi^*_{m\nu'}(\mathbf{y}'\tau')}{\delta v_\nu(\mathbf{y}\tau)}\Bigg]_{\phi_{j\sigma}(t)=\phi^{(0)}_{j\sigma}(t)}. \tag{72}$$

Eq. (72) is within the OEP an exact equation defining the exchange-correlation kernel in terms of functional derivatives of the action functional A_{xc}. After performing the functional derivatives and inserting the stationary orbitals it takes the following form:

$$\sum_{\sigma'\nu'}\int dt'\int d^3r'\int d\tau'\int d^3y'\; \chi^{(0)}_{\sigma\sigma'}(\mathbf{r}t,\mathbf{r}'t')f^{OEP}_{xc\,\sigma'\nu'}(\mathbf{r}'t',\mathbf{y}'\tau')\chi^{(0)}_{\nu',\nu}(\mathbf{y}'\tau',\mathbf{y}\tau)=P_1+P_2 \tag{73}$$

where

$$P_1=\sum_{\sigma'\nu'}\delta_{\sigma\sigma'}\delta_{\nu'\nu}\int dt'\int d^3r'\int d\tau'\int d^3y'\;\theta(t-t')\theta(\tau'-\tau)\sum_{jlmk}\Bigg[$$
$$\Bigg\{\frac{1}{\phi^*_{j\sigma'}(\mathbf{r}'t')}\left(\frac{1}{\phi^*_{m\nu'}(\mathbf{y}'\tau')}\frac{\delta}{\delta\phi_{m\nu'}(\mathbf{y}'\tau')}-\frac{1}{\phi_{k\nu'}(\mathbf{y}'\tau')}\frac{\delta}{\delta\phi^*_{k\nu'}(\mathbf{y}'\tau')}\right)\frac{\delta A_{xc}[\{\phi_{j\sigma}\}]}{\delta\phi_{j\sigma'}(\mathbf{r}'t')}-$$
$$\frac{1}{\phi_{l\sigma'}(\mathbf{r}'t')}\left(\frac{1}{\phi^*_{m\nu'}(\mathbf{y}'\tau')}\frac{\delta}{\delta\phi_{m\nu'}(\mathbf{y}'\tau')}-\frac{1}{\phi_{k\nu'}(\mathbf{y}'\tau')}\frac{\delta}{\delta\phi^*_{k\nu'}(\mathbf{y}'\tau')}\right)\frac{\delta A_{xc}[\{\phi_{j\sigma}\}]}{\delta\phi^*_{l\sigma'}(\mathbf{r}'t')}\Bigg\}$$
$$\times\,\tilde{F}_{\sigma'\,jl}(\mathbf{r}t,\mathbf{r}'t')\tilde{F}^*_{\nu'\,mk}(\mathbf{y}'\tau',\mathbf{y}\tau)\Bigg]_{\phi_{j\sigma}(t)=\phi^{(0)}_{j\sigma}(t)}$$

and

$$P_2 = -i\theta(t-\tau)\delta_{\sigma\nu}\sum_{jlmk}\varphi_{j\sigma}(\mathbf{r})\varphi^*_{l\sigma}(\mathbf{r})\varphi_{m\sigma}(\mathbf{y})\varphi^*_{k\sigma}(\mathbf{y})\int d^3r'\Bigg\{$$

$$e^{-i(\epsilon_{k\sigma}-\epsilon_{m\sigma})(t-\tau)}\Bigg(\frac{\delta_{jk}\,\varphi^*_{m\sigma}(\mathbf{r}')\varphi_{l\sigma}(\mathbf{r}')}{\epsilon_{l\sigma}-\epsilon_{m\sigma}+i\eta}\left[\Big(f_{m\sigma}-f_{k\sigma}\Big)v^{(0)}_{xc\,\sigma}(\mathbf{r}')-u^{(0)}_{xc\,m\sigma}(\mathbf{r}')\right]$$

$$-\frac{\delta_{lm}\varphi^*_{j\sigma}(\mathbf{r}')\varphi_{k\sigma}(\mathbf{r}')}{\epsilon_{k\sigma}-\epsilon_{j\sigma}+i\eta}\left[\Big(f_{m\sigma}-f_{k\sigma}\Big)v^{(0)}_{xc\,\sigma}(\mathbf{r}')+u^{(0)\,*}_{xc\,k\sigma}(\mathbf{r}')\right]\Bigg)$$

$$+e^{-i(\epsilon_{j\sigma}-\epsilon_{l\sigma})(t-\tau)}\Bigg(\frac{\delta_{jk}\,\varphi^*_{m\sigma}(\mathbf{r}')\varphi_{l\sigma}(\mathbf{r}')}{\epsilon_{l\sigma}-\epsilon_{m\sigma}+i\eta}\left[\Big(f_{j\sigma}-f_{l\sigma}\Big)v^{(0)}_{xc\,\sigma}(\mathbf{r}')+u^{(0)\,*}_{xc\,l\sigma}(\mathbf{r}')\right]$$

$$-\frac{\delta_{lm}\varphi^*_{j\sigma}(\mathbf{r}')\varphi_{k\sigma}(\mathbf{r}')}{\epsilon_{k\sigma}-\epsilon_{j\sigma}+i\eta}\left[\Big(f_{j\sigma}-f_{l\sigma}\Big)v^{(0)}_{xc\,\sigma}(\mathbf{r}')-u^{(0)}_{xc\,j\sigma}(\mathbf{r}')\right]\Bigg)\Bigg\}$$

with

$$\tilde{F}_{\sigma'\,jl}(\mathbf{r}t,\mathbf{r}'t') = \phi_{j\sigma'}(\mathbf{r}t)\phi^*_{j\sigma'}(\mathbf{r}'t')\phi_{l\sigma'}(\mathbf{r}'t')\phi^*_{l\sigma'}(\mathbf{r}t)\,. \tag{74}$$

Some simplification of this equation can be achieved for action functionals depending on the orbitals only through the spin density matrices $\Gamma_\sigma(\mathbf{r}t,\mathbf{r}'t')=\sum_j f_{j\sigma}\phi_{j\sigma}(\mathbf{r}t)\phi^*_{j\sigma}(\mathbf{r}'t')$. In this case the OEP integral equations (21) for the stationary xc-potential and the exchange-correlation kernel $f_{\mathrm{xc}\,\sigma\sigma'}$, respectively, read as follows:

$$\int dt'\int d^3r'\int d^3z'\;\kappa^{(0)}_\sigma(\mathbf{r}t;\mathbf{r}',\mathbf{z}',t')\left(v^{(0)}_{xc\,\sigma}(\mathbf{r}'t')\delta(\mathbf{r}'-\mathbf{z}')-\left.\frac{\delta A_{xc}}{\delta\Gamma_\sigma(\mathbf{z}',\mathbf{r}';t')}\right|_{\phi^{(0)}}\right) = 0 \tag{75}$$

and

$$\int dt'\int d\tau'\int d^3r'\int d^3z\int d^3y'\int d^3z'\;\kappa^{(0)}_\sigma(\mathbf{r}t;\mathbf{z},\mathbf{r}',t')\;\kappa^{(0)\,\dagger}_\nu(\mathbf{z}',\mathbf{y}',\tau';\mathbf{y}\tau)$$

$$\times\left[f^{OEP}_{xc\,\sigma\nu}(\mathbf{r}'t',\mathbf{y}'\tau')\delta(\mathbf{z}-\mathbf{r}')\delta(\mathbf{z}'-\mathbf{y}')-\left.\frac{\delta^2A_{xc}}{\delta\Gamma_\nu(\mathbf{y}',\mathbf{z}';\tau')\delta\Gamma_\sigma(\mathbf{r}',\mathbf{z};t')}\right|_{\phi_{j\sigma}(t)=\phi^{(0)}_{j\sigma}(t)}\right]$$

$$= -i\,\theta(t-\tau)\delta_{\sigma\nu}\sum_{jlmk}\varphi_{j\sigma}(\mathbf{r})\varphi^*_{l\sigma}(\mathbf{r})\varphi_{m\sigma}(\mathbf{y})\varphi^*_{k\sigma}(\mathbf{y})$$

$$\times\Bigg\{\Bigg(\Big(f_{m\sigma}-f_{k\sigma}\Big)e^{-i(\epsilon_{k\sigma}-\epsilon_{m\sigma})(t-\tau)}+\Big(f_{j\sigma}-f_{l\sigma}\Big)e^{-i(\epsilon_{j\sigma}-\epsilon_{l\sigma})(t-\tau)}\Bigg)$$

$$\times\Bigg(\frac{\delta_{jk}}{\epsilon_{l\sigma}-\epsilon_{m\sigma}+i\eta}<\varphi_{m\sigma}\mid v^{(0)}_{xc\,\sigma}-\left.\frac{\delta A_{xc}}{\delta\Gamma_\sigma}\right|_{\phi^{(0)}}\mid\varphi_{l\sigma}>$$

$$-\frac{\delta_{lm}}{\epsilon_{k\sigma}-\epsilon_{j\sigma}+i\eta}<\varphi_{j\sigma}\mid v^{(0)}_{xc\,\sigma}-\left.\frac{\delta A_{xc}}{\delta\Gamma_\sigma}\right|_{\phi^{(0)}}\mid\varphi_{k\sigma}>\Bigg)\Bigg\}. \tag{76}$$

Here

$$\kappa^{(0)}_\sigma(\mathbf{r}t;\mathbf{z},\mathbf{r}',t') = i\sum_{jl}\Big(f_{j\sigma}-f_{l\sigma}\Big)\varphi_{j\sigma}(\mathbf{r})\varphi^*_{j\sigma}(\mathbf{z})\;\varphi_{l\sigma}(\mathbf{r}')\varphi^*_{l\sigma}(\mathbf{r})e^{-i(\epsilon_{j\sigma}-\epsilon_{l\sigma})(t-t')}\theta(t-t') \tag{77}$$

describes the density response of a noninteracting system to a nonlocal perturbation,

$$\kappa_{\nu}^{(0)\dagger}(\mathbf{z}',\mathbf{y}',\tau';\mathbf{y}\tau) =$$
$$-i\sum_{mk}\Big(f_{m\sigma}-f_{k\sigma}\Big)\varphi_{m\nu}^{*}(\mathbf{z}')\varphi_{m\nu}(\mathbf{y})\ \varphi_{k\nu}^{*}(\mathbf{y})\varphi_{k\nu}(\mathbf{y}')e^{-i(\epsilon_{k\nu}-\epsilon_{m\nu})(\tau'-\tau)}\theta(\tau'-\tau). \quad (78)$$

is its adjoint and we adopted the notation

$$< \varphi_{j\sigma} \mid v_{xc\,\sigma}^{(0)} - \left.\frac{\delta A_{xc}}{\delta\Gamma_\sigma}\right|_{\phi^{(0)}} \mid \varphi_{l\sigma} >$$
$$= \int d^3r' \int d^3y'\ \varphi_{j\sigma}^{*}(\mathbf{y}')\left(v_{xc\,\sigma}^{(0)}(\mathbf{r}')\delta(\mathbf{r}'-\mathbf{y}') - \left.\frac{\delta A_{xc}}{\delta\Gamma_\sigma(\mathbf{r}',\mathbf{y}')}\right|_{\phi^{(0)}}\right)\varphi_{l\sigma}(\mathbf{r}'). \quad (79)$$

We remark that the KLI approximation scheme amounts to replacing κ^0 by

$$\kappa_{\sigma}^{(0)\,\mathrm{KLI}}(\mathbf{r}t;\mathbf{r}',\mathbf{z}',t') = i\sum_{jk}\Big(f_{j\sigma}+f_{k\sigma}-2\delta_{j\,k}f_{j\sigma}\Big)\varphi_{j\sigma}(\mathbf{r})\varphi_{j\sigma}^{*}(\mathbf{r}')\ \varphi_{k\sigma}(\mathbf{z}')\varphi_{k\sigma}^{*}(\mathbf{r})\delta(t-t') \quad (80)$$

and that the approximation (38) corresponds to omission of the term $-2\delta_{j\,k}f_{j\sigma}$. These manipulations can also be applied on the lhs of the integral equation (76). One is also tempted to neglect the rhs of Eq. (76) because the combination of matrix elements (79) appearing here should be forced to be small by Eq. (75). In this way one gets fully analytic approximations for $f_{\mathrm{xc}\,\sigma\sigma'}$. The resulting expressions are however still quite complicated. In the applications discussed in this paper they perform also not better than the simple formula (39).

ACKNOWLEDGMENTS

We thank Cyrus Umrigar for providing us with the exact KS potential of the He atom. Partial financial support by the Deutsche Forschungsgemeinschaft is gratefully acknowledged.

REFERENCES

1. E. Runge and E.K.U. Gross, Phys. Rev. Lett. **52**, 997 (1984).
2. K.L. Liu and S.H. Vosko, Can. J. Phys. **67**, 1015 (1989).
3. E.K.U. Gross, J.F. Dobson, and M. Petersilka, in *Density Funcional Theory II*, Vol. 181 of *Topics in Current Chemistry*, edited by R.F. Nalewajski (Springer, Berlin, 1996), p. 81.
4. C.A. Ullrich, U.J. Gossmann, and E.K.U. Gross, Phys. Rev. Lett. **74**, 872 (1995).
5. R.T. Sharp and G.K. Horton, Phys. Rev. **90**, 317 (1953).
6. J.D. Talman and W.F. Shadwick, Phys. Rev. A **14**, 36 (1976).
7. J.B. Krieger, Y. Li, and G.J. Iafrate, Phys. Rev. A **45**, 101 (1992).
8. J.B. Krieger, Y. Li, and G.J. Iafrate, Phys. Rev. A **46**, 5453 (1992).
9. J.B. Krieger, Y. Li, and G.J. Iafrate, in *Density Functional Theory*, Vol. 337 of *NATO ASI Series B*, edited by E.K.U. Gross and R.M. Dreizler (Plenum Press, New York, 1995), p. 191, and references therein.
10. T. Grabo and E.K.U. Gross, Chem. Phys. Lett. **240**, 141 (1995).
11. J. Chen, J.B. Krieger, Y. Li, and G.J. Iafrate, Phys. Rev. A **54**, 3939 (1996).
12. E.K.U. Gross, C.A. Ullrich, and U.J. Gossmann, in *Density Functional Theory*, Vol. 337 of *NATO ASI Series B*, edited by E.K.U. Gross and R.M. Dreizler (Plenum Press, New York, 1995), p. 149.
13. C.A. Ullrich, U.J. Gossmann, and E.K.U. Gross, Ber. Bunsenges. Phys. Chem. **99**, 488 (1995).

14. C.A. Ullrich, S. Erhard, and E.K.U. Gross, in *Super Intense Laser Atom Physics IV*, edited by H.G. Muller and M.V. Fedorov (Kluwer, Dordrecht, 1996), p. 267.
15. C.A. Ullrich and E.K.U. Gross, Comments on Atomic and Molecular Physics (1997), in press.
16. C.A. Ullrich, P.G. Reinhard, and E. Suraud, J. Phys. B: At. Mol. Opt. Phys. (1997), to appear.
17. J.C. Slater, Phys. Rev. **81**, 385 (1951).
18. O. Gunnarsson and B.I. Lundqvist, Phys. Rev. B **13**, 4274 (1976).
19. T. Ziegler, A. Rauk, and E.J. Baerends, Theoret. Chim. Acta **43**, 261 (1977).
20. U. von Barth, Phys. Rev. A **20**, 1693 (1979).
21. A. Theophilou, J. Phys. C **12**, 5419 (1979).
22. N. Hadjisavvas and A. Theophilou, Phys. Rev. A **32**, 720 (1985).
23. W. Kohn, Phys. Rev. A **34**, 5419 (1986).
24. E.K.U. Gross, L.N. Oliveira, and W. Kohn, Phys. Rev. A **37**, 2805 (1988).
25. E.K.U. Gross, L.N. Oliveira, and W. Kohn, Phys. Rev. A **37**, 2809 (1988).
26. L.N. Oliveira, E.K.U. Gross, and W. Kohn, Phys. Rev. A **37**, 2821 (1988).
27. A. Nagy, Phys. Rev. A **42**, 4388 (1990).
28. A. Nagy, J. Phys. B **24**, 4691 (1991).
29. A. Nagy, Phys. Rev. A **49**, 3074 (1994).
30. A. Nagy, Int. J. Quantum Chem. Symp. **29**, 297 (1995).
31. M. Levy, Phys. Rev. A **52**, R4313 (1995).
32. P.K. Chattaraj, S.K. Ghosh, S. Liu, and R.G. Parr, Int. J. Quant. Chem, in press.
33. A. Görling, Phys. Rev. A **54**, 3912 (1996).
34. M. Petersilka, U.J. Gossmann, and E.K.U Gross, Phys. Rev. Lett. **76**, 1212 (1996).
35. M. Petersilka and E.K.U. Gross, Int. J. Quantum Chem. Symp. **30**, 1393 (1996).
36. Ch. Jamorski, M.E. Casida, and D.R. Salahub, J. Chem. Phys. **104**, 5134 (1996).
37. C.J. Umrigar and X. Gonze, Phys. Rev. A **50**, 3827 (1994).
38. A. Kono and S. Hattori, Phys. Rev. A **29**, 2981 (1984).
39. E.K.U. Gross and W. Kohn, Adv. Quant. Chem. **21**, 255 (1990).
40. S.H. Vosko, L. Wilk, and M. Nusair, Can. J. Phys. **58**, 1200 (1980).
41. C. Umrigar, A. Savin, and X. Gonze, contribution to present volume.
42. S.J.A. van Gisbergen, V.P. Osinga, O.V. Gritsenko, R. van Leeuwen, J.G. Snijders, and E.J. Baerends, J. Chem. Phys. **105**, 3142 (1996).

CURRENT-DENSITY FUNCTIONAL THEORY OF LINEAR RESPONSE TO TIME-DEPENDENT ELECTROMAGNETIC FIELDS

G. Vignale[1] and Walter Kohn,[2]

[1] Department of Physics, University of Missouri
Columbia, Missouri 65211

[2]Department of Physics, University of California
Santa Barbara, California 93106-4030

INTRODUCTION

Local density approximations are known to be very useful in calculating the ground-state exchange and correlation (xc) energy of many-electron systems [1], and local approximations are gaining in importance for the description of xc effects in time-dependent situations also [2,3]. Most of the time-dependent work has dealt with a *scalar* time-dependent xc potential $v_{xc}(\vec{r},\omega)$ approximated as a local functional of the time-dependent density $n(\vec{r},\omega)$ as described in the chapter "Time-dependent density functional theory" in the introductory section of this Book. Despite much progress in this scalar approach, it will be shown in this Chapter that existing approximations for v_{xc} require modification. In particular, it will be shown that (i) there is no local density approximation for the scalar xc potential at finite frequency (ii) there is, however, a consistent local approximation for a *vector* xc potential $\vec{a}_{xc}(\vec{r},\omega)$ in terms of the dynamic current density $\vec{j}(\vec{r},\omega)$ and its space derivatives, as well as the ground-state density and its space derivatives. This approximation is valid, at a given frequency, for sufficiently slow spatial variations of the ground-state density and of the perturbing dynamic potential. The appropriately modified vector xc potential will be derived here in some detail, thus filling out the brief description published recently [4]. New material will also be presented, providing interpretation of the findings in the simple case of one dimensional inhomogeneity.

The time dependent density functional theory (TDFT) of Runge and Gross [5] provides the starting point for the present development. Their approach is to transform the problem of many interacting particles subjected to a time-dependent external potential $v(\vec{r},t)$, to an equivalent problem of noninteracting electrons, in an effective potential

$$v_{eff}(\vec{r},t) = v(\vec{r},t) + v_H(\vec{r},t) + v_{xc}(\vec{r},t), \tag{1}$$

Electronic Density Functional Theory: *Recent Progress and New Directions*
Edited by Dobson *et al.*, Plenum Press, New York, 1998

which yields the same time-dependent density. The Hartree-potential

$$v_H(\vec{r},t) = \int \frac{n(\vec{r}',t)}{|\vec{r}-\vec{r}'|} d\vec{r}' \tag{2}$$

is known exactly. The "exchange-correlation" (xc) potential $v_{xc}(\vec{r},t)$ - a functional of the density $n(\vec{r},t)$ - needs to be approximated. It is here that a local density approximation (LDA) - whereby $v_{xc}(\vec{r},t)$ is taken to be a *function* of the density at point $\vec{r}$ - would be desirable. The simplest approximation of this kind, known as the "adiabatic local density approximation" (ALDA) [3] assumes that the time-dependent xc potential has the functional form of the *static* LDA, only with a time-dependent density

$$v_{xc}^{ALDA}[n(\vec{r},t);\vec{r},t] = \frac{d}{dn}\epsilon_{xc}^{h}(n(\vec{r},t)), \tag{3}$$

where $\epsilon_{xc}^h(n)$ is the xc energy density of the homogeneous electron gas of density n. Clearly, this is only justifiable when the external potential is slowly varying in time as well as in space.

In a first step towards overcoming the restriction to slow variation in time, Gross and Kohn [2] considered the *linear* density response $n_1(\vec{r},\omega)e^{-i\omega t}$ of a system of interacting electrons in their ground-state to an external time-dependent potential $v_1(\vec{r},\omega)e^{-i\omega t}$. The linearized effective potential has the form

$$v_{eff}(\vec{r},t) = v_{0,eff}(\vec{r}) + v_{1,eff}(\vec{r},\omega)e^{-i\omega t}, \tag{4}$$

where $v_{0,eff}(\vec{r})$ is the static effective potential for the system in equilibrium (including Hartree and xc contributions),

$$v_{1,eff}(\vec{r},\omega) = v_1(\vec{r},\omega) + \int \frac{n_1(\vec{r}',\omega)}{|\vec{r}-\vec{r}'|} d\vec{r}' + v_{1xc}(\vec{r},\omega), \tag{5}$$

is the time-dependent component, to first order in the external potential, and $v_{1xc}(\vec{r},\omega)$ is linear in $n_1(\vec{r},\omega)$,

$$v_{1xc}(\vec{r},\omega) = \int f_{xc}(\vec{r},\vec{r}';\omega) n_1(\vec{r}',\omega) d\vec{r}'. \tag{6}$$

The kernel $f_{xc}(\vec{r},\vec{r}';\omega)$ is a functional of the unperturbed ground state density $n_0(\vec{r})$.

In the spirit of the local density approximation Gross and Kohn (GK) [2] considered the case where both n_0 and n_1 are sufficiently slowly varying functions of $\vec{r}$. As f_{xc} is of short range [6] for a *homogeneous* system, they proposed the following plausible approximation for systems of slowly varying $n_0(\vec{r})$:

$$f_{xc}(\vec{r},\vec{r}';\omega) \sim f_{xc}^h(|\vec{r}-\vec{r}'|,\omega;n_0(\vec{r})). \tag{7}$$

The superscript h refers to a homogeneous electron gas and the function f_{xc}^h is a property of the homogeneous electron gas [2,7]. Because $n_1(\vec{r},\omega)$ is also slowly varying, Eqs. (6) and (7) together lead to the approximation

$$v_{1xc}(\vec{r},\omega) = f_{xc}^h(\omega;n_0(\vec{r}))n_1(\vec{r},\omega), \tag{8}$$

where $f_{xc}^h(\omega;n_0) \equiv \int f_{xc}^h(|\vec{r}-\vec{r}'|,\omega;n_0(\vec{r}))d\vec{r}'$. (Note that the linearized ALDA is obtained from this, putting $\omega = 0$ in f_{xc}^h).

However it was noted later by Dobson [8] that the approximation (8), when applied to an electron gas in a static harmonic potential $\frac{1}{2}kr^2$ and subjected to a uniform electric field, $v_1(\vec{r},\omega) = -\vec{E}\cdot\vec{r}e^{-i\omega t}$, violates the so called harmonic potential theorem (HPT), related to the generalized Kohn's theorem [9], according to which the density follows rigidly the classical motion of the center of mass: $n_1(\vec{r},\omega) = \vec{\nabla} n_0(\vec{r})\cdot\vec{R}_{CM}(\omega)$. The violation is a genuine one: that is, it cannot be dismissed as a manifestation of gradient corrections that are obviously not included in the LDA (see next section, where it is shown that the error occurs in the leading order of a gradient expansion of the exchange-correlation potential). This failure raised serious questions about the validity of the approximation (7). Dobson observed that one could satisfy the HPT by requiring that the GK approximation (7) be applied in a frame of reference moving with the local velocity of the electron fluid. The mathematical implementation of this idea leads to the following approximation [8]:

$$v_{1xc}(\vec{r},\omega) = f^h_{xc}(\omega; n_0(\vec{r}))[n_1(\vec{r},\omega) - \frac{\vec{\nabla} n_0(\vec{r})\cdot\vec{j}_1(\vec{r},\omega)}{i\omega n_0(\vec{r})}] + f^h_{xc}(\omega=0; n_0(\vec{r}))\frac{\vec{\nabla} n_0(\vec{r})\cdot\vec{j}_1(\vec{r},\omega)}{i\omega n_0(\vec{r})}, \tag{9}$$

where the second term arises from the change in density due to a rigid displacement of the local equilibrium density, and the first one from the local compression (or expansion) of the electron fluid (This construction has recently been improved and extended to the nonlinear regime [10].) The most interesting feature of this approximation is that the separation of n_1 into translational and compressional components requires the knowledge of the local fluid velocity $\vec{j}_1/n_0$. Thus, the current density $\vec{j}_1$ makes its first appearance in the theory.

A deeper understanding of the problem with approximation (7) was achieved by examining the behavior of the theory under transformation to an accelerated frame of reference [11,12]. Vignale observed [12] that the covariance of the time-dependent Schrödinger equation under such transformations requires that, if $n'(\vec{r},t) = n(\vec{r}+\vec{x}(t),t)$ where $n(\vec{r},t)$ is an arbitrary time-dependent density, and $\vec{x}(t)$ is an arbitrary time-dependent vector, then the exchange-correlation potentials associated with the densities n and n' must satisfy the symmetry relation

$$v_{xc}[n';\vec{r},t] = v_{xc}[n;\vec{r}+\vec{x}(t),t]. \tag{10}$$

This symmetry is violated by Eq. (7). More precisely, one can deduce from (10) (see Ref. [13] and the next section of this paper) that $f_{xc}(\vec{r},\vec{r}',\omega)$ for a non-uniform system is of long range in space and does not admit a gradient expansion. Thus, a consistent local-density approximation for the time-dependent xc potential in general does not exist.

At this point one should ask whether this surprising result has its root in the physics of the problem, or it is simply a consequence of our choice to represent the xc potential as a functional of the density. In a recent paper [4] we have argued that the second answer is the correct one. We have shown that a consistent local approximation for the time-dependent linear response theory can be constructed after replacing the scalar xc potential by a vector xc potential $\vec{a}_{1xc}(\vec{r},\omega)$, and the density by the current-density. The central result of [4] was a local expression for $\vec{a}_{1xc}(\vec{r},\omega)$ in terms of properties of the homogeneous electron gas. This expression satisfies the HPT and all the other symmetries required by transformations to accelerated frames. The purpose of this paper is to supply more steps in the derivation of this result, and to give a more complete discussion of various conceptual points. The plan is the following: In Section

II we explain in detail the failure of the standard LDA in time-dependent DFT. In Section III we generalize the GK theory by allowing a time-dependent vector potential to be present, and introduce the current-density as the basic variable. In Section IV we derive some exact properties of the tensorial kernel $\mathbf{f}_{xc,ij}(\vec{r},\vec{r}',\omega)$, which replaces the scalar f_{xc} of Eq. (6). In Section V we use these exact properties to construct the exact $\mathbf{f}_{xc,ij}$ in an almost uniform electron gas of slowly varying density. In Section VI we derive the correct form of $\vec{a}_{1xc}$ for finite frequencies, slowly varying densities and perturbing potential. Finally, in Section VII, we apply the general formula for $\vec{a}_{1xc}$ to the special case of a system which is inhomogeneous only in one direction: in this case, the correct scalar f_{xc} can also be calculated, and its long-rangedness is explicitly displayed.

THE DEMISE OF THE LOCAL-DENSITY APPROXIMATION FOR THE TIME-DEPENDENT EXCHANGE-CORRELATION POTENTIAL

We start from the exact result (10), derived in Ref. [12]. Let us take $n(\vec{r},t)$ to be the unperturbed density $n_0(\vec{r})$, and $\vec{x}(t) = x(\omega)e^{-i\omega t}$ to be a *small* periodic displacement with frequency ω. To first order in $\vec{x}$, $n'(\vec{r},t) \simeq n_0(\vec{r}) + \vec{x}(t)\cdot\vec{\nabla}n_0(\vec{r})$. Then, defining $v_{xc0}(\vec{r}) \equiv v_{xc}[n_0;\vec{r}]$ - the static xc potential - and using the definition of f_{xc} in (6) we can write

$$v_{xc}[n';\vec{r},t] = v_{xc0}(\vec{r}) + \int f_{xc}(\vec{r},\vec{r}',t-t')\vec{x}(t')\vec{\nabla}'n_0(\vec{r}')d\vec{r}'dt'$$

and

$$v_{xc}[n;\vec{r}+\vec{x}(t),t] = v_{xc0}(\vec{r}) + \vec{x}(t)\cdot\vec{\nabla}v_{xc0}(\vec{r}).$$

Substituting these in Eq. (10), and using the arbitrariness of $\vec{x}(\omega)$ yields the sum rule

$$\int f_{xc}(\vec{r},\vec{r}';\omega)\vec{\nabla}'n_0(\vec{r}')d\vec{r}' = \vec{\nabla}v_{0xc}(\vec{r}), \tag{11}$$

for all frequencies. Let us now assume that the kernel $f_{xc}(\vec{r},\vec{r}',\omega)$ - a functional of the unperturbed density $n_0(\vec{r})$- admits a gradient expansion. This means that, when $n_0(\vec{r})$ is sufficiently slowly varying, one can expand

$$f_{xc}(\vec{r},\vec{r}',\omega) = f^h_{xc}(\vec{r}-\vec{r}',\omega;n_0(\vec{r})) + \vec{g}^h_{xc}(\vec{r}-\vec{r}',\omega;n_0(\vec{r}))\cdot\vec{\nabla}n_0(\vec{r}) + O((\vec{\nabla}n_0)^2,\nabla\nabla n_0), \tag{12}$$

where f^h_{xc}, $\vec{g}^h_{xc}$ etc. are properties of the homogeneous electron gas at density $n_0(\vec{r})$. The tacit assumption, in writing such an expansion, is that the kernels f^h_{xc}, $\vec{g}^h_{xc}$ ... are of *finite* range [6] in $|\vec{r}-\vec{r}'|$, while the density is slowly varying on the scale of this range. The static xc potential also has a gradient expansion,

$$v_{0xc}(\vec{r}) = \frac{d\epsilon^h_{xc}(n)}{dn}|_{n=n_0(\vec{r})} + O((\vec{\nabla}n_0)^2)\,. \tag{13}$$

Substituting these expansions in Eq. (11), we find, to first order in $\vec{\nabla}n_0(\vec{r})$:

$$f^h_{xc}(\omega,n_0(\vec{r}))\vec{\nabla}n_0(\vec{r}) = \frac{d^2\epsilon^h_{xc}(n)}{dn^2}|_{n=n_0(\vec{r})}\vec{\nabla}n_0(\vec{r}), \tag{14}$$

for all ω's. This is impossible, unless $f^h_{xc}(\omega)$ is independent of frequency and equal to $d^2\epsilon^h_{xc}(n)/dn^2$ (as in the ALDA). Since f^h_{xc} is known to be frequency-dependent, we have found a contradiction showing that the gradient expansion hypothesized in Eq. (12) does not exist. In order for Eq. (11) to be satisfied at all frequencies, $f_{xc}(\vec{r},\vec{r}',\omega)$ *must be of infinite range in* $|\vec{r}-\vec{r}'|$.

This assertion is, at first glance, surprising, in view of the fact that the f_{xc} of a homogeneous electron gas is known to have a finite range. However, a nonuniform system has an extra degree of freedom, namely it can be translated in space as a whole (such a translation would be unobservable in a uniform gas). Imagine now a finite system whose density is uniform in a neighborhood of a point $\vec{r}$, far away from the boundaries of the system. It is evident that the density in a neighborhood of $\vec{r}$ is "blind" to a global rigid displacement of the system. Therefore, in order to produce the correct xc potential, the kernel f_{xc} must be able to probe changes in density at large distances from $\vec{r}$ (comparable to the size of the system). Hence, f_{xc} must be long-ranged.

This conclusion is consistent with previous results. For instance, the "ad-hoc" xc potentials introduced in Refs. [8] and [12] to satisfy the HPT, had to be nonlocal functionals of the density, and the corresponding kernels were long-ranged.

The Dobson functional (9) is particularly interesting because it invokes a knowledge of the current density, which is related to the density by the continuity equation

$$\vec{\nabla}\cdot\vec{j}(\vec{r},\omega)=i\omega n(\vec{r},\omega). \tag{15}$$

What makes the current density interesting in the present context is the fact that (in contrast to the density) it provides a local indicator of a global rigid translation of the system. From Eq. (15) one sees that the *longitudinal* component of the current density has the form of an electrostatic field generated by the (complex) charge distribution $i\omega n(\vec{r},\omega)/4\pi$. Thus, the relation between $\vec{j}$ and n is a long-ranged one. All these considerations suggest that the nonlocality of ordinary TDFT can be imputed to the nonlocal relation between the current and the density. A theory formulated in terms of the current-density should recover short-ranged kernels, and hence be amenable to the local density approximation.

Current density as the basic dynamical variable

Let us consider the linear *current* response $\vec{j}_1(\vec{r},\omega)e^{-i\omega t}$ of interacting electrons in their ground-state to a time-dependent *vector* potential $\vec{a}_1(\vec{r},\omega)e^{-i\omega t}$. This problem includes, as a special case, the scalar potential problem studied by GK, because, by gauge invariance, any scalar potential $v_1(\vec{r},\omega)$ is equivalent to a longitudinal vector potential

$$\vec{a}_1(\vec{r},\omega)=\frac{\vec{\nabla}v_1(\vec{r},\omega)}{i\omega}. \tag{16}$$

The density response can be calculated from the current response using the continuity equation. In addition, the formalism will allow us to calculate the response of the system to a transverse electromagnetic field, which was not permitted in the formulation of GK.

As usual, we express the exact induced current as the response of a non-interacting reference system (the "Kohn-Sham" system) to an effective vector potential $\vec{a}_1^{eff}=\vec{a}_1+\vec{a}_{1H}+\vec{a}_{1xc}$:

$$j_{1i}(\vec{r},\omega)=\int\sum_j\chi_{KS,ij}(\vec{r},\vec{r}',\omega)a_{1j}^{eff}(\vec{r}',\omega)d\vec{r}', \tag{17}$$

where

$$\vec{a}_{1H}(\vec{r},\omega) = \frac{1}{(i\omega)^2}\int \vec{\nabla}\frac{1}{|\vec{r}-\vec{r}'|}(\vec{\nabla}'\cdot\vec{j}_1(\vec{r}',\omega))d\vec{r}' \tag{18}$$

is the longitudinal vector potential corresponding to the dynamic Hartree potential of Eq. (2), and

$$\vec{a}_{1xc}(\vec{r},\omega) = \int \mathbf{f}_{xc}(\vec{r},\vec{r}',\omega)\cdot\vec{j}_1(\vec{r}',\omega)d\vec{r}' \tag{19}$$

is a linear functional of the current. The 3×3 tensor kernel $\mathbf{f}_{xc}$ is defined as

$$\mathbf{f}_{xc,ij}(\vec{r},\vec{r}',\omega) = \chi_{KS,ij}^{-1}(\vec{r},\vec{r}',\omega) - \chi_{ij}^{-1}(\vec{r},\vec{r}',\omega) + \frac{1}{\omega^2}\nabla_i\frac{1}{|\vec{r}-\vec{r}'|}\nabla'_j, \tag{20}$$

where χ and χ_{KS} are the current response tensors [14] of the interacting system and the "Kohn-Sham" noninteracting system respectively. The latter is given as

$$\chi_{KS,ij}(\vec{r},\vec{r}',\omega) = \frac{n_0(\vec{r})}{m}\delta(\vec{r}-\vec{r}')\delta_{ij} + \frac{1}{m^2}\sum_{\alpha,\beta}(f_\alpha - f_\beta)\frac{\psi_\alpha^*(\vec{r})\nabla_i\psi_\beta(\vec{r})\psi_\beta^*(\vec{r}')\nabla'_j\psi_\alpha(\vec{r}')}{\omega-(\epsilon_\beta-\epsilon_\alpha)+i\eta}, \tag{21}$$

where $\psi_\alpha(\vec{r})$ are the solutions of the static Kohn-Sham equation, with eigenvalues ϵ_α. (Note that we have put $e = c = 1$).

It is evident that the tensorial kernel $\mathbf{f}_{xc,ij}$ is the central unknown quantity of the formalism. For a *homogeneous* electron gas of density $\bar{n}$, $\mathbf{f}_{xc}$ is a function of $\vec{r}-\vec{r}'$, which can be Fourier-transformed to $\mathbf{f}^h_{xc}(\vec{k},\omega)$. In the following, we shall be interested in the limit of small wave vector k, which, for fixed frequency ω, means $k << k_F, \omega/v_F$ where k_F is the Fermi momentum, and v_F is the Fermi velocity. In this regime, $\mathbf{f}^h_{xc}$ is a quadratic function of $\vec{k}$ (see Appendix for details), and can be written as the sum of a longitudinal and a transverse part (this follows from the isotropy of the electron gas):

$$\mathbf{f}^h_{xc,ij}(\vec{k},\omega) = \frac{1}{\omega^2}[f^h_{xcL}(\omega,\bar{n})k_ik_j + f^h_{xcT}(\omega,\bar{n})(k^2\delta_{ij}-k_ik_j)]. \tag{22}$$

The factor $1/\omega^2$ has been put in evidence so that the function $f^h_{xcL}(\omega,\bar{n})$ coincides with the function $f^h_{xc}(k,\omega,\bar{n})$ introduced by GK (see Eq. (7)). Both $f^h_{xcL}(\omega,\bar{n})$ and $f^h_{xcT}(\omega,\bar{n})$ can, in principle, be computed from the current response function of the homogeneous electron gas (see Eq. (20)). While some aspects of these functions are already known (see Appendix), and considerable progress has been recently made [7], further work will be needed for a complete evaluation. In particular, it is not obvious that the limit, for $\omega\to 0$, of $f^h_{xcL(T)}(\omega,\bar{n})$ is the same as the limit, for $k\to 0$, of the corresponding *static* functions $f^h_{xcL(T)}(k,\omega=0,\bar{n})$, as tacitly assumed by GK. In this paper we shall assume that the frequency is large enough to satisfy the condition $\omega/v_F >> k$. We shall denote by $f^h_{xcL(T)}(\omega=0,\bar{n})$ the $lim_{\vec{k}\to 0}f^h_{xcL(T)}(\vec{k},\omega=0,\bar{n})$, leaving open the possibility that it might not coincide with $lim_{\omega\to 0}f^h_{xcL(T)}(\vec{k}=0,\omega,\bar{n})$.

From Eq. (22) we can obtain the xc potential in real space for a homogeneous electron gas subjected to a perturbation which is slowly varying on the scales k_F^{-1} ($\sim$ interelectron distance) and v_F/ω ($\sim$ distance travelled by an electron during a period of the perturbing field):

$$\vec{a}^h_{1xc}(\vec{r},\omega) = -\frac{1}{\omega^2}\{\vec{\nabla}[f^h_{xcL}(\omega,\bar{n})\vec{\nabla}\cdot\vec{j}_1(\vec{r},\omega)] - \vec{\nabla}\times[f^h_{xcT}(\omega,\bar{n})\vec{\nabla}\times\vec{j}_1(\vec{r},\omega)]\}. \tag{23}$$

At this point, in the spirit of the local density approximation, it would be tempting to construct the $\vec{a}_{1xc}$ of a slowly varying system simply by substituting $\bar{n} \to n_0(\vec{r})$. This is not sufficient, however, because in an inhomogeneous system, characterized by a length scale q^{-1}, the tensor $\mathbf{f}_{xc,ij}$, has additional contributions, proportional to $k_i q_j$ and $q_i q_j$, which can be of the same order of magnitude as the terms appearing in Eq. (22), but are not captured by the substitution $\bar{n} \to n_0(\vec{r})$ in Eq. (23). In order to construct the correct form of $\vec{a}_{1xc}$, up to second order in the gradients of the density, a far more careful analysis is needed. We begin by establishing, in the next section, some exact properties of the kernel $\mathbf{f}_{xc,ij}$.

EXACT PROPERTIES OF THE TENSORIAL EXCHANGE-CORRELATION KERNEL

Onsager Reciprocity Relation

If the unperturbed system is invariant under time reversal then

$$\mathbf{f}_{xc,ij}(\vec{r},\vec{r}',\omega) = \mathbf{f}_{xc,ji}(\vec{r}',\vec{r},\omega). \tag{24}$$

Proof. The exact expression for the linear current response tensor is

$$\chi_{ij}(\vec{r},\vec{r}',\omega) = \frac{n_0(\vec{r})}{m}\delta(\vec{r}-\vec{r}')\delta_{ij} + \frac{1}{Z}\sum_{\alpha,\beta}(e^{-E_\alpha/T} - e^{-E_\beta/T})\frac{\langle\alpha|j_{pi}(\vec{r})|\beta\rangle\langle\beta|j_{pj}(\vec{r}')|\alpha\rangle}{\omega-(E_\beta-E_\alpha)+i\eta}, \tag{25}$$

where Z is the partition function, T is the temperature, $|\alpha\rangle$ is a complete set of eigenstates of the unperturbed Hamiltonian with eigenvalues E_α, and $\vec{j}_p(\vec{r}) = \sum_n(\vec{p}_n\delta(\vec{r}-\vec{r}_n) + \delta(\vec{r}-\vec{r}_n)\vec{p}_n)/2m$ is the paramagnetic current density operator ($\vec{p}_n$ is the canonical momentum operator of the n-th particle). If the unperturbed Hamiltonian is invariant under time reversal, the wave functions corresponding to the $|\alpha\rangle$'s can all be chosen to be *real*, so that $\langle\alpha|j_{pi}(\vec{r})|\beta\rangle = \langle\beta|j_{pi}(\vec{r})|\alpha\rangle$. Using this equality in Eq. (25) one can verify at once that

$$\chi_{ij}(\vec{r},\vec{r}',\omega) = \chi_{ji}(\vec{r}',\vec{r},\omega). \tag{26}$$

Since both χ and χ_{KS} satisfy this symmetry, it follows, from the definition (20) that $\mathbf{f}_{xc}$ satisfies it too. This proves Eq. (24). If the unperturbed system is *not* time-reversal invariant because of a static magnetic field, then it is easy to verify that Eq. (24) holds with the sign of the magnetic field reversed on the right hand side.

The "Zero-Force" Theorem

This is the tensorial generalization of Eq. (11). It states that

$$\int \mathbf{f}_{xc,ij}(\vec{r},\vec{r}',\omega)n_0(\vec{r}')d\vec{r}' = \frac{\nabla_i\nabla_j v_{0xc}(\vec{r})}{(i\omega)^2}. \tag{27}$$

By virtue of Eq. (24), the transpose of this equation is also valid

$$\int n_0(\vec{r})\mathbf{f}_{xc,ij}(\vec{r},\vec{r}',\omega)d\vec{r} = \frac{\nabla_i'\nabla_j' v_{0xc}(\vec{r}')}{(i\omega)^2}. \tag{28}$$

Proof. Consider the unperturbed system in its ground state with density $n_0(\vec{r})$ in a reference frame in which the center of mass is at rest. The system is assumed to be time-reversal invariant, so that the current density is zero everywhere. Now we look at the system from a linearly accelerated reference frame, whose origin oscillates according to the law $\vec{x}(t) = \vec{x}(\omega)e^{-i\omega t}$. Evidently, to first order in x, the current density in this reference frame is $\vec{j}'(\vec{r},t) = -n_0(\vec{r})d\vec{x}(t)/dt$ and its Fourier amplitude is

$$\vec{j}'(\vec{r},\omega) = i\omega n_0(\vec{r})\vec{x}(\omega). \tag{29}$$

At the same time, the system, in the accelerated frame of reference, sees an external potential

$$v'(\vec{r},t) = v_0(\vec{r}+\vec{x}(t)) + m\vec{r}\cdot\frac{d^2\vec{x}(t)}{dt^2} \tag{30}$$

where $v_0(\vec{r})$ is the static external potential in the inertial frame. The second term in v' corresponds to an "inertial force" appearing in the accelerated frame. Now, if $\vec{x}(\omega)$ is sufficiently small, the current in the accelerated frame can be viewed as the linear response of the system, in its ground state, to a time-dependent perturbing potential $\delta v(\vec{r},t) = v'(\vec{r},t) - v_0(\vec{r})$. Because of gauge-invariance, this scalar potential is physically equivalent to a perturbing vector potential (Eq. 16) $\delta\vec{a}(\vec{r},t) = \delta\vec{a}(\vec{r},\omega)e^{-i\omega t}$, where

$$\delta\vec{a}(\vec{r},\omega) = i\omega m\vec{x}(\omega) + \frac{1}{i\omega}\vec{\nabla}[\vec{x}(\omega)\cdot\vec{\nabla}v_0(\vec{r})]. \tag{31}$$

Thus, we are led to the identity

$$\vec{j}'_i(\vec{r},\omega) = \sum_j \int \chi_{ij}(\vec{r},\vec{r}',\omega)\delta a_j(\vec{r}',\omega)d\vec{r}'. \tag{32}$$

Substituting Eqs. (29) and (31) into this, and using the arbitrariness of $\vec{x}(\omega)$, we get the sum rule

$$\sum_k \int \chi_{ik}(\vec{r},\vec{r}',\omega)[\delta_{kj} + \frac{\nabla'_k\nabla'_j v_0(\vec{r}')}{m(i\omega)^2}]d\vec{r}' = \frac{n_0(\vec{r})}{m}\delta_{ij}. \tag{33}$$

The left hand side of Eq. (33) has the structure of a generalized matrix product, with $\vec{r}$ and $\vec{r}'$ playing the role of continuous indices. Using this formal analogy, we can immediately deduce that

$$\int \chi^{-1}_{ij}(\vec{r},\vec{r}',\omega)n_0(\vec{r}')d\vec{r}' = m\delta_{ij} + \frac{\nabla_i\nabla_j v_0(\vec{r})}{(i\omega)^2}. \tag{34}$$

A similar identity holds for the inverse of the Kohn-Sham response function, since it is the response function of a physical system, namely, N independent electrons moving in a static external potential $v_{0,eff} = v_0 + v_{0H} + v_{0xc}$:

$$\int \chi^{-1}_{KS,ij}(\vec{r},\vec{r}',\omega)n_0(\vec{r}')d\vec{r}' = m\delta_{ij} + \frac{\nabla_i\nabla_j[v_0(\vec{r}) + v_{0H}(\vec{r}) + v_{0xc}(\vec{r})]}{(i\omega)^2}. \tag{35}$$

Subtracting equation (34) from Eq. (35), and using the definition of v_H in (2) and $\mathbf{f}_{xc,ij}$ from (20), we obtain the "zero force" theorem, Eq. (27). The physical content of this identity is that the net force exerted on the system by the exchange-correlation potential must vanish, in agreement with Newton's third law, i.e.

$$\vec{F}(t) = \int [n(\vec{r},t)\left(\frac{\partial\vec{a}_{xc}(\vec{r},t)}{\partial t} - \vec{\nabla}v_{0xc}(\vec{r})\right) + \vec{j}(\vec{r},t)\times(\vec{\nabla}\times\vec{a}_{xc}(\vec{r},t))]d\vec{r} = 0. \tag{36}$$

It is easy to verify that this condition, after linearization with respect to the perturbing field, coincides with Eq. (28).

The Zero-Torque Theorem

This theorem states that

$$\sum_{j,k} \int \epsilon_{ljk} \mathbf{f}_{xc,ij}(\vec{r}, \vec{r}', \omega) n_0(\vec{r}')(r'_k - r_k) d\vec{r}' = \frac{1}{(i\omega)^2} \sum_j \epsilon_{lji} \nabla_j v_{0xc}(\vec{r}), \tag{37}$$

where ϵ_{ijk} is the Levi-Civita tensor. The transpose of this identity, i.e.,

$$\sum_{j,k} \int \epsilon_{ljk}(r'_k - r_k) n_0(\vec{r}') \mathbf{f}_{xc,ji}(\vec{r}', \vec{r}, \omega) d\vec{r}' = \frac{1}{(i\omega)^2} \sum_j \epsilon_{lji} \nabla_j v_{0xc}(\vec{r}), \tag{38}$$

is also valid.

Proof. Start from the unperturbed system in its ground-state, with density $n_0(\vec{r})$ and zero current-density. Let us look at the system from a frame of reference rotating with an angular velocity $\vec{\Omega}(t) = \vec{\Omega}(\omega) e^{-i\omega t}$ about a time-independent axis. To first order in Ω, the current-density in this reference frame is $\vec{j}'(\vec{r}, t) = -n_0(\vec{r}) \vec{\Omega}(t) \times \vec{r}$, and its Fourier amplitude is

$$\vec{j}'(\vec{r}, \omega) = -n_0(\vec{r}) \vec{\Omega}(\omega) \times \vec{r}. \tag{39}$$

Because of the constancy of the rotation axis, the orientation of the rotating frame is specified by an angular displacement vector

$$\vec{\theta}(t) = \int_{-\infty}^{t} \vec{\Omega}(t') dt' \tag{40}$$

directed along the rotation axis. We shall assume that this angular displacement remains small at all times: $|\vec{\theta}(t)| << 1$. Then, to first order in the angular displacement, the electrons see a scalar potential

$$v'(\vec{r}, t) = v_0(\vec{r} + \vec{\theta}(t) \times \vec{r}, t), \tag{41}$$

and a vector potential $\vec{a}'(\vec{r}, t) = -m\vec{\Omega}(t) \times \vec{r}$ (this latter follows from Larmor's theorem). Using Eq. (16) to transform the scalar potential into an equivalent vector potential, one sees that the effect of the rotation is to perturb the system in its ground state with a vector potential

$$\delta\vec{a}(\vec{r}, \omega) = -m\vec{\Omega}(\omega) \times \vec{r} - \frac{1}{(i\omega)^2} \vec{\nabla}[\vec{\nabla} v_0(\vec{r}) \cdot (\vec{\Omega}(\omega) \times \vec{r})]. \tag{42}$$

In the following, we define the quantity

$$\vec{V}(\vec{r}) \equiv \vec{\Omega}(\omega) \times \vec{r}. \tag{43}$$

Now requiring that he current $\vec{j}'$ from Eq. (39) and the perturbing vector potential $\delta\vec{a}$ from Eq. (42) be related by the linear response formula, Eq. (32), we obtain

$$\sum_j \int \chi_{ij}(\vec{r}, \vec{r}', \omega) \left(V_j(\vec{r}') + \frac{1}{m(i\omega)^2} \sum_k \nabla'_j (\nabla'_k v_0(\vec{r}') V_k(\vec{r}')) \right) d\vec{r}' = \frac{n_0(\vec{r})}{m} V_i(\vec{r}). \tag{44}$$

Multiplying both sides of this equation by the inverse response matrix χ^{-1} we obtain

$$\sum_j \int \chi_{ij}^{-1}(\vec{r}, \vec{r}', \omega) \frac{n_0(\vec{r}')}{m} V_j(\vec{r}') d\vec{r}' = V_i(\vec{r}) + \frac{1}{m(i\omega)^2} \sum_k \nabla_i \left([\nabla_k v_0(\vec{r})] V_k(\vec{r}) \right). \tag{45}$$

A similar equation holds for the inverse of the Kohn-Sham response function, with $v_0(\vec{r})$ replaced by $v_{0eff} = v_0(\vec{r}) + v_{0H}(\vec{r}) + v_{0xc}(\vec{r})$:

$$\sum_j \int \chi^{-1}_{KS,ij}(\vec{r},\vec{r}',\omega)\frac{n_0(\vec{r}')}{m}V_j(\vec{r}')d\vec{r}' = V_i(\vec{r}) + \frac{1}{m(i\omega)^2}\sum_k \nabla_i\left([\nabla_k v_{0eff}(\vec{r})]V_k(\vec{r})\right). \tag{46}$$

Subtracting Eq. (46) from Eq. (45), and using the definition of $\mathbf{f}_{xc}$ in Eq. (20), after some straightforward algebraic manipulations we obtain:

$$\sum_j \int \mathbf{f}_{xc,ij}(\vec{r},\vec{r}',\omega)n_0(\vec{r}')V_j(\vec{r}')d\vec{r}' = \frac{1}{(i\omega)^2}\sum_j \left(\nabla_i\nabla_j v_{0xc}(\vec{r})V_j(\vec{r}) + \nabla_j v_{0xc}(\vec{r})\nabla_i V_j(\vec{r})\right). \tag{47}$$

Finally, we go back to Eq. (27) and multiply both sides by $V_j(\vec{r})$. The resulting equation is subtracted from Eq. (47) and we obtain

$$\sum_j \int \mathbf{f}_{xc,ij}(\vec{r},\vec{r}',\omega)n_0(\vec{r}')[V_j(\vec{r}') - V_j(\vec{r})]d\vec{r}' = \frac{1}{(i\omega)^2}\sum_j \nabla_j v_{0xc}(\vec{r})\nabla_i V_j(\vec{r}). \tag{48}$$

Expanding $V_j(\vec{r}) \equiv \sum_{k,l}\epsilon_{jkl}\Omega_k(\omega)r_l$, and using the arbitrariness of the vector $\vec{\Omega}(\omega)$ we arrive at the desired result, Eq. (37). Its transpose (Eq. (38)) follows from the application of Eq. (24). The physical content of this identity is that the net torque exerted on the sytem by the exchange-correlation potential must vanish. It can be proved that Eq. (38) follows directly from this requirement.

THE EXCHANGE-CORRELATION KERNEL FOR A WEAKLY NON-HOMOGENEOUS ELECTRON GAS OF SLOWLY VARYING DENSITY

In this section we construct the tensorial xc kernel for a system in which both the unperturbed static density $n_0(\vec{r})$ and the perturbing field $\vec{a}_1(\vec{r},\omega)$ are slowly varying on the scales of the local k_F^{-1} and v_F/ω. In addition, we assume that the nonuniformity of the density is *weak*. The simplest case is that of a periodically modulated electron gas, where the unperturbed density, given by

$$n_0(\vec{r}) = \bar{n}(1 + 2\gamma\cos(\vec{q}\cdot\vec{r})), \tag{49}$$

is not only slowly varying (i.e., $q << k_F, \omega/v_F$), but, also, *almost uniform*, i.e. $\gamma << 1$. Our objective is to compute the exact $\mathbf{f}_{xc}$ for this system, to first order in γ.

In order to exploit the periodicity of the static density, we introduce the momentum representation of $\mathbf{f}_{xc}$ as follows:

$$\mathbf{f}_{xc}(\vec{k}+m\vec{q},\vec{k},\omega) \equiv \frac{1}{V}\int d\vec{r}\int d\vec{r}'\mathbf{f}_{xc}(\vec{r},\vec{r}',\omega)e^{-i(\vec{k}+m\vec{q})\cdot\vec{r}}e^{i\vec{k}\cdot\vec{r}'}, \tag{50}$$

where m is an integer and V is the volume of the system. The inverse of this relation reads

$$\mathbf{f}_{xc}(\vec{r},\vec{r}',\omega) = \sum_m \int \frac{d\vec{k}}{(2\pi)^3}\mathbf{f}_{xc}(\vec{k}+m\vec{q},\vec{k},\omega)e^{i(\vec{k}+m\vec{q})\cdot\vec{r}}e^{-i\vec{k}\cdot\vec{r}'}. \tag{51}$$

To first order in γ only the diagonal element ($m = 0$) and the off-diagonal elements of $\mathbf{f}_{xc}$ with $m = \pm 1$ are nonvanishing. All the matrix elements with $|m| \geq 2$ require the intervention of at least two Fourier components of the external potential at wave vector $\vec{q}$ and thus are of order γ^2 at least. Furthermore, the leading correction, for small γ, to the diagonal component of $\mathbf{f}_{xc}$, requires the intervention of two Fourier components of the external potential (at wave vectors $\vec{q}$ and $-\vec{q}$), which makes it of order γ^2. Thus, *to first order in* γ we find

$$\mathbf{f}_{xc}(\vec{k},\vec{k},\omega) \simeq \mathbf{f}^h_{xc}(\vec{k},\omega), \tag{52}$$

where $\mathbf{f}^h_{xc}$ is defined in Eq. (22) and calculated at the density $\bar{n}$. The remaining task is to calculate the off-diagonal matrix elements $\mathbf{f}_{xc}(\vec{k}\pm\vec{q},\vec{k},\omega)$. This task is greatly facilitated by the use of the following exact identities.

Onsager Reciprocity Relation in Momentum Space

$$\mathbf{f}_{xc,ij}(\vec{k}+m\vec{q},\vec{k},\omega) = \mathbf{f}_{xc,ji}(-\vec{k},-\vec{k}-m\vec{q},\omega). \tag{53}$$

Proof. It follows immediately from the Onsager relation, Eq. (24), and from the definition (50).

Zero-Force and Zero-Torque Theorems for a Weakly Nonhomogeneous Electron Gas of Slowly Varying Density

To first order in γ,

$$lim_{k\to 0}\mathbf{f}_{xc,ij}(\vec{k}+\vec{q},\vec{k},\omega) = -\frac{\gamma}{\omega^2}[\delta f^h_{xcL}q_iq_j + f^h_{xcT}(q^2\delta_{ij}-q_iq_j)], \tag{54}$$

and

$$lim_{k\to 0}\sum_{j,k}\epsilon_{ljk}\frac{\partial}{\partial k_k}\mathbf{f}_{xc,ij}(\vec{k}+\vec{q},\vec{k},\omega) = -\frac{\gamma}{\omega^2}[\delta f^h_{xcL} - 3f^h_{xcT}]\sum_k \epsilon_{lki}q_k. \tag{55}$$

For notational simplicity, we have suppressed the arguments ω and $\bar{n}$ in writing f^h_{xcL} and f^h_{xcT}. We have also defined

$$\delta f^h_{xcL}(\omega,\bar{n}) \equiv f^h_{xcL}(\omega,\bar{n}) - f^h_{xcL}(\omega=0,\bar{n}). \tag{56}$$

Proof. All we have to do is to substitute in the left hand side of equations (27) and (37) $\mathbf{f}_{xc}$ from Eq. (51) and n_0 from Eq. (49). On the right hand side, we observe that, *to first order in* γ, and for small q,

$$v_{0xc}(\vec{r}) = v^h_{xc}(\bar{n}) + 2\gamma\bar{n}f^h_{xcL}(\omega=0,\bar{n})cos\vec{q}\cdot\vec{r}, \tag{57}$$

where $v^h_{xc}(\bar{n})$ is the (constant) xc potential of the homogeneous electron gas of density $\bar{n}$. Equating the terms of first order in γ on the two sides of (27) and (37) we obtain the above stated identities. We emphasize that these results are only valid for $q << k_F, \omega/v_F$, and to first order in γ.

The Ward Identity

A third limiting form of $\mathbf{f}_{xc,ij}$, to first order in γ, is obtained from the Ward identity

$$lim_{q\to 0}\mathbf{f}_{xc}(\vec{k}+\vec{q},\vec{k},\omega) = \gamma\bar{n}\frac{\partial \mathbf{f}^h_{xc}(\vec{k},\omega,\bar{n})}{\partial\bar{n}}. \tag{58}$$

Proof. This identity can be proved by standard diagrammatic methods similar to the ones introduced in [14]. We prove it here by a more direct physical argument. First, let us prove that

$$lim_{q\to 0}\chi_{ij}(\vec{k}+\vec{q},\vec{k},\omega)=\gamma\bar{n}\frac{\partial\chi^h_{ij}(\vec{k},\omega,\bar{n})}{\partial\bar{n}}. \tag{59}$$

Proof. Consider a homogeneous electron gas of density $\bar{n}$ perturbed by a vector potential $\vec{a}(\vec{k},\omega)$ of wave vector $\vec{k}$ and frequency ω, *and* by a static scalar potential of wavevector $\vec{q}(<< k_F)$ that produces, to first order, the density modulation of Eq. (49):

$$v_0(\vec{q})\sim-\frac{\gamma\bar{n}}{\partial\bar{n}/\partial\bar{\mu}}, \tag{60}$$

where $\bar{\mu}$ is the chemical potential of the homogeneous electron gas. This formula follows from the fact that the static density-density response function of a homogeneous electron gas reduces, for small wavevectors, to $-\partial\bar{n}/\partial\bar{\mu}$ [14]. To first order in $\vec{a}$ and γ, the component of the induced current at wave vector $\vec{k}+\vec{q}$ and frequency ω is given by

$$j_i(\vec{k}+\vec{q},\omega)=\sum_j\chi_{ij}(\vec{k}+\vec{q},\vec{k},\omega)a_j(\vec{k},\omega). \tag{61}$$

Now take the limit $\vec{q}\to 0$. In this limit, $j_i(\vec{k}+\vec{q},\omega)$ must coincide with the first-order change in $j_i(\vec{k},\omega)=\sum_j\chi^h_{ij}(\vec{k},\omega,\bar{n})a_j(\vec{k},\omega)$ due to the application of the static potential $v_0(\vec{q})$ of Eq.(60). But, in the limit of $\vec{q}\to 0$, the application of this potential is equivalent to changing the chemical potential of the uniform electron gas by the amount $\delta\bar{\mu}=-v_0(\vec{0})$. Therefore,

$$\delta j_i(\vec{k},\omega)=-v_0(\vec{0})\frac{\partial}{\partial\bar{\mu}}\sum_j\chi^h_{ij}(\vec{k},\omega,\bar{n})a_j(\vec{k},\omega). \tag{62}$$

Substituting $v_0(\vec{0})$ from Eq. (60) and equating $\delta j_i(\vec{k},\omega)$ with the $q\to 0$ limit of $j_i(\vec{k}+\vec{q},\omega)$, Eq.(61), we obtain the desired result, Eq.(59).

Next, we invert χ to first order in γ:

$$\chi^{-1}(\vec{k}+\vec{q},\vec{k},\omega)\simeq[\chi^h]^{-1}(\vec{k}+\vec{q},\omega)\cdot\chi(\vec{k}+\vec{q},\vec{k},\omega)\cdot[\chi^h]^{-1}(\vec{k},\omega), \tag{63}$$

(the dot represents the usual matrix multiplication) and take the limit $q\to 0$, to find, with the help of Eq. (59),

$$lim_{q\to 0}\chi^{-1}_{ij}(\vec{k}+\vec{q},\vec{k},\omega)=\gamma\bar{n}\frac{\partial[\chi^h]^{-1}_{ij}(\vec{k},\omega,\bar{n})}{\partial\bar{n}}. \tag{64}$$

A similar result holds for the inverse of the Kohn-Sham response tensor:

$$lim_{q\to 0}[\chi^h_{KS}]^{-1}_{ij}(\vec{k}+\vec{q},\vec{k},\omega)=\gamma\bar{n}\frac{\partial[\chi^h_0]^{-1}_{ij}(\vec{k},\omega,\bar{n})}{\partial\bar{n}}, \tag{65}$$

where $[\chi^h_0]_{ij}$ is the response tensor of the *noninteracting* homogeneous electron gas. Combining Eqs. (64) and (65) with the definition

$$\mathbf{f}_{xc,ij}(\vec{k}+\vec{q},\vec{k},\omega)=\chi^{-1}_{KS,ij}(\vec{k}+\vec{q},\vec{k},\omega)-\chi^{-1}_{ij}(\vec{k}+\vec{q},\vec{k},\omega) \tag{66}$$

yields the target result, Eq. (58).

Explicit Form of XC Kernel to First Order in γ

Equations (54),(55),(58), and (53) imply that, for small k and q, $\mathbf{f}_{xc,ij}(\vec{k}+\vec{q},\vec{k},\omega)$ is expressible as a linear combination of $q_iq_j, q^2\delta_{ij}, (k_i+q_i)k_j, k_i(k_j+q_j)$, and $\vec{k}\cdot(\vec{k}+\vec{q})\delta_{ij}$. Equations (54), (55), and (58), completely determine the coefficients of this combination. Thus, to first order in γ, we obtain

$$\mathbf{f}_{xc,ij}(\vec{k}+\vec{q},\vec{k},\omega) = -\frac{\gamma}{\omega^2}\{(\delta f^h_{xcL} - f^h_{xcT})q_iq_j + f^h_{xcT}q^2\delta_{ij} - \bar{n}\frac{\partial f^h_{xcT}}{\partial \bar{n}}\vec{k}\cdot(\vec{k}+\vec{q})\delta_{ij} + A(\bar{n},\omega)(k_i+q_i)k_j - B(\bar{n},\omega)k_i(k_j+q_j)\}, \quad (67)$$

where $A(\bar{n},\omega) \equiv [\bar{n}(2\partial f^h_{xcT}/\partial\bar{n} - \partial f^h_{xcL}/\partial\bar{n}) + 3f^h_{xcT} - \delta f^h_{xcL}]$, and $B(\bar{n},\omega) \equiv [\bar{n}\partial f^h_{xcT}/\partial\bar{n} + 3f^h_{xcT} - \delta f^h_{xcL}]$. The essential point is that Eq. (67) is analytic for small k and q. It is this feature that will enable us to construct a local approximation for $\vec{a}_{1xc}$ in terms of the current density. By contrast, the off-diagonal component $f_{xc}(\vec{k}+\vec{q},\vec{k},\omega)$ in the usual time dependent density functional theory, has a singularity of the form $\vec{k}\cdot\vec{q}/k^2$ when $k\to 0$ at finite q. This small-k singularity reflects the already mentioned long-rangedness of f_{xc}, and the consequent nonlocality of the xc scalar potential as a functional of the density.

THE XC VECTOR POTENTIAL FOR AN ELECTRON GAS OF SLOWLY VARYING DENSITY

Let us now return to the more general problem of determining $\vec{a}_{1xc}$ for a system whose density is *slowly varying* in the sense that $|\vec{\nabla}n_0(\vec{r})|/n_0(\vec{r}) << k_F(\vec{r})$ and $\omega/v_F(\vec{r})$, but may have large global deviations from uniformity. We assume that $\vec{a}_{1xc}$ of such a system can be expanded in a power series of gradients of the local unperturbed density. The most general form of $\vec{a}_{1xc}$ that is linear in $\vec{j}_1$, and contains up to two gradient operators, is then a linear combination of the following terms: $\vec{j}_1, \vec{\nabla}n_0\times\vec{j}_1, \vec{\nabla}\times\vec{j}_1, \vec{\nabla}(\vec{\nabla}\cdot\vec{j}_1), \vec{\nabla}\times(\vec{\nabla}\times\vec{j}_1), \vec{\nabla}(\vec{\nabla}n_0\cdot\vec{j}_1), \vec{\nabla}\times(\vec{\nabla}n_0\times\vec{j}_1), \vec{\nabla}n_0(\vec{\nabla}\cdot\vec{j}_1), \vec{\nabla}n_0\times(\vec{\nabla}\times\vec{j}_1), (\vec{\nabla}n_0\cdot\vec{\nabla})\vec{j}_1, \vec{\nabla}n_0(\vec{\nabla}n_0\cdot\vec{j}_1), \vec{\nabla}n_0\times(\vec{\nabla}n_0\times\vec{j}_1)$ with coefficients that are functions of $n_0(\vec{r})$ and ω. This gradient expansion is applicable, in particular, to the periodically modulated electron gas of Eq. (49), in which case it must yield the same perturbative (in γ) result that one obtains from the combination of Eqs. (19), (22), and (67). Thus, by requiring agreement between the gradient and the perturbative expansions, we are able to determine the coefficients of all the terms appearing in the gradient expansion, except the last two, which are of order $|\vec{\nabla}n_0|^2$ (or γ^2 in the periodically modulated electron gas).

In practice, we proceed as follows: we use Eqs. (19) and (51) to construct the xc vector potential:

$$a_{1xc,i}(\vec{r},\omega) \simeq \sum_j \int f^h_{xc,ij}(\vec{k},\omega)j_j(\vec{k},\omega)e^{i\vec{k}\cdot\vec{r}}d\vec{k} + \sum_{\sigma=\pm 1} e^{i\sigma\vec{q}\cdot\vec{r}}\sum_j\int f_{xc,ij}(\vec{k}+\sigma\vec{q},\vec{k},\omega)j_j(\vec{k},\omega)e^{i\vec{k}\cdot\vec{r}}\frac{d\vec{k}}{(2\pi)^3}, \quad (68)$$

and substitute our results (22) and (67) in this equation. After lengthy, but elementary manipulations, we obtain:

$$\vec{a}_{1xc}(\vec{r},\omega) =$$

$$-\frac{1}{\omega^2}\{\vec{\nabla}[f^h_{xcL}\vec{\nabla}\cdot\vec{j}_1(\vec{r},\omega) - \delta f^h_{xcL}\vec{\nabla}n_0(\vec{r})\cdot\frac{\vec{j}_1(\vec{r},\omega)}{n_0(\vec{r})}] - \vec{\nabla}\times[f^h_{xcT}n_0\vec{\nabla}\times\frac{\vec{j}_1(\vec{r},\omega)}{n_0(\vec{r})}]$$
$$+\delta f^h_{xcL}\vec{\nabla}n_0(\vec{r})\vec{\nabla}\cdot\frac{\vec{j}_1(\vec{r},\omega)}{n_0(\vec{r})} + f^h_{xcT}[(\vec{\nabla}n_0(\vec{r})\cdot\vec{\nabla})\frac{\vec{j}_1(\vec{r},\omega)}{n_0(\vec{r})} - 4\vec{\nabla}n_0(\vec{r})\vec{\nabla}\cdot\frac{\vec{j}_1(\vec{r},\omega)}{n_0(\vec{r})}$$
$$+3\sum_j\nabla_j n_0(\vec{r})\vec{\nabla}\frac{j_{1j}(\vec{r},\omega)}{n_0(\vec{r})}] + 2n_0(\vec{r})[\sum_j\nabla_j f^h_{xcT}\vec{\nabla}\frac{j_{1j}(\vec{r},\omega)}{n_0(\vec{r})} - \vec{\nabla}f^h_{xcT}\vec{\nabla}\cdot\frac{\vec{j}_1(\vec{r},\omega)}{n_0(\vec{r})}]\}, \tag{69}$$

where f^h_{xcL} and f^h_{xcT} are functions of the local density $n_0(\vec{r})$ and the frequency ω.

Clearly, this is a linear combination of the admissible terms listed above, and the coefficients of all the terms of first order in $|\nabla n_0| \sim \gamma$ are necessarily exact. At first sight, there seems to be an arbitrariness in the special choice of the the coefficient of the terms of order γ^2 (i.e., $\vec{\nabla}n_0(\vec{\nabla}n_0\cdot\vec{j}_1)$ and $\vec{\nabla}n_0\times(\vec{\nabla}n_0\times\vec{j}_1)$) implied by Eq. (69). For example, one might think that a term of the form $\vec{\nabla}n_0(\vec{r})\vec{\nabla}\cdot(\vec{j}_1(\vec{r},\omega)/n_0(\vec{r}))$ could have been written, with equal right, in the form $(\vec{\nabla}n_0(\vec{r})/n_0(\vec{r}))\vec{\nabla}\cdot\vec{j}_1(\vec{r},\omega)$ since the difference between the two expression is of order γ^2, which is beyond the accuracy of our perturbative treatment. It turns out, however, that Eq. (69) *is* the correct result, *including the second order terms in* γ, for the following reason.

Let us imagine that the system is subjected to the external potential of Eq. (30), which causes it to translate *rigidly* with the center of mass position given by $-\vec{x}(\omega)e^{-i\omega t}$, so that the current is $j_1(\vec{r},\omega) = i\omega n_0(\vec{r})\vec{x}(\omega)$. For this special motion we know from Eq. (10) that the *exact* time-dependent xc potential, is obtained from the static one by a rigid translation. This means that, in the linear response approximation

$$v_{1xc}(\vec{r},\omega) = \vec{x}(\omega)\cdot\vec{\nabla}v_{0xc}(\vec{r}) = \frac{\vec{j}_1(\omega)}{i\omega n_0(\vec{r})}\cdot f^h_{xcL}(\omega=0,n_0(\vec{r}))\vec{\nabla}n_0(\vec{r}), \tag{70}$$

and, from Eq. (16),

$$\vec{a}_{1xc}(\vec{r},\omega) = \frac{1}{(i\omega)^2}\vec{\nabla}[\frac{\vec{j}_1(\omega)}{n_0(\vec{r})}\cdot f^h_{xcL}(\omega=0,n_0(\vec{r}))\vec{\nabla}n_0(\vec{r})]. \tag{71}$$

It is immediate to verify that our expression (69) does indeed reduce to this exact limit, when $\vec{j}_1(\vec{r},\omega) = i\omega n_0(\vec{r})\vec{x}(\omega)$. On the other hand, any additional term proportional to $\vec{\nabla}n_0(\vec{\nabla}n_0\cdot\vec{j}_1)$ or $\vec{\nabla}n_0\times(\vec{\nabla}n_0\times\vec{j}_1)$ would necessarily violate this exact identity, for a general equilibrium density. We conclude that Eq. (69) is the correct form of the local density approximation for the dynamical $\vec{a}_{1xc}$, consistent with the local density approximation for the ground-state.

Our result simplifies considerably in two limiting cases. (1) The density is slowly varying on the scale set by the wavelength of the perturbation. In this regime we can neglect the terms containing $\vec{\nabla}n_0$, and we recover the former results of GK [5] for the response to a scalar potential, and of Ng [15] for the response to a general vector potential. This result can also be simply obtained from Eq. (23) by the substitution $\bar{n}\to n_0(\vec{r})$. (2) The velocity field j_1/n_0 is constant in space. This is the case when the static external potential is parabolic and the perturbing electric field is uniform, the regime of Kohn's theorem [9] and the HPT [8]. Then all derivatives of j_1/n_0 vanish, and Eq. (69) reduces to Eq. (71).

CASE OF ONE-DIMENSIONAL INHOMOGENEITY

The expression obtained in the previous section for the xc vector potential of an electron gas of slowly varying density is complicated. It becomes much simpler in the case of systems in which the density and the current-density depend only on one coordinate x and the current vector is parallel to the $x-$ axis: $\vec{j}_1(x) = j_1(x)\hat{x}$. The special interest of this case arises from the fact that the current-density is completely specified by the density, because the transverse component is missing, and the equation of continuity can be integrated to give:

$$j_1(x,\omega) = i\omega \int_{-\infty}^{x} n_1(x',\omega)dx'. \tag{72}$$

At the same time, the xc-vector potential, being purely longitudinal, can also be expressed in terms of a scalar potential, according to Eq. (16). Thus, it becomes possible to compute exactly the *scalar* kernel $f_{xc}(x,x',\omega)$ (i.e, $f_{xc}(\vec{r},\vec{r}',\omega)$ averaged over the transverse coordinates y,z, and y',z') defined in Eq. (6), and see clearly what is wrong with approximations such as Eq. (7) and (9).

As a first step, we calculate the xc vector potential from our Eq. (69). Because of the one-dimensional geometry, all the tranverse terms cancel out, and we are left with

$$\begin{aligned} a_{1xc}(x,\omega) = {} & \frac{1}{(i\omega)^2}\{\frac{d}{dx}[f^h_{xcL}(x)\frac{dj_1(x,\omega)}{dx} - \delta f^h_{xcL}(x)\frac{n_0'(x)}{n_0(x)}j_1(x,\omega)] \\ & +\delta f^h_{xcL}(x)n_0'(x)\frac{d}{dx}(\frac{j_1(x,\omega)}{n_0(x)})\}, \end{aligned} \tag{73}$$

where $f^h_{xcL}(x)$ is a shorthand for $f^h_{xcL}(\omega,n_0(x))$ - a function of ω. Now express j_1 in terms of n_1 according to Eq. (72) and construct the equivalent scalar potential

$$v_{1xc}(x,\omega) = -i\omega \int_x^{\infty} a_{1xc}(x',\omega)dx', \tag{74}$$

according to Eq.(16). After some straightforward manipulations we obtain

$$\begin{aligned} v_{1xc}(x,\omega) = {} & f^h_{xcL}(x)n_1(x,\omega) \\ & - \int_{-\infty}^{\infty} [\delta f^h_{xcL}(x)\frac{n_0'(x)}{n_0(x)}\theta(x-x') + \delta f^h_{xcL}(x')\frac{n_0'(x')}{n_0(x')}\theta(x'-x)]n_1(x',\omega)dx' \\ & + \int_{-\infty}^{\infty} dx' n_1(x',\omega)[\int_{-\infty}^{\infty} \theta(x''-x)\left(\frac{n_0'(x'')}{n_0(x'')}\right)^2 \delta f^h_{xcL}(x'')\theta(x''-x')dx'']. \end{aligned} \tag{75}$$

where $\theta(x) = 1$ for $x > 0$, and 0 otherwise. From this we deduce the correct form of $f_{xc}(x,x',\omega)$:

$$\begin{aligned} f_{xc}(x,x',\omega) = {} & f^h_{xcL}(x)\delta(x-x') \\ & -[\delta f^h_{xcL}(x)\frac{n_0'(x)}{n_0(x)}\theta(x-x') + \delta f^h_{xcL}(x')\frac{n_0'(x')}{n_0(x')}\theta(x'-x)] \\ & + \int_{-\infty}^{\infty} \theta(x''-x)\left(\frac{n_0'(x'')}{n_0(x'')}\right)^2 \delta f^h_{xcL}(x'')\theta(x''-x')dx''. \end{aligned} \tag{76}$$

The first term on the right hand side of Eq. (76) is the GK approximation [2]: it is local, but it fails to satisfy the HPT. The first and second term together are equivalent to the

Dobson approximate form, Eq. (9): this satisfies the HPT, but it violates Onsager's reciprocity theorem * i.e., the symmetry under interchange of x and x' (this symmetry can be proved for the scalar f_{xc} in the same way as it was proved for the tensorial $\mathbf{f}_{xc}$ in section IV). The third term on the right hand side of (76) restores the Onsager symmetry, but now the HPT is violated again. The fourth and last term restores the HPT without breaking the Onsager symmetry.

The most striking feature of the scalar f_{xc} is its extreme long-rangedness, evidenced by the θ functions. In contrast to this, the tensorial $\mathbf{f}_{xc}$ extracted from Eq. (73), is local. It is also interesting to look at v_{1xc} as a functional of the current-density. Because of Eq. (74) and the fact that a_{1xc} is a local functional of j_1, we immediately conclude that v_{1xc} is a nonlocal functional of the current density. This shows that it is not possible to construct a consistent local theory in terms of a scalar xc potential, even when the current-density is used as the basic variable. A vector xc potential is required.

CONCLUSION

In conclusion, our analysis uniquely specifies a local current density functional theory of the linear current (and density) response, which becomes exact in the limit of slowly varying unperturbed densities and perturbing potentials. The scale on which the variations must be slow is set by the smaller of the wave vectors k_F and ω/v_F. Therefore, this theory is applicable to the study of high frequency phenomena, such as electromagnetic absorption, for which the adiabatic approximation [3] is in general not justified. A complete local current density response theory, for spatially slowly varying unperturbed density and perturbing field, and *all* frequencies ($0 < \omega < \infty$) remains to be developed. Important generalizations of this theory to include (i) a static magnetic field, and (ii) the spin degrees of freedom, are presently under study.

ACKNOWLEDGMENTS

We acknowledge support from NSF grants No. DMR-9403908 and DMR-9308011. One of us (GV) gratefully acknowledges the hospitality of the Institute of Theoretical Physics and of the Physics Department at UCSB, where this work was initiated. Thanks are due to John Dobson and A. K. Rajagopal for many stimulating comments on the manuscript.

APPENDIX - THE CURRENT-CURRENT RESPONSE TENSOR FOR THE HOMOGENEOUS ELECTRON GAS

In this Appendix we review the basic properties of the current-current response tensor $\chi^h_{ij}(\vec{k},\omega)$ of a uniform electron gas of density n, in the limit of small wave vector k. From a knowledge of this response function, one can extract, using Eq. (20), the functions $f^h_{xc,L(T)}(\omega,n)$, which determine the xc vector potential (Eq. (69)) in systems of slowly varying density.

First of all, it follows from the isotropy of the electron gas that the tensor χ^h_{ij} has

*Recently [10] a revised version of Dobson's construction has been given for the case of one-dimensional inhomogeneity. This nonlinear version reduces to the present theory in the 1D linear limit, and hence avoids the violation of symmetry produced by the original version [8].

the structure

$$\chi^h_{ij}(\vec{k},\omega) = \chi^h_L(\vec{k},\omega)\hat{k}_i\hat{k}_j + \chi^h_T(\vec{k},\omega)(\delta_{ij} - \hat{k}_i\hat{k}_j). \tag{77}$$

($\hat{k}$ is a unit vector in the direction of $\vec{k}$). The longitudinal (transverse) response function is defined as

$$\chi^h_{L(T)}(\vec{k},\omega) = \frac{\bar{n}}{m} + R^h_{L(T)}(\vec{k},\omega), \tag{78}$$

where

$$R^h_{L(T)}(\vec{k},\omega) \equiv -i\int_0^\infty \langle [j_{p,L}(\vec{k},t), j_{p,L}(-\vec{k},0)]\rangle e^{i\omega t}dt. \tag{79}$$

$j_{p,L}(\vec{k}) = \frac{1}{2m}\sum_{n=1}^N (\vec{p}_n\cdot\hat{k})e^{i\vec{k}\vec{r}_n} + e^{i\vec{k}\vec{r}_n}(\vec{p}_n\cdot\hat{k})$ is the *longitudinal* component of the paramagnetic current operator at wave vector $\vec{k}$ ($\vec{r}_n$ and $\vec{p}_n$ are the position and canonical momentum operators of the $n-$th particle) and $j_{p,T}(\vec{k}) = \frac{1}{2m}\sum_{n=1}^N (\vec{p}_n\times\hat{k})e^{i\vec{k}\vec{r}_n} + e^{i\vec{k}\vec{r}_n}(\vec{p}_n\times\hat{k})$ is its *transverse* component. The longitudinal part of χ is related to the density-density response function $\Pi(\vec{k},\omega)$ by the constraints of gauge invariance and the continuity equation:

$$\chi^h_L(\vec{k},\omega) = \frac{\omega^2}{k^2}\Pi(\vec{k},\omega). \tag{80}$$

It is customary to express the response functions $\chi^h_{L(T)}$ of the interacting electron gas in terms of those of the noninteracting electron gas $\chi^{h0}_{L(T)}$ and the "local field factors" $G_{L(T)}(\vec{k},\omega)$ which contain the effect of the interaction (other than the Hartree field):.

$$\chi^h_L(k,\omega) = \frac{\chi^{h0}_L(k,\omega)}{1 - v(k)[1 - G_L(k,\omega)](k^2/\omega^2)\chi^{h0}_L(k,\omega)}, \tag{81}$$

and

$$\chi^h_T(k,\omega) = \frac{\chi^{h0}_T(k,\omega)}{1 + v(k)G_T(k,\omega)(k^2/\omega^2)\chi^{h0}_T(k,\omega)}., \tag{82}$$

where ($v(k) = 4\pi e^2/k^2$). We note that the longitudinal local field factor defined in this manner coincides with the usual local field factor $G(k,\omega)$ used in the theory of the density response [16]. It is easy to show that, in the limit $k\to 0$ the local field factors vanish as k^2. One then defines the functions

$$f^h_{xcL(T)}(\omega,n) \equiv -lim_{k\to 0}v(k)G(k,\omega). \tag{83}$$

Next, inserting in Eqs. (81) and (82) the $k\to 0$, finite frequency limit of the noninteracting longitudinal (transverse) current-current response function, we obtain, to order k^2,

$$\chi^h_L(k,\omega) \sim \frac{n}{m(1-\omega_p^2/\omega^2)}[1 + \frac{12}{5}\frac{E_F^2}{k_F^2}\frac{k^2}{(\omega^2-\omega_p^2)} + \frac{nk^2}{m(\omega^2-\omega_p^2)}f^h_{xcL}(\omega,n)], \tag{84}$$

and

$$\chi^h_T(k,\omega) \sim \frac{n}{m}[1 + \frac{4}{5}\frac{E_F^2}{k_F^2}\frac{k^2}{\omega^2} + \frac{nk^2}{m\omega^2}f^h_{xcT}(\omega,n)], \tag{85}$$

where $\omega_p^2 = 4\pi ne^2/m$ is the square of the plasma frequency. Finally, using Eq. (20), we construct the tensorial xc kernel, and find that, indeed, it has the form stated in Eq. (22).

Some known exact properties of the longitudinal homogeneous kernel $f_{xcL}^h(\omega, n)$ are discussed extensively in [2], and references therein. A simple parametrization forf_{xcL}^h is also provided there. More recently, Böhm *et al* [7] have performed a microscopic calculation of this function, based on physically motivated approximations. Their result differs substantially from the parametrization of Ref. [2].

As for the transverse function $f_{xcT}^h(\omega, n)$, we have thus far obtained a few exact results which can be summarized as follows:

(i) To *first order* in the Coulomb interaction,

$$f_{xcT}^h(\omega, n) = 3\pi e^2/10k_F^2. \tag{86}$$

Note that this is independent of frequency.

(ii) In the limit of $\omega \to \infty$ $f_{xcT}^h(\omega, n)$ can be obtained exactly from the "third moment sum rule" [16], that is, from the calculation of the ground-state expectation value of the double commutator $[j_T(\vec{k}), [j_T(-\vec{k}), H]]$ where H is the hamiltonian of the uniform electron gas. The result is

$$lim_{\omega\to\infty} f_{xcT}^h(\omega, n) = \frac{2}{5} n^{2/3} \frac{d}{dn} \frac{\epsilon_{xc}(n)}{n^{2/3}} + 2n^{1/3} \frac{d}{dn} \frac{\epsilon_{xc}(n)}{n^{1/3}}, \tag{87}$$

where $\epsilon_{xc}(n)$ is the exchange-correlation energy per particle of the uniform electron gas of density n.

As discussed in Section III, the $\omega \to 0$ limit of $f_{xcT}^h(\omega, n)$ is presently unknown. Microscopic calculation of this function along the lines of Ref. [7] are presently underway.

REFERENCES

1. W. Kohn and L. Sham, Phys. Rev. **140**, A1133 (1965).
2. E. K. U. Gross and W. Kohn, Phys. Rev. Lett. **55**, 2850 (1985). See also E. K. U. Gross and W. Kohn, Adv. Quantum Chemistry **21**, 255 (1990).
3. A. Zangwill and P. Soven Phys. Rev. Lett. **45**, 204 (1980); Phys. Rev. B **24**, 4121 (1981).
4. G. Vignale and W. Kohn, Phys. Rev. Lett. **77**, 2037 (1996).
5. E. Runge and E. K. U. Gross, Phys. Rev. Lett. **52**, 997, (1984).
6. In this paper, a function $f(\vec{r}, \vec{r}')$ is said to be of short range if the integral $\int f(\vec{r}, \vec{r}+\vec{s})d\vec{s}$ is finite. This is equivalent to saying that the Fourier transform of f, with wave vector $\vec{k}$, with respect to the separation $\vec{s} = \vec{r}' - \vec{r}$, remains finite in the limit $\vec{k} \to 0$.
7. H.M. Böhm, S. Conti, and M. P. Tosi, J. Phys.: Condensed Matter **8**, 781 (1996).
8. J. F. Dobson, Phys. Rev. Lett. **73**, 2244 (1994).
9. W. Kohn, Phys. Rev. **123**, 1242 (1961), L. Brey *et al* , Phys. Rev. B **40**, 10647 (1989); ibid. **42**, 1240 (1990); S. K. Yip, Phys. Rev. B **43**, 1707 (1991).
10. J. F. Dobson, M. Bünner, and E. K. U. Gross, Phys. Rev. Lett. **79**, 1905 (1997).
11. J. F. Dobson, Proceedings of the NATO ASI on *Density Functional Theory*, edited by E. K. U. Gross and R. M. Dreizler (Plenum, New York, 1994), p. 393.
12. G. Vignale, Phys. Rev. Lett. **74**, 3233 (1995).
13. G. Vignale, Phys. Lett. A **209**, 206 (1995).
14. P. Nozières, *The Theory of Interacting Fermi Systems* (W. A. Benjamin, New York, 1964), Chapter 6.
15. Tai Kai Ng, Phys. Rev. Lett. **62**, 2417 (1989).
16. K. S. Singwi and M. P. Tosi, in *Solid State Physics*, edited by H. Ehrenreich, F. Seitz, and D. Turnbull (Academic, New York, 1981), Vol. 36, p. 177.

EFFECTIVE ACTION IN DENSITY FUNCTIONAL THEORY AND THE BERRY PHASE

A.K. Rajagopal

Naval Research Laboratory
Washington D.C. 20375-5320, USA

INTRODUCTION

This paper addresses two important questions of principle: (1) the contradiction in the traditional time-dependent (td) density functional theory (tdDFT) which seems to require that causality be taken into account explicitly in the variational principle, as pointed out in a critical review article by Gross *et al* [1]; and (2) to obtain the Berry phase that should be present in the time evolution of any given initial state when the system Hamiltonian has time varying parameters such as ion coordinates, magnetic field, etc. The Berry phase within the density functional theory is not expected to be explicitly found at first sight, since phases cancel out in forming the density, and thus are of no consequence in practice. We here show that the resolution of these two issues comes about by reformulating the traditional Dirac-Frenkel td action principle to take explicit account of the time-path in the form suggested by Jackiw and Kerman [2] instead of that used in the development of the tdDFT [1]. (Dr. Robert van Leeuwen has recently informed me that he too has a time-path version of the action principle which I have not seen.) This reformulation leads to an effective action functional which generates appropriate td Green's functions and contains also the Berry phase as an identifiable piece of the action functional. It should be mentioned that the Berry phase is also identifiable in the Dirac-Frenkel action functional [3].

We will first present, in Section 2, the stationary action principle in terms of the td wave function and elsewhere we give the corresponding generalization in terms of td-density matrix [4], which is adopted in the density functional context in [5]. In Section 3, after a brief outline of the theory of the Berry phase, we develop the corresponding theory for tdDFT. Some recent implications of this are also pointed out to show the significance of this development. In Section 4, a brief summary is given.

THE STATIONARY ACTION PRINCIPLE IN tdDFT

We employ the notations in [1] to develop the stationary action principle for the tdDFT following the formulation in [2]. Consider the system Hamiltonian

$$\begin{aligned} \hat{H}(t) &= \hat{T} + \hat{U} + \hat{V}(t) \equiv \hat{H} + \hat{V}(t) \ , \\ \hat{V}(t) &= \int d^3r \nu_{ext}(\vec{r}, t)\hat{n}(\vec{r}); , \\ \hat{n}(\vec{r}) &= \text{density operator} \ . \end{aligned} \tag{1}$$

Here the external potential $\nu_{ext}(\vec{r}, t)$, is taken to be such that it vanishes as $t \to \pm\infty$ and $\hat{n}(\vec{r})$ is the density operator in the Schrödinger picture. Here $\hat{T}$ is the kinetic energy operator for the electrons and $\hat{U}$ represents their mutual Coulomb repulsion [1]. We want to solve the td-Schrödinger equation

$$\begin{aligned} i\hbar\frac{\partial}{\partial t}|\pm, t\rangle = \hat{H}|\pm, t\rangle + \int d^3r \nu_{ext}(\vec{r}, t)\hat{n}(\vec{r})|\pm, t\rangle \ , \\ \text{with} \lim_{t\to\mp\infty} |\pm, t\rangle = |0\rangle, \ \text{with} \ |0\rangle \ \text{the groundstate of} \ \hat{H}. \end{aligned} \tag{2}$$

Note that these are two different boundary conditions for the two equations defining the forward and backward time evolution of the state $|0\rangle$ of the Hamiltonian $\hat{H}$.

Consider the functional

$$\Gamma[\psi_-, \psi_+] = \int_{-\infty}^{+\infty} dt\langle\psi_-, t| \left(i\hbar\frac{\partial}{\partial t} - \hat{H}\right) |\psi_+, t\rangle \tag{3}$$

We require that this functional be stationary when $|\psi_\pm, t\rangle$ are varied arbitrarily and independently but subject to two constraints:

$$\langle\psi_-, t|\hat{n}(\vec{r})|\psi_+, t\rangle = n(\vec{r}, t) \ , (\text{transition density}) \tag{4a}$$

$$\text{and} \ \langle\psi_-, t|\psi_+, t\rangle = 1 \ . \tag{4b}$$

This construction of the functional leads to the generating functional of the connected irreducible N-point functions of density, which will turn out to be a Legendre transform of the action functional.

We impose the boundary conditions

$$\lim_{t\to\mp\infty} |\psi_\pm, t\rangle = |0\rangle \ . \tag{5}$$

These ensure that the variations of the states at the boundary are zero. We will now prove that Γ is a functional of the td-transition density defined in Eq.(4a). The relation of this to the "forward" and "backward" densities defined in terms of "forward" and "backward' states and other associated subtle points are discussed in a more detailed paper which has now appeared [6]. Since the variations are constrained by (4a,4b), two Lagrange multipliers $\nu_{ext}(\vec{r}, t)$ and $w(t)$, are correspondingly introduced and thus we are led to consider the functional

$$\begin{aligned} W'[\psi_-, \psi_+] \equiv \Bigg\{\Gamma[\psi_-, \psi_+] - \int d^3r \int_{-\infty}^{+\infty} dt \, \nu_{ext}(\vec{r}, t)\langle\psi_-, t|\hat{n}(\vec{r})|\psi_+, t\rangle \\ - \int_{-\infty}^{+\infty} dt w(t)\langle\psi_-, t|\psi_+, t\rangle\Bigg\} \ . \end{aligned} \tag{6}$$

We now demand stationarity of W' with respect to variations of $|\psi_\pm, t\rangle$, which lead to the equations, subject to the conditions given by Eq. (5):

$$\left\{ i\hbar\frac{\partial}{\partial t} - \hat{H} - \int d^3r \nu_{ext}(\vec{r}, t)\hat{n}(\vec{r}) \right\} |\psi_+, t\rangle = w(t)|\psi_+, t\rangle \tag{7a}$$

$$\left\{ i\hbar\frac{\partial}{\partial t} - \hat{H} - \int d^3r \nu_{ext}(\vec{r}, t)\hat{n}(\vec{r}) \right\} |\psi_-, t\rangle = w^*(t)|\psi_-, t\rangle \tag{7b}$$

From these we may construct the td equations (2) without the source terms:

$$|+, t\rangle = \left(\exp\frac{i}{\hbar}\int_{-\infty}^{t} dt' w(t') \right) |\psi_+, t\rangle \ , \tag{8a}$$

$$|-, t\rangle = \left(\exp\frac{i}{\hbar}\int_{t}^{\infty} dt' w^*(t') \right) |\psi_-, t\rangle \ , \tag{8b}$$

with two different boundary conditions on the solutions of Eqs. (7a,7b) as above,

$$\lim_{t\to\mp\infty} |\pm, t\rangle = \lim_{t\to\mp\infty} |\psi_\pm, t\rangle = |0\rangle \ . \tag{9}$$

These equations represent forward and backward evolution, reminiscent of the Schwinger-Keldysh time path theory, now stated in terms of the evolution of these states. From (8a,8b) we obtain, upon using the overlap condition (4b) the result

$$\langle -, t|+, t\rangle = \exp\frac{i}{\hbar}\int_{-\infty}^{+\infty} dt\, w(t) \ . \tag{10}$$

From (7a), (4b), and (10), we obtain using (3) and (6) the important result

$$\begin{aligned} W[\psi_-, \psi_+] &\equiv \int_{-\infty}^{+\infty} dt w(t) \\ &= \Gamma[\psi_-, \psi_+] - \int_{-\infty}^{+\infty} dt \int d^3r \nu_{ext}(\vec{r}, t)\hat{n}(\vec{r}, t) \\ &= \int_{-\infty}^{+\infty} dt \langle\psi_-, t| \left\{ i\hbar\frac{\partial}{\partial t} - \hat{H} - \int d^3r \nu_{ext}(\vec{r}, t)\hat{n}(\vec{r}) \right\} |\psi_+, t\rangle. \end{aligned} \tag{11}$$

We may further express the functional W in a form that incorporates explicitly the causality requirement by formally solving Eqs. (2) as follows: we first define $|\pm, t\rangle = (\exp -\frac{i}{\hbar}\hat{H}t)|\tilde{\tilde{\pm}}, t\rangle$. Then Eqs. (2) take simple forms with the density operator now in the Heisenberg representation and using the appropriate boundary conditions etc. we obtain from Eq. (11) the result that **W is a functional of $\nu_{\mathbf{ext}}(\vec{\mathbf{r}}, \mathbf{t})$**. Thus,

$$\exp\frac{i}{\hbar}W[\nu_{ext}] = \langle 0|T\left(\exp -\frac{i}{\hbar}\int d^3r \int_{-\infty}^{+\infty} dt' \nu_{ext}(\vec{r}, t')\hat{n}_H(\vec{r}, t') \right)|0\rangle \ . \tag{12}$$

Here T is the time ordering operator. The various functional derivatives of W thus generate appropriate time ordered products of the density operator in the Heisenberg representation.

From this and Eq. (11) we prove, after some algebra, that

$$\frac{\delta W[\nu_{ext}]}{\delta\nu_{ext}(\vec{r}, t)} = -n(\vec{r}, t) \ . \tag{13}$$

The proof of this statement follows from Eqs. (7a,7b), constancy of the overlap given by Eq.(4b), and the boundary conditions. Eq. (13) shows that the conjugate variable to $\nu_{ext}(\vec{r},t)$ is $n(\vec{r},t)$.

We finally obtain

$$W[\nu_{ext}] + \int d^3r \int_{-\infty}^{+\infty} dt\, \nu_{ext}(\vec{r},t)n(\vec{r},t) = \Gamma[n]$$
$$= \int_{-\infty}^{+\infty} dt\langle\psi_-,t|\left(i\hbar\frac{\partial}{\partial t} - \hat{H}\right)|\psi_+,t\rangle. \quad (14)$$

Thus W and Γ are just Legendre transforms of each other. Γ is a functional of $n(\vec{r},t)$ and a direct calculation using Eq. (13) leads to the result

$$\frac{\delta\Gamma[n]}{\delta n(\vec{r},t)} = \nu_{ext}(\vec{r},t)\ . \quad (15)$$

Removal of constraint is thus equivalent to setting $\nu_{ext} = 0$ and the effective action is stationary with respect to variations in $n(\vec{r},t)$.

Note that this is the density functional version of the generating functional of the Green functions in [2]. We have thus defined an action functional in terms of time-path ordered operators and provided the answer to the first item.

THE BERRY PHASE IN tdDFT

In its simplest form, the Berry phase accompanies any cyclic (not necessarily but including adiabatic) change in a system. For a comprehensive review of the various developments in this field, one may refer to the Resource Letter by Anandan *et al* [7]. Here we consider for simplicity of presentation the time-dependent Schrödinger equation

$$i\hbar\frac{d}{dt}|\psi(t)\rangle = \hat{H}\left(\{R(t)\}\right)|\psi(t)\rangle \quad (16)$$

for a system embedded in a slowly changing environment described by a set of N time-dependent parameters $\{R(t)\} = (R_1(t), R_2(t), \ldots R_N(t))$ with the initial state given by

$$|\psi(t=0)\rangle = |n;\ \{R(t=0)\}\rangle\ . \quad (16')$$

If the Hamiltonian has a nondegenerate spectrum and the parameters are slowly varying, it is well-known that one may use the instantaneous eigenstates of the Hamiltonian, with Eq. (16′) being the initial state of that.

Thus

$$\begin{aligned} \hat{H}\left(\{R(t)\}\right)|n;\{R(t)\}\rangle &= E_n\left(\{R(t)\}\right)|n;\{R(t)\}\rangle \\ \langle m;\{R\}|n;\{R\}\rangle &= \delta_{m,n}\ , \end{aligned} \quad (17)$$

and the solution to Eq.(16) is of the form

$$|\psi(t)\rangle = \exp\frac{i}{\hbar}b_n\left(\{R(t)\}\right)\exp -\frac{i}{\hbar}\int_0^t dsE_n\left(\{R(s)\}\right)|n;\{R(t)\}\rangle\ . \quad (18)$$

Here the Berry phases, $b_n(\{R(t)\})$, are obtained by substituting (18) in (16) and using (17):

$$\begin{aligned} b_n(\{R(t)\}) &= \int_0^t ds \langle n; \{R(s)\}| i\hbar \frac{d}{ds} |n; \{R(s)\}\rangle \\ &= \sum_i \int_{R(0)}^{R(t)} dR_i' \langle n; \{R'\}| i\hbar \partial_{R_i'} |n; \{R'\}\rangle \\ &\equiv \sum_i \int_{R(0)}^{R(t)} dR_i' A_i(n; \{R'\}) \ . \end{aligned} \tag{19}$$

Here we have defined the "gauge" potential, $A_i(n; \{R'\})$. Its gauge transformation property follows when we note that another choice of the state, which differs from the ones used above by a phase factor, leads to an addition of the gradient of the phase to the gauge:

$$\begin{aligned} &\text{If } |n; \{R\}\rangle = \exp \frac{i}{\hbar} \zeta_n(\{R\}) |n; \{R\}\rangle \ , \quad \text{then} \\ &A_i'(n; \{R\}) = A_i(n; \{R\}) - \partial_{R_i} \zeta_n(\{R\}) \ . \end{aligned} \tag{20}$$

A useful result in the computation of these phases is the following, which is derived easily using the above definition:

$$\begin{aligned} \langle m; \{R\}| \frac{\partial \hat{H}(\{R\})}{\partial R} |n; \{R\}\rangle &= (E_n(\{R\}) - E_m(\{R\})) \langle m; \{R\}| \partial_R |n; \{R\}\rangle \\ &\quad + \frac{\partial E_n(\{R\})}{\partial R} \delta_{m,n} \ . \end{aligned} \tag{21}$$

Another useful quantity is the "curvature" or two-form, which appears in the dynamics of the parameters $\{R\}$ (as for example, when the R's are ion-coordinates in the electron-ion system, or in vortex dynamics). It is defined as

$$B_i(n; \{R\}) = \varepsilon_{ijk} \partial_{R_j} A_k(n; \{R\}) \ , \quad \varepsilon_{ijk} \text{ is the Levi – Civita symbol.} \tag{22}$$

Aharanov and Anandan (see [7]) gave a theory which did not depend on the adiabatic scheme used above. For degenerate systems, one obtains a nonabelian gauge structure (because the gauge in Eq. (18) becomes a matrix in the degenerate set of states) for the phases as found by Wilczek and Zee (see [7] for reference). This essentially concludes the basic formulation of the theory of Berry phase.

It is clear from the above description that the above framework is directly applicable to our variational formulation quite generally and as a practical matter to the td-Kohn-Sham equations, which is outlined below.

Following the arguments as in [3], from Eqs. (7a) and (8a), we see that $w(t)$ is just a phase associated with the action of the external potential (compare Eq. (2) with Eqs. (7a,7b)) and in fact it is given by

$$w(t) = \langle \psi_-, t| \left\{ i\hbar \frac{\partial}{\partial t} - \hat{H} - \int d^3 r \nu_{ext}(\vec{r}, t) \hat{n}(\vec{r}) \right\} |\psi_+, t\rangle \ . \tag{23}$$

If the external potential begins to act from some time t_0, when the phase is chosen to be zero, then the phase accumulated till any time t afterwards is given by the integral

$$\gamma(t, t_0) = \int_{t_0}^t dt' w(t') \ . \tag{24}$$

We may now identify the Berry phase as

$$\gamma_B(t, t_0) = \int_{t_0}^{t} dt' \langle \psi_-, t' | i\hbar \frac{\partial}{\partial t'} | \psi_+, t' \rangle \,. \tag{25}$$

This is our general reformulation of the arguments based on the Dirac-Frenkel form given in [3]. In a short Appendix, we give a construction of the forward and backward states in terms of the "instantanous" states defined in Eq. (17) and express the Berry phase as in Eq. (25), given above.

Using the td-Kohn-Sham (KS) orbitals of [1], we observe that the time evolution of the many-particle state may be expressed as the evolution of the product of creation operators associated with KS orbitals, each of which will acquire their own Berry phase so that the total Berry phase turns out to be the sum of the Berry phases associated with the occupied KS orbitals. We may cite the example of the motion of the ion coordinate in the condensed matter system where the Berry phase naturally arises (see for example, [8]) and the same type of arguments hold when applied to the density functional approach to this problem. In detail, the Kohn-Sham reference many electron ground state is given by

$$|GS\rangle = \prod_{n(occ.)} c_n^\dagger |vac.\rangle \tag{26}$$

where $c_n^\dagger$ is the creation operator of the n-th KS orbital (we have here used a generic quantum number, n, to denote the KS orbitals). The electron creation operator is given by

$$\begin{aligned} \psi^\dagger(\vec{r}) &= \sum_n c_n^\dagger \phi_n^*(\vec{r}), \ \{\phi_n\} \text{ complete orthonormal KS system}\,, \\ \text{so that } c_n^\dagger &= \int d^3r \psi^\dagger(\vec{r}) \phi_n(\vec{r}) \,. \end{aligned} \tag{27}$$

Adiabatic motion of ions, for example, produces a Berry phase in each of the solutions of the KS equations: $\phi_n \to e^{i\gamma_n}\phi_n$, and hence the ground state develops a Berry phase γ_B, $|GS\rangle \to e^{i\gamma_B}|GS\rangle$, where $\gamma_B = \sum_{n(occ.)} \gamma_n$. Using Eq. (25) and displaying the ion coordinates explicitly in the KS orbitals in the usual way, we may express the total Berry phase in the form

$$\gamma_B(t, t_0) = \int_{t_0}^{t} dt' \sum \langle \phi_n \left(\vec{r}, \left\{\vec{R}(t')\right\}\right) | \nabla_{\vec{R}} | \phi_n \left(\vec{r} \left\{\vec{R}(t')\right\}\right) \rangle \cdot \dot{\vec{R}}(t') \,. \tag{28}$$

Here the over-dot denotes time derivative of the ion coordinate, and $\{\vec{R}(t)\}$ denotes the set of ion coordinates involved in the adiabatic motion under consideration.

We may also mention here that Gaitan [9] investigated the Berry phase induced in the BCS superconducting ground state by movement of the vortex and showed this by considering the BCS state as the occupation of Bogoliubov quasi particle states. This in conjunction with Eq. (22), leads to the derivation that a Magnus force [10] exists for a vortex line in a superconductor.

Another manifestation of the Berry phase is the change of sign in the electronic part of the wave function when the system undergoes a Jahn-Teller effect [11]. Here under a 2π rotation of the vibrational coordinates, the electronic part of the wave function acquires the above mentioned Berry phase.

Professor Ulf von Barth brought to my attention the work of Resta [12] where the Berry phases associated with Kohn-Sham orbitals are used in developing a new approach to the macroscopic electric polarization in crystalline dielectrics. This theory also provides a powerful computational tool for dealing with Born effective charges, linear and non-linear piezoelectricity, and spontaneous polarization in ferroelectric materials.

In the above considerations, we focussed on the evolution of the state vector of the many body system. In the more general context, one would consider the evolution of the density matrix in place of the state vector, for which also appropriate action principles have been derived, both for general many body system [4], and in the context of functional formalism [5], which incorporates other relevant quantites of interest beyond density. Since the procedures are similar except that one now considers the Liouville-von Neumann equation for the density matrix instead of the Schrödinger equation, besides other forms of the Hamiltonian, we will relegate this development to a separate paper. Since this includes the temperature dependence of the initial state of the system, this development adds to that of Resta [12] in providing a framework to incorporate temperature dependence in the theory of electric polarization.

SUMMARY AND CONCLUSIONS

In conclusion, we may state that a suitable reformulation of the Dirac-Frenkel stationary action principle as was done in [2] in the context of many body theory, provides us with an appropriate and correct framework for developing the action functional in the td-density functional context. This procedure answers the doubt that was raised in [1] if one replaces the functional used there by the one derived here, Eq. (14). Also, the occurrence of the Berry phase in time-dependent phenomena is shown here to be present even though it lay buried in the KS orbitals. The recent use of the Berry phase ideas in understanding the dielectric polarization phenomena by Resta [12] points to the importance of the Berry phases where electronic currents play important role.

ACKNOWLEDGEMENTS

It is a pleasure to place on record my thanks to Professor J. Anandan for his correspondence concerning aspects of the Berry phase and for sending me a copy of his paper before publication. I thank Professor Ulf von Barth for pointing out the relevance of my inquiry into the Berry phase in tdDFT to the preprint of Professor R. Resta. I am thankful to the US National Science Foundation for supporting my trip to Brisbane; to the Office of Naval Research for partial support of the research; and the support and hospitality of Professors John Dobson, Mukunda Das, and Giovanni Vignale.

APPENDIX: DERIVATION OF Eq.(25)

In order to derive the Berry phase from our variational principle, we now employ the instantaneous states in place of Eq. (5) as follows:

$$\lim_{t \to \mp\infty} |\psi_\pm, t\rangle = |0, \{R\}\rangle \tag{A1}$$

where $\{R\}$ are for simplicity taken to be the same parameters at the two limits of time. Now we employ the instantaneous eigenstates defined in Eq. (17) to express the states in the form:

$$|\psi_\pm, t\rangle = \exp\left(-\frac{i}{\hbar}\left[\int_{\mp\infty}^{t} dt'\left(w(t') + E_0\left(\{R(t')\}\right)\right)\right.\right.$$

$$\left.\left.+\left(b_0^{(\pm)}\left(\{R(t)\}\right) - b_0^{(\pm)}\left(\{R(\mp\infty)\}\right)\right)\right]\right)|0, \{R(t)\}\rangle\,. \tag{A2}$$

Here $b_0^{(\pm)}\left(\{R(t)\}\right)$ is the phase to be identified as Berry's phase, such that Eq. (A1) is satisfied. The phases are chosen such that the normalization condition Eq. (4b) holds. This leads to the result

$$\int_{-\infty}^{+\infty} dt\, w(t) = -\int_{-\infty}^{+\infty} dt E_0\left(\{R(t)\}\right)$$

$$+\left(b_0^{(+)}\left(\{R(t)\}\right) - b_0^{(-)}\left(\{R(t)\}\right)\right)$$

$$-\left(b_0^{(+)}\left(\{R\}\right) - b_0^{(-)}\left(\{R\}\right)\right)\,. \tag{A3}$$

Eq. (A2) in Eqs. (7a,7b) determine the phases:

$$b_0^{(\pm)}\left(\{R(t)\}\right) - b_0^{(\pm)}\left(\{R\}\right) = \int_{\mp\infty}^{t} ds\langle 0, \{R(s)\}\,|i\hbar\frac{\partial}{\partial s}|0, \{R(s)\}\rangle\,. \tag{A4}$$

Thus Eq. (A3) is expressed in the familiar form

$$W[\psi_-;\psi_+] \equiv \int_{-\infty}^{+\infty} ds\, w(s)$$

$$= -\int_{-\infty}^{+\infty} ds E_0\left(\{R(s)\}\right) + \int_{-\infty}^{+\infty} ds\langle 0, \{R(s)\}\,|i\hbar\frac{\partial}{\partial s}|0, \{R(s)\}\rangle\,. \tag{A5}$$

It is now easy to verify that the identification given in Eq. (25) is indeed correct and is related to the form given in Eq. (18) specialized to the td-DFT case. Thus we finally obtain the familiar result expressed in our formulation

$$\gamma_B \equiv \int_{-\infty}^{+\infty} ds\langle\psi_-, s|i\hbar\frac{\partial}{\partial s}|\psi_+, s\rangle$$

$$= \int_{-\infty}^{+\infty} ds\left\{w(s) + E_0\left(\{R(s)\}\right)\right\}$$

$$= \int_{-\infty}^{+\infty} ds\langle 0, \{R(s)\}|i\hbar\frac{\partial}{\partial s}|0, \{R(s)\}\rangle \tag{A6}$$

This completes the demonstration that the Berry phase can be determined within the general td-DFT framework. A practical way of obtaining it and the consequences thereof is by using the KS orbitals as outlined in the text.

REFERENCES

1. E.K.U. Gross, J.F. Dobson, and M. Petersilka in *Density Functional Theory II*, edited by R.F. Nalewajski, Vol. 181 of *Topics in Current Chemistry* (Springer, Berlin, 1996). I thank Professor Gross for sending me this article before its publication. The various critical comments in this article were a source of inspiration to complete the present paper.

2. R. Jackiw and A. Kerman, Phys. Lett. A **71**, 158 (1979). This is another version of the Schwinger time-path ordered formulation of his action principle technique. J. Schwinger, J. Math. Phys. **2**, 407 (1962).
3. A.K. Pattanayak and W.C. Schieve, Phys. Rev. E **50**, 3601 (1994). See also A.K. Rajagopal and J. Marshall, Phys. Rev. A **26**, 2977 (1982).
4. O. Eboli, R. Jackiw, and S.Y. Pi, Phys. Rev. D **37**, 3557 (1988). This is an adaptation of the td-variational principle for td-expectation value of an observable by R. Balian and M. Vénéroni, Ann. Phys. (NY) **164**, 384 (1985).
5. A.K. Rajagopal and F.A. Buot, in *Density Functional Theory II*, edited by R.F. Nalewajski, Vol. 181 of *Topics in Current Chemistry* (Springer, Berlin 1996).
6. A.K. Rajagopal, Phys. Rev. A **54**, 3916 (1996).
7. J. Anandan, J. Christian, and K. Wanelik, *Resource Letter: Geometric Phase in Physics*, Am. J. Phys. (1996) , to appear. I thank Professor Anandan for sending a copy of this paper before publication.
8. N. Sukumar, Int. J. Quantum Chem. **56**, 423 (1995).
9. F. Gaitan, Phys. Rev. B **51**, 9061 (1995).
10. P. Ao and D.J. Thouless, Phys. Rev. Lett. **70**, 2158 (1993).
11. F.S. Ham, Phys. Rev. Lett. **58**, 725 (1987).
12. R. Resta. Rev. Mod. Phys. **66**, 899 (1994); see also *Density-Polarization-Functional Theory and Long-Range Correlation in Dielectrics* by R. Resta, Preprint (1996). I thank Professor Resta for providing this preprint before its publication.

HYDRODYNAMICS IN THE THOMAS-FERMI-DIRAC-VON WEIZSÄCKER APPROXIMATION

E. Zaremba[1] and H. C. Tso[2]

[1]Department of Physics
Queen's University
Kingston, Ontario, Canada K7L 3N6

[2]Department of Physics
Dalhousie University
Halifax, Nova Scotia, Canada B3H 3J5

INTRODUCTION

The use of hydrodynamics in the description of quantum systems has a long and diverse history. Perhaps one of the earliest applications [1] was to the stopping power problem which concerns the excitation phenomena taking place when a charged particle passes through matter. At an atomistic level, the moving ion suffers a loss of energy as a result of collisions with atomic electrons. However, individual Rutherford scattering events cannot in fact take place due to the long-range nature of the Coulomb potential which implies that the ion interacts at once with many electrons. Furthermore, the electrons which respond to the passage of the ion, interact with each other, so that the field experienced by a given electron is comprised of the total field due to the ion and the dynamic screening charge of all other electrons. This situation is exceedingly complex and indicates that a quantitative understanding can only be achieved by a consideration of the collective response of the electrons. Presumably this is what Bohr [2] had in mind when he imagined electrons as forming a trailing wake behind the ion. This picture has a natural realization in Bloch's hydrodynamic theory [1] which treats the electrons as a charged fluid, analogous to an ordinary fluid, whose dynamical state is specified in terms of local variables such as density, velocity and pressure. What is far from obvious is the extent to which this picture is meaningful for a degenerate quantum system.

In principle, a rigorous justification of hydrodynamics must follow from a detailed microscopic analysis. Our purpose here is not to pursue this argument, but merely to indicate why such a theory is plausible and worthy of further investigation. Hydrodynamics incorporates conservation laws, each of which has a microscopic basis. The continuity equation, for example, is an exact equation of motion for the density operator $\hat{n}(\mathbf{r}) = \sum_{i=1}^{N} \delta(\mathbf{r} - \mathbf{r}_i)$. In terms of Heisenberg operators, we have

$$\frac{\partial}{\partial t}\hat{n}(\mathbf{r},t) = -\nabla \cdot \hat{\mathbf{j}}(\mathbf{r},t) \tag{1}$$

where the current density operator is defined as (we use $e = m = \hbar = 1$ throughout)

$$\hat{j}_\mu(\mathbf{r}) = \frac{1}{2}\sum_{i=1}^{N}[\hat{p}_{i\mu}\delta(\mathbf{r} - \hat{\mathbf{r}}_i) + \delta(\mathbf{r} - \hat{\mathbf{r}}_i)\hat{p}_{i\mu}]\,. \tag{2}$$

The expectation value of Eq. (1) in some quantum state $|\Psi\rangle$, or in an ensemble of states specified by the density matrix $\hat{\rho}$, gives the continuity equation of hydrodynamics. In a similar way, momentum conservation is incorporated into the equation of motion for $\hat{j}_\mu(\mathbf{r})$,

$$\frac{\partial}{\partial t}\hat{j}_\mu(\mathbf{r},t) = -\hat{n}(\mathbf{r},t)\frac{\partial}{\partial r_\mu}v_{ext}(\mathbf{r},t) - \frac{\partial}{\partial r_\nu}\hat{T}^{kin}_{\mu\nu}(\mathbf{r},t) + \hat{F}^{int}_\mu(\mathbf{r},t)\,. \tag{3}$$

The first term on the right hand side represents the force due to an external potential $v_{ext}(\mathbf{r},t)$. The second term is the kinetic stress tensor

$$\hat{T}^{kin}_{\mu\nu}(\mathbf{r}) = \frac{1}{4}\sum_{i=1}^{N}\{\hat{p}_{i\mu}\delta(\mathbf{r} - \hat{\mathbf{r}}_i)\hat{p}_{i\nu} + \hat{p}_{i\mu}\hat{p}_{i\nu}\delta(\mathbf{r} - \hat{\mathbf{r}}_i) + h.c.\} \tag{4}$$

and the last term is the contribution

$$\hat{F}^{int}_\mu(\mathbf{r}) = -\sum_{i\neq j}\delta(\mathbf{r} - \hat{\mathbf{r}}_i)\frac{\partial}{\partial r_\mu}v(\mathbf{r} - \hat{\mathbf{r}}_j) \tag{5}$$

which arises from the interaction, $v(\hat{\mathbf{r}}_i - \hat{\mathbf{r}}_j)$, between the particles. The expectation value of Eqs. (4) and (5) cannot in general be expressed in terms of the density $n(\mathbf{r},t)$ and current density $j_\mu(\mathbf{r},t)$. Thus a systematic development of the equations of motion results in a hierarchy of equations which increase in complexity as more details of the microscopic dynamics are included. The goal of a hydrodynamic description is to find a physically meaningful truncation of this hierarchy [3,4]. For example, in conventional hydrodynamics one assumes that the properties of the fluid vary sufficiently slowly in space, on the scale of the particle mean free path, and in time, on the scale of internal relaxation processes. In this situation, local equilibrium conditions prevail and the kinetic stress tensor gives rise to a term which depends on the local pressure, while the interaction part is the origin of viscous stresses. However this conventional definition of hydrodynamics is overly restrictive since it precludes application to systems whose equilibrium properties vary rapidly in space and in which the long-range Coulomb interaction is the dominant interaction effect. Our usage of the term 'hydrodynamic' will include any description which makes use of a closed set of local field variables such as density and velocity, irrespective of time scales and the rate of spatial variations.

Unlike conventional hydrodynamics which typically deals with homogeneous fluids, our interest here is in strongly inhomogeneous systems. In the realm of mesoscopic physics, with electrons confined in quantum wells, dots, rings and other structures, the electron density decays to zero continuously in some direction. The example of a metallic surface illustrates the idea. The simplest model of this situation is jellium, in which a uniform positive background is terminated at a plane. In equilibrium, the electron density spills out into the vacuum and varies smoothly from a constant value

$\bar{n}$ in the bulk of the metal to zero far from the 'surface'. As a result of the spatial confinement, a new collective excitation appears, the surface plasmon, at the frequency $\omega_p/\sqrt{2}$, where ω_p is the frequency of the bulk plasmon. A hydrodynamic treatment of this problem was provided many years ago [5,6,7] using a model in which the electron density drops to zero discontinuously at the surface. Since the hydrodynamic equations are partial differential equations in space, appropriate boundary conditions must be specified at this abrupt surface. In the literature, these boundary conditions go by the name of additional boundary conditions (ABC's) and, as discussed by Forstmann and Gerhardts [8], there is some ambiguity in how they are to be defined. Going to more realistic density profiles does not in itself resolve these difficulties [9].

At a microscopic level, mathematical boundaries of course do not exist. The equilibrium electron density varies smoothly in space and a consistent description of the dynamics should build these equilibrium properties into the theory. Only in this way will the density fluctuations associated with collective modes have a smooth and well-defined spatial distribution. In other words, it is preferable even within a hydrodynamic context, not to have 'boundaries' at all. This is part of the motivation for developing the Thomas-Fermi-Dirac-von Weizsäcker (TFDW) hydrodynamics [10,11] to be described in the following section.

TIME-DEPENDENT DFT

The TFDW hydrodynamic theory is essentially a simplified form of time-dependent density functional theory (DFT) [12]. In comparison to the equilibrium theory [13] which has been used so successfully to calculate ground state properties, the time-dependent theory is relatively unexplored and still in its early stages of development. Hydrodynamics is but one of the directions this development can be expected to follow.

To place the hydrodynamic theory in context, it is useful to first introduce some of the main DFT ideas that will be needed [14]. In DFT, the ground state density is determined by minimizing the energy functional

$$E[n] = T_s[n] + \frac{1}{2}\int d\mathbf{r}\int d\mathbf{r}'\,\frac{n(\mathbf{r})n(\mathbf{r}')}{|\mathbf{r}-\mathbf{r}'|} + E_{xc}[n] + \int d\mathbf{r}\,v_{ext}(\mathbf{r})n(\mathbf{r}) \tag{6}$$

where $T_s[n]$ is the kinetic energy of an assembly of noninteracting electrons with density $n(\mathbf{r})$, $E_{xc}[n]$ is the exchange-correlation (xc) energy functional and $v_{ext}(\mathbf{r})$ is a confining potential. Within the Kohn-Sham (KS) scheme [15], the minimization is achieved by solving the single-particle Schrödinger equations

$$-\frac{1}{2}\nabla^2\phi_i(\mathbf{r}) + v_{eff}(\mathbf{r};[n])\phi_i(\mathbf{r}) = \epsilon_i\phi_i(\mathbf{r}) \tag{7}$$

which determine the density according to

$$n(\mathbf{r}) = \sum_{\substack{i\\(occ)}} |\phi_i(\mathbf{r})|^2 . \tag{8}$$

The potential in Eq. (7) is $v_{eff} = v_{ext} + v_H + v_{xc}$, where v_H is the classical Hartree potential and v_{xc} is the xc contribution. v_{eff} is a functional of the density and thus Eq. (7) must be solved self-consistently.

If the external potential is changed by a small amount δv_{ext}, the electron density adjusts accordingly. Within the KS scheme, this response corresponds to that of noninteracting electrons, and is given by [16]

$$\delta n(\mathbf{r}) = \int d\mathbf{r}' \chi_0(\mathbf{r}, \mathbf{r}'; \omega = 0) \delta v_{eff}(\mathbf{r}) \tag{9}$$

where $\chi_0(\mathbf{r}, \mathbf{r}'; \omega = 0)$ is the static (zero-frequency) noninteracting electron density response function of the KS system, and $\delta v_{eff} = \delta v_{ext} + \delta v_H + \delta v_{xc}$. This constitutes a rigorous description of the response since only ground state properties are involved.

Adiabatic Local Density Approximation

An extension of this method to time-dependent perturbations was introduced by several authors [17,18] without formal justification, and is now referred to as the adiabatic local density approximation (ALDA). For a dynamic potential $\delta v_{ext}(\mathbf{r}, t) = \delta v_{ext}(\mathbf{r}) e^{-i\omega t}$, one simply assumes that the response of the electrons is given by Eq. (9) with $\chi_0(\mathbf{r}, \mathbf{r}'; \omega = 0)$ replaced by $\chi_0(\mathbf{r}, \mathbf{r}'; \omega)$. Another way to look at the approximation is to consider the time-dependent generalization of the KS equation

$$i \frac{\partial}{\partial t} \phi_i(\mathbf{r}, t) = -\frac{1}{2} \nabla^2 \phi_i(\mathbf{r}, t) + v_{eff}(\mathbf{r}, t; [n]) \phi_i(\mathbf{r}, t) \tag{10}$$

in which $v_{eff}(\mathbf{r}, t; [n])$ is taken to be the *same* functional of the dynamic density as in the static limit. Using this definition, the dynamic perturbation is

$$\delta v_{eff}(\mathbf{r}, t; [n]) = \delta v_{ext}(\mathbf{r}, t) + \int d\mathbf{r}' \left[\frac{1}{|\mathbf{r} - \mathbf{r}'|} + \left. \frac{\delta v_{xc}(\mathbf{r}; [n])}{\delta n(\mathbf{r}')} \right|_{n_0} \right] \delta n(\mathbf{r}', t) \, . \tag{11}$$

The calculation of the deviation of the KS orbitals to first order in $\delta v_{eff}(\mathbf{r}, t)$ is equivalent to the response function form of the ALDA that is usually used. It is clear from its construction that the essential approximation is the neglect of the intrinsic frequency dependence of the xc potential [12,19] which does not simply arise through the frequency-dependent density. It should also be noted that the scheme is a variant of the conventional RPA (random phase approximation) which includes the dynamic Hartree interaction but ignores xc effects.

The ALDA has been applied in a variety of situations including atomic photoabsorption [18,20], small metal particles [21], metallic plasmons [22], surface response [23,24] and mesoscopic systems [17,25,26], with surprisingly good results. The reasons for this are still unclear as a full explanation requires a deeper understanding of the dynamic xc potential, which at present is not available. Nevertheless, for qualitative, and even quantitative purposes, it is an extremely useful description of the dynamic linear response of the electronic systems of interest. However, the approach does have limitations of a practical nature. First, it is numerically intensive, requiring (i) the self-consistent solution of the KS equations, (ii) the calculation of the noninteracting response function $\chi_0(\mathbf{r}, \mathbf{r}'; \omega)$ and (iii) the solution of an integral equation for $\delta n(\mathbf{r}, \omega)$. Second, it is sometimes difficult to distinguish collective excitations from the single-particle excitations contained in χ_0 [27,28]. And finally, the inclusion of a magnetic field requires an entirely new calculation in which the effect of the magnetic field is built into the KS states [25,26]. These difficulties provide an additional motivation for developing a hydrodynamic description.

TFDW Hydrodynamics

The KS kinetic energy is the reason why the ALDA takes the form that it does. If this energy is replaced by an explicit functional of the density, a different kind of adiabatic theory emerges. To be specific, we will take the kinetic energy in the form of the Thomas-Fermi-von Weizsäcker approximation which gives the energy functional [11,29]

$$E[n] = \int d^d r \left[C_{TF}^d n^{(d+2)/d} + C_{vW}^d \frac{|\nabla n|^2}{n} - C_x^d n^{(d+1)/d} \right] + \frac{1}{2} \int d^d r \int d^d r' \frac{n(\mathbf{r}) n(\mathbf{r}')}{|\mathbf{r} - \mathbf{r}'|} + \int d^d r \, v_{ext}(\mathbf{r}) n(\mathbf{r}) \quad (12)$$

where d is the dimensionality of the system ($d = 2, 3$). The theory also includes a local (Dirac) exchange energy with coefficients $C_x^{2D} = \frac{4}{3}\sqrt{\frac{2}{\pi}}$ [30] and $C_x^{3D} = \frac{3}{4}\left(\frac{3}{\pi}\right)^{1/3}$ [14]. For simplicity we neglect any correlation contribution. The Thomas-Fermi (TF) part of the kinetic energy is defined by the coefficients $C_{TF}^{2D} = \frac{\pi}{2}$ and $C_{TF}^{3D} = \frac{3}{10}(3\pi^2)^{2/3}$, while the von Weizsäcker (vW) kinetic energy has a coefficient $C_{vW}^{3D} = \frac{\lambda}{8}$, with $\lambda = 1$ corresponding to the original vW value [31]. This value gives the total kinetic energy density of a single quantum particle, in which case the TF term is superfluous. However, for a weakly inhomogeneous 3D gas, the TF term is the leading order term in a gradient expansion, and the appropriate vW coefficient is $\lambda = 1/9$ [14]. Both terms are required to represent the density in actual physical situations and an empirical value of $\lambda \simeq 1/4$ seems optimal in some 3D applications [32]. For lack of anything better, a similar value will be used in 2D. It should be noted that a gradient expansion of the kinetic energy density does not exist in 2D and the vW correction cannot be motivated in this way. Nevertheless, a term of this form is necessary if the energy functional is to generate smooth and continuous ground state densities. Clearly further work is required to put this kinetic energy functional on a firmer foundation, and more will be said about this later.

The ground state density of the system is found from the solution of the Euler equation

$$\frac{\delta E[n]}{\delta n(\mathbf{r})} = \mu \,, \quad (13)$$

where the chemical potential μ is chosen to fix the total number of particles, N. Since $n(\mathbf{r})$ is positive definite, we can define the nodeless vW wavefunction $\psi(\mathbf{r}) = \sqrt{n(\mathbf{r})}$ which satisfies the Schrödinger-like equation

$$-\frac{\lambda}{2}\nabla^2 \psi(\mathbf{r}) + \tilde{v}_{eff}(\mathbf{r})\psi(\mathbf{r}) = \mu\psi(\mathbf{r}) \quad (14)$$

with the effective potential (distinguished from the KS potential by a tilde)

$$\tilde{v}_{eff}(\mathbf{r}) = \frac{d+2}{d} C_{TF}^d \psi^{\frac{4}{d}}(\mathbf{r}) - \frac{d+1}{d} C_x^d \psi^{\frac{2}{d}}(\mathbf{r}) + v_H(\mathbf{r}) + v_{ext}(\mathbf{r}) \,. \quad (15)$$

If we define an internal force by

$$\mathbf{F}(\mathbf{r}) \equiv -\nabla \frac{\delta E[n]}{\delta n(\mathbf{r})} = -\nabla \left[\tilde{v}_{eff}(\mathbf{r}) - \frac{\lambda}{2}\frac{\nabla^2 \psi}{\psi} \right] , \quad (16)$$

Eq. (13) can be interpreted as a statement of mechanical equilibrium,

$$\mathbf{F}_0(\mathbf{r}) = -\nabla \frac{\delta E[n]}{\delta n(\mathbf{r})} \bigg|_{n_0} = 0 \,. \quad (17)$$

In the TF approximation ($\lambda = C_x^d = 0$), this becomes

$$\nabla P + n\nabla v_H + n\nabla v_{ext} = 0\,, \tag{18}$$

where $P = \frac{2}{d} C_{TF}^d n^{(d+2)/d}$ is the TF pressure.

Time-dependent phenomena can be considered within this scheme by postulating that the electrons constitute a classical fluid whose dynamics is governed by the hydrodynamic equations [11,10]

$$\frac{\partial n}{\partial t} + \nabla \cdot (n\mathbf{v}) = 0 \tag{19}$$

$$\frac{\partial \mathbf{v}}{\partial t} + (\mathbf{v} \cdot \nabla)\mathbf{v} = \mathbf{F} \tag{20}$$

where $\mathbf{F}(\mathbf{r}, t)$ is the dynamic generalization of Eq. (16),

$$\mathbf{F}(\mathbf{r}, t) = -\nabla \left[\tilde{v}_{eff}(\mathbf{r}, t) - \frac{\lambda}{2} \frac{\nabla^2 \sqrt{n(\mathbf{r}, t)}}{\sqrt{n(\mathbf{r}, t)}} \right]. \tag{21}$$

Since this is an explicit functional of $n(\mathbf{r}, t)$, we have a closed set of dynamical equations for the density and velocity variables. In the static limit, Eq. (20) is consistent with the equilibrium condition, Eq. (17). In other words, the dynamics of the fluid is driven by the difference of the internal force from its equilibrium value. To account for an external magnetic field one can augment the right hand side of Eq. (20) by the magnetic force $\mathbf{v} \times \mathbf{B}$, while the addition of the term $\gamma\mathbf{v}$ introduces a phenomenological damping [11].

It is of interest to note that the hydrodynamic equations can also be obtained for $\lambda = 1$ by adopting a time-dependent generalization of Eq. (14) which is the analogue of the time-dependent KS equation, Eq. (10). Introducing the dynamic vW wave equation

$$i\frac{\partial}{\partial t}\psi(\mathbf{r}, t) = \left[-\frac{1}{2}\nabla^2 + \tilde{v}_{eff}(\mathbf{r}, t) \right] \psi(\mathbf{r}, t)\,, \tag{22}$$

and expressing ψ in the form

$$\psi(\mathbf{r}, t) = \sqrt{n(\mathbf{r}, t)} e^{i\phi(\mathbf{r}, t)} \tag{23}$$

where $n(\mathbf{r}, t)$ is the density and $\nabla\phi(\mathbf{r}, t)$ is identified with the velocity field $\mathbf{v}(\mathbf{r}, t)$, we are again led to Eqs. (19), (20) and (21). The restriction to $\lambda = 1$ is due to the fact that this parameter is introduced to optimize the description of the ground state properties and therefore has no dynamical significance. We will later argue that a complete theory would have $\lambda = 1$, in which case the hydrodynamic equations can indeed be based on Eq. (22).

Linearization

Most of the applications we have in mind concern the response of a system to a weak perturbation such as a time-dependent electromagnetic field. This includes the study of magnetoplasma excitations in mesoscopic systems such as parabolic wells [26], quantum dots [25] and electron rings [29]. The linearization of Eqs.(19, 20) proceeds by writing

$$n(\mathbf{r}, t) = n_0(\mathbf{r}) + \delta n(\mathbf{r}, t) \tag{24}$$

with

$$\delta n(\mathbf{r},t) = 2\psi_0(\mathbf{r})\delta\psi(\mathbf{r},t)\,, \tag{25}$$

and interpreting $\mathbf{v}$ as a first order quantity. We then obtain

$$\frac{\partial \delta n}{\partial t} + \nabla\cdot(n_0\mathbf{v}) = 0 \tag{26}$$

$$\frac{\partial \mathbf{v}}{\partial t} = \delta\mathbf{F} \tag{27}$$

where the linearized force is given by

$$\delta\mathbf{F}(\mathbf{r},t) = -\nabla\left[\delta\tilde{v}_{eff}(\mathbf{r},t) - \frac{1}{\psi_0(\mathbf{r})}\hat{h}\delta\psi(\mathbf{r},t)\right] \equiv -\nabla f(\mathbf{r},t)\,. \tag{28}$$

The operator $\hat{h}$ is the Hamiltonian defining the ground state vW wavefunction,

$$\hat{h} = -\frac{\lambda}{2}\nabla^2 + \tilde{v}_{eff}(\mathbf{r}) - \mu\,. \tag{29}$$

An alternative form of the linearized equations (for $\lambda = 1$) follows from Eq. (22). Writing

$$\psi(\mathbf{r},t) = [\psi_0(\mathbf{r}) + \delta\psi(\mathbf{r},t)]e^{-i\mu t}\,, \tag{30}$$

the dynamic wavefunction fluctuation satisfies

$$i\frac{\partial\delta\psi}{\partial t} = \hat{h}\delta\psi + \hat{g}\left(\delta\psi + \delta\psi^*\right) \tag{31}$$

where $\hat{g}$ is an integral operator with kernel

$$g(\mathbf{r},\mathbf{r}') \equiv \psi_0(\mathbf{r})\left.\frac{\delta\tilde{v}_{eff}(\mathbf{r})}{\delta n(\mathbf{r}')}\right|_{n_0}\psi_0(\mathbf{r}')\,. \tag{32}$$

Since $\delta\psi$ is coupled to its complex conjugate in Eq. (31), harmonic solutions of this equation will have the form

$$\delta\psi(\mathbf{r},t) = u(\mathbf{r})e^{-i\omega t} - v^*(\mathbf{r})e^{i\omega t} \tag{33}$$

where the amplitudes satisfy the nonlocal Bogoliubov equations [33]

$$\begin{aligned}(\hat{\mathcal{L}} - \omega)u(\mathbf{r}) - \hat{g}v(\mathbf{r}) &= 0\\ (\hat{\mathcal{L}} + \omega)v(\mathbf{r}) - \hat{g}u(\mathbf{r}) &= 0\end{aligned} \tag{34}$$

with $\hat{\mathcal{L}} \equiv \hat{h} + \hat{g}$. These equations provide a different way of looking at the dynamics which is reminiscent of the dynamics of a Bose condensate as determined by the time-dependent Gross-Pitaevskii equation [33]. However it should be emphasized that the analogy is purely formal and is not meant to imply that the Fermi system has superfluid properties.

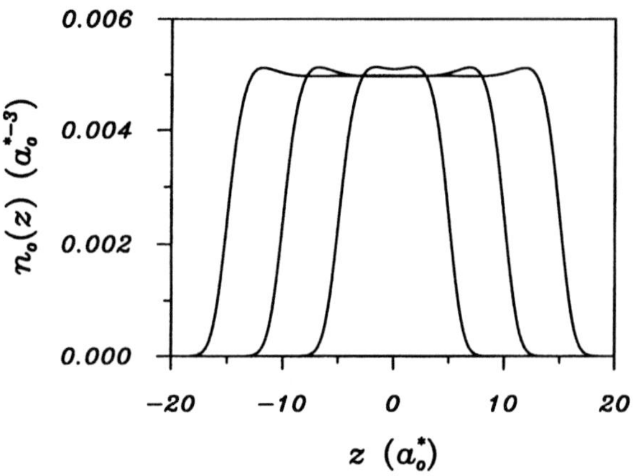

Figure 1. Equilibrium density profiles for a parabolic potential. The increasing width of the electron distribution corresponds to increasing number of electrons [11]. a_0^* is the effective Bohr radius.

AN APPLICATION: PARABOLIC WELLS AND METALLIC SLABS

Parabolic wells can be achieved by controlling the alloy composition in MBE-grown $Al_xGa_{1-x}As$ samples [34]. The technique gives rise to a parabolic variation of the conduction band edge in one spatial dimension and an effective confining potential

$$v_{ext}(z) = \frac{1}{2} m^* \omega_0^2 z^2 \, .$$

This potential is equivalent to that produced by a uniform slab of positive charge of density $\bar{n}$ if the frequency ω_0 is identified with the plasma frequency defined by

$$\omega_p^2 = \frac{4\pi e^2 \bar{n}}{m^*} \, .$$

Electrons provided by remote impurities fall into the potential well and neutralize the effective positive background. One obtains in this way a metallic slab of electrons at the centre of the parabolic well which is the realization of an ideal jellium. The density profile as calculated in the TFDW approximation is shown in Fig. 1 for a range of electron densities.

The dynamics of parabolically confined electrons has been investigated using various techniques [26], including TFDW hydrodynamics [11]. Here, we would like to expand on this earlier work by commenting on the transition from pure parabolic confinement to the neutral metallic slab limit [28,25].

In considering the dynamics of parabolically confined electrons, the centre of mass mode has a privileged status. This mode occurs at exactly ω_0 as a consequence of the generalized Kohn theorem [36,37]. A uniform oscillating electric field typical of long-wavelength radiation couples to the centre of mass degree of freedom of the electron slab. For parabolic confinement, this degree of freedom separates from the internal degrees of freedom, and as a result, only this degree of freedom is excited by the external field. This excitation is a 'sloshing'-type mode in which the equilibrium electron density oscillates rigidly without change in shape: $n(z,t) = n_0(z - \eta_0 cos\omega_0 t)$. Analogous modes are found whenever strict parabolic confinement occurs as in quantum wires (1D), quantum dots (2D) and spherical particles (3D).

The determination of the modes using TFDW hydrodynamics is relatively straightforward. The velocity field can be eliminated from the hydrodynamic equations to yield an equation for the fluctuating vW wavefunction,

$$\lambda\omega^2\delta\psi = \hat{h}(\psi_0 f) = \hat{h}^2\delta\psi + \hat{h}\hat{M}\delta\psi \tag{35}$$

where $\hat{h}$ is defined in Eq. (29) and the operator $\hat{M}$ is given by

$$\begin{aligned}\hat{M}\delta\psi &\equiv \psi_0\delta v_{eff}\\ &= \left[\frac{20}{9}C_{TF}^{3D}\psi_0^{4/3} - \frac{8}{9}C_x^{3D}\psi_0^{1/3}\right]\delta\psi(\mathbf{r}) + 2\int d\mathbf{r}'\frac{\psi_0(\mathbf{r})\psi_0(\mathbf{r}')}{|\mathbf{r}-\mathbf{r}'|}\delta\psi(\mathbf{r}')\end{aligned} \tag{36}$$

Eq. (35) is a general result which can be applied to various physical situations, including strictly 2D geometries with the appropriate modification of the energy functional. Since the slab geometry of interest is translationally invariant in the x–y plane, the modes are propagating waves in the plane and have the form $\delta\psi(\mathbf{r}) = u(z)e^{iqx}$. Substituting this form into Eq. (35), we obtain

$$\left[\hat{h}_q^2(z) - \lambda\omega^2\right]u(z) + \hat{h}_q(z)\int dz'\, M(z,z',q)u(z') = 0 \tag{37}$$

where

$$\hat{h}_q(z) \equiv -\frac{\lambda}{2}\frac{d^2}{dz^2} + \frac{\lambda}{2}q^2 + \tilde{v}_{eff}(z) - \mu \tag{38}$$

and

$$M(z,z',q) = \left[\frac{20}{9}C_{TF}^{3D}\psi_0^{4/3} - \frac{8}{9}C_x^{3D}\psi_0^{1/3}\right]\delta(z-z') + \frac{4\pi}{q}\psi_0(z)e^{-q|z-z'|}\psi_0(z')\,. \tag{39}$$

The method of solution of Eq. (37) depends on whether the ground state potential $\tilde{v}_{eff}(z)$ is *bounded* or *unbounded.* In the latter case, Eq. (37) can be solved easily by making use of an expansion in terms of a complete set of basis functions [11]. The eigenfunctions of $\hat{h}_q(z)$ defined by

$$\hat{h}_q(z)\phi_i(z) = \varepsilon_i\phi_i(z) \tag{40}$$

are a particularly convenient choice. Defining the expansion

$$u(z) = \sum_i c_i\phi_i(z)\,, \tag{41}$$

Eq. (37) reduces to the matrix problem

$$\lambda\omega^2 c_i = \sum_j \varepsilon_i(M_{ij} + \varepsilon_i\delta_{ij})c_j\,. \tag{42}$$

The eigenvalues and eigenvectors of this equation give the normal mode frequencies and densities. We note that ε_i and M_{ij} are both functions of q so that the mode frequencies, $\omega(q)$, are q-dependent. The lowest mode is the 2D plasmon which disperses as $q^{1/2}$ for $q \to 0$. At higher wavevectors it changes over to a surface plasmon of the metallic slab. The next mode at $q = 0$ is the centre of mass mode with frequency ω_0 which also evolves into a surface plasmon at higher wavevectors. The normal mode densities for

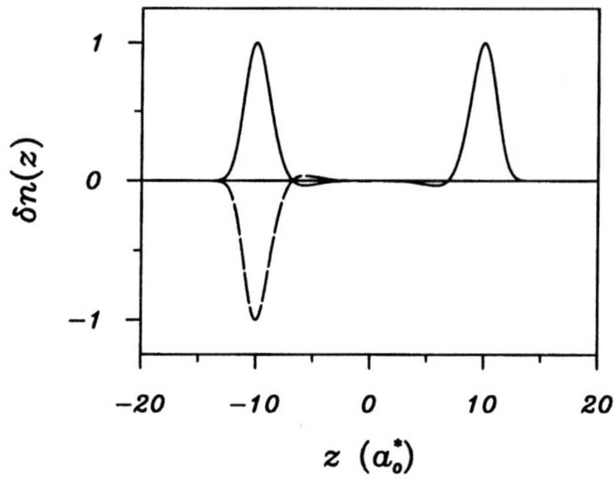

Figure 2. Comparison of the $\mathbf{q} = 0$ density fluctuations of the 2D plasmon (solid) and centre of mass mode (dash).

the 2D plasmon and centre of mass mode at $q = 0$ are shown in Fig. 2 and are seen to be remarkably similar.

We can now ask [28] about the behaviour of these modes as the width of the positive background is reduced, keeping the total number of electrons fixed. At some point the total charge of the slab will be equal to that of the electrons and we reach the situation of a neutral metallic film. Simple classical considerations of the transverse motion of the electrons relative to the positive background indicate that a normal mode of oscillation should occur at the plasma frequency ω_p, which also happens to be the harmonic frequency ω_0. At first sight it might appear that the frequency of the centre of mass mode is pinned at this value. Somewhat unexpectedly, this turns out not to be the case [35].

In Fig. 3 we show the effect on $\tilde{v}_{eff}(z)$ of reducing the width of the positive background toward the neutral limit.

The charge imbalance is specified by the parameter $\Delta\sigma_+ = \sigma_+ - n^{2D}$, where n^{2D} is the areal density of the electrons and σ_+ is the areal density of the positive background; the neutral limit corresponds to $\Delta\sigma_+ = 0$. In this limit, $\tilde{v}_{eff}(z)$ approaches a constant limiting value far from the slab, in contrast to the situation with an excess positive charge which gives rise to a nonzero electric field outside the slab. In the neutral limit, the asymptotic value of the confining potential lies above the chemical potential μ by an amount equal to the workfunction Φ of the slab. The normal situation has $\omega_p > \Phi$ which implies that the plasma frequency is a continuum excitation in contrast to the situation for the centre of mass mode with pure parabolic confinement.

Before considering the neutral limit, it is helpful to look at the evolution of the modes as a function of the excess charge. Fig. 4 shows this evolution for the centre of mass mode and some of the other lower-lying modes. The mode frequencies are relatively insensitive to variations in $\Delta\sigma_+$ until a point is reached where the edge of the positive background begins to penetrate the tail of the equilibrium electron density. As this happens, the higher-lying modes begin to be affected and drop in frequency. As they approach a lower-lying mode, the latter is effectively repelled and an anticrossing behaviour is observed. For example, once the centre of mass mode at ω_0 begins to drop in frequency it is replaced at ω_0 for a short interval by the second-lowest mode. This process continues with each of the higher modes successively passing through ω_0.

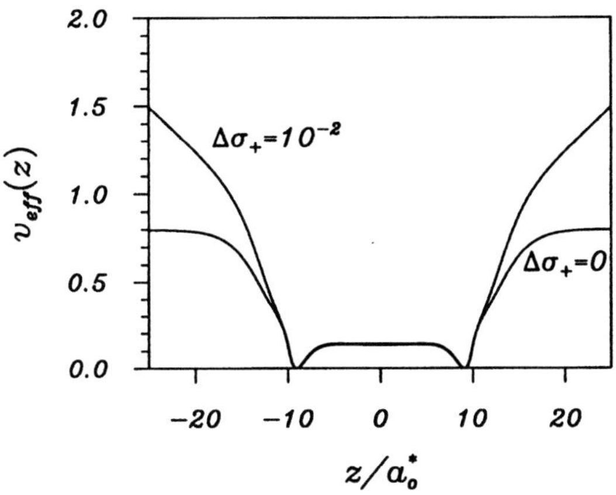

Figure 3. Self-consistent potentials for the neutral metallic film ($\Delta\sigma_+ = 0$) and a slab with a charge excess of $\Delta\sigma_+ = 10^{-2}a_0^{*-2}$. The width of the positive background in the neutral limit is $20a_0^*$.

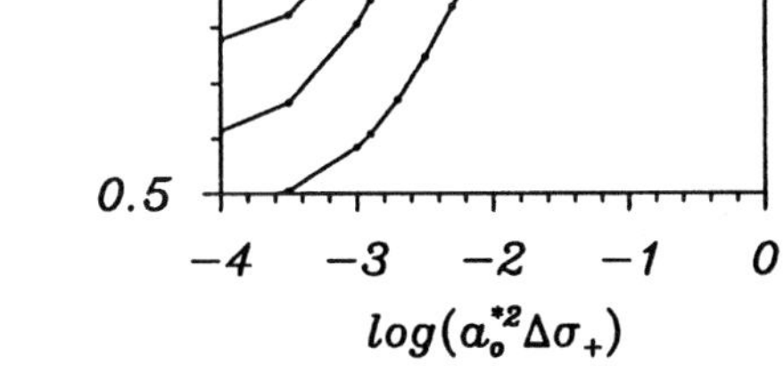

Figure 4. Variation of the mode frequencies with the charge excess $\Delta\sigma_+$.

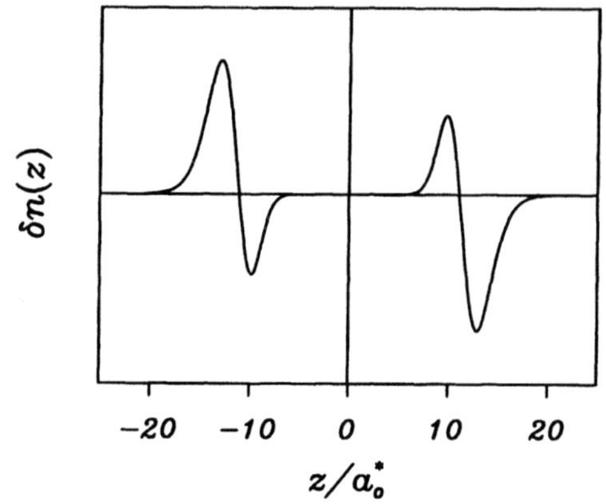

Figure 5. Charge fluctuation at $\mathbf{q} = 0$ of the surface multipole mode.

Although it is not clear from Fig. 4, what is happening is that the lowest centre of mass mode evolves into a mode which is isolated from the rest. All of the remaining higher-lying bulk standing waves evolve into a continuum of modes which extends from a threshold frequency $\omega_{th} = \Phi/\sqrt{\lambda}$ lying slightly above that of the lowest discrete mode. The latter can be thought of as a 'bound' state and, as shown in Fig. 5, has a density fluctuation with the character of a 'surface dipole'. This kind of mode is commonly referred to as a surface multipole mode [33,24,38], in contrast to the plasmon which has monopole character, that is, a surface charge density having a finite integrated weight [33].

The existence of a continuum above ω_{th} in the neutral limit necessitates a different calculational approach [35] from that used in the case of an unbounded potential. Rather than solving the homogeneous equation (37), we consider the response of the electrons to a uniform electric field $E(t) = E_0 e^{-i\omega t}$ perpendicular to the slab. The possible excitations of the electrons are then revealed by a calculation of the average power absorption which is given by the expression

$$\bar{P}(\omega) = -\frac{1}{2} E_0 \omega Im \int dz\, z\delta n(z,\omega)\,, \tag{43}$$

where $\delta n(z,\omega)$ is the induced charge density. As shown in Fig. 6, $\bar{P}(\omega)$ is dominated by a strong peak in the vicinity of the plasma frequency and there is very little absorption at all other frequencies in the continuum above ω_{th}. This result is consistent with the prediction of the simply classical argument to the extent that the electrons respond at the plasma frequency. The novel feature is that the plasma frequency is actually embedded in a continuum of collective excitations.

DISCUSSION AND CONCLUSIONS

The continuum in the TFDW hydrodynamic theory is problematic for two reasons: (i) its physical interpretation is unclear and (ii) the position of the threshold is wrong. For a real metallic film, there of course is a continuum of excitations which extends above a frequency equal to the work function Φ and, in a microscopic theory such as the RPA, this continuum is usually interpreted in terms of *single-particle* excitations. The

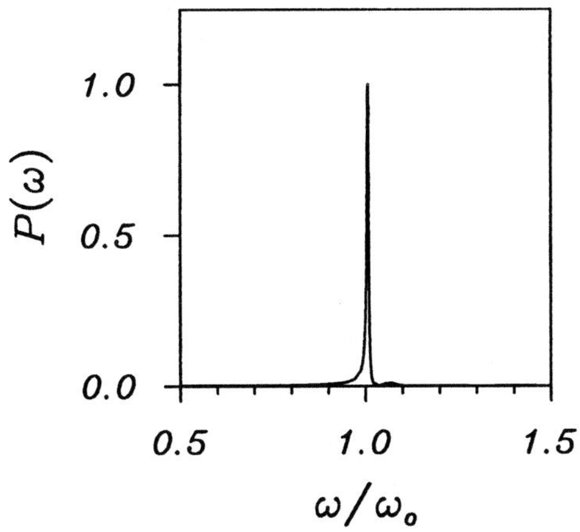

Figure 6. Normalized power absorption as a function of frequency for a neutral metallic film.

fact that the threshold frequency in the TFDW theory is $\Phi/\sqrt{\lambda}$ is the first conundrum. However, this discrepancy can be resolved by simply setting the parameter λ to unity, as opposed to the value $\lambda \simeq 0.25$ suggested earlier. The conclusion that this value is 'optimal' was based on the kinetic energy functional having the restricted form in Eq. (12) in which the usual TF term is retained. An alternative would be to set $\lambda = 1$ and require that any additional contribution to the kinetic energy functional reduce to the TF result in the limit of a uniform electron gas. As discussed by Dreizler and Gross [14], this kind of approximation would at least allow for the vW term to dominate in the limit in which it is believed to be exact, namely, the limit of rapidly varying densities. Functionals of this form have been proposed [39] and are probably a better starting point than the one we have used up until now. Thus there is no difficulty in principle in making a TFDW-like theory consistent with the physical threshold for continuum excitations.

The question of the interpretation of the continuum in the TFDW theory is still to be addressed. Why this is perhaps worrisome is that the TFDW theory deals with density fluctuations which one would normally associate with 'collective' excitations. On the other hand, the excitations above ω_{th} are usually thought of as 'single-particle-like' corresponding to the promotion of an electron from a level below μ to an asymptotic freely-propagating state. This viewpoint is reinforced within the ALDA (or RPA) since the continuum of excitations is built into the noninteracting electron response function χ_0. However these single-particle excitations are in fact renormalized by interactions and are not strictly excitations of a single particle, but rather many-particle excitations which have a spectral weight only slightly different from what the noninteracting electron result would be. From this more holistic point of view, all excitations of a system of interacting electrons are 'collective' in nature and there is no reason *a priori* why a theory could not be developed which treats all possible excitations on an equal footing. We would like to suggest that TFDW hydrodynamics is actually a simple version of such a more general theory.

A formal argument in support of this position follows from a straightforward generalization of the work of Levy *et al* [40]. In their notation, the time-dependent many-electron wavefunction is the solution of

$$i\frac{\partial}{\partial t}\Psi(\overline{N},t) = H(\overline{N},t)\Psi(\overline{N},t) \tag{44}$$

where $\overline{N}$ denotes the coordinates of N electrons. Writing the wavefunction in the form

$$\Psi(\overline{N}, t) = N^{-1/2}\psi(\mathbf{r}_N, t)\Phi(\overline{N-1}; \mathbf{r}_N, t), \tag{45}$$

the single-particle factor will yield the density and current density according to the usual definitions

$$n(\mathbf{r}, t) = |\psi(\mathbf{r}, t)|^2$$

$$\mathbf{j}(\mathbf{r}, t) = \frac{1}{2i}\left[\psi^*(\mathbf{r}, t)\nabla\psi(\mathbf{r}, t) - \nabla\psi^*(\mathbf{r}, t)\psi(\mathbf{r}, t)\right] \tag{46}$$

provided the function Φ satisfies the subsidiary conditions

$$\int d(\overline{N-1})|\Phi(\overline{N-1}; \mathbf{r}, t)|^2 = 1 \tag{47}$$

and

$$\int d(\overline{N-1})\Phi^*(\overline{N-1}; \mathbf{r}, t)\nabla\Phi(\overline{N-1}; \mathbf{r}, t) = 0. \tag{48}$$

The latter two equations are assumed to be true for all times and for all positions $\mathbf{r}$ of the N-th particle. Unlike Eq. (47) which is a necessary consequence of wavefunction normalization, Eq. (48), together with Eq. (46), is more restrictive since it implies that the flow is irrotational. More general flow fields are certainly possible and this limitation of the present analysis should be kept in mind.

Introducing Eq. (45) into Eq. (44), and taking an inner product with respect to $\Phi(\overline{N-1}; \mathbf{r}, t)$, one obtains the following single-particle Schrödinger equation

$$i\frac{\partial}{\partial t}\psi(\mathbf{r}, t) = \left[-\frac{1}{2}\nabla^2 + \overline{v}_{eff}(\mathbf{r}, t)\right]\psi(\mathbf{r}, t), \tag{49}$$

where $\overline{v}_{eff}(\mathbf{r}, t)$ is yet another effective potential defined by

$$\begin{aligned}\overline{v}_{eff}(\mathbf{r}, t) = v_{ext}(\mathbf{r}, t) + \int d\mathbf{r}'\frac{\tilde{n}(\mathbf{r}'; \mathbf{r}, t)}{|\mathbf{r} - \mathbf{r}'|} + \frac{1}{2}\int d(\overline{N-1})\left|\nabla\Phi(\overline{N-1}; \mathbf{r}, t)\right|^2 \\ + \langle\Phi(\overline{N-1}; \mathbf{r}, t)|H(\overline{N-1}, t) - i\frac{\partial}{\partial t}|\Phi(\overline{N-1}; \mathbf{r}, t)\rangle.\end{aligned} \tag{50}$$

Here, $H(\overline{N-1}, t)$ is the Hamiltonian of $N-1$ particles and $\tilde{n}(\mathbf{r}'; \mathbf{r}, t)$ is a $(N-1)$-particle density defined by

$$\tilde{n}(\mathbf{r}'; \mathbf{r}, t) \equiv \langle\Phi(\overline{N-1}; \mathbf{r}, t)|\hat{n}(\mathbf{r}')|\Phi(\overline{N-1}; \mathbf{r}, t)\rangle. \tag{51}$$

Eq. (49) is precisely of the form of Eq. (22) and supports the use of $\lambda = 1$. Furthermore, if we accept Eq. (49) as exact, at least for irrotational flows, then we must conclude that the hydrodynamic equations (19) and (20) with a suitably generalized force are in principle correct. The solution of these equations must include all excitation processes, whether they be 'single-particle-like' or 'collective', and should reproduce the results of alternative microscopic formulations. TFDW hydrodynamics is an approximation to these formally exact equations, and it would appear that the continuum of excitations found for the neutral metallic slab corresponds to the 'single-particle' part of the excitation spectrum, apart from the 'collective' plasmon.

Finally, we return to the question of dissipation. We mentioned earlier that a phenomenological way of introducing this into the TFDW hydrodynamics is to add a term $-\mathbf{v}/\tau$ to the right hand side of Eq. (20). This gives the modes a finite lifetime and adds a width to the collective mode peaks in the power absorption. (This is *not* the origin of the plasmon width shown in Fig. 6!) However, this kind of frictional force cannot account for the single-particle-like excitations which are associated with χ_0 in the ALDA or RPA. In the context of the energy loss of charged particles first mentioned in the Introduction, the losses at low projectile velocities are proportional to velocity and are due to these excitations. Only at higher velocities do collective effects appear when plasmon emission turns on [41]. The low-velocity stopping power is absent in the TFDW theory we have so far advocated and some generalization of it is required. We mention here the possibility of introducing a viscosity into the hydrodynamic equations [42] which was recently used for the stopping power problem [43].

Instead of Eq. (20), we assume a momentum equation of the form

$$n\frac{d\mathbf{v}}{dt} = n\mathbf{F} + \eta\nabla^2\mathbf{v} + (\zeta + \frac{1}{3}\eta)\nabla(\nabla\cdot\mathbf{v}) \tag{52}$$

which, apart from the definition of $\mathbf{F}$, is the usual classical equation for a viscous compressible fluid. The application of Eq. (52) to a uniform electron gas yields the dielectric function

$$\frac{1}{\epsilon(q,\omega)} = 1 + \frac{\omega_p^2}{\omega(\omega + i\nu q^2) - \omega(q)^2} \tag{53}$$

where $\nu = (\zeta + \frac{4}{3}\eta)/\bar{n}$ is the kinematic viscosity and $\omega_p^2 = 4\pi\bar{n}$. The frequency $\omega(q)$ gives the plasmon dispersion in the absence of dissipative effects. For small q and ω, this gives the electron energy-loss function

$$-Im\left(\frac{1}{\epsilon}\right) \simeq \frac{\nu q^2\omega}{\omega_p^2} \tag{54}$$

which is to be compared with the corresponding RPA result [44]

$$-Im\left(\frac{1}{\epsilon_{RPA}}\right) \simeq \frac{\pi}{6}\frac{v_F q\omega}{\omega_p^2}. \tag{55}$$

Instead of the linear dependence on q, viscous hydrodynamics gives q^2, indicating that the combination νq is playing the role of the Fermi velocity v_F in the RPA expression. The linear ω dependence does however allow for the inclusion of single-particle excitations which are needed to explain the low-velocity stopping power. This example illustrates the possibility of developing the hydrodynamic theory beyond the simplest TFDW version. However, it should be emphasized that our introduction of a viscosity is *ad hoc*. A detailed study of the properties of $\overline{v}_{eff}(\mathbf{r},t)$ may provide the needed justification.

We should finally mention some recent work which attempts to rectify the 'adiabatic' limitation of the ALDA by including the intrinsic frequency (or memory) dependence of the xc potential. In particular, the work of Vignale and Kohn [45] has shown that a rigorous formulation of time-dependent LDA is based on the current density, rather than the density as in the ALDA. In TFDW hydrodynamics the kinetic energy functional plays an equally important role. Whether similar conclusions apply to it is still not known.

ACKNOWLEDGMENTS

This work was supported by a grant from the Natural Sciences and Engineering Research Council of Canada.

REFERENCES

1. F. Bloch, Z. Phys. **81**, 363 (1933).
2. N. Bohr, K. Dan. Vidensk. Selsk. Mat.-Fys. Medd. **18**, no. 8 (1948).
3. S. A. Rice and P. Gray, *Statistical Mechanics of Simple Fluids* (Interscience, New York, 1965).
4. L. P. Kadanoff and G. Baym, *Quantum Statistical Mechanics* (Benjamin, New York, 1962).
5. R. H. Ritchie, Progr. Theor. Phys. (Kyoto) **29**, 607 (1963).
6. A. J. Bennett, Phys. Rev. B **1**, 203 (1970).
7. J. Harris, Phys. Rev. B **4**, 1022 (1971).
8. F. Forstmann and R. R. Gerhardts, *Metal Optics Near the Plasma Frequency* (Springer-Verlag, Berlin, 1986).
9. C. Schwartz and W.L. Schaich, Phys. Rev. B **26**, 7008 (1982).
10. S. C. Ying, Nuovo Cimento B **23**, 270 (1974); A. Eguiluz, S. C. Ying, and J. J. Quinn, Phys. Rev. B **11**, 2118 (1975).
11. E. Zaremba and H. C. Tso, Phys. Rev. B **49**, 8147 (1994).
12. E. K. U. Gross, J. F. Dobson and M. Petersilka, in *Density Functional Theory II* edited by R. F. Nalewajski, Springer Series on *Topics in Current Chemistry* (Springer, Berlin, 1996), p.81.
13. P. Hohenberg and W. Kohn, Phys. Rev. **136**, B864 (1964).
14. R.M. Dreizler and E. K. U. Gross, *Density Functional Theory* (Springer-Verlag, Berlin, 1990).
15. W. Kohn and L. J. Sham, Phys. Rev. **140**, A1133 (1965).
16. M. J. Stott and E. Zaremba, Phys. Rev. A **21**, 12 (1980); **22**, 2293 (1980).
17. T. Ando, Solid State Commun. **21**, 133 (1977).
18. A. Zangwill and P. Soven, Phys. Rev. A **21**, 1561 (1980).
19. E. K. U. Gross and W. Kohn, Phys. Rev. Lett. **55**, 2850 (1980); E. K. U. Gross and W. Kohn, Adv. Quantum Chem. **21**, 255 (1990).
20. K. Nuroh, M. J. Stott and E. Zaremba, Phys. Rev. Lett. **49**, 862 (1982).
21. W. Ekardt, Phys. Rev. B **31**, 6360 (1985).
22. K. Sturm, E. Zaremba and K. Nuroh, Phys. Rev. B **42**, 6973 (1990).
23. A. Liebsch, Phys. Rev. B **36**, 7378 (1987).
24. J. F. Dobson and G. H. Harris, J. Phys. C**21**, L729 (1988).
25. D. A. Broido, K. Kempa and P. Bakshi, Phys. Rev. B **42**, 11400 (1990); V. Gudmundsson and R. Gerhardts, Phys. Rev. B **43**, 12098 (1991).
26. J. F. Dobson, Phys. Rev. B **46**, 11163 (1992); J. Dempsey and B. I. Halperin, Phys. Rev. B **47**, 4662 (1993).
27. D. E. Beck, Phys. Rev. B **35**, 7325 (1987).
28. W. L. Schaich and J. F. Dobson, Phys. Rev. B **49**, 14700 (1994).
29. E. Zaremba, Phys. Rev. B **53**, 10512 (1996).
30. A. V. Chaplik, Sov. Phys. JETP **33**, 947 (1971); F. Stern, Phys. Rev. Lett. **30**, 278 (1973).
31. C. F. von Weizsäcker, Z. Phys. **96**, 431 (1935).
32. Y. Tomishima and K. Yonei, J. Phys. Soc. Jpn., **21**, 142 (1966); W. Stich *et al*, Z. Phys. A **309**, 5 (1982); A. Chizmeshya and E. Zaremba, Phys. Rev. B **37**, 2805 (1988).
33. P. Nozières and D. Pines, *Theory of Quantum Fluids*, Vol. II (Addison-Wesley, Redwood City, 1990).
34. M. Sundaram, A. C. Gossard, J. H. English and R. M. Westervelt, Superlatt. Microstruct. **4**, 683 (1988); M. Shayegan, T. Sajoto, M. Santos and C. Silvestre, Appl. Phys. Lett. **53**, 791 (1988).
35. H. C. Tso and E. Zaremba, to be published.
36. L. Brey, N.F. Johnson and B.I. Halperin, Phys. Rev. B **40**, 10647 (1989); S. K. Yip, Phys. Rev. B **43**, 1707 (1991).
37. J. F. Dobson, Phys. Rev. Lett. **73**, 2244 (1994).
38. K. D. Tsuei, *et al*, Surf. Sci. **247**, 302 (1991).
39. A. E. DePristo and J. D. Kress, Phys. Rev. A **35**, 438 (1987).
40. M. Levy, J. P. Perdew and V. Sahni, Phys. Rev. A **30**, 2745 (1984).
41. P. M. Echenique, F. Flores and R. H. Ritchie, Solid State Physics **41**, 229 (1990).
42. R. Kronig and J. Korringa, Physica **10**, 406 (1943).
43. A. Arnau and E. Zaremba, Nucl. Instr. and Meth. in Phys. Res. B **90**, 32 (1994).
44. P. Nozières and D. Pines, *Theory of Quantum Fluids*, Vol. I (Addison-Wesley, Redwood City, 1989).
45. G. Vignale and W. Kohn, Phys. Rev. Lett. **77**, 2037 (1996): see also the chapter by Vignale and Kohn in the present volume.

VAN DER WAALS INTERACTIONS IN DENSITY FUNCTIONAL THEORY

Ylva Andersson, Erika Hult, Henrik Rydberg,
Peter Apell, Bengt I. Lundqvist

Department of Applied Physics
Chalmers University of Technology and Göteborg University
S-412 96 Göteborg, Sweden

David C. Langreth

Department of Physics and Astronomy
Rutgers University
Piscataway, New Jersey 08855-0849, U.S.A.

INTRODUCTION

The history of van der Waals or dispersion forces dates a long way back [1, 2]. The recent book *Van der Waals and Molecular Sciences* [1] gives a detailed account of van der Waals's own contributions and life-long interest in the field. It is interesting to note that this truly quantum-mechanical problem [3, 4, 5] has been addressed by theorists long before the birth of quantum mechanics. The force between atoms, molecules, clusters, complexes, surfaces, and other fragments of matter is dominated by the weak but long-ranged van der Waals interactions at large separations. This is the region that has been primarily addressed. Calculations of the interaction potential between neutral species were first done for molecules [6, 7], leading to the well known asymptotic R^{-6} form of London [5]. The asymptotic z^{-3} form of the interaction potential between a neutral atom and a surface was first identified by Lennard-Jones [8], with subsequent refined treatments of the atom and surface polarizabilities [9, 10]. For the interaction between solid bodies, general formulas have been derived [11], which for flat surfaces a long distance d apart give an interaction energy that varies as d^{-2} [12]. For very large distances, where the limited magnitude of the velocity of light matters, retardation effects are important [13]. Such relativistic effects are physically interesting but beyond the scope of the present work.

In electron systems the attractive van der Waals interaction is always present. Its significance relative to other types of interactions is particularly large for neutral

Electronic Density Functional Theory: *Recent Progress and New Directions*
Edited by Dobson *et al.*, Plenum Press, New York, 1998

non-polar systems, where it, in fact, is the only source of attraction. Examples [14] of physical phenomena that are governed by van der Waals attraction are adhesion and cohesion of less densely packed matter, for example, liquid crystals, layered compounds, polymers, and biomolecular surfaces. The van der Waals interaction is relevant both for microscopic and macroscopic objects. Interactions between atoms and molecules are interesting per se. In addition, in studies of more complex systems, such as molecular solids, liquids, polymers, biomacromolecules, and membranes [15], sometimes called soft matter, a prerequisite is an understanding of such binary interactions. For instance, studies of van der Waals molecules and complexes are sometimes motivated by getting van der Waals interaction potentials as a bridge between microscopic and macroscopic properties [16]. The van der Waals interaction also plays a key role in such modern experimental tools as scanning-force microscopy [17].

Basic experimental data has been scarce in the past, but is now burgeoning. We give a few examples from different areas: The interplay between experiment and theory on dimers of atoms, molecules, and molecular complexes is expanding [18, 16]. The van der Waals force between a ground-state atom and a dielectric wall has recently been obtained in a direct-force measurement [19]. In studies of the forces between macromolecules in liquids, and in particular in water solution, the interplay between the attractive van der Waals forces and the repulsive, basically electrostatic, hydration forces is an area of recent study and progress [20]. A most detailed experimental characterization of the van der Waals potential in the full distance range has been obtained in surface physics, e.g., from the diffractive scattering of beams of light molecules off metal surfaces [21, 22]. There is a hope that the generally limited experimental data should increase substantially in the near future.

On the theoretical side, the magnitude of the van der Waals interaction is very small compared to other contributions to the total energy of a typical system. At the same time, the correlations responsible for the van der Waals effects are subtle ones. These aspects make it difficult to calculate van der Waals interaction energies. A proper and accurate theoretical treatment quickly gets quite demanding with growing size or complexity of the system. To some extent this is reflected in the fact that accurate published calculations on explicit atomic, molecular, or larger objects are relatively few. This scarcity is particularly pronounced when seen alongside of the great number of physical, chemical, and biological systems for which the ubiquitous van der Waals interaction is essential. A simple scheme for calculating the van der Waals interactions, to be used for real systems, would be very desirable.

Density functional theory (DFT) has proved to be an extremely useful tool for many types of calculations. The only approximation needed in a DFT scheme is the one for the exchange and correlation energy. An effect of common local and semi-local approximations for this quantity is that van der Waals interactions are lost. The commonly used local-density (LDA) and generalized-gradient approximations (GGA) have been successful beyond expectation for densely-packed systems [23, 24]. Although DFT embraces the notion of van der Waals interactions in principle, actual usable approximations that make use of this capability have been non-existent.

In this review we describe a practical density functional theory for the calculation of van der Waals interactions. Although its derivation makes approximations that are crude by the most rigorous standards, the resulting density functional scheme is shown by comparison with ab initio calculations to give useful results in a wide variety of circumstances. Because it can be easily used on systems that are way too large or complex to be subjected to the corresponding ab initio calculations, it should have enormous predictive power.

VAN DER WAALS INTERACTIONS IN DENSITY FUNCTIONAL THEORY

Our starting point is the well known *adiabatic-connection formula* for the exchange-correlation energy as defined in DFT [25, 26, 27],*

$$E_{\rm xc}[n] = \frac{e^2}{2} \int\int d^3r\, d^3r'\, n(\mathbf{r}) \frac{1}{|\mathbf{r}-\mathbf{r}'|} n_{\rm xc}(\mathbf{r},\mathbf{r}') \,. \tag{1}$$

The exchange-correlation hole, $n_{\rm xc}$, is expressed as

$$n_{\rm xc}(\mathbf{r},\mathbf{r}'-\mathbf{r}) = n(\mathbf{r}') \int_0^1 d\lambda\, [g_n(\mathbf{r},\mathbf{r}';\lambda) - 1] \,, \tag{2}$$

where λ is a coupling constant multiplying the e^2 in the interparticle Coulomb interaction, and g_n is the pair-correlation function. The latter should be calculated in the presence of a λ-dependent external potential that maintains the density $n(\mathbf{r})$ at its physical value.

Systems where van der Waals forces are important can to a good approximation be divided into subsystems a and b [28]. The Hamiltonian may then be written

$$H = H_a + H_b + V_{ab} \,, \tag{3}$$

where V_{ab} is the Coulomb-interaction operator. We will consider subsystems separated by a vector $\mathbf{R}$, sufficiently large for this to be valid, but still not large enough for retardation effects to be significant ($R \ll c/\omega$, where ω is a characteristic excitation frequency for the fragments).

Using perturbation theory to second order in the interaction V_{ab}, the adiabatic-connection expression Eq. (1) for the exchange-correlation interaction energy can be cast [29, 30] into the form [31]

$$\Delta E_{\rm xc}(\mathbf{R}) = E_{\rm xc}^{\infty} - \int\int\int\int d^3r_1 d^3r_2 d^3r_3 d^3r_4\, V(\mathbf{R}+\mathbf{r}_1-\mathbf{r}_2) V(\mathbf{R}+\mathbf{r}_3-\mathbf{r}_4) \times \int_0^{\infty} \frac{du}{2\pi} \Pi_a(\mathbf{r}_1,\mathbf{r}_3;iu) \Pi_b(\mathbf{r}_2,\mathbf{r}_4;iu) \,, \tag{4}$$

where Π_i is the linear response function giving the density change induced in the subsystem i by an applied potential $v_{\rm applied}$ oscillating with frequency ω:

$$\delta n_i(\mathbf{r},\omega) = \int d^3r'\, \Pi_i(\mathbf{r},\mathbf{r}';\omega)\, v_{\rm applied}(\mathbf{r}',\omega) \,. \tag{5}$$

Since the van der Waals expressions for different model systems can be obtained from the DFT exchange-correlation energy expression, we conclude that the van der Waals interaction is included within DFT [30]. Thus the inability of currently used calculational schemes to account for it is merely a shortcoming of the local and semi-local approximations used for exchange-correlation.

Absence of van der Waals Interactions in Local and Semi-local DFT Schemes

The local-density approximation (LDA) and the various generalized-gradient approximations (GGA's) for the exchange-correlation energy have enabled the success of

*The equations in the text are in Gaussian units, but with $\hbar = 1$. The tables and figures measure energy in Rydberg units and lengths in units of the Bohr radius.

DFT [23, 24]. However, when using LDA and GGA in DFT schemes, the van der Waals interaction is not present. The power-law decay of interaction energies at large separations caused by van der Waals interactions is replaced by an exponential decay. In the light of the locality imposed on the treatment of exchange and correlation within those schemes it is not very surprising that the van der Waals interaction, being non-local correlation, is lost. In the LDA, in particular, the treatment of an electron system as being 'pointwise homogeneous' removes all effects of interactions taking place between points of different density—they are instead treated as if occurring at the same density. In GGA's, the gradient of the electron density is supposed to bring such effects back into the models, but since the exchange and correlation holes are cut off in these schemes, long-ranged effects still are lost.

OUR APPROACH

A number of authors [30, 32, 33, 29, 34, 35, 36] have addressed the problem of restoring the van der Waals interaction in DFT methods. We have developed an approach to this problem that we will describe and prove to be useful by giving physical results for various model systems. It is reassuring that both Refs. [35] and [36] have reached similar conclusions. The latter approach is described in the article by Dobson, Dinte and Wang [37] in this volume.

In the original works on DFT [38, 39] the exchange-correlation energy to second order in the density inhomogeneity is expressed using an exchange-correlation kernel, $K_{\rm xc}(\mathbf{r}-\mathbf{r}', n)$,

$$E_{\rm xc}[n] = \int d^3r\, n(\mathbf{r})\epsilon_{\rm xc}\left(n(\mathbf{r})\right) - \tfrac{1}{2}\int\!\!\int d^3r d^3r'\, K_{\rm xc}(\mathbf{r}-\mathbf{r}'; n)\left[n(\mathbf{r}) - n(\mathbf{r}')\right]^2 . \quad (6)$$

We focus on the non-local interaction between two distant perturbations of the charge in an otherwise uniform electron gas and thereby define a non-local exchange-correlation kernel,

$$E_{\rm xc}^{\rm non-local} = \int\!\!\int d^3r_1 d^3r_2\, K_{\rm xc}^{\rm non-local}(\mathbf{r}_1, \mathbf{r}_2)\delta n(\mathbf{r}_1)\delta n(\mathbf{r}_2)\,, \quad (7)$$

where $\delta n(\mathbf{r})$ is the deviation from the background density at point $\mathbf{r}$. The long-range part of $K_{\rm xc}^{\rm non-local}$ has been calculated by Langreth and Vosko (LV)† [41], Rapcewicz and Ashcroft (RA) [33] and others. RA argue that since the screened interaction is present in this term and therefore screens the interaction between the electrons at $\mathbf{r}_1$ and $\mathbf{r}_2$, the important fluctuations take place within a screening length of the electron. In the slowly-varying electron gas, where the density varies over distances large compared with the screening length, these fluctuations might be assumed to occur at the local densities at $\mathbf{r}_1$ and $\mathbf{r}_2$. The screened interaction is here mediated by plasmon-exchange between $\mathbf{r}_1$ and $\mathbf{r}_2$, an exchange that RA view as occurring in a medium with the effective electron density, $n_{\rm eff} = \sqrt{n(\mathbf{r_1})n(\mathbf{r_2})}$.

†The $1/R^6$ long-range part of $K_{\rm xc}(R)$ is controlled by the non-analyticity in the variable q^2 at small q in its Fourier transform, $K_{\rm xc}(q) = K_{\rm xc}(0) + (\pi e^2/16k_{\rm F}^4)Z(q)q^2$; this is Eq. (4) of LV with the factor of 2 error, previously pointed out [40], corrected. At small q, one can show analytically from LV's expressions that $Z(Q) \to Z(0) - (3\pi^2/8)Q$, where Q is the dimensionless form of q defined by LV. Fourier transformation back to real space gives $\phi(R) \equiv 2K_{\rm xc}(R) \to -(3e^2/4m^2\omega_p^3)(1/R)^6$ in agreement with Eq. (9) in the present chapter. It was not made clear in LV that Eq. (8) was an analytic fit to the small q numerical points, and not the analytic limiting form.

The concept of an effective medium is very useful, because it can be used in attempts to extend (6) beyond its rigorous range of validity. In our case we would like a choice for the effective density, n_{eff}, that gives the correct form for the energy in both the uniform and separated fragment limits. With the application to the interaction between atoms and molecules in mind, we introduce the replacement for the RA effective density by

$$n_{\text{eff}} = \frac{1}{2^{2/3}} \left[\sqrt{n(\mathbf{r}_1) n(\mathbf{r}_2)} \left(\sqrt{n(\mathbf{r}_1)} + \sqrt{n(\mathbf{r}_2)} \right) \right]^{2/3} . \tag{8}$$

Following the analysis of RA, the long-range interaction between two electrons in an electron gas in the high-density, small-momentum limit then is

$$\phi(\mathbf{r}_1, \mathbf{r}_2) = -\frac{3e^4}{2m^2} \frac{1}{\omega_p(n(\mathbf{r}_1))\, \omega_p(n(\mathbf{r}_2))\, \left[\omega_p(n(\mathbf{r}_1)) + \omega_p(n(\mathbf{r}_2)) \right] |\mathbf{r}_1 - \mathbf{r}_2|^6} , \tag{9}$$

with $\omega_p^2(n) = 4\pi n e^2/m$.

The interaction between two separated fragments of matter is obtained by integrating the electron-electron interaction over the electron densities of the two bodies. Finally, 'undoing' the frequency integral over the imaginary frequency iu yields the long-range interaction

$$E_{\text{xc}}^{\text{non-local}} = -\frac{3}{\pi} \int_{V_1} d^3 r_1 \int_{V_2} d^3 r_2 \int_0^\infty du \, \frac{\chi_1(iu; n(\mathbf{r}_1)) \chi_2(iu; n(\mathbf{r}_2))}{|\mathbf{r}_1 - \mathbf{r}_2|^6} \tag{10}$$

between the two macroscopic fragments. Here the dielectric susceptibility is given by

$$\chi(\omega; n(\mathbf{r})) = \frac{1}{4\pi} \left[1 - \frac{1}{\epsilon(\omega; n(\mathbf{r}))} \right] , \tag{11}$$

where $\epsilon(\omega; n)$ is the exact dielectric function at zero wave vector and frequency ω for a uniform electron gas of density n:

$$\epsilon(\omega; n) = 1 - \frac{\omega_p^2(n)}{\omega^2} . \tag{12}$$

Since the χ's are functions of the local density in each of the fragments, Eq. (10) defines that part of the full density functional that obtains when the fragments are disjoint. Eq. (10) is of course close to what one gets exactly [30] from Eq. (1). The approximations are (i) the introduction of finite volumes V_1 and V_2 defined by the cut-off below, (ii) the approximate expression (11), and the introduction of the local density. The approximations (ii) and (iii) are common in the collective descriptions of atoms [42], while (i) is an important innovation introduced by RA [33].

The nature of these approximations deserves some further discussion. A comparison between Eqs. (4) and (10) shows that insofar as these total-energy expressions are concerned, we are assuming that the local polarization in a fragment is given by

$$\mathbf{P}(\mathbf{r}, \omega) = \chi(\omega; n(\mathbf{r}))\, \mathbf{E}_{\text{applied}}(\mathbf{r}, \omega) . \tag{13}$$

This is a doubly local approximation within the fragment. First, it is local in the sense that χ is taken to be the electron gas value at the local density; it is thus a local-density approximation for the dielectric response. Second, it is local in that the electrodynamics is assumed local; this is a separate approximation. Normally one would define local electrodynamics to mean a local relationship between $\mathbf{P}$ and the *total* electric field $\mathbf{E}$:

$$4\pi \mathbf{P}(\mathbf{r},\omega) = [\epsilon(\omega; n(\mathbf{r})) - 1]\, \mathbf{E}(\mathbf{r},\omega)\,. \tag{14}$$

On a macroscopic scale, such a relationship is normally true. However, because fields from the induced polarization can act at macroscopically large distances, the applied field and the polarization do not bear a local relationship to each other. The relationship $\mathbf{E}(\mathbf{r},\omega) = \mathbf{E}_{\text{applied}}(\mathbf{r},\omega)/\epsilon(\omega; n(\mathbf{r}))$ implied by the comparison of (13) and (14) should be expected to be inadequate for inhomogeneous systems of macroscopic size. The exception is the case when the polarization is weak (e.g., due to large ω), in which case the approximation (13) or (10) provides the leading term in an expansion in the size of the dielectric response. Therefore, in the case where one or both of the fragments is macroscopic, one generally needs to treat the electrodynamics exactly, and use (14) instead of (13). This is done simply by substituting into (4) the $\Pi(\mathbf{r},\mathbf{r}',\omega)$ implied by (14), instead of using (10). For a planar surface case the implementation of this procedure is simple. Our application of the better electrodynamics to atoms and molecules is still pending,[‡] although the present results in this case are so good that one could hardly hope for improvement thereby.

The second approximation above, that of a local-density type of approximation for the electrodynamic-response functions, is a venerable one. In the context of van der Waals interactions, an early work [32] from the mid-1970's proposed an approximation of the form

$$4\pi \mathbf{P}(\mathbf{r},\omega) = \int d^3r'\, [\epsilon(\mathbf{r}-\mathbf{r}',\omega; n_{\text{eff}}) - \delta(\mathbf{r}-\mathbf{r}')]\, \mathbf{E}(\mathbf{r}',\omega). \tag{15}$$

and showed that a van der Waals interaction would result. Here $\epsilon(\mathbf{r}-\mathbf{r}',\omega; n_{\text{eff}})$ is the fully non-local electron gas dielectric function at density n_{eff}. Thus the electrodynamics is non-local as it should be, but the actual response function is taken to be given by a local-density type of approximation. Ref. [32] actually proposed a specific approximation for the effective-medium density $n_{\text{eff}}(\mathbf{r},\mathbf{r}')$, while here, for more generality, we leave it unspecified. Here the choice would presumably be less critical than in Eq. (7) where the specification was required to be sensible even over macroscopic distances.

Later, a number of proposals in a context somewhat different from density functional theory, were put forth [44, 45, 46, 47], which in some sense could be regarded as approximations to (15), more like (14). To determine the actual detailed behavior of the fields as a function of frequency and position, this type of approximation has severe limitations [48]. Exact non-local theories should be used for systems simple enough to permit their application. One should therefore be reminded not to over-interpret our results and use them for more than total-energy calculations. A final point in this regard is that the earlier attempts at this type of approximation did not have the benefit of a density-functional motivated cut-off scheme, which has been proven by experience to be effective [49, 50, 33]. The cut-off, which we have found to be crucial for getting good results, presumably mimics some of the non-local aspects. This is discussed in detail in the next section.

‡The procedure requires solving $\nabla \cdot [\epsilon(\omega; n(\mathbf{r}))\mathbf{E}(\mathbf{r})] = 0$, with $\nabla \times \mathbf{E}(\mathbf{r}) = 0$, for each frequency in the presence of an external field (and with $\epsilon(\omega; n(\mathbf{r}))$ replaced with unity in the region specified by the cut-off criterion). This is no problem for spherical atoms, and in general certainly much simpler in the approximation (14), than in the more general situation when ϵ is non-local. Hopefully, when these tests are done, the approximation (13) will prove sufficient in atoms and molecules, because otherwise the potential attractiveness for the use of our functional for the largest systems could be diminished. For an illuminating pedagogical discussion of dielectric screening in the context of atoms and molecules in more general approximation schemes, see Ref. [43].

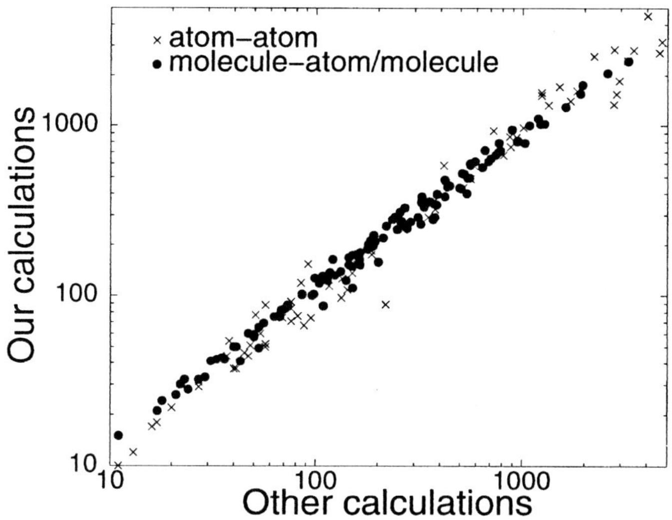

Figure 1. Our results for the van der Waals coefficients for atom–atom, atom–molecule and molecule–molecule interactions, compared with first-principles results.

APPLICATION TO MODEL SYSTEMS

There are three distinct classes of systems that require three successively more sophisticated levels of approximation. These are (i) interaction between two microscopic bodies, (ii) interaction between one microscopic and one macroscopic body, and (iii) interaction between two macroscopic bodies. Each has a simplest prototype problem, which we discuss. For (i), it is the interaction between two atoms or molecules; for (ii), it is the interaction between an atom or molecule and a planar surface; and for (iii), it is two parallel surfaces. The transition between (i) and (ii) requires an improved treatment of the electrodynamics as discussed earlier. The transition between (ii) and (iii) is an even larger step, because second-order perturbation theory in the inter-electron Coulomb interaction [Eq. (4)] fails to give the asymptotic form of the interaction correctly. We treat the prototypes for these three cases in the three sub-sections below.

Two Atoms or Molecules

Our expression (9) for the electron-electron interaction can be identified with the London expression [2, 4] for the van der Waals interaction between two atoms, each having only one important frequency for electron excitations,

$$E_{\mathrm{vdW}}^{\mathrm{London}} = -\frac{3e^4}{2m^2}\frac{Z_a Z_b}{\omega_a \omega_b(\omega_a+\omega_b)}\frac{1}{R^6}\,. \tag{16}$$

Moreover, Eq. (10) applied to two atoms at large separation R becomes

$$E_{\mathrm{xc}}^{\mathrm{non-local}}(R) = -\frac{1}{R^6}\frac{3}{\pi}\int_0^\infty du\, \alpha_1(iu)\alpha_2(iu) = -\frac{C_6}{R^6}\,, \tag{17}$$

which is the standard expression for the van der Waals interaction in terms of atomic polarizabilities [51, 2] (C_6 is the atom–atom van der Waals coefficient). The essence of

Table 1. Some representative results for atom–atom van der Waals coefficients C_6 (Ry atomic units)

	our C_6	C_6 from other calculations	Reference
He–He	4.0	2.9, 3.6, 2.9^a	[52], [53], [54]
Ne–Ne	12	13, 14, 13^a	[52], [53], [54]
Ar–Ar	126	134, 132, 130^a	[52], [53], [54]
Kr–Kr	245	266, 262, 262^a	[52], [53], [54]
Xe–Xe	520	522	[52]
Ar–Xe	253	258	[52]
Ne–Ar	37	41	[52]
Mg–Mg	1513	1237	[55]
Mg–Ca	2592	2224	[55]
H–H	12	13, 13	[52], [55]
H–K	189	218, 209	[52], [55]
Na–Na	1849	3020, 2944, 3080	[52], [56], [55]
H–Ne	10	11	[52]
Li–He	46	45	[52]
K–Ne	136	150	[52]
Na–Kr	486	562	[52]
K–Xe	1327	1338	[52]

^{a}experiment

the approximation is that the exact $\alpha(\omega)$ is replaced by the local approximation

$$\alpha(\omega) = \int_{V_i} d^3r\, \chi(\omega; n(\mathbf{r}))\,, \tag{18}$$

where χ is given by Eq. (11).

When integration over the complex frequency in Eq. (17) is carried out, and the χ_i are expressed using the atomic charge densities, the long-range interaction between the two separated fragments of matter becomes

$$E_{\rm xc}^{\rm non-local} = \frac{6e}{4(4\pi)^{3/2}m^{1/2}} \int_{V_1} d^3r_1 \int_{V_2} d^3r_2 \frac{\sqrt{n_1(\mathbf{r}_1)n_2(\mathbf{r}_2)}}{\sqrt{n_1(\mathbf{r}_1)} + \sqrt{n_2(\mathbf{r}_2)}} \frac{1}{|\mathbf{r}_1 - \mathbf{r}_2|^6}\,, \tag{19}$$

where $n(\mathbf{r}_i)$ is the charge density of fragment i, and the integration is carried out over volumes determined by the cut-off described below.

The replacement of the exact polarizibility by a local approximation, as described above, is reasonable in the interior of a charge distribution. In the outer regions, however, the length scale l_d for density variations gets much shorter than the screening length l_s of the electron gas, causing the approximation to overestimate the response. To cure this we cut off the response in regions where $l_d < l_s$, assuming on good grounds that the true response is small and thus much more accurately approximated by zero than by the local approximation. This argument is, however, not correct at very high frequencies, and imposing the cut-off causes a failure of our approximation to $\alpha(\omega)$ to satisfy the limiting high-frequency form or f-sum rule. The size of this deviation varies considerably among the dozen atoms we have tested it on, but is normally not large (median fractional error ~0.2), since the cut-off occurs well outside the bulk of the atomic electrons.

The electron gas screening length l_s must be inversely proportional to the Thomas-Fermi wave vector $k_{\rm TF}$, given by $k_{\rm TF}^2 = 4me^2(3n/\pi)^{1/3}$. For total-energy calculations the analysis of Ref. [41] suggests that the choice $l_s = (k_{\rm TF}/\sqrt{3})^{-1} = v_F/\omega_p$ is most appropriate, where v_F is the Fermi velocity. It is thus a quantity that is easily evaluated

Table 2. Atomic polarizabilities $\alpha(iu)$ (Ry atomic units). Results from Eq. (18) in left columns and ab initio results from Ref. [53] in right columns. Reprinted from Ref. [34].

u	Ne	Na$^+$	Ar	K$^+$	Kr
0	2.83 2.97	1.15 1.08	14.17 12.15	6.55 5.60	22.27 17.55
1	2.09 2.29	1.01 0.94	5.80 7.22	3.91 4.45	7.56 9.78
2	1.42 1.55	0.81 0.81	3.10 3.64	2.36 2.77	3.71 4.59
4	0.67 0.74	0.47 0.54	1.35 1.28	1.08 1.15	1.69 1.48
6	0.40 0.47	0.34 0.34	0.81 0.61	0.61 0.54	0.94 0.61
8	0.27 0.27	0.27 0.27	0.47 0.27	0.40 0.34	0.67 0.27
10	0.20 0.20	0.20 0.20	0.34 0.13	0.27 0.20	0.47 0.13

from electron densities. It is not quite as obvious how l_d should be defined. This parameter can be identified with the ξ used by Langreth and Mehl [49, 50] to define a wave vector $q = \xi^{-1}$ when going beyond the LDA. We use the expression for q that they arrive at, $q = |\nabla n|/6n$. The cut-off criterion (that the approximation for the response should be set to zero whenever $l_d < l_s$) is fulfilled by multiplying the response function by a step function in real space, which makes sure that no contribution from such regions is included. In practice this means that the integration in Eqs. (19) is carried out only in regions V_1 and V_2 where

$$\left|\frac{\nabla n(\mathbf{r})}{6n(\mathbf{r})}\right| < \sqrt{4me^2}\left(\frac{n(\mathbf{r})}{9\pi}\right)^{1/6}, \tag{20}$$

where individual molecular electron densities $n(\mathbf{r})$ are used. This is precisely the same cut-off criterion as the one derived by RA [33] for van der Waals interactions between fluctuations in the bulk, using somewhat different arguments. We believe that the scaling of the above cut-off with size and nuclear charge for neutral atoms and molecules is correct. The precise numerical definitions of the length scales and hence the overall constant on one side of the inequality (20) are, of course, arbitrary to a limited extent. It is true that the imposition of this cut-off is crucial for obtaining accurate results, and that the predictions are sensitive to the choice made. Therefore one could probably obtain a more accurate functional if the constant in Eq. (20) were optimized by fitting to the experimental database which we have now collected. Nevertheless, in all our calculations (except for sensitivity tests and the case discussed in the next paragraph), we have kept the RA cut-off of Eq. (20), as discussed above.

The method described above, similarly to the local-density approximation for close-packed matter, is based on the total electronic density. For the latter systems, it is well known that the LDA results can be improved by a theory based on the electronic *spin* density [26] for atoms with spin polarization. For our case, it is particularly easy to implement a spin-dependent theory in the case of 100% spin polarization, by adjusting the expression for v_F in l_s for the fact that there are twice as many electron states inside the Fermi-sphere for an unpolarized electron gas as for a polarized one. Then v_F is replaced by $2^{1/3}v_F$ in the cut-off criterion, and (20) is replaced by

$$\left|\frac{\nabla n(\mathbf{r})}{6n(\mathbf{r})}\right| < \sqrt{2me^2}\left(\frac{2n(\mathbf{r})}{9\pi}\right)^{1/6}. \tag{21}$$

For the purposes of calculating the van der Waals coefficient, an atom with one electron outside a closed shell can be considered approximately 100% spin-polarized, because the $\alpha(\omega)$ is dominated by the contribution from the outer shell. Since all the atoms with

Table 3. Some representative results for molecule–molecule van der Waals coefficients C_6 (Ry atomic units).

	our C_6	C_6 other calculations	Reference
HCl–HCl	277	211, 261[a]	[18], [18]
O_2–O_2	137	117.0	[60]
O_2–H_2O	123	106	[60]
H_2O–H_2O	102	97.6, 66.7, 91[a]	[61], [18], [18]
H_2O–CO	139	131.8	[62]
NH_3–H_2O	152	133.4	[62]
NH_3–CO	201	180.4, 207	[62], [63]
CO–CO	189	178, 126, 163[a]	[61], [18], [18]
CO_2–CO_2	265	322, 283, 246, 317[a]	[61], [63], [18], [18]
CH_4–CH_4	268	225, 259[a]	[18], [18]
CH_3OH–CH_3OH	442	362, 444	[18], [18]
C_3H_7OH–C_3H_7OH	1747	1671, 1955	[18], [18]
C_2H_6–NH_3	396	386	[61]
C_2H_2–H_2O	228	191	[64]
C_2H_2–C_3H_7OH	947	890	[64]
C_6H_6–C_2H_2	1102	1186	[64]
C_6H_6–CH_7OH	2057	2586	[64]

[a]experiment

odd electron number which we considered fell into this category, we applied the 100% spin-polarized cut-off (21) to these cases, thus improving the predictions over those from the non-polarized theory. The more general case is more complicated, however, and has not yet been implemented. In particular, all the molecular calculations were made with the form of the theory involving only the total density rather than the spin density, and hence used the cut-off (20).

In Fig. 1 our calculated van der Waals coefficients C_6 for interactions between a large number of atoms and molecules are compared to various first-principles results. Some representative results for atoms are shown numerically in Table 1. Even the values of $\alpha(\omega)$ for imaginary ω are in fair agreement with ab initio calculations, as illustrated in Table 2. Finally we show some representative molecule–molecule calculations in Table 3. We have made calculations for almost all systems that have been considered in first-principles calculations in the literature, and these are all shown in Fig. 1, although not in the tables. The calculations for atom–atom interactions have been made using Hartree-Fock atomic wave functions [34, 57] as input for the electron densities, while we have calculated electron densities for the molecules [58] using DFT-based commercial software [59]. The van der Waals coefficients obtained for the molecules are almost unbelievably consistent with the "better" calculations, and nurture our belief that the method should have good predictive power, when applied to systems too large for such better calculations, which are discussed in the paragraph below, to be feasible.

Systems of stable molecules are termed van der Waals complexes [15]. The last ten years have seen an explosion of calculations and experiments on van der Waals complexes, thanks to a rapid development of theoretical and experimental methods for investigating such systems. Another significant reason is an increasing interest in systems like molecular solids and liquids, polymers, and membranes. The van der Waals interaction we consider here arises from a correlation between fluctuating dipoles. More generally there might also be interactions between a permanent dipole on one molecule and the induced dipole on a second, but this case is not the focus here. This interaction potential is often anisotropic. In the calculations done so far only the isotropic van der Waals coefficient has been considered. Many systems, though,

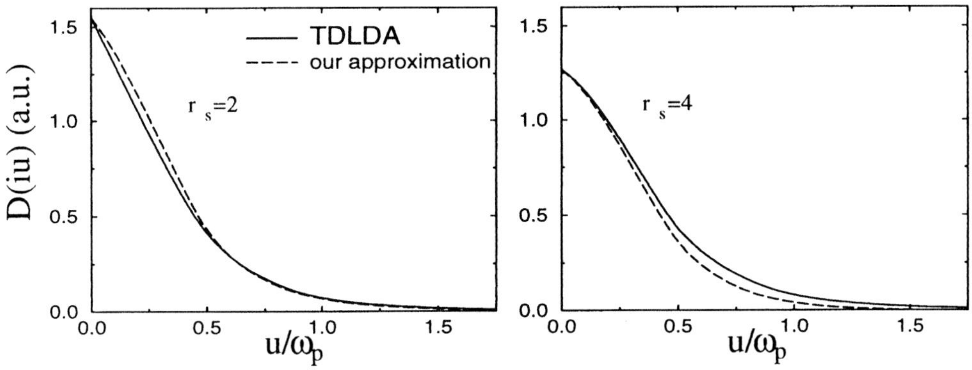

Figure 2. Our approximation for $D(iu) = [\epsilon_b(iu)(\epsilon_b(iu) - 1)/(\epsilon_b(iu) + 1)^2]\, d(iu)$ compared with time-dependent LDA calculations [67]. Reprinted from Ref. [35].

have a low anisotropy, typically about 8% [65]. There are various types of results with which we compare our results for the van der Waals coefficients: From second-order perturbation theory the C_6-coefficient for the interaction between two molecules is obtained as a coupling between their respective polarizabilities taken at imaginary frequencies [66]; Using ab initio methods such as time-dependent coupled Hartree-Fock or many-body perturbation theory, molecular polarizabilities can be obtained with very good accuracy, but this method is applicable only to small molecules, because of an unfavorable scaling of computational effort with system size; finally, C_6-coefficients are sometimes calculated as sums of atom–atom interactions for large molecules.

With increasing complexity of the van der Waals complexes, the possibilities for describing them accurately using first-principles quantum-chemical methods decrease rapidly. This fact puts our approach in a favorable position. It is characteristic for the DFT approach that the price for simplicity is a less detailed insight. The relatively good accuracy (see Table 2) of the atomic polarizabilities calculated from Eq. (18) opens up possibilities for an increased insight, however.

Atom or Molecule Outside a Surface

In the approach applied to two microscopic fragments, the electric field that is caused by the polarization of the medium is taken to be a local function of the applied field. As discussed earlier, this is bound to be an inadequate approximation when one or both of the bodies is macroscopic. Therefore, although still using the same local approximation for response functions, we now treat the electrodynamics within the approximation (14) exactly, rather than approximately according to Eq. (13). For an atom or molecule outside a planar surface, this is implemented simply by using the $\Pi(\mathbf{r}, \mathbf{r}', \omega)$ implied by Eq. (14) directly in Eq. (4). Then, following Zaremba and Kohn [31], one obtains the standard expressions:

$$E_{\text{vdW}} = -\frac{C_3}{(z - Z_0)^3}, \tag{22}$$

where the van der Waals coefficient, C_3, and the displacement, Z_0, of the van der Waals plane from the jellium edge are given by

$$C_3 = \frac{1}{4\pi} \int_0^\infty du\, \alpha(iu)\, \frac{\epsilon_b(iu) - 1}{\epsilon_b(iu) + 1} \tag{23}$$

Table 4. The van der Waals coefficient C_3 and the van der Waals plane position Z_0 (Ry atomic units) for He, Be and H_2 outside jellium. In the second and fourth column results from other calculations are given. The van der Waals plane obtained using polarizabilities as in Ref. [67], Z_0^*, is also given. After Ref. [35].

	r_s	C_3	C_3^{ref}	Z_0	Z_0^{ref}	Z_0^*
He:	2	0.13	0.10[a]	0.83	0.74[b]	0.77
	3	0.086	0.064[a]	0.67	0.64[b]	0.61
	4	0.062	0.045[a]	0.56	0.59[b]	0.53
Be:	2	1.4	1.5[c]	1.2	1.1[c]	
	4	0.96	0.97[c]	0.8	0.79[c]	
H_2:	2	0.36	0.32[a]	0.97	0.85[b]	0.88
	3	0.26	0.22[a]	0.76	0.71[b]	0.68
	4	0.20	0.16[a]	0.63	0.64[b]	0.57

[a] Ref. [31], [b] Ref. [67], [c] Ref. [69]

and

$$Z_0 = \frac{1}{4\pi C_3} \int_0^\infty du \, \alpha(iu) \frac{\epsilon_b(iu) - 1}{\epsilon_b(iu) + 1} \frac{\epsilon_b(iu)}{\epsilon_b(iu) + 1} d(iu) \,, \tag{24}$$

where $\alpha(\omega)$ is the atomic or molecular polarizibility as defined previously, and $\epsilon_b(\omega)$ is the dielectric function deep in the bulk. The quantity $d(\omega)$ is the usual perpendicular surface response function defined by Feibelman [48]. It is proportional to the surface dipole moment per unit area induced by a normal electric field. The normalization is chosen [48] so that $\mathrm{Re}\, d(\omega)$ is the centroid of the induced charge measured relative to the jellium edge, which we take as the origin. We might note that the simplified form of our functional based on (13) rather than (14) gives [68] expressions whose integrands agree with those in (23) and (24) only for the regions where $\epsilon_b - 1$ is small, e.g., large imaginary frequencies. The values thereby obtained for C_3 and Z_0 are therefore not as accurate as those reported here.

Since our original approach gives good results for dimers, we continue to approximate $\alpha(\omega)$ according to Eqs. (18) and (11). Within our approximation (14), it is trivial to calculate the polarization P_z induced by a perpendicular applied field from a charge sheet at infinity. One finds [35]

$$P_z(z, \omega) = \frac{1}{4\pi} \frac{\epsilon(\omega; n(z)) - 1}{\epsilon_b(\omega) + 1} \frac{2\epsilon_b(\omega)}{\epsilon(\omega; n(z))} E_z^{\mathrm{applied}}(\omega) \,, \tag{25}$$

from which $d(\omega)$ is obtained as

$$d(\omega) = \frac{1}{P_z(-\infty, \omega)} \int dz \, z \left(- \frac{dP_z(z, \omega)}{dz} \right) . \tag{26}$$

As discussed above for the atom–atom case, the replacement of the true response function by a local approximation cannot be justified in the tails of electron distributions. Following the same line here, we should multiply the surface response (25) by a step function. For a flat surface we would then multiply the right hand side of Eq. (25) by a factor $\Theta(z_c - z)$. Doing this, however, and taking the static limit shows z_c to equal $d(0)$. Since static image planes $d(0)$ are rather easy to calculate within time independent DFT, we thus choose to replace the cut-off parameter z_c by $d(0)$.

Table 5. Results for some of our homo-nuclear atom–atom van der Waals interactions (unmarked) compared with results we obtain (marked by *) by using first-principles static polarizabilities $\alpha(0)^*$ to define the cut-off radii, r_c^*. C_6^{ref} is from first-principles calculations. (Ry atomic units).

	$\alpha(0)$	$\alpha(0)^*$	r_c	r_c^*	C_6	C_6^*	C_6^{ref}
He	2.1	1.4[a]	1.8	1.7	4	2.6	2.9[b], 3.6[f], 2.9[g]
Ne	2.9	2.8[d], 3.0[f]	2.1	2.1	12.0	12.4	12[b]
Ar	14	11[d], 12[f]	3.5	3.2	126	108	134[b], 132[f], 130[g]
Kr	22	18[f]	4.1	3.8	245	194	266[b]
Be	62	40[d], 37[c]	5.7	4.8	582	326	416[c]
Mg	119	70[d]	7.1	6.0	1513	907	1237[c]
Ca	242	154[c]	9.0	7.6	4500	2810	4010[c]

[a] Ref. [72], [b] Ref. [52], [c] Ref. [55], [d] Ref. [73], [e] Ref. [56], [f] Ref. [53], [g] Ref. [54] (exp't.)

Our approximation for $d(\omega)$ compares favorably with time-dependent LDA calculations [67] for jellium (Fig.2), for the imaginary ω values important for the energy. Thus it is not surprising that our results for the van der Waals plane for jellium, given in Table 4, are in good agreement with more accurate calculations. This comparison is hindered by the fact that the two calculations used different atomic α's; the differing values of C_3 are of course due entirely to this difference. To eliminate this effect for the purpose of testing the surface aspects of our approximation scheme alone, we have calculated the van der Waals plane position using the same atomic polarizabilities as those used in the reference calculations. These results are displayed in the final column of Table 4. Note that the jellium model is used only for the purpose of comparison. To give a realistic account of the surface response, improved models are needed, as discussed later.

As in the atomic case high-frequency sum rules fail to be satisfied exactly [35]. This is due to the small fraction of the electrons in the tail that are not counted. For example, for an $r_s = 2$ jellium surface the value of $d(\infty)$ is not zero as it should be, but rather -0.1 a.u., with still larger deviations for lower electronic densities. Similarly the stress sum rule [70, 71] is not fulfilled, although for $r_s = 2$ the deviation was completely negligible. In the expression (24) these deviations introduce only a minor error, however, because the integral over $\omega = iu$ is heavily weighted at small values of u.

We have also tried the RA type of cut-off (20) used in the atom–molecule case. It gives values for the van der Waals planes that are somewhat too large. We do not understand yet, why the cut-off (20) is better for atoms and molecules than for surfaces. Possibly it relates to the use of the approximation (13) for the first case and (14) for the second. If this is so, the answer will have to await the implementation of (14) in the atomic case. Perhaps, on the other hand, the cut-off for the atomic case should be fixed by the value of $\alpha(0)$ from first-principles calculations. As illustrated by Table 5, the values for the van der Waals coefficient C_6 come out with a reduced but still interesting agreement with the more accurate numbers. A comparison with both the more accurate electrodynamic treatment *and* an $\alpha(0)$ based cut-off would be desirable.

While waiting for good answers to these questions, we feel that one can use the approximations in their present form without unacceptably large errors to calculate the van der Waals coefficients and planes for various combinations of simple metals and rather complex molecules. One should thereby obtain useful results in cases where better ab initio calculations are unavailable and too expensive to perform.

From Eqs. (23) – (25) it is clear that the van der Waals coefficient C_3 depends only

Table 6. The van der Waals reference plane positions Z_0 (atomic units) for H_2 outside the low-indexed Al faces. Values used for d(0) are also given. After Ref. [35].

face	d(0)	Z_0	Z_0^{ref}
111	1.05	0.69	0.61
100	1.32	0.87	0.83
110	2.08	1.32	1.27

$d(0)$ and Z_0^{ref} from Ref. [22, 74]

on the bulk electron density, while the position of the van der Waals plane Z_0 depends in addition on the induced surface charge density. The interactions of an atom with different faces of the same material thus have the same C_3-value but different Z_0-values.

Of course the jellium model is rather unphysical for the high-density (small r_s) surfaces, because the bulk seeks to be at a lower density, and the surface seeks to increase its area (negative surface tension). The result is that electrons spill out, making the selvage layer unphysically thick, and the $d(\omega)$ values unphysically large. As r_s increases, the unphysical spill-out decreases, and the image plane in this model moves closer to the jellium edge. The values of Z_0 follow this trend and also become smaller with increasing r_s. In addition to this spurious density dependence of Z_0, the possibility for dependence on crystalline face is missing entirely.

To test our approximation in a physically correct setting, we have applied it to a case [35], where there are results available that include the effects of the lattice in a realistic way, namely for the interaction between a hydrogen molecule and the low-indexed aluminum surfaces [22, 74]. The densities and the static image-plane for the Al surfaces are taken from a self-consistent DFT-calculation for jellium with laterally averaged pseudopotentials representing the ions in the four outermost layers [22, 74]. From this input we calculated the Z_0 values shown in Table 6. These turn out very close to those obtained by M. Persson [22, 74] in another approximation,§ as also shown in the table.

A simpler way of introducing more realistic features and also face-dependent density profiles is to use a stabilized-jellium model with face corrugation [75, 76]. Hult and Kiejna [77] have recently calculated the van der Waals coefficient and the van der Waals plane for a large number of atoms and molecules outside face-dependent stabilized-jellium surfaces. The stabilized-jellium model gives increasing image-plane positions $d(0)$ with increasing r_s [76], as opposed to the jellium model. This is reflected in the results for the van der Waals plane, where in the more realistic stabilized-jellium model the dependence on r_s is rather weak, while the face-dependence is pronounced. The more open the surface is, the larger the value of Z_0. For fcc-metals the van der Waals plane is located at the smallest distance for the close-packed (111) surface and at the largest one for the open (110) surface. For the bcc-metals the (110) surface gives the smallest van der Waals plane and (111) the largest.

§This calculation [74] used an interpolation formula [71] in the manner of Eq. (2) of Ref. [67]. The parameter λ in the interpolation formula was fixed from the stress sum rule [70, 71], which was generalized to the case of the quasi-one-dimensional pseudopotential model used to get self-consistently calculated electron densities.

Parallel Surfaces

The van der Waals interaction between two surfaces is of interest for adhesion and scanning-force microscopy, among other examples. The standard reference in this case is the Lifshitz theory for dispersion [11], giving $E_{\text{vdW}} = -C_2/d^2$, where d is now the separation between the two surfaces, which are assumed to be sharp. We have derived a modified expression [78], valid also for soft surface profiles, where van der Waals planes are introduced in the same way as in the case of a molecule outside a surface.

For the cases of atom–atom and atom–surface interactions, the second-order perturbation theory formula (4) is applicable and gives the right van der Waals expressions. For the interaction between two parallel surfaces, on the other hand, we cannot treat the Coulomb interaction as a perturbation, since there will be an infinite number of reflections going back and forth between the surfaces. The adiabatic-connection expression (1) can still be used [30], though, giving an interaction at large separations z between two surfaces A and B as [78]

$$E_{\text{vdW}} = -\frac{C_2}{(z - Z_A - Z_B)^2} \,. \tag{27}$$

In this expression the van der Waals coefficient is

$$C_2 = \frac{1}{32\pi^2} \int_0^{\infty} du\, F(iu) \,, \tag{28}$$

where

$$F(\omega) = \int_0^{\infty} \frac{dx\, x^2\, e^{-x}}{\left(\frac{\epsilon_A(\omega)+1}{\epsilon_A(\omega)-1}\right)\left(\frac{\epsilon_B(\omega)+1}{\epsilon_B(\omega)-1}\right) - e^{-x}} \,, \tag{29}$$

and the outward displacements of the van der Waals planes from the respective jellium edges are given by

$$Z_{A/B} = \frac{\int_0^{\infty} du \frac{\epsilon_{A/B}(iu)}{\epsilon_{A/B}(iu)+1}\, d(iu) F(iu)}{\int_0^{\infty} du F(iu)} \,. \tag{30}$$

The expression for C_2 agrees with the standard one [11], while the expression for $Z_{A/B}$ appears to be a new result of Ref. [78]. Since the expressions for C_2 and $Z_{A/B}$ contain the same components as the corresponding atom–surface expressions we can use the same approximation here. Calculations have been done for jellium and the low-indexed aluminum surfaces [30], using the expressions above with surface electron densities and static image plane positions as input. In that process the same cut-off as used previously for surfaces was applied. The value Z_A for an aluminum surface A interacting with a similar aluminum surface B is very close the corresponding values of Z_0 obtained when the surface A interacts with molecules. The displacements $Z_{A/B}$ are larger for the more open surfaces [78].

Experimental data or better calculations for making comparisons with the results we have obtained are lacking. On the other hand, it might be fruitful to incorporate the result (27) into an interpolation formula like that of Ref. [79] in an attempt to obtain a formula useful at closer distances.

SUMMARY AND OUTLOOK

In this review we have described a simple, general, and approximate scheme to account for van der Waals interactions in density functional theory. This should provide the basis for numerous applications on interacting microscopic particles, such as atoms, molecules, and molecular complexes, and on macroscopic objects, such as atoms and molecules interacting with solid surfaces, parallel surfaces, and ultimately more generally shaped macroscopic objects. We give examples of results for the van der Waals interactions at intermediate distances, before the onset of the effect of retardation, which illustrate that a useful accuracy can be achieved. A virtue of the scheme is that to get the forces between two objects, only the electron densities of each object are needed as input. This implies an applicability to systems of growing complexity far beyond what other, more fundamental methods can achieve.

Typically, for most of the experimental results presented so far, the interest is in intermediate-range van der Waals interactions. Here our present results should be adequate. However, some surface experiments, for example, selective-adsorption diffraction and scanning-force microscopy, point at the relevance of the close contact region. In particular, it would be very desirable to account for the region, where the attractive van der Waals interaction goes over into Pauli repulsion at short distances. To provide a general method for this transition region is an important task for the continued research on van der Waals functionals.

ACKNOWLEDGMENT

The authors wish to thank Mats Persson for providing data prior to publication and for fruitful discussions. Adam Kiejna is also gratefully acknowledged. Work at Rutgers was supported in part by NSF grant No. DMR 94-07055. Financial support from the Swedish Natural Science Research Council and the Swedish National Board for Industrial and Technical Development is also acknowledged.

REFERENCES

1. A. Y. Kipnis, B. E. Yavelov, and J. S. Rowlinson, *Van der Waals and Molecular Sciences* (Oxford, New York, 1996).
2. H. Margenau and N. R. Kestner, *Theory of Intermolecular Forces* (Pergamon Press, Oxford, 1969).
3. D. Langbein, *Theory of Van der Waals Attraction* (Springer Verlag, Berlin, 1974).
4. R. Eisenshitz and F. London, Z. Phys. **60**, 491 (1930).
5. F. London, Z. Phys. **63**, 245 (1930).
6. M. Reinganum, Ann. Phys. (Paris) **38**, 649 (1912).
7. S. C. Wang, Z. Phys. **28**, 663 (1927).
8. J. E. Lennard-Jones, Trans. Faraday. Soc. **28**, 333 (1932).
9. J. H. de Boer, Trans. Faraday Soc. **32**, 21 (1936).
10. H. C. Hamaker, Physica **4**, 1058 (1937).
11. E. M. Lifshitz, Sov. Phys. **2**, 73 (1956).
12. I. E. Dzyaloshinskii, E. M. Lifshitz, and L. P. Pitaevskii, Adv. Phys. **10**, 73 (1961).
13. H. B. G. Casimir and D. Polder, Phys. Rev. **73**, 360 (1948).
14. J. Israelachvilii, *Intermolecular and Surface Forces* (Academic, London, 1991).
15. A. Buckingham, P. Fowler, and J. Hutson, Chem. Rev. **88**, 963 (1988).
16. G. Chalasinski and M. M. Szczesinak, Chem. Rev. **94**, 1723 (1994).
17. See, e.g., U. Hartmann, in *Scanning Tunneling Microscopy III*, edited by R. Wiesendanger and H.-J. Guentherodt (Springer, Berlin-Heidelberg, 1993), p. 293.
18. M. Spackman, J. Chem. Phys. **94**, 1295 (1991).
19. A. Landragin, J. Y. Courtois, G. Labeyrie, N. Vansteenkiste, C. I. Westbrook, and A. Aspect, Phys. Rev. Lett. **77**, 1464 (1996).
20. J. Israelachvilii and H. Wennerström, Nature **379**, 219 (1996).
21. S. Andersson, L. Wilzen, and M. Persson, Phys. Rev. B **38**, 2967 (1988).
22. S. Andersson, M. Persson, and J. Harris, Surf. Sci. **360**, L499 (1996).

23. R. O. Jones and O. Gunnarsson, Rev. Mod. Phys. **61**, 689 (1989).
24. K. Burke, J. Perdew, and Y. Wang, "Derivation of a generalized gradient approximation: the PW91 density functional", in this volume.
25. D. C. Langreth and J. P. Perdew, Solid State Commun. **17**, 1425 (1975).
26. O. Gunnarsson and B. I. Lundqvist, Phys. Rev. B **13**, 4274 (1976).
27. D. C. Langreth and J. P. Perdew, Phys. Rev. B **15**, 2884 (1977).
28. J. Harris, Phys. Rev. B **31**, 1770 (1985).
29. J. F. Dobson, in *Topics in Condensed Matter Physics*, edited by M. P. Das (Nova, N. Y., 1994), p. 121.
30. B. I. Lundqvist, Y. Andersson, H. Shao, S. Chan, and D. C. Langreth, Int. J. Quantum. Chem. **56**, 247 (1995).
31. E. Zaremba and W. Kohn, Phys. Rev. B **13**, 2270 (1976).
32. W. Kohn and W. Hanke, Short- and long-wavelength contributions to the exchange-correlation energy of a metal surface, unpublished.
33. K. Rapcewicz and N. W. Ashcroft, Phys. Rev. B **44**, 4032 (1991).
34. Y. Andersson, D. C. Langreth, and B. I. Lundqvist, Phys. Rev. Lett. **76**, 102 (1996).
35. E. Hult, Y. Andersson, B. I. Lundqvist, and D. C. Langreth, Phys. Rev. Lett. **77**, 2029 (1996).
36. J. F. Dobson and B. P. Dinte, Phys. Rev. Lett. **76**, 1780 (1996).
37. J. F. Dobson, B. P. Dinte, and J. Wang, "Van der Waals functionals via local approximations for susceptibilities", in this volume.
38. P. Hohenberg and W. Kohn, Phys. Rev. **136**, B864 (1964).
39. W. Kohn and L. J. Sham, Phys. Rev. **140**, A1133 (1965).
40. D. C. Langreth and S. H. Vosko, in *Advances in Quantum Chemistry*, edited by S. B. Trickey (Academic Press, New York, 1990), Vol. 21, p. 175.
41. D. C. Langreth and S. H. Vosko, Phys. Rev. Lett. **59**, 497 (1987).
42. S. Lundqvist, Ark. Phys. **28**, 399 (1965).
43. A. Zangwill and A. H. Levine, Am. J. Phys. **53**, 1177 (1985).
44. A. Bagchi, N. Kar, and R. G. Barrera, Phys. Rev. Lett. **40**, 803 (1978).
45. K. L. Kliewer, Surf. Sci. **101**, 57 (1980).
46. P. Apell, Physica Scripta **25**, 57 (1982).
47. P. Ahlqvist and P. Apell, Physica Scripta **25**, 587 (1982).
48. P. J. Feibelman, Prog. in Surf. Sci. **12**, 287 (1982).
49. D. C. Langreth and M. J. Mehl, Phys. Rev. Lett. **47**, 446 (1981).
50. D. C. Langreth and M. J. Mehl, Phys. Rev. B **28**, 18090 (1983).
51. C. Mavroyannis and M. J. Stephen, Mol. Phys. **5**, 629 (1962).
52. K. T. Tang, J. M. Norbeck, and P. R. Certain, J. Chem. Phys. **64**, 3063 (1976).
53. G. Mahan, J. Chem. Phys. **76**, 493 (1982).
54. A. Dalgarno and W. Davison, in *Advanced Atomic and Molecular Physics 2*, edited by D. Bates and I. Esterman (Academic, Orlando, 1966).
55. F. Maeder and W. Kutzelnigg, Chem. Phys **42**, 95 (1979).
56. M. Marinescu, H. Sadeghpour, and A. Dalgarno, Phys. Rev. **49**, 982 (1994).
57. E. Clementi and C. Roetti, At. Data Nucl. Data Tables **14**, 177 (1974).
58. Y. Andersson and H. Rydberg, J. Chem. Phys., in press.
59. In *Biosym*, a program package available from Molecular Simulations Inc.
60. W. Rijks, M. van Heeringen, and P. Wormer, J. Chem. Phys. **90**, 6501 (1989).
61. S. van Gisbergen, J. Snijders, and E. Baerends, J. Chem. Phys. **103**, 9347 (1995).
62. W. Rijks and P. Wormer, J. Chem. Phys. **90**, 6507 (1989).
63. R. Amos, N. Handy, P. Knowles, J. Rice, and A. Stone, J. Phys. Chem. **89**, 2186 (1985).
64. A. Kumar and W. Meath, Mol. Phys. **75**, 311 (1992).
65. B. Jhanwar and W. Meath, Chem. Phys **67**, 186 (1982).
66. A. Dalgarno, Adv. Chem. Phys **12**, 143 (1967).
67. A. Liebsch, Phys. Rev. B **33**, 7249 (1986).
68. Y. Andersson, E. Hult, D. C. Langreth, and B. I. Lundqvist, in *Proceedings of the 18th Taniguchi Symposium: Elementary Processes in Excitations and Reactions on Solid Surfaces*, edited by A. Okiji, H. Kasai, and K. Makoshi (Springer, Berlin, 1996), p. 52.
69. C. Holmberg, P. Apell, and J. Giraldo, Physica Scripta **33**, 173 (1986).
70. B. N. J. Persson and P. Apell, Phys. Rev. B **27**, 6058 (1983).
71. B. N. J. Persson and E. Zaremba, Phys. Rev. B **30**, 5669 (1984).
72. D. M. Bishop and J. Pipin, J. Chem. Phys. **97**, 3375 (1992).
73. H. Gollisch, J. Phys. B **17**, 1463 (1984).

74. M. Persson, private communication, 1996.
75. J. P. Perdew, H. Q. Tran, and E. D. Smith, Phys. Rev. B **42**, 11627 (1990).
76. A. Kiejna, Phys. Rev. B **47**, 7361 (1993).
77. E. Hult and A. Kiejna, Surf. Sci. **383**, 88 (1997).
78. Y. Andersson, E. Hult, P. Apell, D. C. Langreth, and B. I. Lundqvist, submitted to Solid State Commun.
79. W. Kohn and A. Yaniv, Phys. Rev. B **20**, 4948 (1979).

VAN DER WAALS FUNCTIONALS VIA LOCAL APPROXIMATIONS FOR SUSCEPTIBILITIES

John F. Dobson, Bradley P. Dinte, and Jun Wang

School of Science
Griffith University
Nathan, Queensland 4111, Australia

INTRODUCTION

The van der Waals * (vdW) or Dispersion interaction between distant sections of a molecular or condensed matter system is a component of the correlation energy. The part of the correlation energy clearly distinguishable as vdW energy is usually much smaller than, say, a covalent bonding energy. Nevertheless the vdW energy and other weak interactions are of great potential importance in biological and pharmaceutical applications, in physisorption studies and in the description of novel chemical species such as complexes containing rare-gas atoms.

This dispersion part is missed, or is given at best unreliably, by the usual local and gradient functionals for the xc energy. Here we discuss and compare various quasi-local schemes which are expected to provide good vdW energies using only a knowledge of the groundstate electron density $n(\mathbf{r})$. The common thread in these various approaches is the use of an exact formula or well-defined approximation for the correlation energy in terms of a susceptibility, followed by an approximation of local-density type for the susceptibility. This approach has already led to useful formulae for the van der Waals energy of well-separated fragments of matter, as will be discussed below. Perhaps more challenging is the inclusion of van der Waals effects, without double-counting, in cases where the interacting systems are not well-separated or have significant overlap of electron density. An energy functional that is uniformly applicable in such cases, as well as in widely-separated cases, will here be termed a "seamless vdW functional". Some seamless vdW functionals will be discussed in the present chapter also.

*Here the name "van der Waals" is used exclusively for the dispersion part of the energy. Unfortunately the usage of this term is not universally agreed. Some workers restrict the term to the $1/R^6$ interaction arising from the *dipolar* approximation to the correlated fluctuations on two small well-separated neutral systems. We make no such restriction here. Other usages occur. In the chemistry literature the term "van der Waals" may refer to the entire interaction in cases where neither strong covalent nor ionic bonds are present. Thus, in the latter terminology, the interactions between permanent and induced static multipoles would be considered part of the vdW energy. These interactions occur already in the Hartree-Fock approximation and are not here considered as part of the vdW interaction.

Electronic Density Functional Theory: *Recent Progress and New Directions*
Edited by Dobson *et al.*, Plenum Press, New York, 1998

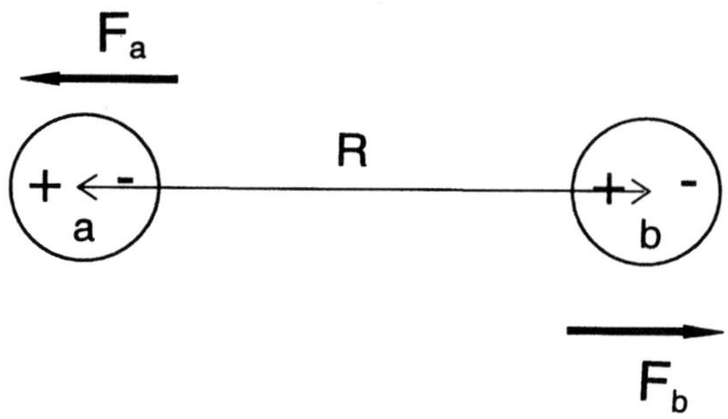

Figure 1. Generation of vdW interaction by spontaneous dipole on atom a followed by induction of dipole on atom b and consequent back-field F_a.

QUALITATIVE PICTURE OF THE VDW INTERACTION

Elementary Derivation of Asymptotic vdW Interaction

In most systems the vdW interaction occurs together with other types of interaction such as covalent bonding forces, or the forces between static electric multipoles. To isolate the vdW phenomenon conceptually, we therefore first consider the special case of two systems ("atoms") each of which is electrically neutral and, when isolated, is spherically symmetric. Each atom therefore has no permanent electric multipoles, and to first order in the inter-atom Coulomb interaction U_{ab} there is thus no interaction energy when two such atoms approach one another without touching. In this situation the vdW interaction is the dominant inter-atom force (it is of second order in U_{ab}, as we shall see).

There are indeed many ways to obtain this vdW interaction, but let us begin with a highly intuitive one which is especially useful in illustrating what the vdW interaction is *not,* and also in guiding approximations to be made when constructing a vdW functional later.

Consider Figure 1, in which the separation R between the atoms is much greater than than the radius of an atom. Because of quantal zero-point motion and/or thermal fluctuations, let the electron density in atom 1 suffer a *spontaneous (temporary, time-dependent)* distortion. Such fluctuations *cannot create charge*, so there is no electric monopole, but the fluctuation will in general have a nonzero dipole moment, say d_a, which for simplicity we assume is oriented as shown. This dipole is the source of an electric field $F_b \approx d_a R^{-3}$ at the position of atom b. The electrons in atom b *respond* by producing a *temporary induced* dipole $d_b = \bar{\alpha}_b F_b = \bar{\alpha}_b d_a R^{-3}$, where $\bar{\alpha}_b$ is the dipole polarizability of atom b. The temporary dipole d_b lives only while driven by the field due to the original dipole fluctuation d_a- i.e. d_b and d_a are *correlated* density fluctuations. Dipole d_b causes a back electric field $F_a \approx -d_b R^{-3} = -\bar{\alpha}_b d_a R^{-6}$ acting on atom a. This causes an electrostatic energy of order

$$E^{(2)} \approx F_a d_a = -\bar{\alpha}_b d_a^2 R^{-6}. \tag{1}$$

Although the expectation $< d_a >$ is zero , the mean-square fluctuation $< d_a^2 >$ is nonzero. To lowest nonvanishing order in U_{ab}, we can estimate $< d_a^2 >$ in the absence of atom b. To achieve this we note that the energy to distort the isolated atom a and produce a dipole d_a is of order $d_a^2/\bar{\alpha}_a$, where $\bar{\alpha}_a$ is the dipole polarizability of atom a.

(Results of this form emerge, for example, from a classical mass-on-spring model of the atom, and equally from a two-level quantal picture). Equating the expectation of this distortion energy to a typical zero point energy, say $\hbar\omega_0$, we deduce $< d_a^2 > \approx \bar{\alpha}_a \hbar\omega_0$. Then the average mutual energy of the correlated fluctuating dipoles from (1) is of order

$$E^{(2)} \approx -(const.)\bar{\alpha}_a \bar{\alpha}_b \hbar\omega_0 R^{-6} \tag{2}$$

Eq. (2) is the result we are seeking: the vdW energy falls off as R^{-6} and is proportional to the two dipole polarizabilites and to a characteristic energy. (The above argument neglects kinetic energy of correlation but is nevertheless qualitatively correct).

Qualitative Points From Elementary Derivation

The above derivation implies the following points:

1. No overlap of electron density is required to produce the vdW interaction. Indeed it would persist even in the following "canonical" vdW situation. Suppose that each atom were surrounded by an impenetrable barrier preventing overlap. The only effect of such barriers would be to stiffen each atom somewhat, resulting in a reduced but nonzero polarizability $\bar{\alpha}$: otherwise the above argument, based on long-ranged coulomb interactions, proceeds as before. It follows also that the vdW interaction has nothing to do with exchange.

2. The vdW interaction arises from *correlations* of distant density fluctuations, and therefore corresponds to a very long-ranged part of the correlation hole.

3. It is not to be confused with the interaction of permanent and induced static multipoles, which are already described in the Hartree approximation.

4. Hartree-Fock theory misses vdW altogether.

5. In the LDA or local-gradient approximations (GGAs), dependence of the correlation energy on the separation R between subsystems only arises via overlap of electron density. † Indeed, the exponential decay of the overlap with R leads, in common GGAs, to faster decay of the R-dependent part of the correlation energy than the correct R^{-6} dependence, at large separation.

6. For a pair of distant systems, the vdW interaction involves two factors of the inter-system Coulomb interaction U_{ab}, and thus should be derivable from second-order perturbation theory in U_{ab}. (The usual second-order Möller-Plesset perturbation theory, applied to the whole system, cannot however be expected to provide good values of the constant C_6 in the asymptotic formula $E^{(2)} = -C_6 R^{-6}$. This is because two orders of perturbation theory are "used up" in obtaining the basic R^{-6} dependence, leaving only a zeroth-order (bare) description of the polarizability $\bar{\alpha}$ of each atom. Higher-order MP theory should, however, be applicable).

†This can be demonstrated by considering the "canonical" vdW situation involving one or more impenetrable barriers described under item 1 above: this is the only situation with literally zero overlap. From the general local form of the LDA and GGA approximations, it can be shown that an R-dependent interaction can only arise via a ***static*** density distortion of ferroelectric sign. This is a Hartree interaction, this being the only part of the LDA/GGA energy functional which connects the densities in the two subsystems. Such a ferroelectric distortion is not the dispersion interaction, and at any rate it vanishes identically for large enough R except possibly in highly degenerate cases.

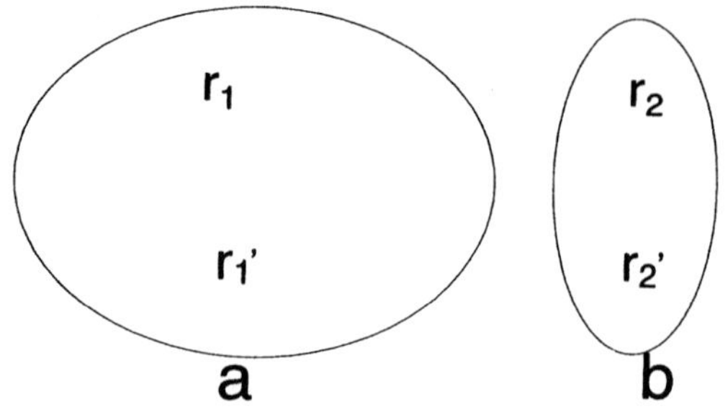

Figure 2. Geometry for second-order interaction between non-overlapping electron densities.

7. The fact that the interacting fluctuations are dipolar and not monopolar is a consequence of charge conservation. This constraint needs to be taken into account explicitly in treatments to be discussed below, in which vdW formulae are derived by approximating response functions.

NON-OVERLAPPING SYSTEMS: SECOND-ORDER PERTURBATIVE RESULTS

Non-Overlapping Systems: Exact Perturbative Results in Terms of χ

The elementary derivation of the R^{-6} vdW interaction given in Eq. (2) of the previous Section can be regarded as a primitive form of second-order perturbation theory, in which we have made the scalar dipole approximation and neglected the kinetic energy contribution. Without making any such approximations, we can apply second-order perturbation theory in the inter-system Coulomb interaction $U_{ab} \equiv e^2/|\mathbf{r}_1 - \mathbf{r}_2|$, leading to the following more general expression for non-overlapping electron densities (see for example [1] or p. 501 of Ref. [2]):

$$E^{(2)} = -\frac{\hbar}{2\pi}\int_a d\mathbf{r}_1 d\mathbf{r}_1' \int_b d\mathbf{r}_2 d\mathbf{r}_2' \frac{e^2}{|\mathbf{r}_1 - \mathbf{r}_2|}\frac{e^2}{|\mathbf{r}_1' - \mathbf{r}_2'|}\int_0^\infty \bar{\chi}_a(\mathbf{r}_1, \mathbf{r}_1', iu)\bar{\chi}_b(\mathbf{r}_2, \mathbf{r}_2', iu)du. \quad (3)$$

Here it is asumed that electrons in system a can be considered distinguishable from those in system b, which is only possible if the electron clouds do not overlap (see Figure 2). $\bar{\chi}_a(\mathbf{r}, \mathbf{r}', \omega)$ and $\bar{\chi}_b(\mathbf{r}, \mathbf{r}', \omega)$ are the exact density-density response functions (in the Kubo sense) of each separate system in the absence of the other . $\bar{\chi}_a$ is defined by the linear density response $n_{1a}(\mathbf{r})\exp(ut)$ of the electrons in system a to an externally applied electron potential energy perturbation $V_1^{ext}(\mathbf{r})\exp(ut)$:

$$n_{1a}(\mathbf{r}) = \int \bar{\chi}_a(\mathbf{r}, \mathbf{r}', iu)V_1^{ext}(\mathbf{r}')d\mathbf{r}' \quad (4)$$

and similarly for $\bar{\chi}_b$. It is important to note that $\bar{\chi}_a$ includes the electron-electron interaction amongst the electrons of system a to all orders, and similarly for $\bar{\chi}_b$. (Note also that, unlike Ref. [1], we have referred the space arguments of $\bar{\chi}_a$ and $\bar{\chi}_b$ in (3) to a common origin.) The form (3) follows from the usual Schrödinger perturbation theory, plus a theorem expressing a single energy-denominator function as a frequency integral over the product of two such functions, the latter forming part of the microscopic susceptibility expressions. (See, e.g., [1].)

Small Non-Overlapping Systems: Density Functional by Approximating Interacting Susceptibility of Each System

We will now use (3) to derive a density functional for the vdW energy, by finding an adequate approximation for the susceptibilities $\bar{\chi}_a$ and $\bar{\chi}_b$, using only the groundstate electron densities $n_a(\mathbf{r})$ and $n_b(\mathbf{r})$ as input. In the spirit of density functional theory, we will approximate $\bar{\chi}_a$ and $\bar{\chi}_b$ by appealing to a simple model of the density-density response of a homogeneous electron gas, suitably modified to approximate the response of the inhomogenous gas in each system. Consider the homogeneous electron gas with groundstate number density n_0, Fermi velocity v_F and plasma frequency $\omega_P = (4\pi n_0 e^2/m)^{1/2}$, perturbed by a time-dependent external scalar potential δV^{ext}. One uses the continuity equation and the Poisson equation plus a linearized Euler ($\mathbf{F} = m\mathbf{a}$) equation with a pressure term $\delta P = -\beta^2 \delta n$, where β^2 is usually [3] given the high-frequency ‡ value $3v_F^2/5$. After space-time Fourier transformation one obtains a density perturbation

$$\delta n(q,\omega) = \bar{\chi}(q,\omega)\delta V^{ext}(q,\omega),$$
$$\bar{\chi}(q,\omega) = \frac{n_0 q^2}{m(\omega^2 - \omega_P^2 - \beta^2 q^2)}.$$

This hydrodynamic response reveals an undamped resonance for plasma waves of wavenumber q and frequency $\omega = (\omega_P^2 + \beta^2 q^2)^{1/2}$. In fact hydrodynamics misses the phenomenon of Landau damping in which a plasma wave decays by creating a particle-hole pair. Conservation of energy and momentum show that this phenomenon sets in for wavenumbers greater than a critical value

$$q_c \approx \omega_P / v_F \equiv q_{TF}.$$

For wavenumbers greater than q_c the Landau damping suppresses the response. The simplest approximation allowing for this would be to set the response to zero for $q > q_c$, and to ignore the diffusion term, i.e. to set $\beta = 0$. This yields

$$\bar{\chi}^{\text{hom}}(q,\omega) \approx \frac{q^2 n_0}{m(\omega^2 - \omega_P^2(n_0))}, \tag{5}$$

where n_0 is the unperturbed electron number density and $\omega_P^2(n) = 4\pi n e^2/m$. Eq. (5) corresponds to a real-space response which is local in the sense of a delta-function second derivative:

$$\bar{\chi}^{\text{hom}}_{local}(\mathbf{r},\mathbf{r}',\omega) = \frac{n_0}{m(\omega_P^2(n_0) - \omega^2)} \nabla^2 \delta^3(\mathbf{r}-\mathbf{r}'). \tag{6}$$

How should we make the corresponding local approximation for an inhomogeneous system, in which the uniform unperturbed density n_0 is replaced by a given inhomogeneous groundstate electronic density $n(\mathbf{r})$? For the present application it will be necessary to take into account the following two exact constraints:

‡The low-frequency or hydrostatic value $\frac{1}{3}v_F^2$ fails to give the correct dispersion relation of plasmons in the uniform electron gas.

(A) **Charge conservation**: because a time-dependent potential perturbation can move electronic charge around but not create or destroy it, we must have

$$\int \bar{\chi}(\mathbf{r},\mathbf{r}',iu)d\mathbf{r} = 0 \quad \forall \mathbf{r}',\ u, \tag{7}$$

and

(B) **Reciprocity:**

$$\bar{\chi}(\mathbf{r},\mathbf{r}',iu) = \bar{\chi}(\mathbf{r}',\mathbf{r},-iu)\ (u\ \ real). \tag{8}$$

Eq. (8) is readily proved from the general Kubo form for the imaginary-frequency response in terms of the interacting many-body eigenstates. Because of (8), (7) also holds with the arguments $\mathbf{r}$ and $\mathbf{r}'$ reversed in the susceptibility. Furthermore, in systems with time-reversal invariance, it is also true that

$$\bar{\chi}(\mathbf{r},\mathbf{r}',iu) = \bar{\chi}(\mathbf{r}',\mathbf{r},iu)\ (u\ \ real). \tag{9}$$

One simple way to ensure (7) is to express $\bar{\chi}$ as a spatial gradient of a function which vanishes at infinity, thus ensuring that χ integrates to zero. Furthermore the *form*

$$\bar{\chi}(\mathbf{r},\mathbf{r}',iu) = \nabla_r \cdot \nabla_{r'} F(\mathbf{r},\mathbf{r}',iu) \tag{10}$$

where F is symmetric in $\mathbf{r}$ and $\mathbf{r}'$ and vanishes at infinity, will guarantee both charge conservation and reciprocity even where F includes density gradient terms. As a first step in implementing these ideas for a general inhomogeneous system, we seek the simplest local-density aproximation for $\bar{\chi}$ consistent with the lowest-order homogeneous result (6). The form (10) essentially mandates the following choice, if F is to be a scalar:

$$\bar{\chi}_{local}^{inhom}(\mathbf{r},\mathbf{r}',\omega) = \nabla_r \cdot \nabla_{r'}\left[\frac{1}{m}\frac{n(\mathbf{r})\delta^3(\mathbf{r}-\mathbf{r}')}{\omega^2 - \omega_P^2(n(\mathbf{r}))}\right] \tag{11}$$

In (11), the square bracket is the function F introduced in eq. (10), and it is essential, for charge conservation, that $n(\mathbf{r})$ does not occur outside the differentiations. While the density arguments in (11) appear at first sight to be reciprocity-violating because they involve only $\mathbf{r}$ and not $\mathbf{r}'$, they occur right against a delta function and so do not in fact violate reciprocity. Indeed $n(\mathbf{r})$ in (11) could be replaced by (e.g.) $[n(\mathbf{r})+n(\mathbf{r}')]/2$ or $[n(\mathbf{r})n(\mathbf{r}')]^{1/2}$ with no effect on the formula (11), when it is integrated with a further function as in (3) and integrations by parts are performed. The symmetric form of the derivatives in (11) is, however, crucial as we shall see: while the operators ∇_r^2 and $-\nabla_r \cdot \nabla_{r'}$ are equivalent in the uniform case, this is not so for the inhomogeneous case in which only the latter form guarantees reciprocity.

Note also that (11) is a "doubly local" approximation in that it uses homogeneous-gas data based on the local density, and also that the uniform-gas response is approximated locally in the sense of a spatial delta function $\delta^3(\mathbf{r}-\mathbf{r}')$. The latter approximation is satisfactory for the nonoverlapping systems treated here, but it will be inadequate, leading to infinities, for the overlapping systems to be studied below

Representing $\nabla_r \cdot \nabla_{r'}$ by $\sum_\mu (\partial/\partial r_\mu)(\partial/dr'_\mu)$ and applying (11) in (3) to approximate $\bar{\chi}_a$ in terms of $n_a(\mathbf{r}_1)$, (and also $\bar{\chi}_b$ in terms of $n_b(\mathbf{r}_2)$), we obtain the approximate vdW energy

$$E^{(2)} = -\frac{\hbar}{2\pi m^2}\int d\mathbf{r}_1 d\mathbf{r}_1' d\mathbf{r}_2 d\mathbf{r}_2' \frac{\partial^2}{\partial r_{1\alpha}\partial r_{2\beta}}\left(\frac{e^2}{|\mathbf{r}_2 - \mathbf{r}_1|}\right)\frac{\partial^2}{\partial r_{1\alpha}'\partial r_{2\beta}'}\left(\frac{e^2}{|\mathbf{r}_2' - \mathbf{r}_1'|}\right)$$
$$\times \int_0^\infty \left[\frac{n_a(\mathbf{r}_1)\delta^3(\mathbf{r}_1 - \mathbf{r}_1')}{-u^2 - \omega_P^2(n_a(\mathbf{r}_1))}\right]\left[\frac{n_b(\mathbf{r}_2)\delta^3(\mathbf{r}_2 - \mathbf{r}_2')}{-u^2 - \omega_P^2(n_b(\mathbf{r}_2))}\right]du$$

Here we have already integrated by parts on each space variable, and have used the Einstein summation convention in the α and β summations. Performing the second derivatives explicitly on the coulomb potentials and then using the delta functions to remove two integrations, we have

$$E^{(2)} = -\frac{\hbar e^4}{2\pi m^2}\int d\mathbf{r}_1 d\mathbf{r}_2 \sum_{\alpha,\beta}\left\{\frac{3r_{12\alpha}r_{12\beta} - \delta_{\alpha\beta}r_{12}^2}{r_{12}^5}\right\}^2$$
$$\times \int_0^\infty \left[\frac{n_a(\mathbf{r}_1)}{-u^2 - \omega_P^2(n_a(\mathbf{r}_1))}\right]\left[\frac{n_b(\mathbf{r}_2)}{-u^2 - \omega_P^2(n_b(\mathbf{r}_2))}\right]du$$
$$= -\frac{\hbar e^4}{2\pi m^2}\int_a d\mathbf{r}_1 \int_b d\mathbf{r}_2 \frac{6}{r_{12}^6} n_a(\mathbf{r}_1) n_b(\mathbf{r}_2)\frac{\pi/2}{(\omega_1 + \omega_2)\omega_1\omega_2}$$
$$= -\frac{3\hbar}{32\pi^2}\int_a d\mathbf{r}_1 \int_b d\mathbf{r}_2 \frac{1}{r_{12}^6}\frac{\omega_1\omega_2}{(\omega_1 + \omega_2)}$$
$$= -\frac{3\hbar e}{2(4\pi)^{3/2} m^{1/2}}\int_a d\mathbf{r}_1 \int_b d\mathbf{r}_2 \frac{1}{r_{12}^6}\frac{\sqrt{n_a(\mathbf{r}_1)n_b(\mathbf{r}_2)}}{\left(\sqrt{n_a(\mathbf{r}_1)} + \sqrt{n_b(\mathbf{r}_2)}\right)} \tag{12}$$

where $\omega_1 \equiv \omega_{P1} = (4\pi e^2 n_a(\mathbf{r}_1)/m)^{1/2}$ and similarly for ω_2 in terms of $n_b(\mathbf{r}_2)$. It is interesting that the integrand in (12) is proportional to the *harmonic mean*, $\omega_1\omega_2/(\omega_1 + \omega_2)$, of the two local plasma frequencies. If the separation R greatly exceeds the dimension of either system, so that we can take $1/R^6$ out of the integration, and if the plasma frequencies ω_1 and ω_2 are taken as independent of position inside each system, then the third-last line of Eq. (12) reduces to $E^{(2)} = -(3/2)\hbar e^4 \pi^{-1} m^{-2} R^{-6} N_1 N_2 (\omega_1\omega_2)^{-1}(\omega_1 + \omega_2)^{-1}$ which is sometimes termed the London formula (see, e.g. Eq. 1.15 of Ref. [4].)

We can also understand the present approximation as a scalar, local-density approximation to the nonlocal linear dipole polarizability $\bar{A}$. This is defined such that

$$\delta P_i(\mathbf{r},\omega) = \sum_{j=1}^{3}\int \bar{A}_{ij}(\mathbf{r},\mathbf{r}',\omega)\delta E_j^{ext}(\mathbf{r}',\omega)d^3r' + O((E^{ext})^2) \tag{13}$$

where $\delta\mathbf{P}(\mathbf{r},\omega)$ is the perturbation to the polarization (dipole moment per unit volume) at $\mathbf{r}$, such that $\delta n = e^{-1}\nabla\cdot\delta\mathbf{P}$. If the external electric field perturbation $\delta\mathbf{E}^{ext}$ is derivable from a scalar potential δV^{ext} then (13) can be written, with summation over i and j understood,

$$\delta n(\mathbf{r},\omega) = \frac{1}{e}\frac{\partial}{\partial r_i}\delta P_i = \frac{1}{e^2}\int \frac{\partial}{\partial r_i}\left[\bar{A}_{ij}(\mathbf{r},\mathbf{r}')\frac{\partial}{\partial r_j'}\delta V^{ext}(\mathbf{r}')\right] \tag{14}$$

Using integration by parts then comparing (14) with (4) we find

$$\bar{\chi}(\mathbf{r},\mathbf{r}',\omega) = -\frac{1}{e^2}\frac{\partial^2}{\partial r_i \partial r_j'}\bar{A}_{ij}(\mathbf{r},\mathbf{r}',\omega) \tag{15}$$

The use of the polarizability, here termed $\bar{A}$, has in fact been common in the chemical literature over the years [5].

By making a local scalar approximation for $\bar{A}$, namely

$$\bar{A}_{ij}(\mathbf{r}, \mathbf{r}', \omega) \approx \delta_{ij}\bar{\alpha}(\mathbf{r})\delta^3(\mathbf{r} - \mathbf{r}'), \tag{16}$$

then putting (15) into (3) and performing integration by parts twice we obtain a local expression for the second-order vdW interaction between nonoverlapping systems :

$$E^{(2)} = -\frac{\hbar}{2\pi} \int_a d\mathbf{r}_1 \int_b d\mathbf{r}_2 \frac{6e^{-4}}{\mid \mathbf{r}_1 - \mathbf{r}_2 \mid^6} \int_0^\infty \bar{\alpha}_a(\mathbf{r}_1, iu)\bar{\alpha}_b(\mathbf{r}_2, iu)du. \tag{17}$$

(17) is somewhat more general than (12), though it is still a type of local approximation. (Note that the quantity $\bar{\alpha}$ used here is termed "χ " in the companion chapter by Andersson *et al* on vdW interactions: our $\bar{\chi}$ is denoted "π" by these authors).

From (15), the Ansatz (11) is now clearly equivalent to the assumption of a local scalar polarizability taken from the $q \to 0$ hydrodynamic response of a uniform electron gas :

$$\bar{\alpha}(\mathbf{r}, \omega) \approx \frac{e^2 n(\mathbf{r})}{m(\omega_P^2(n(\mathbf{r}) - \omega^2)}, \tag{18}$$

and (17) then reproduces (12) . An advantage of the direct local approximation of the nonlocal polarizability $\bar{A}$ (rather than $\bar{\chi}$) is that charge conservation, as expressed by (7), is automatically satisfied by virtue of (15) provided that $\bar{A}$ vanishes at the edges of the system.

Furthermore if the separation R of the two systems sufficiently exceeds their size, we can replace r_{12}^{-6} by R^{-6} in (17), obtaining

$$E^{(2)} \approx -C_6 R^{-6},$$
$$C_6 = \frac{3\hbar}{\pi} \int_0^\infty \bar{\alpha}_a(iu)\bar{\alpha}_b(iu)du \tag{19}$$

where $\bar{\alpha}_a = \int_a \bar{\alpha}_a(\mathbf{r})d\mathbf{r}$ is the total polarizability of system a, and similarly for b. Eq. (19) is a well-known approximate formula (see e.g. Eq. (1.18) of Ref. [4]). Eq. (19) is also essentially the result (2) obtained heuristically in the first part of this chapter, except that the zero-point oscillation frequency ω_0 in (2) has been replaced by an integration over imaginary frequency in (19).

Another way of understanding why an Ansatz $\chi_a(\mathbf{r}, \mathbf{r}', \omega) \approx \nabla^2 F$ is unacceptable (in contrast to the successful form $\nabla_r \cdot \nabla_{r'} F$ used in (11) above) is to note that the ∇^2 form yields a zero net polarizability $\bar{\alpha}_{ij} = e^2 \int \mathbf{r}_i \mathbf{r}'_j \bar{\chi}(\mathbf{r}, \mathbf{r}', \omega) d\mathbf{r} d\mathbf{r}' = 0$ for each subsystem: this follows by two partial integrations. Since the elementary argument given above shows that the vdW interaction is proportional to the dipole polarizabilities of the two subsystems, it is therefore not surprising that the Ansatz $\bar{\chi}(\mathbf{r}, \mathbf{r}', \omega) \approx \nabla^2 F$ gives a zero vdW interaction, as previously noted.

Formulae (12) and (17) are the main results of the present Section [6]. An identical result to (12) was obtained independently by Andersson, Langreth and Lundquist [7] from an argument rather different from ours but related to that of Rapcewicz and Ashcroft [8]. The latter authors represented the vdW interaction between points $\mathbf{r}$ and $\mathbf{r}'$ in two different systems in terms of that from an effective electron gas whose density they chose to be the geometric mean, $\bar{n} = \sqrt{n_a(\mathbf{r})n_b(\mathbf{r}')}$. Andersson *et al* modified the geometric mean density, replacing it by a more complicated average chosen to reproduce the known London result (19) for the case of two point polarizable dipoles.

Ref. [7] uses (12) to make extensive numerical predictions for the constant C_6 defined in equation (19) above. The results for rare-gas pairs, as well as a variety of other pairs, are very good considering the simplicity of the approximation (12) (compared with a near-exact many-body computation of the dipole polarizabilities $\bar{\alpha}_a$, $\bar{\alpha}_b$ which would otherwise be needed to obtain C_6). A very important caveat is that, in this approach, it is necessary to cut off the spatial integrations somewhere in the tail of the electron distribution of each atom: otherwise the prediction for C_6 from (12) is considerably too large. The cutoff criterion used in [7] was that the local wavenumber $k(\mathbf{r}) \equiv |\nabla n| / n$ exceeds the local estimate of the Thomas-Fermi wavenumber, $q_{TF}(\mathbf{r}) \equiv \omega_P(n(\mathbf{r}))/v_F(n(\mathbf{r}))$. This is loosely related to the criterion for the onset of Landau damping of plasma waves given above.

Larger Non-Overlapping Systems: Density Functional by Approximating KS Susceptibility of Each System and Screening Explicitly

Direct local-density approximation of the interacting susceptibilities $\bar{\chi}_a$, $\bar{\chi}_b$ of the separate systems is not necessarily the best local-density approximation, particularly if one or both of the systems is large. Hult *et al* [9] proposed a simple double-local-density approximation for a dielectric constant, followed by solution of the screening equation (self-consistent Maxwell equations). This approach is closely related to a more general procedure proposed some years ago [10] and involving local density approximation for a *bare* susceptibility, followed by solution of a Poisson screening equation . This connection will be discussed further in the section on "seamless" vdW functionals below. For now we merely note that sensible results were obtained by Hult *et al* for various physisorbtion coefficients of noble gases on jellium. As with earlier functionals for non-overlapping systems, a low-density cutoff was required, and the optimal choice for this cutoff was found to be different from the optimal value for atoms: see also the article by Andersson *et al* in the present volume.

Prospects for Improvement of the Cutoff Procedure in vdW Density Functionals for Nonoverlapping Systems

One advantage of the present method of derivation is that, by looking at the detailed physics of $\bar{\chi}$, one can presumably find an improved approximation to $\bar{\chi}_a(\mathbf{r}, \mathbf{r}', \omega)$ when $\mathbf{r}$ and/or $\mathbf{r}'$ are in the tunneling tails of the atomic density distribution, and therefore avoid *ad hoc* cutoff procedures. For example, in (17) one does not need to use the uniform-gas-based approximation for $\bar{\alpha}$. Some progress is currently being made with this [11], by the use of the WKB approximation for tunneling one-body wavefunctions in classically forbidden regions. It is also worth drawing attention to the newly proposed "Airy Gas" (see the chapter by W. Kohn in the present volume) as a possible model for calculation of susceptibilities $\bar{\chi}$ within tunelling regions outside electron concentrations. By these means one may hope to provide a more natural cutoff for calculations based on (12), (17) and similar formulae.

SEAMLESS VDW FUNCTIONALS ALLOWING FOR OVERLAP

All of the discussion above concerned the vdW interaction between two systems with negligible overlap of electron density. For these cases it turned out to be relatively easy to generate explicit, highly nonlocal approximate vdW functionals, dependent simultaneously on the groundstate density in the two interacting susbsystems. In these cases second-order perturbation theory in the inter-system coulomb interaction U_{ab} essentially produced the vdW energy. To achieve an explicit density functional it was only necessary to approximate the density-density response χ of each subsystem in terms of the groundstate density, the basic R^{-6} form emerging naturally provided only that one maintained charge conservation and reciprocity.

The assumption of zero overlap severely limits the usefulness of the functionals, however, because many interesting cases involve both the vdW interaction and significant overlap. In the present Section we will exhibit some functionals designed to cope both with the distant vdW interaction and with the other forms of interaction such as chemical bonding. These new functionals have the same form regardless of the number or separation of any identifiable subsystems, and in this sense they are "seamless" - there is no separation distance at which two different forms of functional are "stitched together". The philosophy, previously described in [10], is to express the xc energy exactly in terms of an interacting response function. The interacting response function is then obtained via a screening integral equation, starting from a local density approximation for a bare or intermediate response function.

The Adiabatic-Connection/Fluctuation-Dissipation (ACF/FDT) Formula for the Exact Correlation Energy

As discussed in the chapter by Dobson on "Time-dependent density functional theory" in the introductory section of the present Volume (see also [12,10]) the adiabatic connection formula [13,14] and the fluctuation-dissipation theorem combine to give the following exact expression (termed here "ACF/FDT") for the xc energy of an inhomogeneous electronic system:

$$E_{xc} = \frac{1}{2}\int_0^1 d\lambda \int d\mathbf{r} \int d\mathbf{r}' \frac{e^2}{|\mathbf{r}-\mathbf{r}'|}\{-\hbar\pi^{-1}\int_0^\infty \chi(\lambda,\mathbf{r},\mathbf{r}',iu)du - n(\mathbf{r})\delta(\mathbf{r}-\mathbf{r}')\}. \tag{20}$$

Here there is no division into subsystems. $\chi(\lambda, \mathbf{r}, \mathbf{r}', iu)$ is the interacting susceptibility or density-density response function of all electrons as defined in Eq. (4) above, but computed for electrons with reduced Coulomb pair interaction $\lambda e^2/|\mathbf{r}-\mathbf{r}'|$. They are assumed to be moving in an external one-electron potential $V_\lambda(\mathbf{r})$ chosen at each value of λ to maintain the interacting density at the true ($\lambda = 1$) groundstate density $n(\mathbf{r})$.

In (20) we are not relying on any finite order of perturbation theory, in contrast to formulae such as (3) where most of the labor of obtaining the vdW interaction was achieved by second order perturbation theory. In (20) the long-ranged vdW correlation energy must be obtained directly from a long-ranged correlation tail in χ. It is therefore not surprising, that, as will be shown below, a direct local-density approximation for χ in (20) produces only the local-density xc energy E^{xc}_{LDA}, and therefore misses the asymptotic vdW interaction completely.

Charge conservation in χ ensures xc hole normalization

As discussed earlier, the conservation of charge is expressed in terms of χ by the exact requirement

$$\int \chi(\lambda, \mathbf{r}, \mathbf{r}', \omega)d^3r = 0 \ \forall \ \mathbf{r}'. \tag{21}$$

Now (20) is based on the fluctuation- dissipation theorem in the form

$$-\frac{\hbar}{\pi}\int_0^\infty \bar{\chi}(\mathbf{r},\mathbf{r}',iu)du = <(\hat{n}(\mathbf{r})-n(\mathbf{r}))(\hat{n}(\mathbf{r}')-n(\mathbf{r}'))>_\lambda$$
$$= <\hat{n}(\mathbf{r})\hat{n}(\mathbf{r}')> -n(\mathbf{r})n(\mathbf{r}')$$
$$= n_2(\mathbf{r},\mathbf{r}') + \delta(\mathbf{r}-\mathbf{r}')n(\mathbf{r}) - n(\mathbf{r})n(\mathbf{r}') \qquad (22)$$

where n_2 is the pair density. Thus if (21) holds we can integrate (22) with respect to $\mathbf{r}$ to obtain

$$0 = \int n_2(\mathbf{r},\mathbf{r}')d^3r + n(\mathbf{r}') - Nn(\mathbf{r}'). \qquad (23)$$

In the DFT context it is common practice to write the pair density n_2 in terms of the xc hole density n_{xc}:

$$n_2(\mathbf{r},\mathbf{r}') = n(\mathbf{r}')(n(\mathbf{r}) + n_{xc}(\mathbf{r}|\mathbf{r}')).$$

Eq. (23) can then be written

$$\int n_{xc}(\mathbf{r}|\mathbf{r}')d^3r = -1 \qquad (24)$$

which is the expression of exchange-hole normalization, known [13,14,15] to be important in the construction of electron density functionals. Thus for reasons of hole normalization it is essential to ensure that any approximation for $\bar{\chi}$, to be used with (20) in constructing a vdW functional, satisfies (21).

LDA for χ in ACF/FDT Will Not Yield vdW

Suppose that we approximate the interacting response $\bar{\chi}$ in (20) by a local density Ansatz, e.g.

$$\chi(\lambda,\mathbf{r},\mathbf{r}',\omega) = \chi^{\text{hom}}(\lambda,\bar{n}=n(\mathbf{r}),|\mathbf{r}-\mathbf{r}'|,\omega) \qquad (25)$$

where $\chi^{\text{hom}}(\lambda,\bar{n},R,\omega)$ is the response at separation R in a homogeneous electron gas of density $\bar{n}$. Then (20) becomes

$$E_{xc} = \int n(\mathbf{r})\epsilon_{xc}(n(\mathbf{r}))d^3r \qquad (26)$$

where

$$\epsilon_{xc}(n) = \frac{1}{2n}\int_0^1 d\lambda \int d^3R\frac{e^2}{R}\{-\hbar\pi^{-1}\int_0^\infty \bar{\chi}^{\text{hom}}(\lambda,n,\mathbf{R},iu)du - n\delta^3(\mathbf{R})\}$$

is the xc energy per particle in the homogeneous electron gas. (The asymmetric form (25) was chosen as the simplest form conserving charge in the sense of Eq. (21). Essentially similar conclusions follow from the form $\bar{\chi} = \nabla_r.\nabla_{r'}F$ provided that F depends locally on density).

(26) is precisely the LDA energy. Thus, since we know the LDA does not contain the vdW interaction, we conclude one cannot obtain a vdW density functional from a local approximation for $\bar{\chi}$ in (20). Nevertheless useful vdW approximations to (20) can be obtained from a knowledge of the groundstate density. To see this we need to discuss bare and interacting response functions in the following paragraphs. Basically, we will show that a vdW functional can be obtained from a local-density approximation for a *bare* susceptibility, followed by solution of a spatially nonlocal integral equation to obtain the interacting susceptibilty which is then substituted into Eq. (20).

Exact TDKS Equation: Definition of Three Density-Density Susceptibilies

Consider electrons interacting via a reduced Coulomb potential $\lambda e^2/|\mathbf{r}-\mathbf{r}'|$. Time dependent density functional theory [16,17] gives the exact linear response $n_{1\lambda}(\mathbf{r},\omega)$ of the electron number density to an external potential perturbation $v_1^{ext}(\mathbf{r})\exp(-i\omega t)$ as

$$n_{1\lambda}(\mathbf{r},\omega)=\int \chi_{KS}(\mathbf{r},\mathbf{r}',\omega)v_{1\lambda}^{TDKS}(\mathbf{r}')d^3r' \equiv (\chi_{KS}*v_{1\lambda}^{TDKS})\big|_{\mathbf{r}}. \tag{27}$$

See also the introductory chapter "Time-dependent density functional theory" in the present volume. In (27), χ_{KS} (often termed χ_s [17,18]) is the density-density response kernel of *independent* electrons moving in the time-dependent external potential plus the groundstate KS potential $V_0^{eff}(\mathbf{r})$. $v_{1\lambda}^{TDKS}$ (often termed simply v_{1s} [17,18]) is the linearized effective potential perturbation composed of external, time-dependent-Hartree-like and exchange-correlation contributions,

$$v_{1\lambda}^{TDKS}=v_1^{ext}+v_{1\lambda}^{TDH}+v_{1\lambda}^{xc} \tag{28}$$

where

$$v_1^{TDH}(\mathbf{r},\omega)=\int \frac{\lambda e^2}{|\mathbf{r}-\mathbf{r}'|}n_{1\lambda}(\mathbf{r}',\omega)d^3r' \equiv U_\lambda * n_{1\lambda}, \tag{29}$$

$$v_{1\lambda}^{xc}(\mathbf{r},\omega)=\int f_{xc\lambda}(\mathbf{r},\mathbf{r}',\omega)n_{1\lambda}(\mathbf{r}',\omega)d^3r' \equiv f_{xc}*n_{1\lambda}. \tag{30}$$

Since the linearized potentials v_1^{TDH} and $v_{1\lambda}^{xc}$ in Eqs. (28)-(30) are linear in the density perturbation $n_{1\lambda}$, it follows that there exist linear relations between $n_{1\lambda}$ and any combination of these potentials. In particular, corresponding to the three potentials v_1^{ext}, $v_{1\lambda}^{TDKS}$ and $v_{1\lambda}^{class}\equiv v_1^{ext}+v_{1\lambda}^{TDH}$, one can define three response functions ("susceptibilities") $\bar{\chi}_\lambda$, χ_{KS} and $\hat{\chi}_\lambda$, each of which yields the exact linear interacting density perturbation $n_{1\lambda}$, :

$$n_{1\lambda}=\bar{\chi}_\lambda * v_1^{ext} \tag{31}$$
$$n_{1\lambda}=\chi_{KS}*v_{1\lambda}^{TDKS} \tag{32}$$
$$n_{1\lambda}=\hat{\chi}_\lambda * v_{1\lambda}^{class}. \tag{33}$$

The first of these susceptibilities, $\bar{\chi}_\lambda$, has been introduced already in Eq. (4). $\bar{\chi}_\lambda$ contains all the physics of electron-electron interactions including the long-ranged correlation effects, while its driving field v_1^{ext} contains none of these effects. As already explained, a direct local approximation for $\bar{\chi}_\lambda$ is not useful for deriving vdW functionals. Note also that $\bar{\chi}_\lambda$ is the density-density response function of the interacting system as defined by Kubo.

The second density-density response, χ_{KS}, introduced in (27) and (32), is the response of independent electrons moving in the groundstate Kohn-Sham potential V_0^{KS}, and contains neither time-dependent Hartree nor time-dependent xc physics. In terms of the eigenfunctions ϕ_i and eigenvalues ϵ_i of the usual static Kohn-Sham theory, χ_{KS} is given exactly [17] by

$$\chi_{KS}(\mathbf{r},\mathbf{r}',\omega)=\sum_{ij}\frac{f_i-f_j}{\epsilon_i-\epsilon_j-\hbar\omega}\phi_i^*(\mathbf{r})\phi_j^*(\mathbf{r}')\phi_i(\mathbf{r}')\phi_j(\mathbf{r}). \tag{34}$$

where f_i is the zero-temperature Fermi factor. For $\chi_{KS}(\mathbf{r},\mathbf{r}',\omega)$ to be nonzero, an independent electron must be able to propagate from $\mathbf{r}$ to $\mathbf{r}'$ under the influence the groundstate KS potential V_0^{KS} alone. In particular,

if $\mathbf{r}$ *and* $\mathbf{r}'$ *are separated by an impenetrable barrier as in the "canonical" vdW case discussed earlier,* χ_{KS} *is strictly zero* (35)

This will be termed the "no-flow" condition. The no-flow condition can be proved directly from (34). Because of the no-flow condition, in the vdW context χ_{KS} is the most "local" of the three density response functions discussed here. We argue that a suitably constrained local approximation for χ_{KS} is a good approximation in this context. From this, with (20) and a screening equation to be discussed below, we will obtain a vdW functional. The dynamic xc kernel f_{xc} can either be set to zero, in which case we will have an approximation to the RPA xc energy: or f_{xc} can be approximated in a suitable local fashion to be discussed below. In that case we have a vdW functional beyond the RPA. This approach appears to be a flexible way to derive a seamless vdW functional.

The third density-density response, defined in (33), is $\hat{\chi}$, an "intermediate" or "direct" response. Its driving field is $v_{1\lambda}^{class} \equiv v_1^{ext} + v_{1\lambda}^{TDH}$, the "total" potential of classical electrodynamics. (The word "classical" is not used here in the sense of $\hbar \to 0$, but rather in the sense that $-q\nabla v_{1\lambda}^{class}$ gives the force on an *infinitesimal* test charge q as in the classical electrodynamic definition of electric field. Because electrons are not infinitesimally charged, and hence cause a finite disturbance in the charge distribution of their neighbours, we need the xc term over and above $v_{1\lambda}^{class}$ in the total TDKS field on the right-hand-side of (28).) Referring to (33), although the driving field $v_{1\lambda}^{class}$ contains the Hartree part of the time-dependent electron-electron interaction, a description of the time-dependent xc effects has to be included in $\hat{\chi}$. To the extent that these time-dependent xc effects can be treated locally, a local approximation for $\hat{\chi}$ will be sensible, and we believe that this (with (20) plus a numerical solution of a "screening" equation given below) yields a useful vdW density functional of the groundstate density $n(\mathbf{r})$. This functional is very close to one proposed long ago by Kohn and Hanke [19] but not so far published. (See below)

Dyson-type "Screening" Equations: relations between density response functions

From the definitions (31)-(33) and the TDKS equations (27) and (32) we can derive the following "screening" equations linking the various response functions:

$$\bar{\chi}_\lambda = \chi_{KS} * (\mathbf{I} + (U_\lambda + f_{xc\lambda}) * \bar{\chi}_\lambda) \tag{36}$$
$$\hat{\chi}_\lambda = \chi_{KS} * (\mathbf{I} + f_{xc\lambda} * \hat{\chi}_\lambda) \tag{37}$$
$$\bar{\chi}_\lambda = \hat{\chi} * (\mathbf{I} + U_\lambda * \bar{\chi}_\lambda). \tag{38}$$

When the convolutions $(*)$ are written out in full, these are recognized as integral equations in position space.

For example, having obtained $\chi_{KS}(\mathbf{r},\mathbf{r}',iu)$ (either exactly from (34), or e.g. by a local approximation based on the homogeneous gas), and f_{xc} (e.g. by an LDA) we solve the space integral equation (36) to obtain $\bar{\chi}$. Notice that this is a highly nonlocal integral equation because of the coulomb potential U, and this will introduce a long-ranged part in $\bar{\chi}$ even where none exists in χ_{KS}.

How a conserving LDA for χ_{KS} Can Yield a Seamless vdW Functional

The steps required in constructing a seamless vdW density functional are thus the following.

Step (a): Use the groundstate density $n(\mathbf{r})$ to construct local approximations for

(a_1) the independent-electron susceptibility χ_{KS} and

(a_2) the dynamic xc kernel f_{xc},

Step (b): Substitute χ_{KS} and f_{xc} into the screening equation (36), then solve this real-space integral equation for the interacting susceptibility $\bar{\chi}$. Step (b) is the time-limiting step of the present scheme.

Step (c): The final step is to substitute $\bar{\chi}$ into the ACF/FDT formula (20) to yield the xc energy of the inhomogeneous system. Some economies are possible, by rearranging the details of steps (b) and (c).

Note that the only information used concerning the inhomogeneous system is its groundstate density $n(\mathbf{r})$ used in step (a). Thus this scheme is a true density functional for the xc energy.

The energy produced is very closely related to the RPA (ring-diagram) energy of the inhomogeneous system. Indeed if $\chi_{KS}(\mathbf{r},\mathbf{r}',\omega)$ (the open bubble of diagrammatic perturbation theory) is evaluated exactly rather than approximated locally, and if f_{xc} is set to zero, the scheme gives exactly the RPA energy.

Further details of steps (a) and (b) are given below. In the case of non-overlapping systems we also show below that, when this scheme is expanded in powers of the inter-system Coulomb interaction U_{ab}, it yields the Zaremba-Kohn (ZK) formula (3) used earlier in this article to treat such cases. The exact interacting susceptibilities $\bar{\chi}_a$, $\bar{\chi}_b$ in the ZK formula are, however, replaced by approximate versions obtained by screening the approximate local version of χ_{KSa} and χ_{KSb}.

Before expanding on steps (a_1), (a_2) and (b) we briefly sketch some very similar alternative schemes.

Alternative scheme based on a local approximation for $\hat{\chi}$

In this scheme the separate local approximations for χ_{KS} and f_{xc} are replaced by a direct local approximation for the intermediate susceptibility $\hat{\chi}$ defined in Eq. (33):

(a') Use $n(\mathbf{r})$ to construct the intermediate susceptibility $\hat{\chi}$.

(b') Substitute $\hat{\chi}$ into the alternative screening equation (38) then solve this for $\bar{\chi}$.

(c) Substitute $\bar{\chi}$ into (20) to obtain E_{xc}.

This scheme is identical to the previous if we neglect the dynamic xc kernel f_{xc}: then, from (37), $\hat{\chi} \equiv \chi_{KS}$.

If $\hat{\chi}$ is generated from a local-density approximation for a conductivity $\hat{\sigma}$ then this is essentially the scheme of Hanke and Kohn [19].

Optimizing the local approximation for χ_{KS}.

In the present Section we address item ($a1$) of the above prescription for a seamless functional, followed by item (a_2) in the following subsection. Some important conditions to be satisfied by χ_{KS} have been discussed above either for χ_{KS} or for the interacting $\bar{\chi}$, and are as follows:

$$\int \chi_{KS}(\mathbf{r}, \mathbf{r}', iu) d^3r = 0 \quad \forall \; \mathbf{r}' \; (charge\ conservation) \tag{39}$$

$$\chi_{KS}(\mathbf{r}, \mathbf{r}', iu) = \chi_{KS}(\mathbf{r}', \mathbf{r}, iu) \quad (reciprocity) \tag{40}$$

$$\chi_{KS}(\mathbf{r}, \mathbf{r}', iu) = 0 \; (no - flow : \mathbf{r}, \mathbf{r}' separated\ by\ wall) \tag{41}$$

$$\alpha \equiv \int \mathbf{r}\mathbf{r}' \chi_{KS}(\mathbf{r}, \mathbf{r}', iu) d^3r d^3r' \neq \mathbf{0} \, (sensible\ dipole\ polarizability\ of\ finite\ system) \tag{42}$$

The aim is to simulate the independent-electron response $\chi_{KS}(\mathbf{r}, \mathbf{r}', iu)$ of an inhomogeneous electron gas, using information such as the known response

$$\chi_{KS}^{\text{hom}}(\bar{n}, |\mathbf{r} - \mathbf{r}'|, iu) = \frac{1}{(2\pi)^3} \int \chi^{Lindhard}(\bar{n}, |\mathbf{q}|, iu) e^{i\mathbf{q}\cdot(\mathbf{r}-\mathbf{r}')} d^3q$$

of the uniform electron gas of independent electrons (the space transform of the well-known Lindhard function [20].) It is surprisingly hard to generate an approximate χ_{KS} using only χ_{KS}^{hom} and a knowledge of the groundstate density $n(\mathbf{r})$, while still satisfying the conditions (39) -(42). For example the Ansatz

$$\chi_{KS}(\mathbf{r}, \mathbf{r}', iu) \approx \chi_{KS}^{\text{hom}}(\bar{n}, |\mathbf{r} - \mathbf{r}'|, iu) \tag{43}$$

with an average density

$$\bar{n} = \frac{n(\mathbf{r}) + n(\mathbf{r}')}{2} \tag{44}$$

satisfies (40) but not (39) nor (41). The Ansatz (43) with the unsymmetrical density argument

$$\bar{n} = n(\mathbf{r}')$$

satisfies (39) (because $\int \chi_{KS}^{\text{hom}}(\bar{n}, R, iu) d^3R = 0$) but it does not satisfy (40) nor (41) nor (42).

Fortunately there is a simple way to achieve all of the above conditions. By analogy with the Ansatz (11) (made in an earlier section of the present Chapter for the *interacting* response), we propose the following form for the independent-particle response:

$$\chi_{KS}(\mathbf{r}, \mathbf{r}', \omega) = \nabla_r \cdot \nabla_{r'} F_0(\bar{n}(\mathbf{r}, \mathbf{r}'), |\mathbf{r} - \mathbf{r}'|, iu) \tag{45}$$

where

$$F_0(\bar{n}, R, \omega) = \frac{1}{(2\pi)^3} \int \frac{1}{q^2} \chi_{KS}^{\text{hom}}(\bar{n}, q, \omega) e^{i\mathbf{q}\cdot\mathbf{R}} d^3q \tag{46}$$

is, up to a factor of $-e^2$, the bare polarizability of the uniform electron gas. The first point to notice about (45) is that, unlike (11), it does NOT contain a delta function constraining $\mathbf{r}$ to equal $\mathbf{r}'$. Indeed (46) is finite with a derivative discontinuity at $\mathbf{r} = \mathbf{r}'$ (i.e. at $R = 0$), and (45) goes as $(const)/|\mathbf{r} - \mathbf{r}'|$ as $\mathbf{r} \to \mathbf{r}'$. One can show [11] that the most singular part of the exact inhomogeneous $\chi_{KS}(\mathbf{r}, \mathbf{r}', iu)$ shares this behavior, and further that the screening equation (36) causes the same behaviour in the interacting susceptibility $\bar{\chi}(\lambda, \mathbf{r}, \mathbf{r}', iu)$. The consequence of a derivative delta function in χ_{KS} would be a similar behavior of $\bar{\chi}(\lambda, \mathbf{r}, \mathbf{r}', iu)$ and hence an infinite value of the xc energy from the Adiabatic Connection/FDT formula (20), from integration by parts using $\nabla^2 |\mathbf{r} - \mathbf{r}'|^{-1} = 4\pi\delta^3(\mathbf{r} - \mathbf{r}')$. This is in sharp contrast to the simple delta function approximation used above (Eq. (11)) to obtain a vdW functional for well-separated charge distributions. In that case second order perturbation theory had already been used to obtain the essential vdW interaction, leaving much greater flexibility in approximating χ. Indeed in (3) the values of χ for $\mathbf{r} \to \mathbf{r}'$ are never sampled because a cutoff is used ensuring that the systems do not overlap, in contrast to the present "seamless" theory where overlap must be treated explicitly. For this reason, many of the approaches involving doubly local approximations [7,9] proposed for separate systems are unsuitable for overlapped cases without further modification. (We note, however, that approaches based on the long-ranged part of the Coulomb interaction (e.g. [19]) can avoid this difficulty).

The form (45) automatically satisfies the charge conservation condition (39) provided that the average density $\bar{n}(\mathbf{r}, \mathbf{r}')$ vanishes at the edge of the system, as can be seen by integration by parts (Green's theorem). Even the simple average density (44), inserted into (45), results in satisfaction of (39), (40) and (42).

Because we no longer have a factor $\delta^3(\mathbf{r} - \mathbf{r}')$ in our Ansatz for χ_{KS}, we now have the possibility of a spurious long-ranged Friedel tail allowing transport between two isolated systems in violation of the no-flow condition (41). To satisfy (41), we may require an average density $\bar{n}$ which is sensitive to the values of the groundstate density at points between $\mathbf{r}$ and $\mathbf{r}'$. Suppose that $\mathbf{r}$ and $\mathbf{r}'$ lie in two different subsystems separated by an impenetrable wall, as in the "canonical" vdW situation discussed in a previous Section. This impenetrable wall manifests itself in the groundstate density by the fact that $n(\mathbf{r})$ vanishes at least somewhere on every path joining $\mathbf{r}$ and $\mathbf{r}'$. Thus to enforce the no-flow condition, we require the average density $\bar{n}$ to vanish when the groundstate density vanishes at some point or other on every path joining $\mathbf{r}$ to $\mathbf{r}'$. The following formulae have this property:

$$\bar{n}^{(IGA)}(\mathbf{r}, \mathbf{r}') = Max \exp\left[\frac{\int_{\mathbf{r}}^{\mathbf{r}'} \ln(n(\mathbf{r}''))dl}{\int_{\mathbf{r}}^{\mathbf{r}'} dl}\right] \tag{47}$$

$$\bar{n}^{(IHA)}(\mathbf{r}, \mathbf{r}') = Max\left\{\frac{\int_{\mathbf{r}}^{\mathbf{r}'} (n(\mathbf{r}''))^{-1}dl}{\int_{\mathbf{r}}^{\mathbf{r}'} dl}\right\}^{-1} \tag{48}$$

where the "max" is taken over all paths in space joining $\mathbf{r}$ to $\mathbf{r}'$. Clearly both these formulae give zero whenever the density is strictly zero along a finite length of the line integation, as will be the case when two subsystems, containing $\mathbf{r}$ and $\mathbf{r}'$ respectively, are each surrounded by a rigid wall. On the other hand, both formulae yield $\bar{n} = n_0$ when $\mathbf{r}$ and $\mathbf{r}'$ lie inside an infinite uniform electron gas of density n_0. For many geometries it will probably suffice to evaluate the line integrals in (47) and (48) along a straight line joining $\mathbf{r}$ to $\mathbf{r}'$. There are certainly some geometries, however, in which the line integral neeeds to be evaluated along the "most metallic" path giving the maximum average density as in (47) and (48). For example, if $\mathbf{r}$ and $\mathbf{r}'$ are near the two tips of a horseshoe-shaped molecule, a straight-line path will give a near-zero value for $\bar{n}$ because of the vacuum region between the tips. In reality, however, substantial electron transport may be possible from $\mathbf{r}$ to $\mathbf{r}'$ around the curve of the molecule.

In the case of systems with one dimensional spatial inhomogeneity and an infinite number of electrons (e.g. jellium surfaces or $Ga_{1-x}Al_xAs$ quantum wells) considerable evidence already exists [21,22] for the efficacy of approximations like (45), (46), (47) or (48). For finite systems or insulating solids, it may be preferable to modify this approach to incorporate a HOMO-LUMO gap, in a manner suggested by the Penn model [23] of dielectric response.

By the above means we obtain an approximation for χ_{KS} of an inhomogeneous electron distribution using only the groundstate density (plus possibly an estimate of the HOMO-LUMO gap). Ultimately we will substitute this into the screening equation (36), and then use the resulting interacting susceptibility $\bar{\chi}$ in the ADF/FDT formula (20) for the correlation energy. If we ignore the dynamic xc kernel f_{xc} in the screening equation, this will give an approximation to the RPA correlation energy, which is known [10] to contain the R^{-6} vdW interaction between isolated systems (albeit with the true polarizability of each system replaced in the London formula by the RPA polarizability.) While this may be acceptable for some small systems, the RPA correlation energy is seriously deficient in systems such as the uniform electron gas with large size and closely spaced energy levels. Therefore for such systems it is not acceptable to ignore f_{xc} in the screening equation (36). In the following section we therefore discuss approximations for f_{xc}.

Inclusion of the Exchange-Correlation Kernel f_{xc}

In an inhomogeneous system $f_{xc\lambda}(\mathbf{r},\mathbf{r}',iu)$ depends in a complicated fashion on λ, $\mathbf{r}$, $\mathbf{r}'$ and u. Local-density approximations for f_{xc} are certainly not new. The adiabatic LDA [24,25] has

$$f_{xc\lambda}(\mathbf{r},\mathbf{r}',iu) \approx \delta^3(\mathbf{r}-\mathbf{r}')f_{xc\lambda}^{ALDA}(n) \equiv \delta^3(\mathbf{r}-\mathbf{r}')\frac{d^2(n\epsilon_{xc\lambda}^{\rm hom}(n))}{dn^2}.$$

f_{xc}^{ALDA} scales with λ in a simple fashion because $\epsilon_{xc\lambda}^{\rm hom}(n) = H(\lambda)e(na_B^3(\lambda))$ where $e(\rho)$ is a dimensionless function of one independent variable, $H(\lambda) = \lambda^2 me^4/\hbar^2$ is the Hartree energy unit for the case of reduced Coulomb interaction $\lambda e^2/r_{12}$, and $a_B(\lambda) = \hbar^2/(\lambda me^2)$ is the corresponding Bohr radius. It follows that $f_{xc\lambda}$ scales with λ as follows:

$$f_{xc\lambda}^{ALDA}(n) = \lambda^{-1}f_{xc}^{ALDA}(\lambda^{-3}n). \tag{49}$$

Better approximations for $f_{xc\lambda}(\mathbf{r},\mathbf{r}',iu)$ [17,26,27] are functions of frequency and sometimes of wavenumber as well [28] (i.e. they have nonzero spatial range). When f_{xc} is used to predict, e.g., plasmon dispersion, its detailed dependence on (real) frequency is of the essence. Here we have rather a different goal, namely the calculation of a whole-system correlation energy (Eq. (20)) which is a highly integrated quantity. Accordingly we propose, as a first approximation, a frequency-independent local kernel

$$f_{xc\lambda}(\mathbf{r},\mathbf{r}',iu) \approx \delta^3(\mathbf{r}-\mathbf{r}')f_{xc\lambda}^{av}(\bar{n}=n(\mathbf{r})). \tag{50}$$

Here $f^{av}_{xc\lambda}(\bar{n})$ is an "energy averaged" xc kernel defined so that, when substituted into the screening equation (36) and the ACF/FDT formula (20) for a *homogeneous* electron gas of density $\bar{n}$, it yields the numerically-exact homogeneous correlation energy $\epsilon_c(\bar{n})$ as given by Ceperley and Alder [29]. The λ dependence of f^{av}_{xc} will be assumed to be the same as that of $f^{ALDA}_{xc\lambda}(n)$ as given by Eq. (49). With this assumption, the requirement of reproducing the Ceperley-Alder xc energy at each density is sufficient to determine $f^{av}_{xc\lambda}(\bar{n})$ completely [10,30]. Note that the pair distribution resulting from such a constant f_{xc} via the screening equations and the fluctuation-dissipation theorem is weakly divergent as $\mathbf{r} \to \mathbf{r}'$, but this does not lead to a divergent energy in (20). If one wants to obtain the correct coulomb cusp at the origin of the pair distribution it will be necessary to use a nonlocal f_{xc} kernel. The coulomb cusp has very little effect on the xc energy, however.

The benefit of including $f^{av}_{xc\lambda}(\bar{n})$ in the screening equation is that our vdW functional now gives the numerically exact xc energy when applied to the uniform electron gas: with f_{xc} omitted, the vdW functional will reproduce only the RPA (ring-diagram) energy per particle of the homogeneous gas.

Relation to Methods Based on the Conductivity or Polarizability

In Eqs. (31)-(33) we defined three forms of density-density response function $\bar{\chi}_\lambda$, χ_{KS} and $\hat{\chi}_\lambda$ linking density perturbations to potentials. Similarly, if magnetic fields are absent or neglected, we can define three (in general nonlocal) conductivity quantities linking currents to electric fields:

$$j_\alpha(\mathbf{r},\omega) = \sum_{\beta=1}^{3} \int \bar{\sigma}_{\alpha\beta}(\mathbf{r},\mathbf{r}',\omega) E^{ext}_{1\beta}(\mathbf{r}',\omega) d^3r' \quad i.e. \quad \mathbf{j} = \bar{\sigma} * \mathbf{E}^{ext} \tag{51}$$

$$\mathbf{j} = \sigma_{KS} * \mathbf{E}^{TDKS} \tag{52}$$

$$\mathbf{j} = \hat{\sigma} * \mathbf{E}^{TDH} \tag{53}$$

Here $\mathbf{E}^{ext}(\omega)\exp(-i\omega t)$ is the electric field generated by sources external to the electronic system.

$$\mathbf{E}^{TDH} = \mathbf{E}^{ext} - \nabla \int \frac{-en_1(\mathbf{r}',\omega)}{|\mathbf{r}-\mathbf{r}'|} d^3r' \tag{54}$$

is the field including a Hartree-like contribution. Thus $\mathbf{E}^{TDH}$ is in fact the "total" field of classical electrodynamics, and thus it is $\hat{\sigma}$, not $\bar{\sigma}$ nor σ_{KS}, which is usually called "the" conductivity. $\mathbf{E}^{TDKS}$ is a field including an xc component, analogous to v_1^{TDKS} in (28).

Further, since the current $\mathbf{j}$ is the time derivative of polarization density $\mathbf{P}$ we can relate the three conductivities to corresponding nonlocal polarizability quantities:

$$\bar{\sigma} = -i\omega\bar{\mathbf{A}}, \ \sigma_{KS} = -i\omega\mathbf{A}_{KS}, \ \hat{\sigma} = -i\omega\hat{\mathbf{A}}$$

Consider cases where the fields are each derivable from a potential so that the density-density response is an adequate description. We then have from continuity and the assumption of vanishing boundary terms in integration by parts (Green's theorem)

$$\begin{aligned} n_1(\mathbf{r},\omega) &= \frac{-1}{i\omega e}\nabla\cdot\mathbf{j}_1(\mathbf{r},\omega) \\ &= \sum_{\alpha,\beta}\int [\frac{1}{i\omega e^2}\frac{\partial}{\partial r_\alpha}\frac{\partial}{\partial r'_\beta}\bar{\sigma}_{\alpha\beta}(\mathbf{r},\mathbf{r}',\omega)] V_1^{ext}(\mathbf{r}') \end{aligned} \tag{55}$$

with similar results for $\hat{\sigma}$ and σ_{KS}. It follows by comparing (55) and (31),(32), (33) that

$$\bar{\chi}(\mathbf{r},\mathbf{r}',\omega) = \frac{-1}{e^2}\sum_{\alpha,\beta}\frac{\partial}{\partial r_\alpha}\frac{\partial}{\partial r'_\beta}\bar{A}_{\alpha\beta}(\mathbf{r},\mathbf{r}',\omega) = \frac{-i}{\omega e^2}\sum_{\alpha,\beta}\frac{\partial}{\partial r_\alpha}\frac{\partial}{\partial r'_\beta}\bar{\sigma}_{\alpha\beta}(\mathbf{r},\mathbf{r}',\omega) \tag{56}$$

$$\chi_{KS}(\mathbf{r},\mathbf{r}',\omega) = \frac{-1}{e^2}\sum_{\alpha,\beta}\frac{\partial}{\partial r_\alpha}\frac{\partial}{\partial r'_\beta}A_{KS\alpha\beta}(\mathbf{r},\mathbf{r}',\omega) = \frac{-i}{\omega e^2}\sum_{\alpha,\beta}\frac{\partial}{\partial r_\alpha}\frac{\partial}{\partial r'_\beta}\sigma_{KS\alpha\beta}(\mathbf{r},\mathbf{r}',\omega) \tag{57}$$

$$\hat{\chi}(\mathbf{r},\mathbf{r}',\omega) = \frac{-1}{e^2}\sum_{\alpha,\beta}\frac{\partial}{\partial r_\alpha}\frac{\partial}{\partial r'_\beta}\hat{A}_{\alpha\beta}(\mathbf{r},\mathbf{r}',\omega) = \frac{-i}{\omega e^2}\sum_{\alpha,\beta}\frac{\partial}{\partial r_\alpha}\frac{\partial}{\partial r'_\beta}\hat{\sigma}_{\alpha\beta}(\mathbf{r},\mathbf{r}',\omega). \tag{58}$$

Note that the double-gradient structure, used above in order to ensure reciprocity and charge conservation in the construction of approximate χ quantities, occurs naturally in these equations. In this sense, approximation of conductivity or polarizability quantities is a more "forgiving" procedure than direct approximation for density-density quantities.

In particular, if we make a diagonal plasmon-pole approximation for $\bar{\sigma}$,

$$\bar{\sigma}_{\alpha\beta}(\mathbf{r},\mathbf{r}',\omega) \approx \delta_{\alpha\beta}\frac{e^2\delta^3(\mathbf{r}-\mathbf{r}')}{m[\omega^2-\omega_P^2(n(\mathbf{r}))]},$$

then we recover via (56) the simple Ansatz (11), which is known to lead to a useful vdW functional for small nonoverlapping systems, after substitution into the perturbative formula (3).

If instead we make a uniform-gas-based approximation for σ_{KS},

$$\sigma_{KS\alpha\beta}(\mathbf{r},\mathbf{r}',\omega) \approx \delta_{\alpha\beta}\sigma_{KS}^{\hom}(\bar{n}(\mathbf{r},\mathbf{r}'),|\mathbf{r}-\mathbf{r}'|,\omega),$$

where $\sigma_{KS}^{\hom}(\bar{n},R,\omega)$ is a diagonal approximation to the bare conductivity of the uniform gas, then we retrieve the Ansatz (45). This is suitable for insertion into the screening equation (36), followed by use of the Adiabatic Connection/Fluctation Dissipation formula (20). This yields the "seamless" vdW functional suitable for overlapping systems, as described already in the Sections above. A prescription for the treatment of f_{xc}, as in the previous Section, is required to complete the functional. Note that a local (but not doubly local) approximation to σ_{KS} is thus a good alternative way to ensure conservation laws in the ACF/FDT approach to vdW functionals.

It is also possible to make a direct local density approximation for $\hat{\sigma}$, the "conductivity" as usually defined in classical electrodynamics, a quantity which already contains information otherwise provided by f_{xc}. Substitution into the screening equation (38) and use of the ACF/FDT formula (20) yields a vdW functional. This alternative approach is essentially that proposed by Hanke and Kohn [19]. The principal difference from the approach proposed by the present authors is that f_{xc} is not separately considered in the $\hat{\sigma}$ approach, but the corresponding effects must be included in the uniform-gas data used to approximate $\hat{\sigma}$. Kohn and co-workers [19,31] have also explored an interesting scheme in which the bare Coulomb interaction is split into short- and long-ranged parts.

Relation to Methods Based on the Dielectric Function

An alternative way to write the screening equations (31)-(33) is to note that the total dynamic charge density is the sum of external and induced parts:

$$-en_1^{tot} = -en_1^{ext} - en_1^{ind}.$$

Using Gauss's law for n_1^{tot}, and continuity for n_1^{ind}, then expressing the current via Eq. (53) we obtain

$$\frac{1}{4\pi}\nabla.\mathbf{E}_1^{TDH} = \rho^{ext} + \frac{1}{i\omega}\nabla\cdot\sigma * \mathbf{E}_1^{TDH}. \tag{59}$$

Here " $*$ " represents spatial convolution. $\mathbf{E}^{TDH}$ is the "total" field of standard electrodynamics, defined in Eq. (54). (59) can be written

$$\nabla\cdot(\epsilon * \mathbf{E}^{TDH}) = 4\pi\rho^{ext} \tag{60}$$

where

$$\epsilon_{\alpha\beta}(\mathbf{r},\mathbf{r}',\omega) = \delta_{\alpha\beta}\delta^3(\mathbf{r}-\mathbf{r}') - \frac{4\pi}{i\omega}\sigma_{\alpha\beta}(\mathbf{r},\mathbf{r}',\omega) \equiv (\mathbf{I} - \frac{4\pi}{i\omega}\sigma)_{\alpha\beta} \tag{61}$$

is the dielectric function.

Hult *et al* (see Ref. [9] and the chapter by Andersson et al in the present Volume) introduced a doubly local approximation for ϵ :

$$\epsilon_{\alpha\beta}(\mathbf{r},\mathbf{r}',\omega) \approx \delta_{\alpha\beta}\delta(\mathbf{r}-\mathbf{r}')[1 - \frac{\omega_P^2(\mathbf{r})}{\omega^2}] \tag{62}$$

where $\omega_P^2(\mathbf{r}) = 4\pi e^2 n(\mathbf{r})/m$. They then used (60) to obtain an interacting response function of a metal surface for substitution into the Zaremba-Kohn formula (3). In this way they could predict, with surprisingly good accuracy, a number of properties of adsorbates on metal surfaces. From (61) it follows that the Ansatz (62) of Hult *et al* is equivalent to the assumption of a doubly local conductivity

$$\sigma_{\alpha\beta}(\mathbf{r},\mathbf{r}',\omega) \approx \delta_{\alpha\beta}\delta(\mathbf{r}-\mathbf{r}')\frac{e^2 n(\mathbf{r})}{i\omega m}. \tag{63}$$

A very similar but more sophisticated approximation for σ was proposed by Hanke and Kohn [32]. Furthermore, the Ansatz (62) is also equivalent to one for an intermediate polarizability $\hat{\mathbf{A}}$:

$$\hat{A}_{ij}(\mathbf{r},\mathbf{r}',\omega) = \delta_{ij}\delta(\mathbf{r}-\mathbf{r}')e^2 n(\mathbf{r})/m$$

and hence, via an equation analagous to (15), to an Ansatz for $\hat{\chi}$.

Thus if we used a non-delta function approximation for ϵ, but based on local density values, we could rederive the seamless vdw functional proposed in earlier sections of the present chapter. Work based on ϵ to date [9] has, however, used a doubly local approximation for ϵ and is therefore restricted to non-overlapping systems. It also ignored the contribution of f_{xc} to ϵ.

Perturbative Expansion of the Screening Equations Yields the Previous Results for Well-Separated Systems

The "seamless" functionals proposed above are all approximations to the ACF/FDT formula (20) which is more general than the perturbative result (3). By suitably expanding the screening equation (36), however, we can show that (20) reduces to (3) in the limit of two distant systems with no overlap. The only way to achieve strictly zero overlap is to have an infinite barrier potential separating the two systems a and b in (3). Formula (3) is problematical if there is any overlap at all. For nonoverlapping systems we can then separate space into two regions, R_a and R_b separated by the barrier. The

KS susceptibility $\chi_{KS}(\mathbf{r}, \mathbf{r}', \omega)$ is zero when $\mathbf{r} \in R_a$ and $\mathbf{r}' \in R_b$ or vice versa (this is the "no-flow condition" (41)), so

$$\chi_{KS} = \chi_{KSaa} + \chi_{KSbb} \tag{64}$$

where $\chi_{KSaa}(\mathbf{r}, \mathbf{r}', \omega)$ is zero unless both $\mathbf{r}$ and $\mathbf{r}'$ lie in R_a, and similarly for χ_{KSbb}. The coulomb interaction U can similarly be split into three parts,

$$U = U_{aa} + U_{ab} + U_{bb} \,. \tag{65}$$

The particles are NOT treated as distinguishable here, which is essential for the representation of a seamless vdW functional: rather, the split in (64) and (65) is based on the values of the space variables $\mathbf{r}$ and $\mathbf{r}'$. We then expand the screening equation (36) to second order in U_{ab}, and substitute the resulting $\bar{\chi}$ into (20). A subtle but important point here is that the λ integration in (20) must be done on all parts of U simultaneously: otherwise the procedure does not represent a "seamless" functional. This renders the proof quite nontrivial, but it is nevertheless shown in [10] that this process yields the perturbative formula (3), in which no λ integration remains. The proof in [10] omits f_{xc} and so yields the RPA version of the susceptibilites $\bar{\chi}_a$ and $\bar{\chi}_b$ in (3). A proof including f_{xc} can presumably be constructed, noting that any contribution to f_{xc} connecting systems a and b will be of second or higher order in U_{ab}. The proof will again be nontrivial.

Special difficulties for small systems

For small systems, two additional problems arise with the seamless uniform-gas-based vdW density functional proposed above, in common with most uniform-gas-based theories.

- We need to correct for spurious orbital self-interaction. This can be expressed by saying that, in a 1-electron system,

  $$f_{xc\lambda}(\mathbf{r}, \mathbf{r}', \omega) = -\lambda e^2 \left|\mathbf{r} - \mathbf{r}'\right|^{-1}.$$

  With reference to the screening equation (36), this means that $U_\lambda + f_{xc} = 0$, so that there is actually no electron-electron screening when $N = 1$: correspondingly $\bar{\chi}_\lambda \equiv \chi_{KS}$ so that the correlation energy $E_c = E_{xc} - E_x$ is strictly zero. For small values of $N > 1$, corrections of Fermi-Amaldi type may be possible here.

- It seems likely that χ_{KS} in small systems with finite energy-level spacing will be over-estimated by approximations based on the uniform electron gas, which has continuous energy levels. Similar physics can be invoked to explain the over-estimation of the correlation energy of small systems by the LDA. It remains to be seen how well this tendency is corrected when both E_c and E_x are calculated together in a seamless functional of the present type, as is the case when they are calculated together in the usual LDA. It is possible that a Penn-type model [23,33] may help here.

Test Cases for the Seamless vdW Density Functional

In view of the preceding points, the first systems on which we are testing the seamless functional are jellium slabs as described below. These systems are immune from difficulties involving orbital self-interaction, and furthermore the problem of finite level spacing is not severe for the wider slabs, allowing one to separate this from the effect of strong inhomogeneity (occurring at the jellium edge). The simplicity of jellium slab geometry also permits comparison of the new density functionals against "benchmark" calculations in which exact numerical calculation of $\chi_{KS}(z, z', q_{||}, iu)$ is performed. The systems which we are testing are as follows.

- A single slab of charge-neutral jellium, infinite in the x and y directions and having thickncss L in the z direction. Study of the energy as a function of L will allow a new nonlocal prediction of the jellium surface energy. In this system there is also a van der Waals energy proportional to L^{-2}: this can be understood as the contribution from the zero-point plasmon energy arising from the *coupling* between surface plasmons on the two surfaces. This contribution to the energy is absent in the LDA.

- Two slabs of jellium, each of thickness L, separated by a distance D. The energy will be followed as a function of D from the large-separation limit (with a D^{-2} vdW contribution in the case of thick slabs), through to complete overlap. The region $D \approx \lambda_{TF}$ of significant overlap, characterized by a separation on the order of the Thomas-Fermi wavelegth, is not reliably treated by any existing functional.

This work is currently underway.

SUMMARY

Local-density approximations for an *appropriately chosen* density-density susceptibility χ, conductivity σ, polarizability α or dielectric function ϵ can lead to vdW density functionals of the groundstate density $n(\mathbf{r})$ in two distinct ways.

1. For two systems without overlap of electron density, second-order perturbation theory leads to the expression (3) giving the vdW energy in terms of the interacting susceptibility $\bar{\chi}$ of each system separately. Direct local-density approximation of each $\bar{\chi}$, with great care regarding constraints (see (7), (8), (11)), leads to a simple, explicit, but highly nonlocal vdW expression (12) involving only the groundstate density $n(\mathbf{r})$. The same expression can be derived, with less sensitivity to constraints, by obtaining $\bar{\chi}$ via direct local density approximation for interacting polarizability $\bar{\alpha}$ (see Eq. (17) or conductivity $\bar{\sigma}$. Many different approximations can be made here because the second-order perturbation formula (3) has already introduced the essential physics of the vdW interaction. In particular this approach does not rely on long-ranged properties of $\bar{\chi}$. The drawback of the existing functionals in this class [8,7,6,9] is that they require an integration cutoff in the low-density tails. The optimal cutoff procedure appears to differ between atomic and surface geometries [9]. This is difficult to justify in detail from the arguments used to date [8,7] which attempt to model the whole system, including the vacuum gap, as some sort of average uniform gas. The approach in [6], however, offers hope of a more reliable cutoff procedure based directly on a study of the density-density response in the tunneling tails of electron distributions (see also Eq. (17)).

2. For the general case with (or without) overlap of electron density, the Adiabatic Connection/Fluctuation-Dissipation formula (20) relates the exchange-correlation

energy exactly to the interacting density-density response $\bar{\chi}(\lambda, \mathbf{r}, \mathbf{r}', \omega)$ of the combined system. However, to obtain the vdW interaction via (20) we must explicitly construct the system-specific long-ranged tail in $\bar{\chi}$. Thus direct local-density approximations for $\bar{\chi}$ fail to generate vdW functionals. Instead, we can use the groundstate density $n(\mathbf{r})$ to construct a local approximation for an inhomogeneous *independent*-electron susceptibility χ_{KS}, and then generate the appropriate long-ranged tail in $\bar{\chi}$ via the long-ranged coulomb interaction in the screening equation (36). (In order to satisfy conservation laws, it is useful to construct χ_{KS} indirectly as a double gradient of a locally-approximated bare polarizability, conductivity or dielectric quantity: there are a variety of essentially equivalent approaches). Substitution of $\bar{\chi}$ into (20) then yields the xc energy E_{xc}. Use of χ_{KS} in this fashion constitutes an approximation to the inhomogeneous RPA correlation energy, but can be taken beyond this level via a local approximation for the dynamic xc kernel f_{xc} in the screening equation (36), or via approximations for an intermediate quantity $\hat{\chi}$ which already contains the physics of f_{xc} (see Eq.(38)). By suitable approximation of f_{xc}, this approach is able [10] to generate the exact correlation energy in the limit of the homogeneous gas, while retaining the vdW properties in the relevant inhomogeneous limit and using the inhomogeneous groundstate electron density $n(\mathbf{r})$ as input. This is the "seamless" type of vdW functional one has been seeking, and one particular version of it is summarized by equations (47), (50), (45), (46), (36) and (20). Numerical tests are currently underway.

ACKNOWLEDGMENTS

This work was supported by a Large Research Grant from the Australian Research Council. It is a pleasure to acknowledge discussions with Walter Kohn, Werner Hanke, David Langreth, Andreas Savin and Bernard Silvi.

REFERENCES

1. E. Zaremba and W. Kohn, Phys. Rev. B **13**, 2270 (1976).
2. R. McWeeny, *Methods of Molecular Quantum Mechanics*, 2nd ed. (Academic Press, London, 1989).
3. G. Barton, Rep. Prog. Phys. **42**, 964 (1979).
4. J. Mahanty and B. W. Ninham, *Dispersion Forces* (Academic Press, London, 1976).
5. K. L. C. Hunt, J. Chem. Phys **80**, 393 (1984).
6. J. F. Dobson and B. P. Dinte, Phys. Rev. Lett. **76**, 1780 (1996).
7. Y. Andersson, D. C. Langreth, and B. I. Lundqvist, Phys. Rev. Lett. **76**, 102 (1996).
8. K. Rapcewicz and N. W. Ashcroft, Phys. Rev. B **44**, 4032 (1991).
9. E. Hult, Y. Andersson, B. I. Lundqvist, and D. C. Langreth, Phys. Rev. Lett. **77**, 2029 (1996).
10. J. F. Dobson, in *Topics in Condensed Matter Physics*, edited by M. P. Das (Nova, New York, 1994), p. 121.
11. J. F. Dobson, unpublished.
12. J. Harris and A. Griffin, Phys. Rev. B **11**, 3669 (1975).
13. D. C. Langreth and J. P. Perdew, Sol. State Commun. **17**, 1425 (1975).
14. O. Gunnarsson and B. I. Lundqvist, Phys. Rev. B **13**, 4274 (1976).
15. J. P. Perdew, Physica B **172**, 1 (1991).
16. E. Runge and E. K. U. Gross, Phys. Rev. Lett. **52**, 887 (1984).
17. E. K. U. Gross and W. Kohn, Phys. Rev. Lett. **55**, 2850 (1985).
18. E. K. U. Gross, J. F. Dobson, and M. Petersilka, in *Density Functional Theory II*, Vol. 181 of *Topics in Current Chemistry*, edited by R. F. Nalewajski (Springer, Berlin, 1996), Chap. 2, pp. 81 – 172.
19. W. Kohn and W. Hanke, private communication.

20. J. Lindhard, Kgl. Danske Vidensk. Mat-Fys. Meddr. **28**, (1954).
21. J. F. Dobson, NATO ASI Series B, Physics **337**, 393 (1995).
22. J. F. Dobson and G. H. Harris, J. Phys. C: Solid State Phys. **21**, 6127 (1987).
23. J. C. Inkson, *Many-body Theory of Solids* (Plenum, New York, 1984).
24. T. Ando, Z. Phys. B **26**, 263 (1977).
25. A. Zangwill and P. Soven, Phys. Rev. A **21**, 1561 (1980).
26. N. Iwamoto and E. K. U. Gross, Phys. Rev. B **35**, 3003 (1987).
27. G. Vignale and W. Kohn, Phys. Rev. Lett. **77**, 2037 (1996).
28. B. Dabrowski, Phys. Rev. B **34**, 4989 (1986).
29. D. M. Ceperley and B. J. Alder, Phys. Rev. Lett. **45**, 566 (1980).
30. J. F. Dobson and G. H. Harris, unpublished.
31. W. Kohn and Y. Meir, (Preprint, CECAM meeting September 1996).
32. W. Kohn and W. Hanke, unpublished.
33. J. F. Dobson and A. Savin, unpublished.

BOUND-FREE CORRELATION POTENTIALS FOR SCATTERING THEORY

Robert K. Nesbet

IBM Almaden Research Center
650 Harry Road
San Jose, CA 95120, USA

INTRODUCTION

Density functional theory [1,2] (DFT) models a system of N interacting electrons by an effective one-electron Hamiltonian with a local exchange-correlation potential function. There is no competitive method available for realistic calculations on condensed matter, large molecules, or molecules containing heavy atoms. For atoms or small molecules, DFT is much more efficient than alternative methods, and often gives results of useful accuracy [3]. Many applications of this theory to bound states have used the local-density approximation (LDA), based on the interacting homogeneous electron gas, which does not provide an adequate model of an electron outside the charge distribution responsible for the interaction energy. LDA is not expected to be a correct model for the nonlocal polarization response involved in bound-free correlation and the van der Waals interaction [3].

Although screening of the short-range Coulomb singularity is important to the success of the LDA for bound systems, it has recently been recognized that a model of longer-range electronic correlation should use a very different aspect of the physics of an electron gas. This motivates a local-density theory of polarization response, which has recently been applied with considerable success to van der Waals interaction coefficients for rare gas atoms and molecular hydrogen [4]. It has been shown [5] that fundamental principles of charge conservation and reciprocity, when applied to modify an earlier Ansatz for local-density response theory [6], are consistent with the practical formula tested by Andersson *et al* [4].

Correct treatment of the polarization response is essential for accurate calculations of low-energy electron scattering. Cross sections depend strongly on the polarization potential that takes the form $-\alpha/2r^4$ at large distance r from the scattering center, where α is the static dipole polarizability. The formal theory of such interactions is that of bound-free electronic correlation, described in general by a nonlocal, energy-dependent optical potential deduced from an N+1-electron thory of continuum states [7]. It will be shown here that the local-response theory [4,5] can be adapted to provide a generalized polarization potential or bound-free correlation potential. The local response theory corresponds to a classical electrostatic model of an object (atom or molecule) with uniform static polarizability, bounded by a closed response surface. This classical model poses some questions of internal consistency, which are resolved here by modifying a postulate of the local-response theory. It is shown that an internally consistent dielectric model of bound electrons requires reducing the mean excitation energy, which is $\hbar\omega_p$ in the jellium model with plasma frequency ω_p, by a factor $3^{-1/2}$, while reducing atomic response radii by $3^{-1/3}$. The argument leading to this modification will be presented here. This modification removes another apparent inconsistency of the plasma response model, which requires response radii for atoms to be outside classical turning radii. The plasma model is clearly not appropriate to a classically forbidden region of the potential.

NEW OUTLOOK ON CORRELATION POTENTIALS

Except at resonance energies, a scattered electron makes no contribution to the local density of the $N+1$-electron scattering system. This implies that exact exchange can be computed relatively easily for the scattered wave function, using a fixed target wave function. The static potential and asymptotic potentials valid at large separation are obtained from the fixed target wave function or from empirical values of multipole moments and polarizabilities. Standard DFT potentials should be valid for the scattered electron inside the target system. Hence application of DFT to electron scattering depends primarily on defining a model of the bound-free correlation potential valid in the region extending from outside the target system to the range in which asymptotic potentials are accurate.

A new justification for separating the exchange and correlation potentials has recently been provided by the reference-state density functional theory (RDFT) [8]. Here the Kohn-Sham postulate, requiring reference-state electron density to be identical with that of a true ground state correlated wave function, is replaced by the postulate that the reference state should have maximum projection on the true wave function. This has no effect in LDA, but gives a somewhat different interpretation to the various contributions to the energy functional. All terms other than the correlation energy are defined as mean values in the reference state [8]. Thus an exact exchange calculation in the electron-scattering problem is consistent with the general RDFT functional. A closed formula is given in the RDFT for the correlation energy functional, expressed as a nonlocal interaction between virtual excitations of the system. In appropriate geometry, the second-order perturbation approximation to this formula describes the van der Waals interaction. Hence the physical arguments and classical models discussed here for the closely-related polarization potential should also be relevant to a generalized treatment of the RDFT correlation energy functional and correlation potential.

LOCAL RESPONSE THEORY

The underlying idea of the local-response theory can be stated very simply as a two-parameter parametrization of the frequency-dependent oscillator strength of each electronic density volume element in an extended system. One of these parameters is fixed by the oscillator-strength sum rule, and it appears very natural to determine the other parameter from the density-dependent plasma frequency of the jellium model. The charge conservation condition, essentially $\int \nabla\cdot\delta\mathbf{P}d^3\mathbf{r} = 0$, follows from integrating a charge displacement over the given charge density if all charge is enclosed within an outer boundary. This property implies that for a finite system the lowest order solid-harmonic term in the response potential corresponds to an induced electric dipole field.

The local response theory assumes that dynamical properties of an electron distribution are determined by local values of the plasma frequency. The remaining element in a universal theory of polarization response is a low-density cutoff, required for a meaningful theory of the static polarizability. As discussed by Andersson *et al* [4], a correction is required because the effective electron wave number can exceed the value for which the collective plasma excitation of an electron gas merges into the background of particle-hole excitations. This correction was implemented [4] by setting the local polarizability to zero according to a criterion based on this property of the electron gas, discussed earlier by Langreth and Mehl [9]. A length scale is associated with the relative value of the density gradient. As an element in an empirically valid model of bound-free correlation, it is proposed here to fix this cutoff by requiring static atomic polarizabilities to have known empirical values. This leaves only the local mean excitation energy to be determined in a consistent physical model.

The local-response theory is expressed here in terms of a parametrized classical electrostatic model, determining local values of a frequency-dependent dielectric constant from the local electron density. The low-density cutoff defines a response surface surrounding any target atom or molecule, such that a classical dielectric model is valid within this surface, and polarization response vanishes outside. Cutoff boundaries associated with each surface atom are determined by requiring the model to give correct results for the asymptotic dipole polarization potential. In this model, the generalized polarization potential or bound-free correlation potential is computed at any point outside the response surface by solving a classical electrostatic problem.

A complex-valued function $\alpha(z)$ is defined for any polarizable entity such that the frequency-dependent electric dipole polarizability is given by $\alpha(\omega)$ at real frequency ω. This function is a Stieltjes integral over the oscillator-strength distribution function $df(\epsilon)$,

$$\alpha(z) = \int_0^\infty \frac{df(\epsilon)}{\epsilon^2 - z^2}. \tag{1}$$

Here ϵ is an excitation energy in Hartree units, giving α in units a_0^3, where a_0 is the Bohr radius. A principal representation of an oscillator strength distribution is defined so that $\alpha(z)$ is approximated by the finite sum

$$\alpha(z) \simeq \sum_{a=1}^{n} \frac{f_a}{\epsilon_a^2 - z^2}. \tag{2}$$

Physical properties related to polarization response [13] are easily expressed in terms of a principal representation in this form, and elegant computational procedures facilitate ab initio computation of the representation parameters [10,11]. Integrals over the oscillator strength distribution are replaced by a generalized Gaussian quadrature formula determined by requiring $2n$ values of the inverse energy moments

$$\mu_k = \int_0^\infty \epsilon^{-k} df(\epsilon) \tag{3}$$

to be given exactly by n pairs of abscissa and weight parameters (ϵ_a, f_a), such that

$$\mu_k = \sum_{a=1}^{n} f_a \epsilon_a^{-k}, \quad k = 0, \ldots, 2n-1. \tag{4}$$

The simplest possible mathematical model of a polarization response requires two parameters, to determine the first two inverse-energy moments μ_0 and μ_1. In the local response model, for each electron density volume element, the oscillator strength sum rule determines

$$\mu_0 = \int_0^\infty df(\epsilon) = n(\mathbf{r})d^3\mathbf{r}, \tag{5}$$

and the plasma excitation energy determines

$$\mu_1 = \int_0^\infty \epsilon^{-1} df(\epsilon) = (\hbar\omega_p)^{-1} n(\mathbf{r})d^3\mathbf{r}. \tag{6}$$

A principal representation of the oscillator strength for each volume element is defined by the two parameters $\epsilon_1 = \hbar\omega_p$ and $f_1 = n(\mathbf{r})d^3\mathbf{r}$, both determined by the local density. On expressing Eq. (2) in this representation in terms of the electronic charge and mass it becomes

$$\alpha(z) \simeq \frac{e^2}{m}\frac{f_1}{\omega_p^2 - z^2} = \frac{e^2}{m}\frac{n(\mathbf{r})d^3\mathbf{r}}{(4\pi e^2/m)n(\mathbf{r}) - z^2}. \tag{7}$$

The van der Waals interaction between two separated polarizable charge distributions is characterized by the coefficient [12]

$$C_6(A, B) = \frac{3}{\pi}\int_0^\infty \alpha_A(iu)\alpha_B(iu)du, \tag{8}$$

defined by an integral over imaginary frequency. If the polarization response of each charge distribution is approximated by a principal representation defined by parameters (ϵ_a, f_a), Eq. (8) takes the form [13], in Hartree atomic units,

$$C_6(A, B) = \frac{3}{\pi}\sum_a\sum_b \frac{f_{aA}f_{bB}}{\epsilon_{aA}\epsilon_{bB}}\int_0^\infty \frac{\epsilon_{aA}\epsilon_{bB}du}{(\epsilon_{aA}^2 + u^2)(\epsilon_{bB}^2 + u^2)} = \frac{3}{2}\sum_a\sum_b \frac{f_{aA}f_{bB}}{\epsilon_{aA}\epsilon_{bB}(\epsilon_{aA} + \epsilon_{bB})}, \tag{9}$$

using the definite integral [14]

$$\int_0^\infty \frac{abdu}{(a^2 + u^2)(b^2 + u^2)} = \frac{\pi}{2}\frac{1}{a + b}. \tag{10}$$

Substituting Eq. (7) for each volume element, the local response theory gives

$$C_6(A, B) = \frac{3}{2}\frac{e}{m^{1/2}}\frac{1}{(4\pi)^{3/2}}\int d^3\mathbf{r}_A \int d^3\mathbf{r}_B \frac{n_A^{1/2}(\mathbf{r}_A)n_B^{1/2}(\mathbf{r}_B)}{n_A^{1/2}(\mathbf{r}_A) + n_B^{1/2}(\mathbf{r}_B)}. \tag{11}$$

This is the formula proposed and tested by Andersson et al [4], and derived from an Ansatz with correct symmetry and charge conservation properties by Dobson and Dinte [5].

POLARIZATION POTENTIAL

The electrostatic energy of a polarizable charge distribution and an infinitesimal test charge δQ at a fixed external point $\mathbf{r}$ varies with the magnitude of the test charge.

The first-order energy $\partial E/\partial Q$ is the static potential acting on δQ. The second-order energy defines the polarization potential

$$v_{pol}(\mathbf{r}) = \frac{1}{2}\partial^2 E/\partial Q^2, \tag{12}$$

acting at point $\mathbf{r}$. The asymptotic form is easily shown to be $-\alpha(0)/2r^4$ in classical theory. In quantum theory, if $\mathbf{r}$ is outside a sphere that encloses an atom or molecule with internal coordinate $\mathbf{r}'$, the perturbing Coulomb potential $\delta Q/|\mathbf{r}-\mathbf{r}'|$ can be expanded as $\delta Q \sum_\ell (r'^\ell/r^{\ell+1}) P_\ell(\cos\theta)$. The leading term in transition matrix elements for a neutral polarizable charge distribution is the dipole term, with $\ell = 1$ and proportional to r^{-2}. The dominant energy term for large r is $-\alpha(0)/2r^4$, in agreement with Eq. (12). Using the static polarizability, the polarization potential is a bound-free correlation potential for elastic electron scattering. For inelastic scattering with energy transfer $\hbar\omega$, the frequency-dependent polarizability $\alpha(\omega)$ replaces $\alpha(0)$.

The bound-free correlation potential for an electron far outside the response surface of a target system is computed in the local response theory by integrating Eq. (7) inside the response surface determined by cutoff parameters. This procedure must be modified for shorter-range interactions, because it cannot be assumed that the internal electric field is uniform. A classical model will be developed here in which the nonuniformity of the local electric field is taken into account. This provides a theoretical framework for computation of shorter-range interactions. Because the frequency-dependent polarizability given by Eq. (7) varies with local density, this argument indicates that the local electric field and polarization should be computed even in the long-range limit except for static interactions. This was not considered by Andersson *et al* [4] in their calculations of van der Waals coefficients. Several conceptual points required to establish the classical model will be discussed here.

Jackson [15] derives the formula

$$w = -\frac{1}{8\pi}(\epsilon - \epsilon_0)\mathbf{E}\cdot\mathbf{E}_0 = -\frac{1}{2}\mathbf{P}\cdot\mathbf{E}_0 \tag{13}$$

for the energy density of a dielectric placed in an electric field $\mathbf{E}_0$ (electrostatic potential ϕ_0) with fixed sources. Here $\mathbf{P}$ is the classical polarization, and $\mathbf{E}$ is the electric field vector taking polarization into account. The local dielectric constant ϵ reduces to $\epsilon_0 = 1$ in the absence of polarization. When integrated by parts over the volume τ enclosed by a response surface σ, Eq. (13) gives

$$\Delta E = -\frac{1}{2}\int_\tau \phi_0 \nabla\cdot\mathbf{P}d\tau + \frac{1}{2}\int_\sigma \phi_0\mathbf{P}\cdot\hat{\mathbf{n}}d\sigma. \tag{14}$$

If the fixed field is that of an infinitesimal test charge which induces polarization $\mathbf{P}$ in a given polarizable charge distribution, this formula defines the second-order response energy. The ratio of this energy to the squared magnitude of the test charge is $\frac{1}{2}\partial^2 E/\partial Q^2$. This ratio can be interpreted as an effective local potential acting on the test charge, defining a generalized polarization potential. Eq. (14) is half the Coulomb energy of the test charge interacting with the induced polarization charge. Because polarization preserves total charge, the net induced charge must integrate to zero for any geometry.

In classical potential theory [15], the electric field is modified inside a polarizable medium by an amount that depends on the dielectric constant. If the local response theory is taken literally as a model of a classical dielectric, the local dielectric constant is determined by the local polarizability through the Clausius-Mossotti equation,

$$\alpha = \frac{3}{4\pi}\frac{\epsilon - 1}{\epsilon + 2}. \tag{15}$$

Table 1. Atomic data

Atom	$\alpha(a_0^3)^a$	IP(eV)	$r_{mod}(a_0)^b$	$r_\alpha(a_0)$	$r_{cl}(a_0)$	$3^{-1/3}r_\alpha(a_0)$
H	4.500	13.60	2.440	2.381	2.348	1.650
He	1.380	24.59	1.838	1.606	1.391	1.114
Ne	2.667	21.56	2.040	2.000	1.642	1.387
Ar	11.09	15.76	3.490	3.216	2.410	2.229
Kr	16.78	14.00	4.058	3.692	2.738	2.559
Xe	27.07	12.13		4.331	3.184	3.003

[a]Bell and Kingston [16]
[b]Andersson *et al* [4]

Use of this equation in the present context must be justified. The classical theory assumes its validity on a scale in which individual molecules are small [15]. Here the scale relationships are reversed, assuming uniformity in a volume element within an atom. However, on this microscale, an infinitesimal spherical volume element can be considered in which the true electric field vector $\mathbf{E}$ includes a contribution $\mathcal{E} = -(4\pi/3)\mathbf{P}$ from the charge displacement due to local polarization $\mathbf{P}$. Since the local driving field for polarization must exclude this term, the defining relationship for local polarizability α is

$$\mathbf{P} = \alpha(\mathbf{E} - \mathcal{E}) = \alpha(\mathbf{E} + \frac{4\pi}{3}\mathbf{P}). \tag{16}$$

The Clausius-Mossotti formula follows when this equation is expressed in terms of the dielectric constant [15].

Due to exact cancellation of numerator and denominator in Eq. (7), the local polarizability $\alpha(iu)$ is $1/4\pi$ in the static limit or the limit of high local density, decreasing to zero for low density if $u > 0$. From Eq. (15), the local dielectric constant ϵ decreases from 5/2 to 1. In the static limit, $(\epsilon - 1)/(\epsilon + 2) = 1/3$, and the uniform polarization of a response sphere of radius a at the coordinate origin induced by a test charge δQ at distance $r \gg a$ is $-(\delta Q/4\pi)\mathbf{r}/r^3$. If the test charge is on the positive $z-$axis the induced surface charge at points with polar angle θ is$-(\delta Q/4\pi r^2)\cos\theta$. Half the Coulomb interaction energy of this dipolar surface charge with the external point charge is $-a^3\delta Q^2/6r^4$. This gives the standard dipole polarization potential after substituting $a^3 = 3\alpha(0)$.

This last formula defines a response radius for hydrogen or any spherical atom. Table 1 lists response radii r_{mod} corresponding to $\alpha(0)$ computed [4] using the length scale parameter ξ of Langreth and Mehl [9], compared with r_α determined by empirical values of $\alpha(0)$ [16]. The formula for the length scale contains a parameter given nominally as $f = 1/6$ which was fitted [9] to an empirical value $f = 0.15$ from atomic DFT calculations. The data in Table 1 support the idea that this parameter could be determined from empirical values of $\alpha(0)$.

CLASSICAL PLASMA MODEL

An essential assumption of the model of local response is that polarization response can be parametrized by plasma resonance excitation of "jellium", described as an electron gas embedded in uniform positive charge. In a semiclassical model, an infinitesimal displacement $\delta\mathbf{x}$ of a local charge density $-en(\mathbf{r})$ produces polarization $\delta\mathbf{P} = -en\delta\mathbf{x}$. The induced electric field $\delta\mathcal{E} = -4\pi\delta\mathbf{P}$ provides a restoring force. The classical equation of motion is

$$m\delta\ddot{\mathbf{x}} = -e\delta\mathcal{E} = -4\pi ne^2\delta\mathbf{x}. \tag{17}$$

This is an oscillator equation with frequency parameter given by

$$\omega_p^2 = \frac{4\pi n e^2}{m}, \tag{18}$$

where ω_p is the classical plasma frequency. The semiclassical excitation energy is $\hbar\omega_p$.

The classical derivation of the plasma frequency given in Eq. (18) uses a relationshop between a displacement of the electron gas and the restoring force, equivalent to an induced electric field $\delta\mathcal{E} = -4\pi\delta\mathbf{P}$. The resulting formula for the plasma frequency is well-established and plays an essential role in dynamical screening of the Coulomb interaction. However, this is inconsistent with $\delta\mathcal{E} = -(4\pi/3)\delta\mathbf{P}$, appropriate to the physical model of a dielectric, used in deriving the Clausius-Mossotti formula. This inconsistency can be removed either by changing the Clausius-Mossotti formula or by changing the local mean excitation energy parameter, so that excitation and polarization are described by the same physical model. If the plasma excitation model is selected, the plasma excitation energy is retained, but Eq. (16) takes the modified form

$$\mathbf{P} = \alpha(\mathbf{E} - \mathcal{E}) = \alpha(\mathbf{E} + 4\pi\mathbf{P}). \tag{19}$$

This replaces the Clausius-Mossotti formula by

$$\alpha = \frac{1}{4\pi}\frac{\epsilon - 1}{\epsilon}. \tag{20}$$

Eq. (7) is unaltered, so that the static polarizability per unit volume is $1/4\pi$ inside the response surface. But now Eq. (20) implies that $\epsilon = \infty$.

This actually simplifies the classical potential problem. Since there is no true electric charge inside the response surface, the classical displacement vector $\mathbf{D} = \epsilon\mathbf{E}$ is finite throughout. Infinite ϵ requires $\mathbf{E}$ to vanish and the potential function to reduce to a constant in the interior. The method of images gives a solution in closed form for a test charge δQ at an external point $\mathbf{r}$ interacting with a sphere of radius a and infinite dielectric constant centered at the coordinate origin. Dividing the potential function by $2\delta Q$, the classical polarization potential at $\mathbf{r}$ is

$$v_{pol}(\mathbf{r}) = -\frac{1}{2}\frac{a^3}{r^2(r^2 - a^2)}. \tag{21}$$

Since $\alpha(0) = a^3/3$ in this model, this result disagrees with the standard asymptotic formula derived from quantum mechanics.

Eq. (21) shows that the classical model itself has a singularity when the test charge approaches the response boundary. This indicates that polarization response must be modified near the cutoff boundary in a valid model of close-in interactions.

The plasma model has a more subtle implication that also appears to be incorrect. For an atom with ionization potential IP, the outer turning radius r_{cl} of the classically allowed region for a bound s-electron is a solution of the equation $1/r + \alpha(0)/2r^4 = IP$. Values obtained from this equation are included in Table 1. The cutoff radii r_{mod} or r_α given in Table 1 by the plasma model for atoms are systematically greater than r_{cl}. The underlying jellium model assumes electronic states with positive kinetic energy, in analogy to a classical charge distribution. This model should not be valid in a classically forbidden region, requiring negative kinetic energy.

CLASSICAL DIELECTRIC MODEL

An alternative internally consistent classical model is obtained by modeling the restoring force for local polarization displacement in terms of the induced electric field

$\delta\mathcal{E} = -(4\pi/3)\delta\mathbf{P}$ assumed in deriving the Clausius-Mossotti formula. This reduces the restoring force and the square of the classical excitation frequency by 1/3, compared with the plasma response model, while Eq. (16) is unchanged. Denoting this dielectric response frequency by ω_d, Eq. (7) is changed to

$$\alpha(z) \simeq \frac{e^2}{m} \frac{f_1}{\omega_d^2 - z^2} = \frac{e^2}{m} \frac{n(\mathbf{r})d^3\mathbf{r}}{(4\pi e^2/3m)n(\mathbf{r}) - z^2}. \tag{22}$$

In the static limit, the polarizability per unit volume takes the constant value $3/4\pi$ inside the response surface. If a is the radius of this surface, spherical for a closed-shell atom, then the integrated static polarizability is $\alpha(0) = a^3$, which can be used to determine a from empirical static polarizabilities.

From the Clausius-Mossotti formula, consistent with this model, uniform $\alpha = 3/4\pi$ implies $\epsilon = \infty$, as in the internally consistent version of the plasma response model. The closed-form expression Eq. (21) remains valid for the point and sphere model. However, because $\alpha(0) = a^3$ with this choice of the physical model, it now gives the correct result. The atomic response radius r_α determined by the empirical static polarizability is reduced by a factor $3^{-1/3}$ compared with the plasma response model. As indicated in Table 1, this change resolves the problem of using a response model in a classically forbidden region. Since the collective excitation energy is reduced by $3^{-1/2}$, it merges with the particle-hole continuum at a correspondingly smaller value of momentum transfer q, indicating that this rescaling also affects the intrinsic length scale [9].

REFERENCES

1. P. Hohenberg and W. Kohn, Phys. Rev. **136**, B864 (1964).
2. W. Kohn and L. J. Sham, Phys. Rev. **140**, A1133 (1965).
3. R. O. Jones and O. Gunnarsson, Rev. Mod. Phys. **61**, 689 (1989).
4. Y. Andersson, D. C. Langreth and B. I. Lundqvist, Phys. Rev. Lett. **76**, 102 (1996).
5. J. F. Dobson and B. P. Dinte, Phys. Rev. Lett. **76**, 1780 (1996).
6. K. Rapcewicz and N. W. Ashcroft, Phys. Rev. B **44**, 4032 (1991).
7. R. K. Nesbet, *Variational Methods in Electron-Atom Scattering Theory* (Plenum, New York, 1980).
8. R. K. Nesbet, J. Phys. B **29**, L173 (1996).
9. D. C. Langreth and M. J. Mehl, Phys. Rev. Lett. **47**, 446 (1981).
10. P. W. Langhoff, Chem. Phys. Lett. **22**, 60 (1973).
11. R. K. Nesbet, Phys. Rev. A **14**, 1065 (1976).
12. C. Mavroyannis and M. J. Stephen, Mol. Phys. **5**, 629 (1962).
13. A. Dalgarno and W. D. Davison, Adv. At. Mol. Phys. **2**, 1 (1966).
14. I. S. Gradshteyn and I. M. Ryzhik, *Table of Integrals, Series, and Products*, edited by A. Jeffrey (Academic Press, New York, 1980). Eq. 3.264(2), p300.
15. J. D. Jackson, *Classical Electrodynamics*, 2nd edition (Wiley, New York, 1975). Chapter 4, pp136-167.
16. R. J. Bell and A. E. Kingston, Proc. Phys. Soc.(London) **88**, 901 (1966).

Part IV

INVITED CHAPTERS ON SPECIAL SYSTEMS

EDGE ELECTRONIC STRUCTURE: THE AIRY GAS

W. Kohn

Department of Physics, University of California
Santa Barbara, California 93106, U.S.A.

INTRODUCTION

Both Thomas Fermi theory and the local density approximation of density functional theory take as their starting point the properties of a uniform electron gas. This is qualitatively inappropriate in "edge" regions where the density changes from a bulk-like character to evanescence. I shall report here on work in progress by F. Picus and myself in which we propose to amend both TF theory and the local density approximation of DFT in edge regions. We approximate the effective potential near the edges by linear potentials. In 1D, $v_{eff}(x) \approx \mu - F(x - x_o)$ where μ is the chemical potential x_o is the edge point, and $F = -v'(x_o)$. This leads to Schrödinger equations of the Airy form, which combined with single occupation up to μ, leads to the concept of the Airy gas.

NATURE OF EDGE REGIONS

"Edge" regions are those where the electronic density drops from large interior values towards zero or near zero. Traditionally the emphasis has been on the *interior* regions, where, in first approximations, one regards the electrons as a locally uniform electron gas. Examples are Thomas Fermi (TF) theory [1,2] or the local density approximation (LDA) of density functional theory (DFT). This point of view is appropriate when in effective single particle descriptions (such as in Hartree theory or Kohn-Sham (KS) theory [3]) the local kinetic energy, $\epsilon - v_{eff}(r)$, of the highest occupied orbital is positive, as it is in a uniform gas. In edge regions, however, the local kinetic energy changes from positive (running waves) to negative (evanescent waves) and thus the traditional viewpoint of the locally uniform electron gas is qualitatively inappropriate.

To be specific we define the electronic edge by the equation

$$v_{eff}(r) = \mu \tag{1}$$

where $v_{eff}(r)$ is the exact KS potential, which reproduces the exact physical density $n(r)$, and μ is the chemical potential, half way between the highest occupied and lowest unoccupied levels. For the purpose of this talk I limit myself to a 1D system. Without loss of generality, I can take $\mu = 0$ and the position of an edge-point as $x_o = 0$. Then, near the edge point we can write

$$\mu - v_{eff}(x) = -Fx + 0(x^2) \qquad F > 0. \tag{2}$$

THE AIRY GAS

This leads to a consideration of what I call the Airy gas: Non-interacting electrons, obeying the Schroedinger equation

$$\left(-\frac{\hbar^2}{2m}\frac{d^2}{dx^2} - v(x)\right)\varphi_n(x) = \epsilon_n \varphi_n(x)\,, \tag{3}$$

where

$$\begin{aligned} v(x) &\equiv -Fx \qquad && -\infty < x < L\,, \\ &\equiv +\infty && x \geq L\,, \end{aligned} \tag{4}$$

and the boundary conditions

$$\varphi(-\infty) = \varphi(L) = 0\,. \tag{5}$$

$\varphi_n(x)$ and ϵ_n are, respectively, the normalized eigenfunctions and eigenvalues; and, for purposes of orthonormality, there is a (non-physical) hard wall at L, which eventually becomes infinite. All states with $\epsilon_n \leq 0$ are singly occupied, say by spin-up electrons. The Schroedinger equation (3) can, by scaling and a coordinate shift, be converted to the Airy equation

$$(-\frac{d^2}{dx^2} + x)Ai(x) = 0 \tag{6}$$

which is why we call this gas the Airy gas.

The Airy gas has one characteristic length scale, and one energy scale,

$$\ell \equiv (2m/\hbar^2 F)^{1/3}\,; \qquad \epsilon = (\hbar^2 F^2/2m)^{1/3}. \tag{7}$$

ℓ measures the thickness of the edge in the sense that for

$$x << \ell \qquad n(x) << n(0)\,, \tag{8}$$

and for

$$x >> \ell \qquad n(x) \approx n_{TF}(x) \tag{9}$$

where $n_{TF}(x)$ is the bulk density as given approximately by TF theory. (Note that, for $x < 0$, $n_{TF}(x) \equiv 0$.) ϵ is of the order of the energy/electron in the edge region [4]. The extension of these ideas to edges of 2D or 3D systems has been carried out. Note also that previous work [4] has considered linear potentials, but these publications emphasize bulk regions where the linear potential gives a correction to TF theory. In the present paper, the interest is in the edge region.

We propose to amend both traditional TF theory and DFT in the edge region by taking as our starting point the Airy gas rather than the uniform electron gas. Details will be published elsewhere.

This work was supported in part by NSF DMR 9630452.

REFERENCES

1. L. H. Thomas, Proc. Camb. Phil. Soc. **23**, 542 (1927).
2. E. Fermi, Z. Physik **42**, 73 (1928).
3. W. Kohn and L. J. Sham, Phys. Rev. **140** A1133 (1965).
4. R. Baltin, Z. Neturforsch. **27a**, 1176 (1972); R. Baltin and N.M. Marsh, Phys. Rev. A **37**, 3947 (1988).

DENSITY FUNCTIONALS FOR ENERGIES AND EIGENVALUES: LOCAL MASS APPROXIMATION

G. E. Engel[1] and Warren E. Pickett[2]

[1]George Mason University, Computational Science Institute
Fairfax, VA 22030
[2]Naval Research Laboratory, Washington, DC 20375

INTRODUCTION

The considerable success of density functional theory (DFT) in describing the properties of molecules and solids [1] has been built on the Kohn-Sham [2] (KS) prescription for separating the energy functional into separate parts. The KS approach leads to a one-body problem in which electrons with the free electron mass move in an effective potential (which is determined self-consistently). The result is a set of dispersion relations ϵ_i widely referred to as the density functional spectrum, and their associated single-particle wavefunctions ψ_i. These are obtained in the process of minimizing the energy functional with respect to the density, and the density ρ and the energy $E[\rho]$ are the fundamental properties of the many-body system that are guaranteed by the Hohenberg-Kohn theorem [3].

There has been continuing interest over the years in improving the density and the energy that result from the various approximations to the energy functional (see [4,5,6] and elsewhere in this volume). The band gap in insulators, and in correlated electron systems the spectrum in general, is also unsatisfactory [1,7,8], but because there is very little of a fundamental nature in DFT about eigenvalues, improving the spectrum has seemed to be less expedient. It is not widely recognized that it is also possible to seek improved band structures and wavefunctions directly, without simply accepting the outcome of a particular improvement in the energy functional.

The band gap in semiconductors arising from the KS procedure is commonly referred to as the "(true) density functional band gap." In the following section we emphasize that the conventional approach gives what must be recognized as only the *Kohn-Sham* band structure and gap, and that DFT itself allows great flexibility in manipulating the eigenvalue spectrum. We obtain and implement a new method [9], a *modified Kohn-Sham* procedure, that leads to an improved band structure for Si and Ge, without degradation of the energetic properties. Although our derivation ultimately follows a route analogous to the KS self-consistency loop, the discussion we provide serves to illustrate the flexibility that exists within the general DFT approach, and particularly in the separation of terms in the functional.

Electronic Density Functional Theory: *Recent Progress and New Directions*
Edited by Dobson *et al.*, Plenum Press, New York, 1998

To understand our new approach, recall that the problem of the semiconducting gap in KS theory is connected to the discontinuity with respect to particle number of both the KS kinetic energy functional $T_s[n]$ and the exchange-correlation functional $E_{xc}[n]$ [7,8], whose definition (see below) is tied to T_s. The KS separation of kinetic and other contributions leads to one-electron equations that are mathematically simple. It is possible, however, that a separation of the energy functional in a different way may alleviate the intrinsic problem in the KS approach that arises because the discontinuity is accounted for (exactly) in T_s but not at all in the common approximations to E_{xc}. We choose a different separation that is physically motivated in the homogeneous limit, and extend it to inhomogeneous systems, with promising results. A related generalization of the KS prescription has been presented recently by Seidl *et al* [10]

The method, its implementation, and selected results have been presented elsewhere [9], to be referred to as EP. It was also shown that the linear response around the homogeneous limit remains reasonable compared to LDA, and the relation of the new functional to self-energies was investigated. In this paper we concentrate on the motivation, and present the method in terms of a different and somewhat more physically based notation that may lead to better understanding, or perhaps will stimulate extensions of the method. An overview of the results will be presented to give a realistic indication of the success of the approach.

GENERALIZING THE KOHN-SHAM METHOD

We consider the usual time-independent problem of N electrons interacting via the Coulomb interaction $v(\mathbf{r}-\mathbf{r}')$ in an external potential V_{ext}. Denoting the ground state kinetic energy by $T[\rho]$ and the potential energy by $U[\rho]$, the energy functional can be written, trivially, as

$$E[\rho] = F[\rho] + \int d\mathbf{r}\rho(\mathbf{r})V_{\text{ext}}(\mathbf{r}) \tag{1}$$

$$= \Big(T[\rho] + U[\rho] \Big) + E_{\text{ext}}[\rho]. \tag{2}$$

This relation in terms of exactly separated kinetic and potential energies is useful for formal manipulations (*e.g.* establishing relations using coupling constant integrations) but is not useful in obtaining good approximations. The KS procedure [2] consists of introducing the density ρ expressed in terms of a set of occupied one-particle orbitals ψ_i as

$$\rho(\mathbf{r}) = \sum_i^{\text{occ}} |\psi_i(\mathbf{r})|^2. \tag{3}$$

The unknown functional $F[\rho]$ is then partitioned in a different way,

$$F[\rho] \equiv T_s[\rho] + E_{\text{H}}[\rho] + E_{\text{xc}}[\rho], \tag{4}$$

where the KS kinetic energy T_s in atomic units $e = \hbar = 1$ (we retain the mass in the equations, for reasons to become obvious) is expressed as

$$T_s[\rho] \equiv \sum_i^{\text{occ}} \int \frac{1}{2m}|\nabla\psi_i(\mathbf{r})|^2 \equiv \int \tau_s(\mathbf{r};\rho), \tag{5}$$

where τ_s is a local kinetic energy density of the inhomogeneous system (*not* the same at the kinetic energy density per particle of the homogeneous electron gas (HEG), t_s, to be introduced below). The Hartree energy is given by

$$E_{\mathrm{H}}[\rho] \equiv \frac{1}{2} \int \int d\mathbf{r} d\mathbf{r}' \frac{\rho(\mathbf{r})\rho(\mathbf{r}')}{|\mathbf{r} - \mathbf{r}'|}. \tag{6}$$

The local density approximation (LDA) consists of approximating the unknown KS exchange-correlation functional $E_{\mathrm{xc}}[\rho]$ in terms of the exchange-correlation energy per particle $\epsilon_{\mathrm{xc}}(\rho)$ of a HEG of density ρ,

$$E_{\mathrm{xc}}[\rho] \approx \int d\mathbf{r} \rho(\mathbf{r}) \epsilon_{\mathrm{xc}} (\rho(\mathbf{r})) . \tag{7}$$

Modification for the Homogeneous System

In this KS procedure, the exchange-correlation functional E_{xc} is simply the remaining energy after T_s, E_{H}, and E_{ext} have been separated out. In particular, E_{xc} contains contributions from both kinetic and potential energies. The standard Kohn-Sham partitioning of the density functional, whatever the precise form of the exchange-correlation functional $E_{\mathrm{xc}}[\rho]$, always results in a constant exchange-correlation potential V_{xc}, which is the total potential. It follows that the spectrum is always that of a free electron gas, and the valence band width is given by the Fermi-energy $\epsilon_{\mathrm{F}} = \frac{1}{2m} k_{\mathrm{F}}^2$. A change in the valence band width can never be reproduced within such a scheme. Hartree-Fock theory, on the other hand, predicts a widening of the valence band by a factor of $1.5 - 2$ for typical metallic densities, contrary to experiment, indicating that correlation must be included together with exchange to obtain the proper bandwidth.

This realization, and the fact that both T_s and E_{xc} have related discontinuities with respect to particle number, suggest that some other repartition of their sum might be used to advantage, perhaps to reduce the band gap error and to produce a better eigenvalue spectrum in general. Heuristically, one might consider scaling the kinetic energy functional by an effective mass m^* instead of the bare electron mass m, and then redefining the exchange-correlation energy functional,

$$F[\rho] = T_s^*[\rho] + E_{\mathrm{xc}}^*[\rho] = \sum_i^{\mathrm{occ}} \int \frac{1}{2m^*} |\psi_i(\mathbf{r} : \rho)|^2 d\mathbf{r} + E_{\mathrm{xc}}^*[\rho], \tag{8}$$

and then trying to approximate the new exchange-correlation functional E_{xc}^*.

For the HEG, the orbitals are known (plane waves) regardless of the mass, and the kinetic energy T_s^* is just $\frac{m}{m^*}$ times the corresponding KS single-particle kinetic energy T_s, whose density dependence is known. This difference can be subtracted from the KS term E_{xc}, giving

$$E_{xc}^*[\rho] = E_{xc}[\rho] - \left(\frac{m}{m^*(\rho)} - 1\right) T_s[\rho] \tag{9}$$

where T_s is just $\frac{3}{5} k_F^2 / 2m$ per particle.

To determine m^*, we note that *GW* calculations for the HEG reported by Mahan and Sernelius [11] lead to a reduction in bandwidth $\Delta\epsilon$ compared to the free-electron value. Apparently the experimentally observed [12] reduction in valence bandwidths of the free-electron like metals is largely due to many body correlation effects, which are treated realistically in the *GW* approximation. We choose therefore to fix $m^*(\rho)$ to match the *GW* results, which we find are fit well by the analytic form

$$\frac{\Delta\epsilon}{\epsilon_{\mathrm{F}}} = A\, e^{-br_s} - 1 = \frac{m}{m^*(\rho)} - 1, \tag{10}$$

with A=1.0917, b=0.0527. For practical purposes this fit is equivalent to the linear fit by EP, [9] but this form is more reasonable in the large r_s limit, where $1/m^*$ approaches zero rather than going negative. With the new functional of Eq. (8), the energy of the HEG is still given exactly, but in addition the bandwidth narrowing given by the *GW* calculations is also reproduced. Thus the eigenvalue spectrum has been improved with no degradation of the energies that result.

Inhomogeneous Systems

We now consider the new functional for an arbitrary system to be given by a generalization of Eq. (8), with appropriate extension to apply to inhomogeneous densities. We write the modified kinetic energy functional as

$$T_s^*[\rho] = \sum_i^{\text{occ}} \int \psi^*(\mathbf{r})\ \widehat{\mathcal{T}}[\rho]\ \psi(\mathbf{r})d\mathbf{r} \tag{11}$$

where the kinetic operator is generalized to the Hermitian form with a density-dependent effective mass

$$\widehat{\mathcal{T}}[\rho] = -\nabla \cdot \frac{1}{2m^*\Big(\rho(\mathbf{r})\Big)}\nabla. \tag{12}$$

We write the basic functional in the form

$$\begin{aligned} F[\rho] &= T_s^*[\rho] + E_{\text{H}}[\rho] + E_{\text{xc}}^*[\rho] & (13)\\ &= T_s^*[\rho] + E_{\text{H}}[\rho] + \Big(E_{\text{xc}}[\rho] + E_{\text{l}}[\rho]\Big), & (14)\end{aligned}$$

where the new term is a local density term that corrects the kinetic energy for the HEG, as in Eq. (9):

$$E_{\text{l}} \equiv -\int d\mathbf{r}\Big(\frac{m}{m^*\Big(\rho(\mathbf{r})\Big)} - 1\Big)\rho(\mathbf{r})t_s\Big(\rho(\mathbf{r})\Big). \tag{15}$$

Predictions of this functional will be presented below. We first turn to the interpretation of the resulting single particle equation.

SINGLE PARTICLE EQUATIONS

Variation of this LMA functional with respect to the wavefunctions ψ_i yields the modified Schrödinger equation

$$\widehat{H}_{\text{eff}}[\rho]\psi_i(\mathbf{r}) = \epsilon_i\psi_i(\mathbf{r}) \tag{16}$$

with

$$\begin{aligned}\widehat{H}_{\text{eff}}[\rho] \equiv &-\nabla \cdot \frac{1}{2m^*\Big(\rho(\mathbf{r}))\Big)}\nabla + V_{\text{ext}}(\mathbf{r}) + V_{\text{H}}(\mathbf{r})\\ &+V_{\text{xc}}(\mathbf{r}) + V_{\text{lma}}^*(\mathbf{r};\rho) + V_{\text{l}}(\mathbf{r}),\end{aligned} \tag{17}$$

where V_{H} and V_{xc} are the usual Hartree and exchange-correlation potentials, and $V_{\text{l}} \equiv \delta E_{\text{l}}/\delta\rho$ is a simple local density form. The new nonlocal potential is

$$V_{\text{lma}}^*(\mathbf{r}) \equiv -\frac{d\ \log\ m^*}{d\rho}\frac{1}{2m^*(\rho)}\sum_i^{\text{occ}}\Big|\nabla\psi_i(\mathbf{r})\Big|^2 \equiv \lambda^*\Big(\rho(\mathbf{r})\Big)\tau_s(\mathbf{r};\rho)/\rho(\mathbf{r}), \tag{18}$$

where r_s is defined in Eq. (5). It is useful to introduce some notation involving the density derivative of m^*, and we have chosen the dimensionless positive quantity

$$\lambda^*(\rho) \equiv -\frac{m}{m^*(\rho)}\frac{d\ \log\ m^*}{d\ \log\ \rho} = \frac{Abr_s}{3}\ e^{-br_s}, \tag{19}$$

where $r_s=(3/4\pi\rho)^{1/3}$ is the usual electron gas parameter, and we have made use of the fit given in Eq. (10). $\lambda(\rho)$ thus given acquires its maximum value of $\lambda^{max} \approx 0.134$ at $r_s = 1/b \approx 19$; note that this point is well beyond the range where it was fit. Eqs. (16)-(18) together with Eq. (3) have to be solved selfconsistently.

We will show below that this LMA widens the bandgap of Si and Ge compared to the LDA result (but that the change in diamond is negligible). Since the eigenvalue spectrum is a product of the one-electron equation, to begin to understand the changes the new terms must be interpreted and understood. Neglecting for the moment differences between the LMA density and the LDA density, we introduce the "corrections" $\widehat{H}^{LMA} = \widehat{H}^{LDA} + \delta\widehat{H}$, $\psi^{LMA} = \psi^{LDA} + \delta\psi$, and $\epsilon^{LMA} = \epsilon^{LDA} + \delta\epsilon$. To lowest order, the change in eigenvalue is $\delta\epsilon = <\psi^{LDA}|\delta\widehat{H}|\psi^{LDA}>$.

The contribution to $\delta\epsilon$ due to the kinetic energy in $\delta\widehat{H}$,

$$-\Big[\frac{1}{2m^*\big(\rho(\mathbf{r})\big)} - \frac{1}{2m}\Big]\nabla^2\psi(\mathbf{r}) + \frac{1}{2m}\lambda^*\big(\rho(\mathbf{r})\big)\nabla\big(\log\ \rho(\mathbf{r})\big)\cdot\nabla\psi(\mathbf{r}), \tag{20}$$

contains the simple mass-enhanced kinetic energy minus the free-mass value, plus a new term that involves the gradient of the density. Unlike the density gradient arising in the Generalized Gradient Approximation [5,6] as a correction to the LDA, the present density gradient is intrinsic to the local mass concept and occurs in only first order, because the mass enhancement varies with density, and the density is not constant. This density gradient term may affect valence orbitals, which are likely to be decreasing where the density is decreasing, differently than it does conduction orbitals which are more likely to be increasing, or decreasing less rapidly, where the density falls off. Such an effect can directly change the energy gap in the same direction in all insulators. The mass enhancement term, on the other hand, may produce a different shift of eigenvalues in systems whose mean density is large (corresponding to $r_s \leq 1.6$, the value where the mass enhancement changes sign, see Fig. 1) than for lower density systems.

The contribution to $\delta\epsilon$ from the potential terms in $\delta\widehat{H}$ involves several terms. The most interesting change is in the new potential V^*_{lma} of Eq. (18). This term is a positive definite local potential directly related to the kinetic energies of all occupied states. This potential is large where the product of the density and the kinetic energy is large, which will have the effect of expanding atomic densities. The result should be in the direction to increase lattice constants, a trend that would help to compensate for the overbinding (too large binding energies; too small bond lengths) by the LDA. It is not immediately clear, however, whether this effect dominates or perhaps is compensated by other changes. From the results (below) it is apparent that there is no clear tendency toward larger volumes in general.

LINEAR RESPONSE WITHIN THE LMA

For weakly inhomogeneous systems one can study the linear response described by the LMA analytically. Since a full discussion has been presented by EP [9] we simply summarize the results here, compared to the corresponding quantity within the LDA and a recent quasi-exact quantum Monte Carlo calculation of Moroni, Ceperley, and

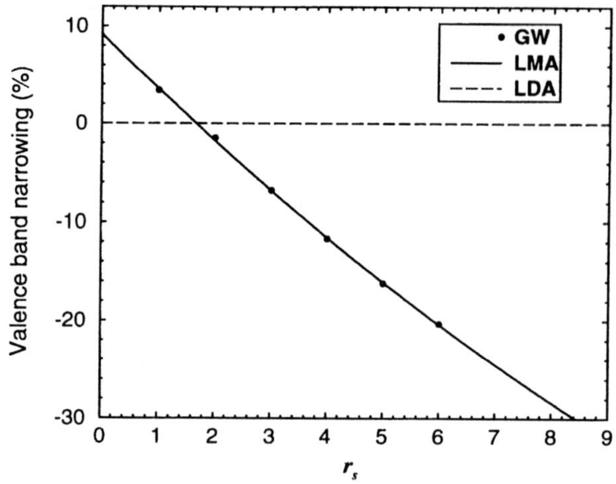

Figure 1. Relative valence band narrowing $\Delta\epsilon/\epsilon_F$ using the *GW* results of Mahan and Sernelius and the fit as described in the text (solid line).

Senatore [13]. The Fourier transform of the linear density response $\delta\rho(\mathbf{r})/\delta V_{\text{ext}}(\mathbf{r}')$ can be written written in the form

$$\chi(q; r_s) = \frac{\chi_0(q; r_s)}{1 - v(q)(1 - G(q; r_s))\chi_0(q; r_s)}, \tag{21}$$

where $v(q) \equiv 4\pi/q^2$ is the Fourier transform of the Coulomb interaction $v(r) = 1/r$ and χ_0 is the Lindhard response function of a non-interacting electron gas,

$$\chi_0(q; r_s) = -\frac{k_F}{\pi^2}\left[\frac{1}{2} + \frac{1 - x^2}{4x}\ln\left|\frac{1+x}{1-x}\right|\right], \tag{22}$$

where $x = q/(2k_F)$. $G(q; r_s)$ is the local field factor, of which there has been much study for the HEG.

The result is shown in Fig. 2. Since the LMA, like the LDA, gives the correct energy of the homogeneous gas at all densities, it satisfies the limit $q \to 0$ exactly (by construction). For $q > 2k_F$, G^{LDA} and G^{LMA} both become progressively different from the exact form. For many materials, this matters little: as evidenced by the success of the LDA, this q region of $G(q; r_s)$ is rarely explored by the relatively smooth density variations of the valence electrons. Thus it may be anticipated that for systems within the linear response regime of the homogeneous gas, LMA total energies will be very close to the LDA energies, and this expectation is borne out for the alkali metals (see below). We find that even for the covalent semiconductors diamond, Si, and Ge, the new functional is at least as successful in predicting lattice constants and bulk moduli as the LDA, while simultaneously changing the band structure significantly.

COMPARISON OF LDA AND LMA STRUCTURAL AND ELECTRONIC PROPERTIES

We have performed plane-wave pseudopotential calculations for the alkali metals Li and Na in the bcc structure and for the semiconductors Si, Ge, and C in the diamond structure in order to compare the LMA to the usual LDA (using the Ceperley-Alder form of the exchange-correlation energy [14]). The method of calculation and convergence criteria can be found in EP [9].

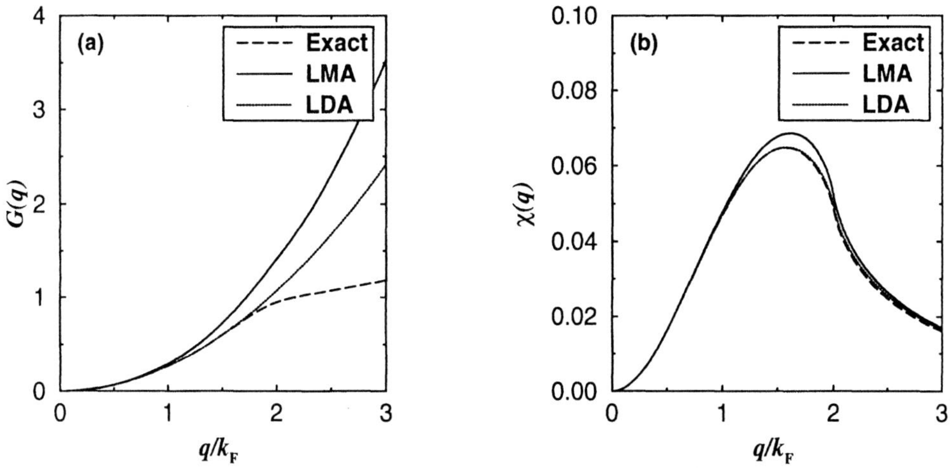

Figure 2. (a) Comparison of LMA and LDA local field corrections with the exact one of Moroni *et al* (b) Density response functions $\chi(q)$ obtained from each of the local field corrections.

Ground State Predictions

The LDA yields surprisingly poor results for the structural properties of alkali metals. The lattice constant of Na is underestimated by about 5 % by pseudopotential LDA schemes, [15] and is improved very little by all-electron calculations [16,17]. Moreover, different parametrizations of the exchange-correlation energy of the homogeneous electron gas can result in sizeable differences in the structural properties and even change the predicted ground state structures (bcc or fcc). (See EP [9] for full references.) Part of this sensitivity is due to the very soft lattices of the alkali metals, which makes small energy changes translate into relatively large differences in volume.

The development of the LMA functional gives no reason to expect that structural properties should be more accurately described than by the LDA. We find that LMA and LDA structural properties differ very little. Table 1 summarizes the results: for the metals and semiconductors we have tested, the LDA and LMA lattice constants agree to within 1 % and the bulk moduli to within 5 %. For semiconductors, the LMA lattice constants are larger and closer to experiment than the LDA values; for Li and Na, they are smaller.

"Excitations"

The eigenvalues, bandgaps, and bandwidths that we report were calculated at the experimental lattice constants shown in Table 1. Within the LDA, the valence bandwidth for Na is virtually unchanged by the presence of the lattice compared to its free-electron value. For Li, the presence of the lattice reduces the valence band width by 27 % compared to the free-electron value (4.75 eV).

The LMA leads to a reduction of 13 % for Li and 11 % for Na with respect to the LDA values. The corresponding bandwidth for the HEG is (from the GW calculations) 8% for $r_s = 3.25$ and 11% for $r_s = 3.93$, respectively. Although Li has a higher electron density than Na, the relative bandwidth reduction is larger in Li than in Na, contrary to what might be expected from Eq. (10). This is easily understood if we assume that the absolute distortion of the valence bandwidth due to the lattice is the same within

Table 1. Lattice constants a and bulk moduli B calculated within the Local Mass Approximation, compared to the LDA predictions and to experiment for metallic bcc Li and Na and insulating diamond, Si, and Ge.

	a (Å)			B (Mbar)		
	LDA	LMA	Expt.	LDA	LMA	Expt.
Li	3.308	3.275	3.480	0.138	0.146	0.127
Na	4.025	3.989	4.220	0.081	0.085	0.073
C	3.533	3.565	3.567	4.51	4.38	4.43
Si	5.384	5.406	5.429	0.96	0.92	0.99
Ge	5.545	5.579	5.652	0.80	0.71	0.77

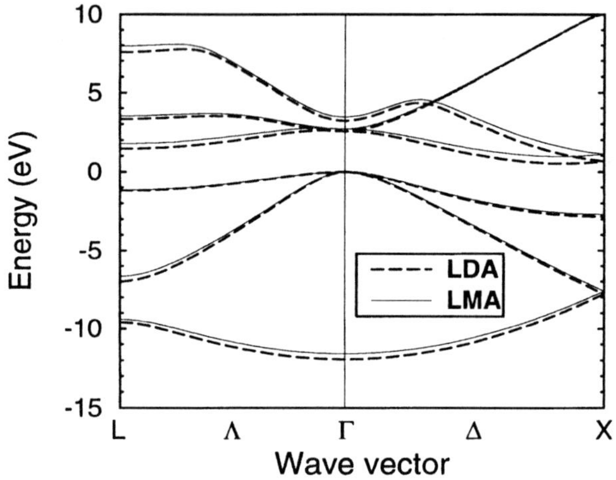

Figure 3. Comparison of the band structure of Si along the (111) and (100) high symmetry directions.

the LDA and LMA, *i.e.*, the valence band in both cases is about 1.30 eV narrower than in a homogeneous gas of the same average density. Since the LMA bandwidth of a homogeneous system with $r_s = 3.25$ is about 4.38 eV, we can then estimate the LMA bandwidth in the presence of the lattice potential to be about 3.08 eV, which is close to the value 3.02 eV of the full calculation. While the LMA clearly improves some aspects of the bandstructures in simple metals, it does not improve the unsatisfactory description of the shape of the Fermi surface within the LDA. The reason, as discussed in EP, [9] seems to be the inability of the LMA (at least as formulated here) to model the *dynamical* mass enhancement. This topic presents a fertile area for extension of the ideas of this paper.

For the semiconductors Si and Ge, whose bandstructures along L-Γ-X are shown in Figs. 3 and 4, we find that the LMA results in a substantial increase of the indirect band gap. For Si the increase is from 0.5 eV to 1.0 eV, which is a substantial improvement. For Ge the indirect LMA gap ($\Gamma \rightarrow$ L) is 0.5 eV; the corresponding LDA gap is 0.1 eV. The increase in the direct band gap is less dramatic, from 0.0 eV to 0.3 eV. Overall, there are two main effects: the conduction bands are shifted upward with respect to the valence bands, but somewhat non-uniformly, and the valence bandwidths are

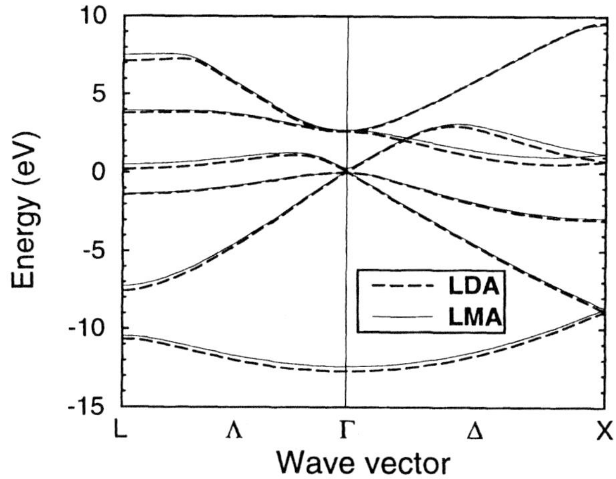

Figure 4. Comparison of the band structure of Ge along the (111) and (100) high symmetry directions.

decreased slightly but are still in reasonable agreement with experiment. For most of the conduction bands, the shift from the LDA bands is larger near the edge of the Brillouin zone than it is around the zone center.

Diamond structure carbon shows a different behavior: the bandstructures predicted by the LDA and LMA are almost identical. This result reveals that the LMA, which builds in self-energy corrections for the HEG, does not give a good representation of "self-energy effects" in strongly inhomogeneous insulators such as carbon. Diamond differs from Si and Ge in two distinctive ways. First, as alluded to above, diamond is more inhomogeneous. For the LMA there is another important difference: the mean density of diamond is essentially that where the effective mass we have incorporated is equal to the free electron mass, so the average "effective mass" correction may be expected to be very small for diamond, as the eigenvalue differences are found to be. Nonetheless it is encouraging that the LMA accurately predicts structural properties even for semiconductors and insulators and moves the excitation spectra of Si and Ge in closer agreement with experiments.

A comparison of cohesive energies will determine whether the LMA yields absolute total energies with a similar degree of accuracy as the LDA. To avoid ambiguities in the cohesive energies related to the use of LDA pseudopotentials and to the neglect of nonlinear core corrections, one of the authors has implemented the LMA within an all-electron Gaussian molecular orbital code. Preliminary results [18] for Si_{20} clusters indicate that for geometries optimized within the LDA, the LMA cohesive energies in Si are about 20 % smaller in magnitude than the LDA cohesive energies, thereby substantially correcting the well-known overbinding observed in LDA calculations. It is possible that this improvement may be fortuitous, and further investigations are under way to see if the improvement is systematic also for other materials.

DISCUSSION

We have introduced the LMA in a semiempirical fashion as a means of reproducing the valence-band narrowing in simple metals. The form we have chosen is just one choice of a general form of nonlocal functional [9]

$$E_{\rm nl}[\rho] \equiv \sum_i^{\rm occ} \int\int d\mathbf{r}d\mathbf{r}'\, \psi_i^*(\mathbf{r})\, \Sigma(\mathbf{r},\mathbf{r}';[\rho])\, \psi_i(\mathbf{r}'), \tag{23}$$

where Σ is a symmetric nonlocal operator which may depend upon the density. With this form, the density dependence may enter explicitly, through $\Sigma[\rho]$, but it also arises implicitly and very nonlocally through the density dependence of the orbitals $\psi[\rho]$, as in the conventional KS kinetic energy functional. In addition, Σ may include implicit density dependence, such as through the orbitals.

Choosing Σ as the Fock exchange operator provides an example of the more general type of functional, (see [10]) and suggests a hybrid Hartree-Fock/density functional in which the exchange is described by the nonlocal Fock operator and the correlation energy is added in a local density fashion. The LMA functional implemented in this paper includes only an explicit local density dependence of Σ. Density dependence of the operator Σ renders the variation of the energy functional in Eq. (13) with respect to the orbitals ψ_i more complicated than in the usual KS scheme. This is a substantial complication for the Hartree-Fock formalism. However, when the density dependence of Σ is explicit, as in the LMA, the additional computational demand is insignificant. The strategy outlined above is readily generalized to spin-density dependent functionals replacing the usual local spin density approximation (LSDA).

The presentation here has emphasized that the search for density functionals beyond the LDA should include not only functionals with explicit density dependence, but should be extended to nonlocal functionals whose dependence on the density is implicit through the Kohn-Sham orbitals. The additional freedom gained by the introduction of a nonlocal orbital-dependent term can be used to absorb some of the discontinuity in T_s, and thereby to reduce some of the nonanalyticities of the usual $E_{\rm xc}[\rho]$ as a function of ρ. One objective should be to construct density functionals where the discontinuity of the exchange-correlation potential upon addition of a single electron in semiconductors and insulators is minimized, since this would optimize the predicted energy gap.

The use of a scalar density dependent mass is only the crudest approximation for self-energy effects. A more general nonlocal operator Σ might, and probably should [9], also contain angular momentum dependent terms. Related corrections should also be kept in mind in constructing optimized effective potentials for Hartree-Fock like theories.

ACKNOWLEDGMENTS

The plane-wave code used in this study was adapted from a program by R. J. Needs. The norm-conserving Kerker pseudopotentials were generated using a code by E. Shirley. This work was supported by the U. S. Office of Naval Research.

REFERENCES

1. For a review, see R. O. Jones and O. Gunnarsson, Rev. Mod. Phys. **61**, 689 (1989).
2. W. Kohn, and L. J. Sham, Phys. Rev. **140**, A1133 (1965).
3. P. Hohenberg and W. Kohn, Phys. Rev. **136**, 864 (1964).
4. H. Eschrig, K. A. Kikoin, and V. G. Kohn, Solid State Commun. **56**, 773 (1985).
5. J. P. Perdew, J. A. Chevary, S. H. Vosko, K. A. Jackson, M. R. Pederson, D. J. Singh, and C. Fiolhais, Phys. Rev. B **46**, 6671 (1992).

6. J. P. Perdew, K. Burke, and M. Ernzerhof, Phys. Rev. Lett. **77**, 3865 (1996).
7. L. J. Sham and M. Schlüter, Phys. Rev. Lett. **51**, 1888 (1983); Phys. Rev. B **32**, 3883 (1985).
8. J. P. Perdew and M. Levy, Phys. Rev. Lett. **51**, 1884 (1983).
9. G. E. Engel and W. E. Pickett, Phys. Rev. B**54**, 8420 (1996).
10. A. Seidl, A. Görling, P. Vogl, J. A. Majewski, and M. Levy, Phys. Rev. B **53**, 3764 (1996).
11. G. D. Mahan and B. E. Sernelius, Phys. Rev. Lett. **62**, 2718 (1989); G. D. Mahan, *Many-Particle Physics*, 2nd edition, (Plenum, New York, 1990), p. 477.
12. P. H. Citrin, G. K. Wertheim, T. Hashizume, F. Sette, A. A. MacDowell, and F. Comin, Phys. Rev. Lett. **61**, 1021 (1988).
13. S. Moroni, D. M. Ceperley, and G. Senatore, Phys. Rev. Lett. **75**, 689 (1995).
14. D. M. Ceperley and B. J. Alder, Phys. Rev. Lett. **45**, 556 (1980), parametrized by J. P. Perdew and A. Zunger, Phys. Rev. B**23**, 5048 (1981).
15. S. G. Louie, S. Froyen, and M. L. Cohen, Phys. Rev. B **26**, 1738 (1982).
16. V. L. Moruzzi, J. F. Janak, and A. R. Williams, *Calculated Electronic Properties of Metals* (Pergamon, New York, 1978).
17. M. Sigalas, N. C. Bacalis, D. A. Papaconstantopoulos, M. J. Mehl, and A. C. Switendick, Phys. Rev. B **42**, 11637 (1990).
18. G. E. Engel and M. Pederson (unpublished).
19. D. M. Bylander and L. Kleinman, Phys. Rev. B **41**, 7868 (1990).

ENSEMBLE DENSITY FUNCTIONAL THEORY FOR INHOMOGENEOUS FRACTIONAL QUANTUM HALL SYSTEMS

O. Heinonen, M.I. Lubin, and M.D. Johnson

Department of Physics, University of Central Florida
Orlando, FL 32816-2385

INTRODUCTION

The fractional quantum Hall effect (FQHE) occurs in a two-dimensional electron gas (2DEG) in a strong magnetic field oriented perpendicular to the plane of the electrons [1]. The effect was discovered as a transport anomaly. In a transport measurement it is noted that at certain strengths $B^*(n)$, which depend on the density n of the 2DEG, current can flow without any dissipation. That is, there is no voltage drop along the flow of the current. At the same time, the Hall voltage perpendicular to both the direction of the current and of the magnetic field is observed to attain a quantized value for a small, but finite, range of magnetic field or density, depending on which quantity is varied in the experiment. The effect is understood to be the result of an excitation gap in the spectrum of an infinite 2DEG at these magnetic fields. A convenient measure of the density of a 2DEG in a strong magnetic field is given by the filling factor $\nu = 2\pi\ell_B^2 n$, with $\ell_B = \sqrt{\hbar c/(eB)}$ the magnetic length. The filling factor gives the ratio of the number of particles to the number of available states in a magnetc sub-band (Landau level), or, equivalently, the number of particles per flux quantum $\Phi_0 = hc/e$. The quantum Hall effect was first discovered [2] at integer filling factors. In this integer quantum Hall effect, the energy gap is nothing but the kinetic energy gap $\hbar\omega_c = \hbar eB/(m^*c)$. Later, the fractional quantum Hall effect (FQHE) was discovered [3] at certain rational filling factors of the form $\nu = p/q$, with p and q relative primes, and q odd. In the FQHE, the excitation gap is a consequence of the strong electron-electron correlations. Therefore, any computational approach to the quantum Hall effect must accurately treat the electron correlations in order to capture the FQHE *at all.* We emphasize that this fact makes any density functional approach to the FQHE qualitatively different from 'standard' applications to atomic, molecular or condensed matter systems. In such standard applications, the electron correlations usually give a quantitatively important correction of the order of perhaps 10 – 20% but do not usually *qualitatively* change the results in a fundamental way. Furthermore, standard applications of density functional theory do not require ensemble density functional theory. In such applications, the densities of the systems under consideration can typically be obtained from a single Slater determinant of Kohn-Sham orbitals. Our work represents (to the best of our

Electronic Density Functional Theory: *Recent Progress and New Directions*
Edited by Dobson *et al.*, Plenum Press, New York, 1998

knowledge) the first practical applications of ensemble density functional theory to a system for which this is *not* possible.

Our understanding of the origin of the FQHE started with Laughlin's seminal paper of 1983 [4], which dealt with the simplest fractions $\nu = 1/m$, with m an odd integer. At these values of ν, there are on the average m magnetic flux quanta $\Phi_0 = hc/e$ per electron. Laughlin constructed a variational wavefunction for spin-polarized systems in strong magnetic fields, strong enough that the splitting $\hbar\omega_c$ between the magnetic subbands, or Landau levels, can be taken to be infinite. The wavefunction can then be constructed from single-particle states entirely within the lowest Landau level. Laughlin wrote the variational wavefunction as

$$\Psi_m \propto \prod_{i,j} (z_i - z_j)^m \exp\left[-\frac{1}{4}\sum_k |z_k|^2\right] \tag{1}$$

where $z_j = x_j + iy_j$ is the coordinate of the jth electron in complex notation. This wavefunction is an eigenstate of angular momentum. Laughlin demonstrated that the system having the wavefunction Eq. (1) is an incompressible liquid with $\nu = 1/m$, m odd, with an energy gap to excitations, and that the elementary excitations are fractionally charged quasi-holes or quasi-particles of charge $e^* = \pm e/m$. The origin of the energy gap can be understood in the so-called pseudo-potential representation of the electron-electron interactions [5]. Here, the electron-electron interaction $V(\mathbf{r}_i - \mathbf{r}_j)$ between electrons i and j is decomposed into strengths V_ℓ in relative angular momentum channels $\ell = 0, 1, 2, \ldots$, of the two electrons. For any realistic interaction $V(\mathbf{r}_i - \mathbf{r}_j)$, it turns out that $V_0 > V_1 > V_2 > \ldots$. Consider the case $\nu = 1/3$. In this case, the lowest angular momentum pseudo-potential that enters into the $m = 3$ Laughlin description is V_1, the interaction energy of two electrons of unit relative angular momentum (in units of $\hbar$). (Only odd relative angular momenta are permissible for spin-polarized electron wavefunctions, since they have to be anti-symmetric under interchange of electron coordinates.) The Laughlin wavefunction is a very cleverly constructed highly correlated state which completely excludes unit relative angular momentum between any two electrons, and is furthermore the only state which satisfies this property at $\nu = 1/3$. Therefore, any excited state at this filling factor must contain some electrons with unit relative angular momentum. The energy gap is due to the cost of this, and hence is of order V_1. Figure 1 depicts the exchange-correlation energy per particle for infinite, homogeneous FQHE systems *vs.* filling factor. The cusps at filling factors $\nu = 1/3, 2/5, 3/5, 2/3, 4/3$, and $\nu = 7/5$ have been included to scale. Note that these cusps are barely visible on this scale, yet they are responsible for all the physics of the FQHE!

FINITE SYSTEMS, AND SPINS

We have outlined above how the electron-electron interactions in an infinite, homogeneous system produce the excitation gap. It is important to note that these gaps are only for excitations in the *bulk* of the system. When a system is bounded there *must* be gapless excitations located at the boundaries of the system [6,7]. Since all experimental systems are finite and inhomogeneous, the low-energy properties probed by experiments are necessarily determined by the gapless edge excitations. Advances in semiconductor nanofabrication technologies have lead to the possibility of manufacturing systems which are extremely inhomogeneous, and in practice dominated by edges. As an example, recent experiments have even been performed on tiny quantum dots,

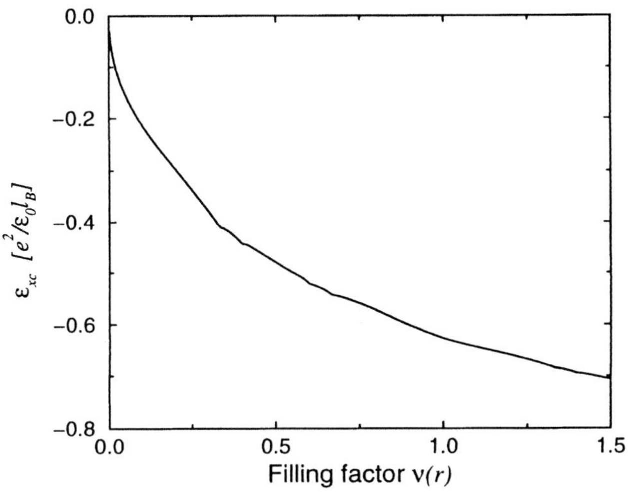

Figure 1. The ground state energy ϵ_{xc} per particle of an infinite, homogeneous, spin-polarized FQHE system is depicted as a function of filling factor. The cusps at $\nu = 1/3, 2/5, 3/5, 2/3, 4/3$, and $\nu = 7/5$ are included. In this plot, the smooth part of the exchange-correlation energy was obtained from the Levesque-Weiss-MacDonald interpolation formula Eq. (7).

containing about 30 electrons [8,9]. Also, edge structures in inhomogeneous systems have been studied directly using capacitance spectroscopy [10], time-resolved measurements of edge magnetoplasmons (the gapless modes along the boundary) [11], and by surface acoustic wave techniques capable of resolving very small inhomogeneities in the electron density [12].

So far, we have limited the discussion to spin-polarized system. At first look it seems reasonable that in the strong magnetic fields used in experiments on QHE systems the Zeeman splitting $g\mu_B B$, where g is the effective Landé factor and μ_B the Bohr magneton, is large enough that the high-energy spin direction is energetically inaccessible. However, two factors conspire to make the Zeeman splitting very low in GaAs systems. First of all, spin-orbit coupling in the GaAs conduction band effectively lowers the Landé factor to $g \approx 0.44$. Second, the low effective mass, $m^* \approx 0.067 m_e$, further reduces the ratio of Zeeman energy to cyclotron energy to about 0.02, compared to its value of unity for free electrons. For magnetic fields of about 1 – 10 T, the Coulomb energy scale of the electron-electron interactions, $e^2/(\epsilon_0 \ell_B)$, where $\epsilon_0 \approx 12.4$ is the static dielectric constant, is of the same order as the cyclotron energy. As a first approximation, one should then set the Zeeman energy to zero, rather than infinite, since it is two orders of magnitude smaller than the other energy scales. As a consequence, the spin-degree of freedom is governed by the electron-electron interactions, rather than by the Zeeman energy. This dramatically changes the nature of the low-energy bulk single-particle excitations near filling factors of the form $\nu = 1/m$, with m odd, from single-particle spin-flips to charge-spin textures. In these objects, loosely called skyrmions, the spin density varies smoothly over a distance of several magnetic lengths, so that the system can locally take advantage of the exchange energy. Skyrmions in the quantum Hall effect were first predicted by Sondhi *et al* [13]. Recent experiments [14] measuring the spin-polarization near $\nu = 1$ have provided evidence for this. According to calculations by Karlhere *et al* [15], the skyrmion excitations in an infinite homogeneous system may lead to charge-spin textures appearing at the edges of QHE systems, which would dramatically change our understanding of QHE edges.

In order to accurately understand the experiments and inhomogeneous FQHE systems in general, we must have a way of accurately calculating their properties, including the spin degrees of freedom. Certain aspects of inhomogeneous FQHE systems have been studied by different techniques. For example, field theories can be constructed to study the low-energy limit of the gapless edge excitations for spin-polarized systems [16], or the spin texture on the edge of an infinite Hall bar [15]. However, effective field theories typically do not give accurate quantitative results, other than for *e.g.* critical exponents near fixed points. So-called composite fermion methods have been used for non-interacting composite fermions [17] and in a Hartree approximation to study finite spin-polarized FQHE systems [18,19,20]. In this approach, the Chern-Simons term, arising from the singular gauge transformation, is replaced by its smooth spatial average, and the composite fermion mass, which is not well known, has to be put in by hand. The electron-electron interaction can then treated in a self-consistent Hartree approximation. The hope is then that the most important aspects of the electron-electron correlations are included in this approximation. Near $\nu = 1$, at which the Slater determinant Ψ_1 is the exact ground state, it makes sense to use the Hartree-Fock approximation, and the stability of a spin-polarized quantum dot at $\nu = 1$ as a function of confining potential has been studied in this approximation [21,22]. The edge structure of spin-polarized systems has also been studied using semiclassical methods [23,24], in which the electron-electron interaction is included at the Hartree level and it is furthermore assumed that all potentials vary on a length scale much larger than ℓ_B. Beenakker [23], and Chklovskii, Shklovskii and Glazman [24] demonstrated that the edge of an integer quantum Hall system, in which the correlation energies between the electrons can be ignored, consists of a sequence of compressible and incompressible strips. The widths of the incompressible strips is determined by the length over which the effective confining potential (external plus Hartree) varies an amount equal to the energy gap $\hbar\omega_c$. The origin of the compressible and incompressible strips are the energy gaps, which are the kinetic energy gaps $\hbar\omega_c$ in the case of the integer quantum Hall effect. But it is easy to generalize the argument to include the energy gaps causing the FQHE [24]. The conclusion is then that there should be compressible and incompressible strips, with the density of each incompressible strip fixed at the value of an FQHE fraction. The width of each incompressible strip is then fixed by the length over which the effective confining potential varies an amount equal to the energy gap of the FQHE fraction corresponding to the density of that strip.

Quantum dots including the spin degree of freedom have also been studied by direct numerical diagonalizations [25]. These calculations demonstrated the importance of the spin degree of freedom. At the present, numerical diagonalizations are limited to systems with of the order of 10 electrons.

It is highly desirable to have a computational approach which accurately includes electron-electron correlations and spin degrees of freedom, and which can handle inhomogeneous systems with on the order of 10^2–10^3 electrons. One such approach which is in principle valid for any interacting electron system is density functional theory (DFT) [26,27,28]. There have been some attempts to apply density functional theory to the FQHE. Ferconi and Vignale [29] applied current density functional theory [30] to small, parabolically confined quantum Hall systems and showed that the current density functional theory gave good results for the ground state energy and spin polarization near $\nu = 1$. However, the energy gaps due to correlation effects were not included in that calculation. Ferconi, Geller and Vignale [31] also recently studied spin-polarized FQHE systems within the spirit of the DFT using an extended Thomas-Fermi approximation at low, but non-zero, temperatures. In this, the kinetic energy was treated as a local functional, as in the standard Thomas-Fermi approximation, while the exchange-correlation energy was included in a local density approximation (LDA). This extended Thomas-Fermi approximation is valid in the limit of very slowly varying confining potential. Ferconi, Geller and Vignale focused on the incompressible and compressible strips at an edge of an FQHE system, and obtained results in agreement with the predictions by Chklovskii, Shklovskii, and Glazman [24].

We have developed for the fractional quantum Hall effect an ensemble DFT scheme within the local density approximation, and have applied it to circularly symmetric quantum dots [32,33]. In our approach, the kinetic energy is treated exactly, and the density represented by Kohn-Sham orbitals. The results are in good agreement with results obtained by semiclassical [23,24,31], Hartree-Fock [21,22] (for cases where the correlations do not play a major role), and exact diagonalization methods [25]. Our calculations for spin-polarized systems show that the exchange and correlation effects of the FQHE are very well represented by the LDA and that our approach provides a computational scheme to model large inhomogeneous FQHE systems. We note that there exist previous formal DFTs for strongly correlated systems, in particular for high-temperature superconductors [34], and DFT calculations of high-T_c materials [35] and transition-metal oxides [36]. However, ours are, to the best of our knowledge, the first practical LDA-DFT calculations of a strongly correlated system in strong magnetic fields, and demonstrate the usefulness of the LDA-DFT in studying large inhomogeneous FQHE systems. Recently, we have generalized our DFT approach to include spin degrees of freedom in an approximation in which the spin-quantization axis is parallel to the external magnetic field. While such an approximation cannot capture skyrmion-like excitations, in which the spin quantization axis is tumbling in space, it does demonstrate the existence of spin structures near edges of inhomogeneous systems consistent with numerical diagonalizations [25] and Hartree-Fock calculations [15]. We are presently working on extending the DFT approach to allow for the possibility of a tumbling spin quantization axis.

ENSEMBLE DENSITY FUNCTIONAL THEORY APPROACH FOR SPIN-POLARIZED SYSTEMS

In typical DFT calculations of systems of N_{el} electrons, the standard Kohn-Sham (KS) scheme [37] is implemented, in which the particle density $n(\mathbf{r})$ is expressed in terms of a Slater determinant of $N \geq N_{\mathrm{el}}$ KS orbitals, $\psi_\alpha(\mathbf{r})$. These obey an effective single-particle Schrödinger equation $H_{\mathrm{eff}}\psi_\alpha = \epsilon_\alpha \psi_\alpha$, which is solved self-consistently by occupying the N_{el} KS orbitals with the lowest eigenvalues ϵ_α, and iterating. This scheme works well in practice for systems for which the true ground state electron density can be represented by a single Slater determinant of single-particle wavefunctions. However, when the KS orbitals are degenerate at the Fermi energy (which we identify with the largest ϵ_α of the occupied orbitals) there is an ambiguity in how to occupy these degenerate orbitals. There exists an extension of DFT which is formally able to deal with this situation. This extension is called ensemble DFT [27,28], and in it, the density of the system is represented by an ensemble of Slater determinants of KS orbitals. However, while it can be shown using ensemble DFT that such a representation of the density is rigorous, it cannot be shown *how* the degenerate KS orbitals at the Fermi energy should be occupied, *i.e.*, there has not been available a practical computational scheme for ensemble density functional theory.

We begin by first demonstrating that the ground state densities of fractional quantum Hall systems cannot in general be obtained from a single Slater determinant of Kohn-Sham orbitals. Hence, an ensemble representation has to be used. Consider a uniform fractional quantum Hall system at filling factor $\nu = 1/3$. This density can be obtained by forming a density matrix of the form

$$\hat{D} = \frac{1}{3}\sum_{i=1}^{3} |\Psi_i\rangle\langle\Psi_i|. \tag{2}$$

Here, $|\Psi_i\rangle$ are the three possible degenerate Slater determinants obtained by filling every third single-particle orbital in the lowest Landau level. Using momentum eigenfunctions, we can write in occupation-number representation $|\Psi_1\rangle = \{100100100\ldots\}$, $|\Psi_2\rangle = \{010010010\ldots\}$, and $|\Psi_3\rangle = \{001001001\ldots\}$. The corresponding ensemble density is then

$$n_D(\mathbf{r}) = \mathrm{Tr}\left\{\hat{D}\hat{n}(\mathbf{r})\right\} = \frac{1}{3}\sum_{i=1}^{3} n_i(\mathbf{r}) = \frac{1}{3 \times 2\pi\ell_B^2}, \tag{3}$$

that is, the filling factor is fixed at $\nu = 1/3$.

We have thus constructed the correct ground state density using fractional occupations of degenerate single-particle orbitals. One can show [38], using a generalization of the Hohenberg-Kohn theorem to degenerate ground states, that the single-particle orbitals used here are the correct KS orbitals, and that we then *must* use fractional occupancies of these orbitals. Hence the ground state density of the system cannot be obtained from a single Slater determinant of KS orbitals, and ensemble DFT must be used.

Although ensemble DFT has been developed formally, there are in practice few examples of applications and calculations using ensemble DFT for ground state calculations. A significant aspect of our work is that we have developed an ensemble scheme which is practical and useful for the study of the FQHE. In ensemble DFT, any physical density $n(\mathbf{r})$ can be represented by $n(\mathbf{r}) = \sum_{mn} f_{mn}|\psi_{mn}(\mathbf{r})|^2$, where f_{mn} are occupation numbers satisfying $0 \leq f_{mn} \leq 1$, and the orbitals ψ_{mn} satisfy the equation

$$\left\{\frac{1}{2m^*}\left[\mathbf{p} + \frac{e}{c}\mathbf{A}(\mathbf{r})\right]^2 + V_{\mathrm{ext}}(\mathbf{r}) + V_{\mathrm{H}}(\mathbf{r}) + V_{\mathrm{xc}}(\mathbf{r},\mathbf{B})\right\}\psi_{mn}(\mathbf{r}) = \epsilon_{mn}\psi_{mn}(\mathbf{r}), \tag{4}$$

where $\nabla \times \mathbf{A}(\mathbf{r}) = \mathbf{B}(\mathbf{r})$. In equation (4), $V_{\mathrm{H}}(\mathbf{r})$ is the Hartree interaction of the 2D electrons, and, as usual, $V_{\mathrm{xc}}(\mathbf{r},\mathbf{B})$ is the exchange-correlation potential, defined as a functional derivative of the exchange-correlation energy $E_{\mathrm{xc}}[n(\mathbf{r}),\mathbf{B}]$ of the system with respect to density: $V_{\mathrm{xc}}(\mathbf{r},\mathbf{B}) = \frac{\delta E_{\mathrm{xc}}[n(\mathbf{r}),\mathbf{B}]}{\delta n(\mathbf{r})}$. (We will hereafter not explicitly indicate the parametric dependence of V_{xc} and E_{xc} on $\mathbf{B}$.) For the case of the FQHE, we know that the exchange-correlation potential will be crucial, as it contains all the effects of the electron correlations which cause the FQHE in the first place, and a major part of the DFT application is to come up with an accurate model of E_{xc} and so of V_{xc}. Leaving this question aside for a moment, and assuming that we have succeeded in doing so, the practical question is then how to determine the KS orbitals and their occupancies in the presence of degeneracies. We have devised an empirical scheme, which produces a set of occupancies for the KS orbitals which satisfy some minimum requirements, namely (a) the scheme converges to physical densities (to the best of our knowledge) for FQHE systems, (b) it reproduces finite temperature DFT distributions at finite temperatures, and (c) it reproduces the standard Kohn-Sham scheme for systems whose densities can be represented by a single Slater determinant.

In our scheme, we start with input occupancies and single-particle orbitals and iterate the system N_{eq} times using the KS scheme. The number N_{eq} is chosen large enough (about 20–40 in practical calculations) that the density is close to the final density after N_{eq} iterations. If the density of the system could be represented by a single Slater determinant of the KS orbitals, we would now essentially be done. However, in this system there are now in general many degenerate or near-degenerate orbitals at the Fermi energy. After each iteration, the Kohn-Sham scheme chooses to occupy the N_{el} orbitals with the lowest eigenvalues, corresponding to making a distinct Slater determinant of these orbitals. But there will be small fluctuations in the density between each iteration, which cause a different subset of these (near) degenerate orbitals to be occupied after each iteration. This corresponds to constructing different Slater determinants after each iteration, and the occupation numbers f_{mn} of these orbital are zero or unity more or less at random after each iteration. This means that the computations will never converge. However, the *average* occupancies, *i.e.*, the occupancies averaged over many iterations, become well defined and approach a definite value, *e.g.*, 1/3 for orbitals localized in a region where the local filling factor is close to $\nu = 1/3$. Therefore, we use these average occupancies to construct an ensemble by accumulating running average occupancies $\langle f_{mn} \rangle$ after the initial N_{eq} iterations

$$\langle f_{mn} \rangle = \frac{1}{(N_{it} - N_{eq})} \sum_{i=N_{eq}+1}^{N_{it}} f_{mn,i}, \tag{5}$$

where $f_{mn,i}$ is the occupation number (0 or 1) of orbital ψ_{mn} after the ith iteration, and use these to calculate densities. Thus, our algorithm essentially picks a different (near) degenerate Slater determinant after each iteration, and these determinants are all weighted equally in the ensemble. It is clear that this scheme reduces to the KS scheme for which the density can be represented by a single Slater determinant of KS orbitals (for which the KS scheme picks only the one Slater determinant which gives the ground state density) for N_{eq} large enough. Moreover, we have numerically verified that a finite-temperature version of our scheme converges to a thermal ensemble at finite temperatures down to temperatures of the order of $10^{-3}\hbar\omega_c/k_B$. We have also performed some Monte Carlo simulations about the ensemble obtained by our scheme. In these simulations, we used a Metropolis algorithm to randomly change the occupation numbers about our converged solution, keeping the chemical potential fixed. The free energy of the new set of occupation numbers was calculated self-consistently. If the free energy decreased, this set was kept, and if the free energy increased, the set was kept if a random number was smaller than $\exp[-\Delta F/k_B T^*]$, where ΔF is the change in free energy, and T^* a fictitious temperature. The results were that to within numerical accuracy our ensemble DFT scheme gives the lowest free energy. As a condition for convergence, we typically demanded that the difference between the input and output ensemble densities, $n_{in}(r)$ and $n_{out}(r)$, of one iteration should satisfy

$$\frac{1}{N_{el}} \int_0^{\infty} |n_{in}(\mathbf{r}) - n_{out}(\mathbf{r})| \, d\mathbf{r} < 5 \times 10^{-4}. \tag{6}$$

Practical density functional theory calculations hinge on the availability of good approximations for the exchange-correlation potential V_{xc}, which enters in the effective Schrödinger equation for the KS orbitals. The simplest, and probably the most commonly used, approximation is the local density approximation (LDA). In this approximation, the exchange-correlation energy is assumed to be a *local function* of density, so that the total exchange-correlation energy consists of contributions from the local density of the system. Thus, in this approximation one writes $E_{xc}/N = \int d\mathbf{r} \epsilon_{xc}(\nu) n(\mathbf{r})$, where $\epsilon_{xc}(\nu)$ is the exchange-correlation energy per particle in a *homogeneous* system of constant density $n = \nu/(2\pi\ell_B^2)$ and filling factor ν. In other words, in the LDA one assumes that the system is locally homogeneous, *i.e.*, the system can locally be approximated to have the energy per particle of an infinite, homogeneous system of the local density. This approximation obviously makes sense if the density of the system varies on a very long length scale, while it could be questionable for systems in which the density varies on some microscopic length scale. However, experience has shown that the LDA often works surprisingly well, even for systems in which the electron density is strongly inhomogeneous [26]. In the FQHE, the length scale of exchange-correlation interactions and density fluctuations is given by the magnetic length ℓ_B due to the Gaussian fall-off of any single-particle basis in which the interacting Hamiltonian is expanded. The densities are relatively smooth on this length scale, which gives us additional hope that the LDA will work well for the FQHE, too. In addition, the cusps in the exchange-correlation energy will suppress density fluctuations, so in this sense one can actually expect the basic physics of the FQHE to make the LDA a good approximation.

We construct our exchange-correlation energy by writing

$$\epsilon_{\rm xc}(\nu) = \epsilon_{\rm xc}^{\rm LWM}(\nu) + \epsilon_{\rm xc}^{\rm C}(\nu). \tag{7}$$

Here, $\epsilon_{\rm xc}^{\rm LWM}(\nu)$ is a smooth interpolation formula (due to Levesque, Weiss and MacDonald [39]) between ground state energies at some rational fillings. The second term, $\epsilon_{\rm xc}^{\rm C}(\nu)$, is all-important for the study of the FQHE. This term contains the cusps in the ground state energy which cause the FQHE. Here we have used a simple model which captures the essential physics. We model $\epsilon_{\rm xc}^{\rm C}(\nu)$ by constructing it to be zero at values of $\nu = p/q$ which display the FQHE. Near $\nu = p/q$, $\epsilon_{\rm xc}^{\rm C}(\nu)$ is linear and has at $\nu = p/q$ a discontinuity in the slope related to the chemical potential gap $\Delta\mu = q(|\Delta_p| + |\Delta_h|)$. Here $\Delta_{p,h}$ are the quasiparticle (hole) creation energies which can be obtained from the literature [42,41] at fractions $\nu = p/q$. Farther away from $\nu = p/q$, $\epsilon_{\rm xc}^{\rm C}(\nu)$ decays to zero. Finally, in the LDA $V_{\rm xc}(r)$ is obtained from $\epsilon_{\rm xc}(\nu)$ as $V_{\rm xc}(r) = \left.\frac{\partial[\nu\epsilon_{\rm xc}(\nu)]}{\partial\nu}\right|_{\nu=\nu(r)}$ at constant B. In our calculations, we restrict ourselves to include only the cusps at $\nu = 1/3, 2/5, 3/5$ and $\nu = 2/3$ (and the analogous ones at 4/3, 7/5, 8,5 and 5/3), which are the strongest fractions. These are some of the fractions of the form $\nu = \frac{p}{(2p\pm1)}$ generated by the so-called V_1-model, in which only the pseudo-potential V_1 is included.

A technical difficulty arises in the LDA: the discontinuities in $V_{\rm xc}(r)$ in the LDA give rise to a numerical instability. The reason is that an arbitrarily small fluctuation in charge density close to an FQHE fraction gives rise to a finite change in energy. Imagine that the local filling factor $\nu(r)$ in some neighborhood of a point r is very close to, but less than, say, 1/3 after one iteration. In this neighborhood, the local exchange-correlation potential will then form a potential well with sharp barriers at the points around r where $\nu(r) = 1/3$. During the next iteration, charge will then be poured into this well. As a result, the local filling factor will after this iteration exceed 1/3, and in this neighborhood $V_{\rm xc}$ now forms a potential barrier of finite height. So in the next iteration, charge is removed from this neighborhood, and so on. We can see that this leads to serious convergence problems. To overcome this, we made the compressibility of the system finite, but very small, corresponding to a finite, but very large, curvature instead of a point-like cusp in $\epsilon_{\rm xc}$ at the FQHE fractions. In other words, instead of having a step-like discontinuity $\Delta\mu$ in the chemical potential, it rises smoothly an amount $\Delta\mu$ over an interval γ in the filling factor. What we found worked very well in practice was to have the discontinuity in chemical potential occur over an interval of filling factor γ of magnitude 10^{-3}. This corresponds to a sound velocity of about 10^6 m/s in the electron gas, which is three orders of magnitude larger than

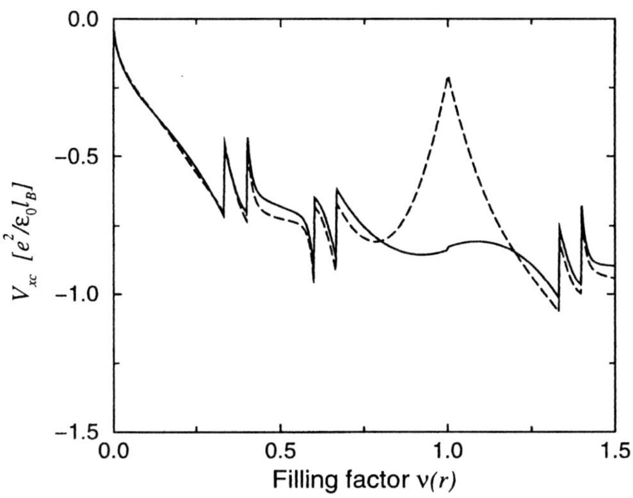

Figure 2. Exchange-correlation potentials $V_{xc\uparrow}$ (solid line) and $V_{xc\downarrow}$ (dashed line) for a spin-polarized system as function of filling factor in units of $e^2/(\epsilon_0 \ell_B)$ for $0 \le \nu \le 1.5$. The increase in V_{xc} at an FQHE filling factor occurs over a range of filling factor of 0.002.

the Fermi velocity of a 2D electron gas at densities typical for the FQHE. In general, the finite compressibility does not lead to any spurious physical effects so long as the energy of density fluctuations on a size of the order of the systems size is larger than any other relevant energy in the problem. The only noticeable effect is that incompressible plateaus, at which the density would be perfectly constant were the compressibility zero, will have density fluctuations on a scale of γ. Figure 2 depicts V_{xc} for a spin-polarized system used in our calculations as a function of filling factor.

APPLICATIONS TO SPIN POLARIZED QUANTUM DOTS

We have self-consistently solved the KS equations Eqs. (4) for a spin-polarized quantum dot in a parabolic external potential, $V_{ext}(r) = \frac{1}{2}m^*\Omega^2 r^2$, by expanding the KS orbitals $\psi_{mn}(\mathbf{r}) = e^{im\phi}\varphi_{mn}(r)$ in the eigenstates of $H_0 = \frac{1}{2m^*}\left(\mathbf{p} + \frac{e}{c}\mathbf{A}(\mathbf{r})\right)^2$. We use the cylindrical gauge, $\mathbf{A}(\mathbf{r}) = \frac{1}{2}Br\hat{\phi}$, and include the four lowest Landau levels $(n = 0, \ldots, 3)$. We chose the static dielectric constant $\epsilon_0 = 12.4$, appropriate for GaAs, and a confining potential of strength [8] $\hbar\Omega = 1.6$ meV.

The use of our LDA-DFT scheme is illustrated by a study of the edge reconstruction of the quantum dot as a function of magnetic field strength. As is known from Hartree-Fock and exact diagonalizations [21,22,25,18,19], for strong confinement the quantum dot forms a maximum density droplet in which the density is uniform at $\nu = 1$ in the interior, and falls off rapidly to zero at $r \approx \sqrt{2N_{el}}\ell_B = r_0$. As the magnetic field strength increases, a "lump" of density breaks off, leaving a "hole" or deficit at about $r = r_0$. This effect is due to the short-ranged attractive exchange interaction: it is energetically favorable to have a lump of density break off so that the system can take advantage of the exchange energy in the lump. As B is further increased, the correlations will cause incompressible strips with densities $\nu = p/q$ to appear [23,24,43,31] on the edges, and incompressible droplets to form in the bulk at densities $\nu = p/q$. Figure 3 depicts various stages of edge reconstruction obtained by us as the magnetic field strength is increased. The value of B for which the exchange lump appears compares very well with the value found by De Chamon and Wen [22] in Hartree-Fock and numerical

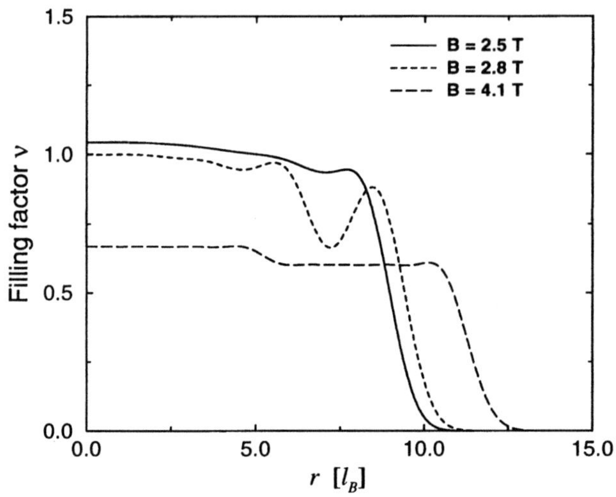

Figure 3. Edge reconstruction of a quantum dot as the magnetic field strength is increased. Plotted here is the local filling factor $\nu(r)$ for a parabolic quantum dot with $\hbar\Omega = 1.6$ meV and 40 electrons. For magnetic field strengths $B < 2.5$ T the dot forms a maximum density droplet, and for $B \approx 2.8$ T, an exchange hole is formed. For stronger magnetic fields, incompressible regions form, separated by compressible strips.

diagonalizations. At higher fields still, incompressible strips appear at the edges, and incompressible droplets are formed in the bulk. Figure 4 depicts the eigenvalues of the KS orbitals for $N_{\text{el}} = 40$, and $B = 4.10$ T. The dashed line indicates the chemical potential of the system. This figure then shows that all KS orbitals in the bulk are in fact degenerate. It may at first seem paradoxical that the eigenvalues are degenerate on an incompressible strip, since, according to the picture by Chklovskii, Shklovskii, and Glazman [24], on such a strip the density is constant, while the total potential varies (since the electrons cannot screen the external potential). If the total potential varies, then ought not the the eigenvalues of the KS orbitals localized on that strip vary, too, since these then in general are subjected to different potential energies? The problem with this argument as applied to DFT is that it ignores the effect of the exchange-correlation potential. As the external and Hartree potentials vary across the strip, the exchange-correlation potential varies across its discontinuity so as to completely screen out the external and Hartree potentials. The discontinuity in V_{xc} does *not* mean that this potential is fixed at the lower limit of its discontinuity while the density is fixed at an incompressible strip. What it does mean, is that V_{xc} is free to achieve any value across its discontinuity so as to completely screen out the external and Hartree potentials. In this way, it is perhaps better to think of incompressibility as the limit of a finite compressibility approaching zero. A strip can then remain incompressible with constant density so long as V_{xc} can screen the external and Hartree potentials, so the width of the incompressible strip is given by the distance over which the external plus Hartree potentials varies an amount given by the energy gap associated with the density at that strip. Also, all bulk KS states are degenerate at the chemical potential. When a single particle is added, the chemical potential simply increases a small amount, and all KS orbitals are again degenerate at the chemical potential.

We have also tested the accuracy of our ensemble DFT-LDA approach by comparing our results for a six-electron system in a confining parabolic potential with the numerical diagonalization results by Yang, MacDonald and Johnson [25]. Figure 5 depicts angular momentum *vs.* magnetic field strength for this system. For better comparison

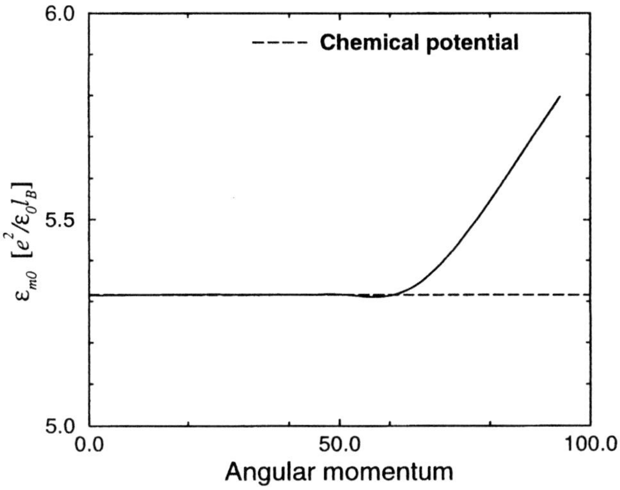

Figure 4. Eigenvalues of the lowest-Landau level Kohn-Sham orbitals for $N_{\rm el} = 40$ and $B = 4.10$ T as a function of angular momentum quantum number. The chemical potential is indicated by the dashed line.

with the numerical diagonalizations, we used here only basis states in the lowest Landau level ($n = 0$). There are clear plateau structures in the angular momenta, and readily identifiable transitions, such as the initial instability of the maximum density droplet at about 2.8 T, and a formation of a $\nu = 1/3$ droplet at about 5.3 T. The agreement is in general very good, in particular considering that (a) the DFT calculations are not constructed to give quantized angular momentum, (b) there are no adjustable parameters in our approach, and (c) the numerical diagonalizations used the full Coulomb interaction, while our DFT approach only included the cusps at $\nu = 1/3, 2/5, 3/5$, and $\nu = 2/3$. The Levesque-Weiss-MacDonald interpolation formula used here tends to overestimate the magnitude of the exchange-correlation potential around $\nu = 1/2$, which increases the values at which transitions occur. For example, the $\nu = 1/3$ droplet formation occurs at 5.29 T in the numerical diagonalization, but at about 5.5T in the DFT-LDA calculation. The overestimation of the exchange-correlation potential also leads to decreased angular momenta at the different plateaus. For example, while the $\nu = 1/3$ droplet has angular momentum 45 in the numerical diagonalization, examination of the density profiles shows that the formation of a $\nu = 1/3$ droplet occurs at an angular momentum of about 40 in our calculations. We are presently working on improving the exchange-correlation energy for better agreement with numerical diagonalization. Initial calculations give highly accurate results for the maximum density droplet instability, and the formation of a $\nu = 1/3$ droplet.

INCLUDING SPIN DEGREES OF FREEDOM

As mentioned above, due to the small effective Landé g-factor in GaAs, the spin degree of freedom plays an important role in quantum Hall systems, in spite of the large applied magnetic fields. We have begun to generalize our ensemble density functional approach to include the spin degree of freedom. In general [43], spin density functional theory has to be based on the single-particle density matrix $\rho_{\sigma\sigma'}(\mathbf{r}) = \langle\hat{\psi}^\dagger_\sigma(\mathbf{r})\hat{\psi}_{\sigma'}(\mathbf{r})\rangle$, where $\hat{\psi}_\sigma(\mathbf{r})$ is the annihilation operator for an electron of spin σ at position $\mathbf{r}$. However, in the presence of a uniform external magnetic field $\mathbf{B} = B\hat{z}$, the z-component of total

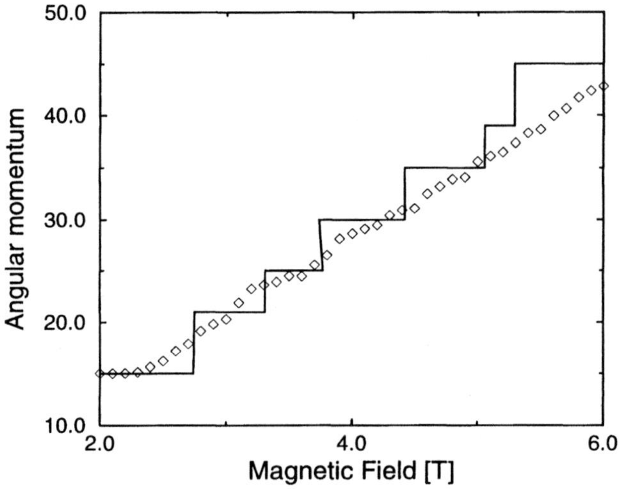

Figure 5. Angular momentum *vs.* magnetic field strength for a six-electron droplet in a parabolic potential with $\hbar\Omega = 2.0$ meV. The solid line is from numerical diagonalizations [25], and the diamonds are from our DFT-LDA calculations.

electron spin, $\hat{S}_z$ commutes with the Hamiltonian, and it is a reasonable approximation to take $\rho_{\sigma\sigma'}(\mathbf{r})$ to be diagonal in the spin indices, $\rho_{\sigma\sigma'}(\mathbf{r}) = n_\sigma(\mathbf{r})\delta_{\sigma\sigma'}$, with $n_\sigma(\mathbf{r})$ the up- and down-spin densities. (Note, however, that this restriction has to be lifted in order to study skyrmion-like spin textures, which complicates the formalism a great deal.) We now obtain two sets of KS equations, one for each spin direction:

$$\left\{\frac{1}{2m^*}\left[\mathbf{p} + \frac{e}{c}\mathbf{A}(\mathbf{r})\right]^2 + V_{\text{ext}}(\mathbf{r}) + V_{\text{H}}(\mathbf{r}) + V_{\text{xc},\sigma}(\mathbf{r}, \mathbf{B}) + \sigma g^* \mu_0 B\right\} \psi_{mn\sigma}(\mathbf{r}) = \epsilon_{mn\sigma}\psi_{mn\sigma}(\mathbf{r}). \tag{8}$$

Here, $\sigma = \pm 1$, μ_0 is the Bohr magneton, and in the local spin density approximation (LSDA) the exchange-correlation potentials are

$$V_{\text{xc},\sigma}(\mathbf{r}, \mathbf{B}) = \frac{\partial}{\partial n_\sigma}\left[n\epsilon_{\text{xc}}(n_\uparrow, n_\downarrow, B)\right]\Big|_{n_\sigma = n_\sigma(\mathbf{r})}. \tag{9}$$

It is more convenient to work with filling factor $\nu(\mathbf{r}) = 2\pi\ell_B^2 n(\mathbf{r})$ and spin polarization $\xi = (n_\uparrow - n_\downarrow)/(n_\uparrow + n_\downarrow)$, in terms of which we have

$$\begin{aligned} V_{\text{xc},\uparrow} &= \frac{\partial}{\partial \nu}(\nu\epsilon_{\text{xc}}) + (1-\xi)\frac{\partial}{\partial \xi}\epsilon_{\text{xc}}, \\ V_{\text{xc},\downarrow} &= \frac{\partial}{\partial \nu}(\nu\epsilon_{\text{xc}}) - (1+\xi)\frac{\partial}{\partial \xi}\epsilon_{\text{xc}}, \end{aligned} \tag{10}$$

where now the exchange-correlation energy per particle in a homogeneous system with a filling factor ν and polarization ξ has to be approximated. Except for a few data points obtained by small system numerical diagonalizations [44], this is largely unkown. In order to obtain a useful approximation, we start by considering only the exchange energy $E_x[\nu_\uparrow, \nu_\downarrow]$. Since the exchange interaction only couples electrons with parallel spins, we have

$$E_x[\nu_\uparrow, \nu_\downarrow] = \frac{1}{2}E_x[\nu_\uparrow, \nu_\uparrow] + \frac{1}{2}E_x[\nu_\downarrow, \nu_\downarrow]. \tag{11}$$

In two dimensions, the exchange energy scales as $n^{3/2}$, so using Eq. (11), we can correctly interpolate between a fully polarized systems ($\xi = 1$) and a completely un-polarized one ($\xi = 0$) by writing

$$\epsilon_x[\nu, \xi] = \epsilon_x(\nu, \xi = 1) + [\epsilon_x(\nu, \xi = 0) - \epsilon_x(\nu, \xi = 1)]\, f(\xi), \tag{12}$$

where

$$f(\xi) = \frac{(1+\xi)^{3/2} + (1-\xi)^{3/2} - 2\sqrt{2}}{2 - 2\sqrt{2}}. \tag{13}$$

We then use the *same* interpolation for the correlation energy (excluding the cusps for a moment), as was first done by von Barth and Hedin [43] for the three-dimensional electron gas in zero magnetic field, and write

$$\begin{aligned} \epsilon'_{\mathrm{xc}}(\nu, \xi) &= \epsilon^{\mathrm{LWM}}_{\mathrm{xc}}(\nu, \xi = 1) + [\epsilon_{\mathrm{xc}}(\nu, \xi = 0) - \epsilon_{\mathrm{xc}}(\nu, \xi = 1)]\, f(\xi) \\ &\equiv \epsilon^{\mathrm{LWM}}_{\mathrm{xc}}(\nu, \xi = 1) + \delta\epsilon_{\mathrm{xc}}(\nu) f(\xi). \end{aligned} \tag{14}$$

The function $\delta\epsilon_{\mathrm{xc}}(\nu)$ can the be obtained by calculating the energy difference between polarized and un-polarized systems using data obtained from small system numerical diagonalizations [44]. An accurate value for the exchange-correlation energy for $\xi = 0$ and $\nu = 1$ is not available, but it is a useful approximation to assume that the different spin directions are completely uncorrelated at $\xi = 0$ and $\nu = 1$, which gives $\epsilon_{\mathrm{xc}}(\nu = 1, \xi = 0) = \epsilon_{\mathrm{xc}}(\nu = 1/2, \xi = 0)$. This approximation also gives a good value for the exchange enhancement, which is the energy required to flip a spin in a polarized system at $\nu = 1$.

So far, we have constructed a function $\epsilon'_{\mathrm{xc}}(\nu, \xi)$ which gives a smooth interpolation for the exchange-correlation energy for any value of ν and ξ. What is left is to add the cusps to this function. We already have a good approximation for this at $\xi = 1$. We now need to extend this approximation to arbitrary values of ξ. Very little is known about the cusps, *i.e.*, the energy gaps, for arbitrary polarizations. It is known that there *is* a gap for un-polarized systems at fillings $\nu = 2/5$, $\nu = 3/5$, and $\nu = 2/3$. The gap, and thus the cusps, occur at very special 'magic' configurations at which the system can take advantage of a particularly low correlation energy. Therefore, it seems plausible that for a given value of ν, say $\nu = 2/5$, there cannot be an energy gap for any value of ξ between 0 and 1. In order to incorporate this assumption into a usable approximation, we interpolate our cusp energy constructed for polarized systems, $\epsilon^{\mathrm{C}}_{\mathrm{xc}}(\nu)$, to arbitrary polarizations by multiplying it by a function $g(\xi)$ which is unity at $\xi = 0$ and $\xi = 1$, and vanishes away from these values of polarization. All together, then, we have

$$\epsilon_{\mathrm{xc}}(\nu, \xi) = \epsilon^{\mathrm{LWM}}_{\mathrm{xc}}(\nu) + \delta\epsilon_{\mathrm{xc}}(\nu) f(\xi) + \epsilon^{\mathrm{C}}_{\mathrm{xc}}(\nu) g(\xi). \tag{15}$$

Figure 2 depicts $V_{\mathrm{xc}\sigma}$ (here for $\xi = 1$) used in our calculations.

We have applied this spin ensemble DFT to study the phase diagram of a maximum density droplet. For large values of the Landé-factor g, the maximum density droplet is fully polarized, and as the magnetic field is increased, there is an instability to forming a spin-polarized exchange-hole. But for small values of g, the instability is towards forming a spin structure at the edge. The value of g separating the spin-polarized and spin-structured instabilities, $\widetilde{g} = g\mu_B B/(e^2/\epsilon_0 \ell_B) \approx 0.05$, is in good agreement with the value found for $\widetilde{g} \approx 0.03$ found from numerical diagonalizations of parabolic dots by Yang, MacDonald and Johnson [25], and consistent with the value obtained in calculations by Kivelson *et al* [15], who used a Hartree-Fock approximation in which the spin axis was allowed to tumble. They obtained a value $\widetilde{g} = 0.17$ for an infinite Hall bar.

CONCLUSION AND SUMMARY

In conclusion, we have shown that ensemble density functional theory can be applied to the FQHE. This opens the door to doing realistic calculations for large systems. We believe that our results are also significant in that they are the first LDA-DFT calculations of a strongly correlated system in a strong magnetic field, and they are (to the best of our knowledge) the first practical ensemble DFT calculations. We find excellent agreement between our ensemble DFT calculations and numerical diagonalizations and Hartree-Fock calculations. Preliminary calculations including spin degrees of freedom are consistent with numerical diagonalizations and Hartree-Fock calculations. We are presently working on improving the exchange-correlation energy as a function of density and polarization. Finally, the spin DFT presented here cannot be used to study skyrmion-like structures, in which the spin density vector is tumbling in space. Work is presently underway, together with J. Kinaret (Chalmers University of Technology) to develop such a theory.

The authors would like to thank M. Ferconi, M. Geller and G. Vignale for helpful discussions and for sharing their results prior to publication, K. Burke and E.K.U Gross for useful comments about the DFT, and M. Levy for a stimulating discussion. O.H. would like to thank Chalmers Institute of Technology, where part of the numerical work was done. This work was supported by the NSF through grants DMR93-01433 and DMR96-32141.

REFERENCES

1. *The Quantum Hall effect*, edited by R.E. Prange and S.M. Girvin (Springer, New York 1987).
2. K. von Klitzing, G. Dorda, and M. Pepper, Phys. Rev. Lett. **45**, 494 (1980).
3. D.C. Tsui, H.L. Störmer, and A.C. Gossard, Phys. Rev. Lett. **48**, 1559 (1982).
4. R.B. Laughlin, Phys. Rev. Lett. **50**, 1395 (1983).
5. F.D.M. Haldane, Phys. Rev. Lett. **51**, 605 (1983).
6. B.I. Halperin, Phys. Rev. B **25**, 2185 (1982).
7. See, for example, A.H. MacDonald, in *Quantum Transport in Semiconductor Microstructures*, edited by B. Kramer (Kluwer Academic, 1996).
8. P.L. McEuen, E.B. Foxman, Jari Kinaret, U. Meirav, M.A. Kastner, Ned S. Wingreen, and S.J. Wind, Phys. Rev. B**45**, 11 419 (1992).
9. O. Klein, C. de C. Chamon, D. Tang, D.M. Abusch-Magder, U. Meirav, X.-G. Wen, M.A. Kastner, and S.J. Wind, Phys. Rev. Lett. **74**, 785 (1995).
10. N.B. Zhitenev, M. Brodsky, R.C. Ashoori, M.R. Melloch, SISSA report cond-mat/9601157.
11. N.B. Zhitenev, R.J. Haug, K. von Klitzing, and K. Eberl, Phys. Rev. Lett. **71**, 2292 (1993); G. Ernst, N.B. Zhitenev, R.J. Haug, K. von Klitzing, and K. Eberl, Surf. Sci. **361/362**, 102 (1996)
12. J.M. Shilton *et al*, J. Phys. Condens. Mat. **8**, L337 (1996); A. Knäbchen and Y.B. Levinson, SISSA report cond-mat/9608074.
13. S.L. Sondhi, A. Karlhede, S.A. Kivelson and E.H. Rezayi, Phys. Rev. B **47**, 16 419 (1993).
14. S.E. Barrett, D. Dabbagh, L.N. Pfeiffer and K.W. West, Phys. Rev. Lett. **74**, 5112 (1995).
15. A. Karlhede, S.A. Kivelson, K. Lejnell and S.I. Sondhi, Phys. Rev. Lett. **77**, 2061 (1996).
16. X.-G. Wen, Phys. Rev. B **44**, 5708 (1991).
17. J.K. Jain and T. Kawamura, Europhys. Lett. **29**, 321 (1995).
18. L. Brey, Phys. Rev. B **50**, 11 861 (1994).
19. D. B. Chklovskii, Phys. Rev. B **51**, 9895 (1995).
20. E. Goldman and S. R. Renn, unpublished.
21. A.H. MacDonald, S.R.E. Yang and M.D. Johnson. Aust. J. Phys. **46**, 345 (1993).
22. C. De C. Chamon and X.G. Wen, Phys. Rev. B**49**, 8227 (1994).
23. C.W.J. Beenakker, Phys. Rev. Lett. **64**, 216 (1990).
24. D.B. Chklovskii, B.I. Shklovskii, and L.I. Glazman, Phys. Rev. B**46**, 4026 (1992).
25. See, for example, S.-R. Eric Yang, A.H. MacDonald, and M.D. Johnson, Phys. Rev. Lett. **71**, 3194 (1993).

26. W. Kohn and P. Vashista in *Theory of the Inhomogeneous Electron Gas*, edited by S. Lundqvist and N. March (Plenum, New York, 1983).
27. *Density Functional Theory: an Approach to the Quantum Many-Body Problem*, by R.M. Dreizler and E.K.U Gross (Springer, Berlin 1990).
28. *Density-Functional Theory of Atoms and Molecules*, by R.G. Parr and W. Yang (Oxford University Press, New York, 1989).
29. M. Ferconi and G. Vignale, Phys. Rev. B **50**, 14 722 (1994).
30. G. Vignale and M. Rasolt, Phys. Rev. Lett. **59**, 2360 (1987); Phys. Rev. B **37**, 10 685 (1988).
31. M. Ferconi, M. Geller and G. Vignale, Phys. Rev. B **52**, 16 357 (1995).
32. O. Heinonen, M.I. Lubin, and M.D. Johnson, Phys. Rev. Lett. **75**, 4110 (1995).
33. O. Heinonen, M.I. Lubin and M.D. Johnson, Int. J. Quant. Chem.: Quant. Chem. Symposium **30**, 231 (1996).
34. E.K.U Gross, L.N. de Oliveira and W. Kohn, Phys. Rev. A **37**, 2805 (1988); Phys. Rev. A **37**, 2809 (1988).
35. W. Pickett, Rev. Mod. Phys. **61**, 433 (1989).
36. A. Svane and O. Gunnarsson, Phys. Rev. Lett. **65**, 1148 (1990); Z. Szotek, W. Temmerman and H. Winter, Phys. Rev. B **47**, 4029 (1993).
37. W. Kohn and L.J. Sham, Phys. Rev. **140**, A1133 (1965).
38. M.I. Lubin, O. Heinonen, and M.D. Johnson, Phys. Rev. B **56**, 10373 (1997).
39. D. Levesque, J.J. Weiss and A.H. MacDonald, Phys. Rev. B **30**, 1056 (1984).
40. R. Morf and B.I. Halperin, Phys. Rev. B **33**, 221 (1986).
41. N. d'Ambrumenil and R. Morf, Phys. Rev. B **40**, 6108 (1989).
42. B.Y. Gelfand and B.I. Halperin, Phys. Rev. B **49**, 1994.
43. U. von Barth and L. Hedin, J. Phys. C **5**, 1629 (1972).
44. For a comprehensive reference, see *The Quantum Hall Effects*, by T. Chakraborty and P. Pietiläinen (Springer Verlag, Berlin 1995).

APPLICATIONS OF SELF-INTERACTION CORRECTIONS TO LOCALIZED STATES IN SOLIDS

W.M. Temmerman[1], A. Svane[2], Z. Szotek[1], and H. Winter[3]

[1]Daresbury Laboratory, Warrington WA4 4AD, United Kingdom

[2]Institute of Physics and Astronomy, University of Aarhus
DK-8000 Aarhus C, Denmark

[3]Forschungszentrum Karlsruhe, INFP, Postfach 3640
Karlsruhe, Germany

I. INTRODUCTION

The local-spin-density (LSD) approximation to density functional theory provides a simple and rather successful scheme for describing many interacting electrons in a solid [1]. There exist, however, important cases where electron correlations are too strong to be properly treated within LSD. One prominent example is the high-temperature superconductors, for which LSD fails to produce the antiferromagnetic and semiconducting ground states of the generic La_2CuO_4 and $YBa_2Cu_3O_6$ compounds [2]. Another example are the $3d$ transition metal oxides MnO, FeO, CoO and NiO, which are Mott insulators characterized by localized d-electrons [3,4]. These materials have antiferromagnetic order and large energy gaps although the metal d-shell is incompletely filled. The LSD approximation does reveal the magnetic ordering but with somewhat too small magnetic moments and vanishing (for FeO and CoO) or very small (for MnO and NiO) gaps [5]. In addition, the persistence of the magnetic moments above the Néel temperature is difficult to explain in the Slater picture of magnetism inherent in the LSD band picture. The failure of LSD in producing the correct gaps may be traced to the fact that the LSD eigenenergies do not have built in the large on-site Coulomb repulsion, which characterizes the separation between occupied and unoccupied states [4,6,7]. A third example of inadequacy of the LSD is in describing tri-valent cerium compounds, which are characterized by each Ce atom having a localized $f-$electron. Cerium is the first element of the Periodic Table to accomodate an $f-$electron. This $4f-$electron is peculiar in being spatially localized with a radial extent much smaller than that of the $5s$ and $5p$ semicore states, yet having an energy in the region of the valence $6s$ and $5d$ electrons. In cerium metal the isostructural $\alpha \rightarrow \gamma$ phase transition is believed to occur as a consequence of a change in bonding properties of the $f-$electrons. In the (high–volume) γ phase the $f-$electrons behave as non–bonding localized moments with a Curie–Weiss type susceptibility, while the (low–volume) α-phase is characterized by the f-electrons taking more active part in the cohesion, either

Electronic Density Functional Theory: *Recent Progress and New Directions*
Edited by Dobson *et al.*, Plenum Press, New York, 1998

by hybridizing into the Bloch states, as envisaged in the Mott transition model [8], or by forming a complicated Kondo lattice groundstate [9]. In an LSD calculation the total energy minimum is found in the region of the α phase volume. Upon expansion of the lattice, the $\alpha \rightarrow \gamma$ transition is signalled by the onset of magnetization around the observed lattice parameter for the γ phase [10]. The calculated pressure is much too negative, however, since the f-electrons still contribute significantly to the cohesion in the spin-polarized phase. The loss of cohesion upon localization has been successfully described within the LSD in cases where the f-shell is half filled, as in americium [11].

The concept of an electron being 'localized' is not particularly well-defined. In a solid a periodic array of deep core states are equally well described by a set of atom centered Wannier states or by **k**-dependent Bloch functions. A unitary transformation connects the two representations, but the physical picture of a core state is that of a localized wavefunction. The observation of multiplet effects in photoemission experiments on 3d monoxides and rare-earth systems seems to suggest that the local description is most appropriate for such systems. On the other hand the great success of modern solid state theory in explaining most conventional metals, semiconductors and insulators leaves little doubt that the Bloch picture of normal valence electrons is the more fruitful one.

The celebrated Hubbard model [12] is often used to describe materials where localization phenomena occur. The appreciable Coulomb correlations on particular atoms, say on a Ni d^8 ion in NiO, induce an orbital polarization [4], which may well be described already in a mean-field treatment of the Hubbard model, as has been shown in the LDA+U approach [7,13]. However, the 'localizing' Hubbard term is usually too large to facilitate an accurate description of delicate localization-delocalization transitions, as for example observed in rare-earth and actinide systems.

The self-interaction corrected local spin density (SIC–LSD) approximation is another scheme by which to improve the LSD for Coulomb correlated systems. The basic assumption in the SIC–LSD is that 'localized' electron states experience a different potential from that of the normal delocalized valence electrons. One may argue for such a differentiation by thinking in terms of Wigner delay times of electrons on a particular atom [14]. If an electron resides on a given atom for a long time the local electronic structure of the atom accomodates to the presence of the added electron, while a fast electron has no influence on the effective structure of that atom. This picture is implemented by assuming a fast ('delocalized') electron to experience an effective potential as given by the LSD approximation, while a slow ('localized') electron experiences the LSD potential corrected for the self-interaction of the electron in question. Taking again the d^8 configuration of a Ni ion in NiO as an example, one of the 8 d-electrons residing on a particular atomic site experiences Coulomb interactions with 7 other d-electrons. On the other hand, a conduction electron injected into a perfect NiO crystal will be made out of excited states, which ride on top of an array of Ni d^8 ions, and therefore Coulomb interacts with all the 8 d-electrons. That the LSD is adequate for describing the 'fast' electron is validated by the great success of this approach in describing the bonding properties of conventional metals. This picture is still very idealized, since the transition from localized to delocalized must be a gradual one. The added electron in NiO still may show some localized characteristics, and the ground state electrons do sometimes show **k**-dependent dispersions. One may still hope, though, to be able to distinguish phases of solids being either predominantly 'localized' or predominantly 'delocalized', which will be the aim of the present work.

In Section II of this chapter the general formalism of the SIC–LSD approximation will be discussed, and in Section III two implementations of the approach into ab-initio linear-muffin-tin orbitals (LMTO) band structure codes will be outlined. Section IV will present applications of the SIC–LSD approach to the $3d$ transition metal monoxides and La_2CuO_4, to the $\alpha \to \gamma$ phase transition in cerium, and to the sequence of localization-delocalization transitions observed in the cerium pnictides. Finally, Section V contains the conclusions and outlook for the present work.

II. THE SIC–LSD FORMALISM

The need to correct approximate electronic structure schemes for spurious self-interactions has been discussed by many authors [15-17]. In the SIC–LSD approximation [17] one starts from the LSD total energy functional, but corrects for the self-Coulomb and self-exchange-correlation energy of each of the occupied electron states, which erroneously is included in the LSD approximation. Specifically,

$$\tilde{E}^{SIC}[\{\psi_\alpha\}] = \sum_\alpha <\psi_\alpha| - \Delta|\psi_\alpha> + U[n] + E_{xc}^{LSD}[\bar{n}]$$
$$+ V_{ext}[n] - \sum_\alpha \{U[n_\alpha] + E_{xc}^{LSD}[\bar{n}_\alpha]\} \ . \tag{1}$$

Here, $\tilde{E}^{SIC}$ is a functional of a set of N occupied orthonormal single electron wave-functions ψ_α. $\bar{n}$ is the total spin density of the system, $\bar{n}(\mathbf{r}) = (n^\uparrow(\mathbf{r}), n^\downarrow(\mathbf{r})), n(\mathbf{r}) = n^\uparrow(\mathbf{r}) + n^\downarrow(\mathbf{r})$ and $\bar{n}_\alpha$ is the spin density of the α'th state, $\bar{n}_\alpha(\mathbf{r}) = (|\psi_\alpha(\mathbf{r})|^2, 0)$, $(0, |\psi_\alpha(\mathbf{r})|^2)$, for spin up and spin down states, respectively. $U[n]$ and $E_{xc}^{LSD}[\bar{n}]$ are the Coulomb and the exchange-correlation energies of the electron gas, respectively, while $V_{ext}[n]$ denotes the interaction energy with the lattice of ions:

$$U[n] = \int\int \frac{n(\mathbf{r})n(\mathbf{r}')}{|\mathbf{r} - \mathbf{r}'|} d^3r\, d^3r' \tag{2}$$

$$E_{xc}^{LSD}[\bar{n}] = \int \epsilon_{xc}^{hom}(\bar{n}(\mathbf{r}))\, n(\mathbf{r})\, d^3r \tag{3}$$

$$V_{ext}[n] = \int V_{ext}(\mathbf{r})\ n(\mathbf{r})\, d^3r \ . \tag{4}$$

Throughout we use atomic Rydberg units ($\hbar = 2m_e = e^2/2 = 1$). In Eq. (3), $\epsilon_{xc}^{hom}(\bar{n})$ is the exchange-correlation energy density of a homogeneous electron gas at spin density $\bar{n}$ [18,1], for which we adopt the parametrization as given by Vosko, Wilk and Nusair [19].

The last term in Eq. (1) represents the self-interaction correction, where for each occupied orbital ψ_α the Coulomb and exchange-correlation energies of the corresponding spin density $\bar{n}_\alpha$ is subtracted. If the orbitals in the kinetic energy term are taken to be Kohn-Sham orbitals and the self-interaction is omitted we would have the LSD energy functional E^{LSD}.

The SIC–LSD approximation restores the property, which the true density functional has, that for *any* single-electron density $\bar{n}_\alpha$:

$$U[n_\alpha] + E_{xc}[\bar{n}_\alpha] = 0 \ . \tag{5}$$

This does not hold for the LSD approximation.

$\tilde{E}^{SIC}$ as defined in Eq.(1) is a functional of the set $\{\psi_\alpha\}$ of occupied orbitals. By a reformulation an equivalent SIC functional of the total density alone may be constructed [17,20], the minimization of which in principle may be achieved with the Kohn-Sham procedure. To see this we use the Levy constrained minimization [21] to define the auxiliary functional [17,20]:

$$\Delta^{SIC}[\bar{n}] = \min_{\{\psi_\alpha\}} \sum_\alpha \{< \psi_\alpha| - \Delta\, |\psi_\alpha > - U[n_\alpha] - E_{xc}^{LSD}[\bar{n}_\alpha]\}\Big|_{\bar{n}}\ , \tag{6}$$

where the minimization is constrained to be over those sets $\{\psi_\alpha\}$ of N orthonormal one-electron wavefunctions whose spin densities add up to the total $\bar{n}$. Defining the density functional

$$E^{SIC}[\bar{n}] = \Delta^{SIC}[\bar{n}] + V_{ext}[n] + U[n] + E_{xc}^{LSD}[\bar{n}], \tag{7}$$

one observes that the minimal value of $\tilde{E}^{SIC}$ in Eq.(1) and E^{SIC} in Eq. (7) coincide, and the density by which the minimization of (7) is attained is also the density of the minimum of (1). The proper SIC exchange and correlation energy is now implicitly defined by Eq. (6), which we write as

$$\Delta^{SIC}[\bar{n}] = T[\bar{n}] + E_{xc}^{SIC-LSD}[\bar{n}] - E_{xc}^{LSD}[\bar{n}]. \tag{8}$$

$T[\bar{n}]$ is here the usual kinetic energy term of density functional theory, defined similarly to Δ^{SIC}:

$$T[\bar{n}] = \min_{\{\psi_\alpha\}} \sum_\alpha < \psi_\alpha| - \Delta\, |\psi_\alpha >\Big|_{\bar{n}}\ , \tag{9}$$

and E_{xc}^{LSD} is given by Eq. (3). Introducing Eq. (8) in (7) shows that this is indeed the proper definition of $E_{xc}^{SIC-LSD}$. The $E_{xc}^{SIC-LSD}$ functional is, however, not explicitly defined, so the Kohn-Sham procedure seems difficult to pursue. Recently, it has been shown how the Hartree-Fock approximation may be solved the Kohn-Sham way [22]. A similar procedure should be applicable to SIC–LSD. The Kohn-Sham procedure will lead to the determination of the set of orbitals by which the minimization in Eq. (9) is obtained at the ground state density, which in general will be a different set of orbitals from that by which the minimum of Eq. (6) (and hence Eq. (1)) is attained. In a periodic solid, the former set may always be chosen as Bloch orbitals, which is not the case for the latter set. In the following we will consider two practical ways of minimizing Eq. (1).

The SIC–LSD approximation has been widely studied in atoms [17,23,24], molecules [25,26], clusters [27-29], impurities [30-32] and solids [33-47]. In particular, for periodic solids the SIC–LSD constitutes an extension of LSD in the sense that the self-interaction correction is only finite for electron states, which are spatially localized. This means that if the single-particle states are restricted to be of Bloch type, $\tilde{E}^{SIC}$ coincides with E^{LSD}, i. e., the LSD minimum is also a local minimum of $\tilde{E}^{SIC}$. The question is, whether a set of single-particle states, not all of Bloch form, exists such that $\tilde{E}^{SIC}$ has a lower energy when evaluated with these orbitals. This line of thought was pursued in an investigation of a body-centered cubic array of hydrogen atoms [47], where a discontinous Mott transition was predicted as a function of lattice spacing, precisely as a consequence of the Bloch-like (LSD) local minimum and the localized (SIC–LSD) minimum shifting relatively to each other. Consequently, the concept of localization in the SIC–LSD scheme becomes very literal, the 'localized' states being really spatially confined.

Localization phenomena have been investigated by the SIC–LSD approach in the 3d transition metal oxides, [35-39], in cerium [42-44] and cerium pnictides [45], in praseodymium [41,46] and in americium [40]. Section IV will review some of these results.

III. MINIMIZATION OF THE SIC–LSD ENERGY FUNCTIONAL

The aim is to minimize (1) under the constraint that the orbitals ψ_α be orthonormal:

$$<\psi_\alpha \mid \psi_{\alpha'}>=\delta_{\alpha\alpha'} \ . \tag{10}$$

Varying $\tilde{E}^{SIC}$ with respect to ψ^*_α leads to the SIC equations:

$$\left(H_0+V^{SIC}_\alpha\right)\psi_\alpha=\sum_{\alpha'}^{occ.}\lambda_{\alpha\alpha'}\psi_{\alpha'} \ , \tag{11}$$

where $\lambda_{\alpha\alpha'}$ are Lagrange multipliers associated with the constraints (10). H_0 is the LSD Hamiltonian:

$$H_0=-\Delta+V_H(\mathbf{r})+V_{ext}(\mathbf{r})+V^{LSD}_{xc,\sigma_\alpha}(\bar{n}(\mathbf{r})) \tag{12}$$

$$V_H(\mathbf{r})=2\int\frac{n(\mathbf{r}')}{|\mathbf{r}-\mathbf{r}'|}\,d^3r' \tag{13}$$

$$V^{LSD}_{xc,\sigma}(\bar{n}(\mathbf{r}))=\frac{\delta E^{LSD}_{xc}[\bar{n}]}{\delta n^\sigma(\mathbf{r})} \tag{14}$$

and V^{SIC}_α is the self-interaction potential for orbital α:

$$V^{SIC}_\alpha(\mathbf{r})=-2\int\frac{n_\alpha(\mathbf{r}')}{|\mathbf{r}-\mathbf{r}'|}\,d^3r'-V^{LSD}_{xc,\sigma_\alpha}(\bar{n}_\alpha(\mathbf{r})) \ . \tag{15}$$

It may be explicitly checked that V^{SIC}_α vanishes for an extended state. As stated above, this implies that any set of orbitals solving the Kohn-Sham equations of the LSD energy functional will also solve Eq. (11), thus providing a local minimum of $\tilde{E}^{SIC}$. Here we will be interested in, whether another set of orbitals that solve Eq. (11) exists and has a lower total energy. Such (localized) states will be centered on some particular position, i. e., the index α will be a composite index, $\alpha\equiv\mathbf{T}\nu$, where $\mathbf{T}$ is a Bravais lattice vector, and ν is an additional (site and symmetry) index. Consequently, the periodicity of the lattice is lost in the Hamiltonian in Eq. (11), due to the non-vanishing V^{SIC}_α, and Bloch's theorem cannot be applied for the one-electron states. The overall crystal symmetry is not destroyed, though, since identical copies of the localized states are situated at Bravais lattice positions, so that

$$\psi_{\mathbf{T}\nu}(\mathbf{r})=\psi_{\mathbf{0}\nu}(\mathbf{r}-\mathbf{T}). \tag{16}$$

The λ-matrix in Eq. (11) follows by using the constraint (10) in Eq. (11), i.e., multiplying by $\psi^*_{\alpha'}$ and integrating yields

$$\lambda_{\alpha\alpha'} = <\psi_{\alpha'} \mid H_0 + V_\alpha^{SIC} \mid \psi_\alpha> \quad . \tag{17}$$

The λ-matrix is Hermitian [25]

$$\lambda_{\alpha\alpha'} = \lambda^*_{\alpha'\alpha} \tag{18}$$

or, equivalently

$$<\psi_{\alpha'} \mid \left(V_\alpha^{SIC} - V_{\alpha'}^{SIC}\right) \mid \psi_\alpha>= 0 \; . \tag{19}$$

This follows from the stability of the SIC energy functional with respect to infinitesimal unitary mixing of the orbitals

$$\begin{aligned} &E^{SIC}\left[..,\psi_\alpha + \epsilon\psi_\beta, ..., \psi_\beta - \epsilon^*\psi_\alpha, ...\right] - \\ &\quad E^{SIC}\left[..., \psi_\alpha, ..., \psi_\beta, ...\right] = 2\mathrm{Re}(\epsilon^* r_{\alpha\beta}) + O(\epsilon^2) \end{aligned} \tag{20}$$

with

$$r_{\alpha\beta} = \lambda_{\alpha\beta} - \lambda^*_{\beta\alpha} \; . \tag{21}$$

The relation (19) has been named the localization condition [25] and is a consequence of the state dependence of the Hamiltonian in (11). Alternatively, (18) follows from the complex conjugate of (11) together with the variation of $\tilde{E}^{SIC}$ with respect to ψ_α. To solve Eq. (11) we will adopt one of two approaches, either the steepest descent method or the unified Hamiltonian approach.

To minimize $\tilde{E}^{SIC}$ by the steepest descent method [48] we imagine having in the m'th iteration the set of orbitals $\psi_\alpha^{(m)}$, which satisfy Eq. (10) but do not yet minimize $\tilde{E}^{SIC}$, the orbitals of the $(m+1)$'th iteration are defined by stepping along the gradient of E^{SIC}:

$$\frac{\delta E^{SIC}}{\delta \psi^*_\alpha} = \left(H_0 + V_\alpha^{SIC}\right)\psi_\alpha \; . \tag{22}$$

This step does not preserve orthonormality. Therefore, it is necessary to orthonormalize to all occupied states, i.e., we take

$$\psi_\alpha^{(m+1)} = \psi_\alpha^{(m)} + \delta\psi_\alpha^{(m)} \tag{23}$$

with

$$\delta\psi_\alpha^{(m)} = -x\,\hat{Q}\left(H_0 + V_\alpha^{SIC}\right)\psi_\alpha^{(m)}, \tag{24}$$

where we have defined the projection operator onto unoccupied states

$$\hat{Q} = 1 - \sum_\alpha^{occ.} \hat{P}_\alpha \tag{25}$$

with

$$\hat{P}_\alpha = \mid \psi_\alpha><\psi_\alpha \mid \tag{26}$$

The parameter $x > 0$ in Eq. (24) is chosen at convenience. H_0, V_α^{SIC} and $\hat{Q}$ in Eqs. (24–26) are evaluated with the orbitals of the m'th iteration.

Using Eq. (17), formula (24) may also be written

$$\delta\psi_\alpha^{(m)} = -x\left[\left(H_0 + V_\alpha^{SIC}\right)\psi_\alpha^{(m)} - \sum_{\alpha'}^{occ.}\lambda_{\alpha\alpha'}\psi_{\alpha'}^{(m)}\right] \tag{27}$$

so that convergence of iterations in Eq. (24) leads to Eq. (11).

The λ-matrix may not be Hermitian after iteration m, i.e., according to Eq. (20), the energy may further be lowered by the *simultaneous* change of states α and α':

$$\begin{aligned}\tilde{\psi}_\alpha &= \psi_\alpha - y\,r_{\alpha\alpha'}\psi_{\alpha'}\\ \tilde{\psi}_{\alpha'} &= \psi_{\alpha'} + y\,r^*_{\alpha\alpha'}\psi_\alpha = \psi_{\alpha'} - y\,r_{\alpha'\alpha}\psi_\alpha\end{aligned} \tag{28}$$

with $r_{\alpha\alpha'}$ given in Eq. (21). From here on the iteration index m will be suppressed. $y > 0$ is again chosen at convenience. Considering this unitary mixing for all pairs $\alpha\alpha'$, we get

$$\tilde{\psi}_\alpha = \psi_\alpha - y\sum_{\alpha'} r_{\alpha\alpha'}\psi_{\alpha'}\ . \tag{29}$$

Finally, since steps (24) and (29) preserve orthonormality only to first order in x and y, it is necessary - to avoid errors piling up - to bring the orbitals back onto the orthonormality manifold. To do this, note that if

$$<\psi_\alpha\,|\,\psi_{\alpha'}> -\delta_{\alpha\alpha'} = t_{\alpha\alpha'} \tag{30}$$

then the orbitals

$$\overline{\psi}_\alpha = \psi_\alpha - \frac{1}{2}\sum_{\alpha'} t_{\alpha'\alpha}\psi_{\alpha'} = \psi_\alpha + \frac{1}{2}\hat{Q}\psi_\alpha \tag{31}$$

fulfill

$$<\overline{\psi}_\alpha\,|\,\overline{\psi}_{\alpha'}> -\delta_{\alpha\alpha'} = O(t^2)\ . \tag{32}$$

The unitary step in (29) and the orthonormality step in (31) may be performed interchangeably with the gradient step in (24). At convergence, all steps must vanish. In the steepest descent approach the states, which are not self-interaction corrected are most easily obtained from the periodic Hamiltonian $\hat{Q}H_0\hat{Q}$, filling band states to a Fermi level defined by the appropriate number of non-corrected states. The localized states are eigenstates of $\hat{Q}H_0\hat{Q}$ with eigenvalue 0, since $\hat{Q}\psi_\alpha = 0$.

In the unified Hamiltonian [24] approach the following operator is constructed:

$$H_u = \sum_\alpha \hat{P}_\alpha H_\alpha \hat{P}_\alpha + \sum_\alpha(\hat{P}_\alpha H_\alpha \hat{Q} + \hat{Q}H_\alpha\hat{P}_\alpha) + \hat{Q}H_0\hat{Q} \tag{33}$$

$$= H_0 - \sum_{\alpha,\alpha'}{}'\hat{P}_\alpha H_0 \hat{P}_{\alpha'} + \sum_\alpha \hat{P}_\alpha V_\alpha^{SIC}\hat{P}_\alpha + \sum_\alpha(\hat{P}_\alpha V_\alpha^{SIC}\hat{Q} + \hat{Q}V_\alpha^{SIC}\hat{P}_\alpha), \tag{34}$$

where the primed summation signifies that $\alpha \neq \alpha'$. Further, $H_\alpha = H_0 + V_\alpha^{SIC}$, and H_0 and V_α^{SIC} are given by Eqs. (12) and (15), and the projection operators $\hat{Q}$ and $\hat{P}_\alpha$ are defined in Eqs. (25) and (26). H_u has the property that a solution ψ_α to Eq. (11) is also an eigenstate of H_u (with eigenvalue $\lambda_{\alpha\alpha}$). The remaining eigenstates of H_u may be interpreted as unoccupied electron states, and are simply the LSD eigenstates in the subspace orthogonal to the occupied states.

H_u has the periodicity of the lattice, so that crystal momentum provides a good quantum number, and H_u may be diagonalized for each **k**-vector by standard band-structure methods. From Eq. (34) it is evident that only those states, which are self-interaction corrected need be considered in the sums. The occupied non-localized states are simply determined by the LSD Hamiltonian in the subspace orthogonal to the space of localized states. A problem arises in the self-consistency requirement in the unified Hamiltonian approach, since the SIC-potential necessarily must be evaluated with the density of the localized state. Concerning localized states, the diagonalization of H_u gives only a Bloch-combination of the localized states, and to define the transformation back to localized states we introduce the transformation matrix M:

$$\psi_{\mathbf{T}\nu} = \sum_{\mathbf{k},\nu'} e^{i\mathbf{kT}} M_{\nu,\nu'}(\mathbf{k}) \psi_{\mathbf{k}\nu'} \tag{35}$$

Here, **k** runs over the full ν'th band. To get the most optimally localized Wannier functions, $\psi_{\mathbf{T}\nu}$, the transformation matrix M is best constructed in the three following steps. First we define the matrix $M^{(1)}$ by

$$M^{(1)}_{\nu\nu'}(\mathbf{k}) = N_\nu(\mathbf{k}) < \psi_{\mathbf{k}\nu'} | \psi_{\mathbf{k}=0\nu} >_{\Omega_0} \tag{36}$$

with the constant

$$N_\nu(\mathbf{k}) = (\sum_{\nu'} | < \psi_{\mathbf{k}\nu'} | \psi_{\mathbf{k}=0\nu} >_{\Omega_0} |^2)^{-1/2} \tag{37}$$

The subscript Ω_0 indicates that in the overlap matrix elements only the central unit cell is integrated over. In the second step we apply a Schmitt-orthogonalization procedure to the states $\psi^{(1)}_{\mathbf{T}\nu}$, obtained by inserting $M^{(1)}$ into Eq. (35), thus generating states $\psi^{(2)}_{\mathbf{T}\nu}$. Finally, we perform a series of unitary transformations on pairs of states $\psi^{(2)}$ until the localization criterion (Eq. (19)) is fulfilled. The last two steps are equivalent to a multiplication of the matrix $M^{(1)}$ with two other (numerically determined) matrices, say $S^{(1)}$ and $S^{(2)}$, thus leading to the final transformation matrix M.

The inverse transformation of (35) reads

$$\psi_{\mathbf{k}\nu'} = \sum_{\mathbf{T}\nu} e^{-i\mathbf{k}\cdot\mathbf{T}} M^{-1}_{\nu',\nu}(\mathbf{k}) \psi_{\mathbf{T}\nu} \tag{38}$$

When this transformation is applied to Eq. (11) the following wave equation is obtained for the Bloch states $\psi_{\mathbf{k}\nu}$:

$$[H_0 + V^{SIC}_{\mathbf{k}\nu}]\psi_{\mathbf{k}\nu} = \sum_{\nu'} \varepsilon_{\nu\nu'}(\mathbf{k}) \psi_{\mathbf{k}\nu'} \tag{39}$$

with the **k**-dependent SIC-potential, $V^{SIC}_{\mathbf{k}\nu}$, defined by

$$V^{SIC}_{\mathbf{k}\nu} \psi_{\mathbf{k}\nu} = \sum_{\mathbf{T}\nu'} e^{-i\mathbf{kT}} M^{-1}_{\nu\nu'}(\mathbf{k}) V^{SIC}_{\mathbf{T}\nu'} \psi_{\mathbf{T}\nu'} \tag{40}$$

and

$$\varepsilon_{\nu\nu'}(\mathbf{k}) = \sum_{\mathbf{T}\nu_1\nu_2} e^{i\mathbf{kT}} M^{-1}_{\nu\nu_1}(\mathbf{k}) \lambda_{0\nu_1,\mathbf{T}\nu_2} M_{\nu_2\nu'}(\mathbf{k}) \tag{41}$$

To proceed, the electron wavefunctions are expanded in a suitably complete set of basis functions. The solution of the problem within the unified Hamiltonian approach then becomes a standard matrix diagonalization problem, and the steepest descent steps (24), (29) and (31) become simple updates of the expansion coefficients, given by standard matrix and vector operations. In the present work the linear-muffin tin orbitals method (LMTO) [49-53] was used, with the additional atomic sphere approximation (ASA), i. e., the atom centered spheres are taken to overlap slightly to cover a volume equal to the actual crystal volume, and the potentials and charge densities are assumed spherically symmetric inside the spheres. Here we only give a short account of the implementation and refer the reader interested in these more technical parts to the references [44] and [41].

Let

$$\psi_\alpha = \sum_i a_i^\alpha \, \chi_i \,, \tag{42}$$

where the indices α and i label the occupied states, $\alpha \equiv \mathbf{T}\nu$, and the degrees of freedom of the basis set, $i \equiv (\tau + \mathbf{R})L$, respectively, where $\mathbf{R}$ is a lattice vector, τ an atomic position vector within the primary unit cell, and $L = (\ell, m)$ specify the angular character of the muffin-tin orbital $\chi_{(\tau+\mathbf{R})L}$. The steepest descent steps (24), (29) and (31) then read

$$\delta a^\alpha = -x \, (\mathbf{O}^{-1} - \Pi) \cdot (\mathbf{H} + \mathbf{V}^\alpha) \cdot a^\alpha \,, \tag{43}$$

$$\delta a^\alpha = -y \sum_{\alpha'} r_{\alpha\alpha'} \, a^{\alpha'} \,, \tag{44}$$

and

$$\delta a^\alpha = \frac{1}{2} \, (\mathbf{O}^{-1} - \Pi) \cdot \mathbf{O} \cdot a^\alpha \,, \tag{45}$$

respectively. Here, $\mathbf{O}$, $\mathbf{H}$ and $\mathbf{V}^\alpha$ are the overlap, the Hamiltonian, and the SIC-potential matrices: $\mathbf{O}_{ij} =< \chi_i|\chi_j >$, $\mathbf{H}_{ij} =< \chi_i|H_0|\chi_j >$, and $\mathbf{V}^\alpha_{ij} =< \chi_i|V_\alpha^{SIC}|\chi_j >$. Π is given by $\Pi_{ij} = \sum_{\alpha'}^{occ.} \delta_{\sigma_\alpha \sigma_{\alpha'}} \cdot a_i^{\alpha' *} \, a_j^{\alpha'}$, i. e., only projection on states with the same spin as ψ_α is considered. In a periodic lattice, $\mathbf{O}$, $\mathbf{H}$, and Π are translationally invariant. In the latter case this follows from the translational invariance of the a's (combining (16) and (42)):

$$a^{\mathbf{T}\nu}_{(\tau+\mathbf{R})L} = a^{\mathbf{0}\nu}_{(\tau+\mathbf{R}-\mathbf{T})L} \,. \tag{46}$$

Only the SIC potential matrix $\mathbf{V}^\alpha$ is not translationally invariant. It is advantageous to switch to $\mathbf{k}$-space: We assume $\mathbf{R}$ running over a finite cluster (N unit cells with periodic boundary clusters, where N must be so large as to ensure that the SIC states vanish outside this region). Let $\mathbf{k}$ enumerate the corresponding reciprocal space, i. e., the $N \times N$ matrix $\mathbf{U_{kR}} \equiv \frac{1}{\sqrt{N}} e^{i\mathbf{k}\cdot\mathbf{R}}$ is unitary: $UU^+ = U^+U = 1$. Then Eqs. (43)-(45) hold for the Fourier transformed quantities as well, except for the product $\mathbf{V}^\alpha a^\alpha$, which must first be evaluated in real space and Fourier transformed afterwards. Specifically, let

$$b^{\mathbf{T}\nu}_{(\tau+\mathbf{R})L} =< \chi_{(\tau+\mathbf{R})L}|V^{SIC}_{\mathbf{T}\nu}|\psi_{\mathbf{T}\nu} > \,. \tag{47}$$

Then, with

$$b^\nu_{\tau L}(\mathbf{k}) = \frac{1}{\sqrt{N}} \sum_{\mathbf{R}} e^{i\mathbf{k}\cdot(\mathbf{R}-\mathbf{T})} \, b^{\mathbf{T}\nu}_{(\tau+\mathbf{R})L} \,, \tag{48}$$

we get

$$\delta a^{\nu}(\mathbf{k}) = -x\left(\mathbf{O}^{-1}(\mathbf{k}) - \Pi(\mathbf{k})\right) \cdot \left[\mathbf{H}(\mathbf{k}) \cdot a^{\nu}(\mathbf{k}) + b^{\nu}(\mathbf{k})\right], \tag{49}$$

$$\delta a^{\nu}(\mathbf{k}) = -y \sum_{\nu'} r_{\nu\nu'}(\mathbf{k})\, a^{\nu'}(\mathbf{k}), \tag{50}$$

and

$$\delta a^{\nu}(\mathbf{k}) = \frac{1}{2}\left(\mathbf{O}^{-1} - \Pi\right) \cdot \mathbf{O} \cdot a^{\nu} \tag{51}$$

for the gradient-step, the unitary-mixing step and the orthonormalizing step, respectively. The r-matrix in $\mathbf{k}$-space in (50) is simply given by:

$$r_{\nu\nu'}(\mathbf{k}) = \sum_{\tau L} [a^{\nu'*}_{\tau L}(\mathbf{k}) \cdot b^{\nu}_{\tau L}(\mathbf{k}) - b^{\nu'*}_{\tau L}(\mathbf{k}) \cdot a^{\nu}_{\tau L}(\mathbf{k})]\,. \tag{52}$$

In the unified Hamiltonian scheme the expansion coefficients of the Bloch states are obtained by diagonalization of the following $\mathbf{k}$-dependent Hamiltonian matrix

$$\begin{aligned} <\chi_i(\mathbf{k})|H_u|\chi_j(\mathbf{k})> &= <\chi_i(\mathbf{k})|H_0|\chi_j(\mathbf{k})> \\ &- \sum_{\nu\nu'}{}' <\chi_i(\mathbf{k})|\psi_{\mathbf{k}\nu}><\psi_{\mathbf{k}\nu}|H_0|\psi_{\mathbf{k}\nu'}><\psi_{\mathbf{k}\nu'}|\chi_j(\mathbf{k})> \\ &+ \sum_{\nu} <\chi_i(\mathbf{k})|\psi_{\mathbf{k}\nu}><\psi_{\mathbf{k}\nu}|V^{SIC}_{\mathbf{k}\nu}|\psi_{\mathbf{k}\nu}><\psi_{\mathbf{k}\nu}|\chi_j(\mathbf{k})> \\ &+ \sum_{\nu\mu}(<\chi_i(\mathbf{k})|\psi_{\mathbf{k}\nu}><\psi_{\mathbf{k}\nu}|V^{SIC}_{\mathbf{k}\nu}|\psi_{\mathbf{k}\mu}><\psi_{\mathbf{k}\mu}|\chi_j(\mathbf{k})> \\ &+ <\chi_i(\mathbf{k})|\psi_{\mathbf{k}\mu}><\psi_{\mathbf{k}\mu}|V^{SIC}_{\mathbf{k}\nu}|\psi_{\mathbf{k}\nu}><\psi_{\mathbf{k}\nu}|\chi_j(\mathbf{k})>) \end{aligned} \tag{53}$$

In this expression $\chi_i(\mathbf{k})$ is a lattice Fourier transform of the muffin-tin orbital $\chi_{(\tau+\mathbf{R})L}$, and the labels i, j stand for combined angular momentum- and atomic site-indices, $i = \tau L$. The index ν runs over all SIC-corrected Bloch states, wheras the index μ covers all non-SIC-corrected Bloch states. Diagonalizing the matrix in Eq. (53) is equivalent to solving Eq. (39) with a self-consistent determination of the Lagrange parameters $\varepsilon_{\nu\nu'}$. All the bands are obtained from one matrix diagonalization, and therefore the charge density contribution of both the localized and the delocalized states is obtained in $\mathbf{k}$-space.

IV. RESULTS AND DISCUSSION

In this section we present results of some applications of the above methodology to systems where LSD fails to provide the correct physical character of the ground state. We start with the application to $3d$ transition metal (TM) oxides and the high T_c superconductor parent compound La_2CuO_4. These are then followed by the applications to $\alpha \rightarrow \gamma$ transition in cerium and cerium monopnictides.

Transition Metal Oxides And La_2CuO_4

The TM oxides are type II antiferromagnetic (AFII) wide gap insulators, with magnetic order along the (111) direction. Although the LSD approximation predicts them

Table 1. The LSD and SIC–LSD energy band gaps and spin magnetic moments compared to experimental gaps and total moments. In the columns marked 'LSD' and 'SIC–LSD', the first set of numbers corresponds to the unified Hamiltonian diagonalization [36,39] and the second set to the steepest–descent minimization [35,38].

	Band gap (eV)			Magnetic moment (μ_B)		
compound	LSD	SIC–LSD	expt.	LSD	SIC–LSD	expt.
VO	0.0	0.0	0.0	0.0	0.0	0.0
CrO	0.0	- ,1.01	-	2.99	- ,3.49	-
MnO	1.45,0.8	3.57,3.98	3.6-3.8^a	4.49,4.39	4.49,4.64	4.79^b,4.58^c
FeO	0.0	3.25,3.07	-	3.46,3.42	3.55,3.54	3.32^d
CoO	0.0	2.51,2.81	2.5^e	2.38,2.33	2.59,2.53	3.35^f,3.8^d,3.8^g
NiO	0.4,0.2	2.66,2.54	4.3^h,4.0^i	1.06,1.04	1.49,1.53	1.77^b,1.64^j,1.9^c
CuO	0.0	1.00,1.43	1.37^k	0.0	0.64,0.65	0.65^l
La_2CuO_4	0.0	2.1,1.04	2.0^m	0.0	0.66,0.47	0.4^n

a: Ref. [55]. b: Ref. [56]. c: Ref. [57]. d: Ref. [58]. e: Ref. [59]. f: Ref. [60]. g: Ref. [61]. h: Ref. [62]. i: Ref. [63]. j: Ref. [64]. k: Ref. [65]. l: Ref. [66]. m: Ref. [67]. n: Ref. [68].

to be magnetic, the magnetic moments are in general too small. Concerning the energy gaps, they are either too small or zero. In the case of La_2CuO_4, the LSD electronic band structure calculations fail in predicting its antiferromagnetic and semiconducting ground state [2], but describe the cohesive properties and phonon frequencies rather well [54]. The reason why LSD fails for La_2CuO_4 and TM oxides lies in the mean–field character of the Kohn–Sham equations, where both valence and conduction states are described by the same potential. This is corrected for in the SIC–LSD approximation where every occupied orbital sees a different potential, while all conduction states respond to the same LSD potential. Therefore, the SIC–LSD approximation provides the mechanism by which the occupied and unoccupied states are split by the on–site Coulomb interaction U, leading to wide band gaps also for localized systems with occupations, which differ from half filling.

The SIC–LSD solutions consist of localized $3d$–like orbitals on every TM atom (Cu in case of La_2CuO_4), and with the oxygen $2p$ states forming extended Bloch solutions. Assuming also the oxygen $2p$ states to be localized turns out to be energetically unfavourable, leading to an increase in the total energy in comparison to the band–like oxygen $2p$ solutions. Arai and Fujiwara [37] analyzed that this result may be an artifact of the ASA (the use of spherically symmetric SIC potentials). The optimal choice of localized states was accomplished by reaching the absolute energy minimum and maximizing the spin magnetic moment, in accord with Hund's first rule.

In Table 1, the *ab initio* SIC–LSD results for the band gaps and magnetic moments for VO, CrO, MnO, FeO, CoO, NiO, CuO, and La_2CuO_4 are compared with the LSD results and experimental results, when available. The results corresponding to both the steepest–descent minimization [35,38], and the unified Hamiltonian diagonalization [36,39] are given, when available. Since in the SIC approximation the effects of non–spherical contributions to the potential are expected to be more important than for LSD, these non–spherical terms were taken into account for TM oxides, but only in the case of the steepest–descent minimization. However, the Wigner–Seitz cells were still approximated by spheres. For VO, it is unfavourable to form a localized V $3d$ state, and only the usual LSD minimum is found, i.e., VO is predicted to be a metal, in accord with experiment. The later $3d$ monoxides, (hypothetical)CrO, MnO, FeO, CoO, NiO, and CuO, are correctly predicted to be wide gap insulators. In contrast, the LSD approximation gives no band gap for FeO, CoO, and CuO, and for MnO and NiO the

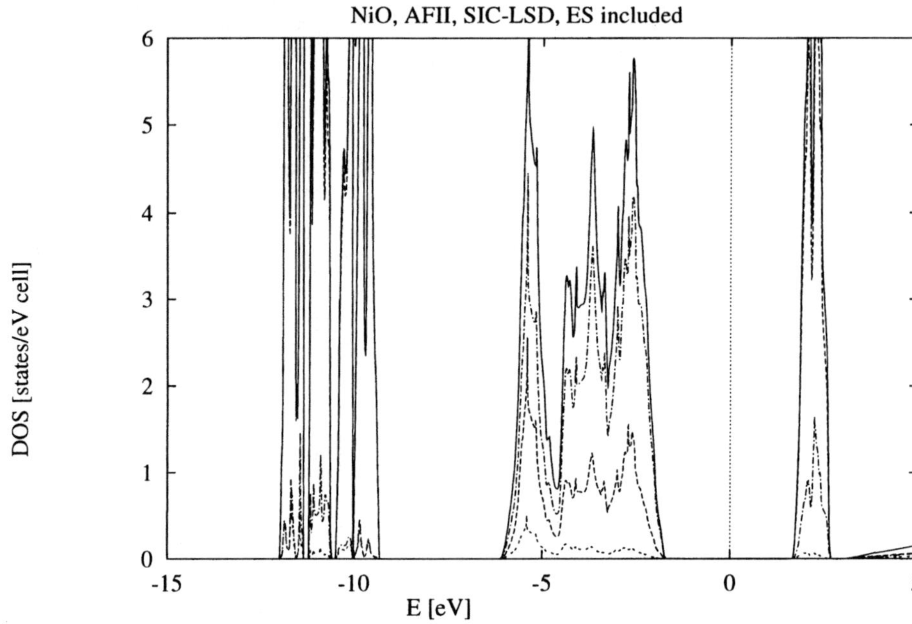

Figure 1. Total density of states for NiO in the SIC–LSD approximation (solid curve) decomposed into Ni (dashed curve), oxygen (chain curve), and interstitial sphere (dotted curve) contributions.

LSD gaps are much too small. It is reassuring that the SIC–LSD results using the two different minimization schemes are very similar. In principle, both implementations of the SIC–LSD methodology should give exactly the same answers for fully converged basis sets. The small deviations between the two calculations seen in Table 1 are caused by the minor details of the implementations of the LMTO method.

Among all systems studied NiO seems to be a special case, because the band gap is still too small in comparison with experiment. Recently, we have performed new calculations for this system with an improved basis set. This has led to a larger band gap of 3.35 eV, and a spin magnetic moment of 1.61 μ_B, as compared to the values from Table 1. The corresponding density of states (DOS) is presented in Fig. 1, where the decomposition of the total DOS into Ni, O and empty (interstitial) sphere contributions is also shown.

The nature of the band gap of TM oxides has attracted much interest in the past. Recently, the understanding has greatly increased due to the comparison of many–body model calculations and spectroscopic experiments [62,69,6]. For NiO it is found that the highest hole state has most of its weight on the oxygen atoms; i.e., it has mainly $3d^8L^{-1}$ character with some admixture of $3d^7$ character, where L^{-1} stands for a ligand (O $2p$) hole. Since the lowest electron state is a $3d^9$ state, NiO is an insulator of charge–transfer type [62,69,6]. The LSD approximation predicts that both the highest occupied and the lowest unoccupied states have primarily $3d$ character. The character of the states is therefore of the same type as in a Mott–Hubbard insulator, although the LSD approximation in other respects is more similar to the Slater picture. In the SIC–LSD calculation, as seen in Fig. 1, the lowest unoccupied states have primarily $3d$ character as in most other theories. The occupied part of the $3d$ DOS has most of its weight below the O $2p$ band, but there is also some $3d$ weight mixed into this band. The highest hole states in SIC–LSD, therefore have a mixture of $3d^8L^{-1}$ and $3d^7$ character, as in the model calculation, and NiO turns out a charge–transfer insulator.

In fact, the SIC–LSD seems to exaggerate the charge–transfer character of the gap, while the LSD approximation greatly overestimates the Mott–Hubbard character, as compared to the model calculation.

The SIC–LSD approximation gives good magnetic moments for all systems listed in Table 1, while LSD fails to give a finite magnetic moment for CuO and La_2CuO_4, and too small moment for NiO. For FeO and CoO the orbital moment plays a substantial role. The orbital contribution to the total magnetic moment was estimated to be $1.00\mu_B$, $1.13\mu_B$, $0.31\mu_B$, and $0.14\mu_B$, for the sequence of FeO, CoO, NiO and CuO [35], which seems to explain the discrepancies between the calculated spin magnetic moments in Table 1 and the experimentally derived moments, with the exception of FeO.

Summarizing this subsection, the SIC–LSD approximation is seen to produce large antiferromagnetic moments and significant charge-transfer insulating gaps in the TM oxides and La_2CuO_4, while the LSD approximation fails badly for this class of strongly correlated materials.

Cerium $\alpha \rightarrow \gamma$ Transition

The electronic structure of cerium has been widely studied over the years both experimentally and theoretically. Several phases of Ce metal exist [70] of which in particular the α and γ phases attracted a lot of attention. Both the α and γ phases have the face-centered-cubic (fcc) structure, but with a significant volume difference. The α phase is a Pauli–enhanced paramagnet, while the γ phase is characterized by localized magnetic moments with a Curie–Weiss behavior of the susceptibility. At low temperature and zero pressure the (low–volume) α phase is stable, while at room temperature the (high–volume) γ phase is more stable, but at a pressure of $\sim$8 kbar it transforms back to the α phase. This transformation is accompanied by a volume collapse of 14.8% and quenching of the magnetic moment.

Several theoretical models have been proposed to explain this first order isostructural transition that has a phase boundary terminating at a critical point, a feature which makes Ce unique among elemental metals. The promotional model [71] advocates a promotion of the Ce $4f$ electron to the $5d$ conduction band state to explain the transition. This model is, however, difficult to reconcile with the cohesive properties of Ce [72] as well as with positron annihilation [73] and photoemission [74] data, neither of which indicate a dramatic change in the $f-$electron population during the transition. In addition, band structure calculations are not able to reproduce this picture, which would require a very narrow $f-$band in the vicinity of the Fermi level [75,76]. In the Mott transition model [72] the $f-$electrons are considered to be localized in the γ phase but delocalized in the α phase. The localization of $f-$electrons in the γ phase is favored by the large on-site Coulomb interaction, dominating over the $f-$hybridization energies. When volume is decreased the $f-f$ and $f-(spd)$ hopping integrals increase and eventually the band formation energies, gained by building Bloch states, outweigh the increase in Coulomb energy, whereupon delocalization is favored. In the Kondo Volume Collapse (KVC) model [9] both the γ phase and the α phase are considered to consist of localized $f-$electrons. The phase transition is assumed to be governed by spin screening of the localized $f-$electrons by the surrounding (*spd*) conduction electrons. In this model, the γ phase consists of essentially unscreened localized moments, while the α phase consists of screened moments, and the change in Kondo energy of $\sim$ 60 meV provides the small energy quantity, a parameter of the model, needed to describe the p–V phase diagram. The Mott transition model and the KVC model have several features in common. Both of them agree that the γ phase consists of local-

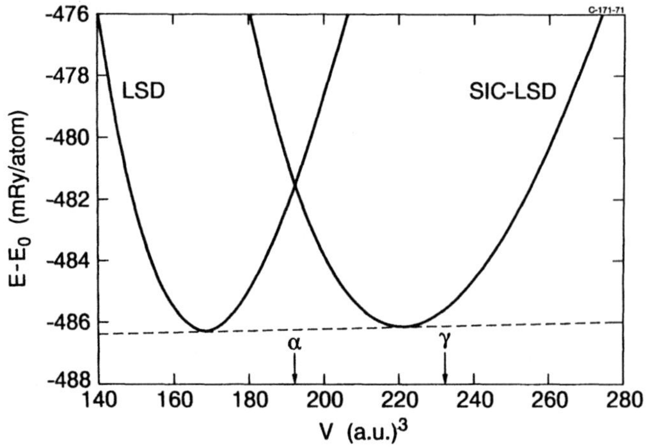

Figure 2. Cohesive energy of Ce (in mRy/atom) as a function of atomic volume (in a_0^3/atom). The curve marked 'SIC–LSD' corresponds to the calculation with one localized f−electron per Ce atom, while the curve marked 'LSD' corresponds to itinerant f-electrons. The common tangent marks the phase transition.

ized f−electrons and that the α phase is characterized by the f−electrons playing a more active role in the cohesion. However, they disagree concerning whether this *'entanglement'* of the f−electrons with the sea of $non - f$ conduction electrons can be described simply in terms of Bloch states, as assumed in the Mott transition model, or a more complicated wave function, beyond a single Slater determinant, would be more appropriate to describe the α phase.

The LSD calculations for Ce [75,76] do find an onset of spin polarization around the experimental γ phase volume, however with a too small loss of cohesion to produce a discontinuous phase transition. Recently, Eriksson *et al* [77], by implementing the orbital–polarization formalism, involving orbital and spin quenching, have reproduced certain features of the $\gamma \to \alpha$ transition. Here we will demonstrate that the SIC–LSD approximation provides a scheme for a unified description of the α and γ phases of cerium [42-44]. For clarity, below we will only report the results obtained with the steepest–descent minimization [43].

The calculated total energy of Ce as a function of volume is shown in Fig. 2. The two different curves shown are calculated assuming all f−electrons to be itinerant (curve marked LSD) and treating one f−electron per Ce atom as localized (curve marked SIC–LSD), respectively. For both sets of calculations the same (double) set of LMTO basis functions was used, assuming $s-$, $p-$, $d-$ and $f-$functions for two different linearization energies E_ν, so that orbitals of both 5s and 6s character as well as 5p and 6p character are included, in addition to 5d and 4f states.

In the LSD approximation a non-magnetic Ce ground state is found for the Wigner–Seitz radii $S \leq 3.85$ a.u., whereas a ferromagnetic state is found for $S > 3.85$ a.u., in accord with previous work [75-79]. The LSD minimum of the total energy is located in the non–magnetic region at $S = 3.42$ a.u., with a bulk modulus of $B = 484$ kbar. In the SIC–LSD calculation a localized f−state of T_{2u} symmetry gives the lowest total energy and produces the curve marked 'SIC–LSD' in Fig. 2. The minimum of the curve is found at $S = 3.74$ a.u., with a bulk modulus of $B = 310$ kbar. The difference between the two energy minima in Fig. 2 is only 0.1 mRy. Since both curves represent local minima of the same energy functional, Eq. (1), a phase transition is predicted with the transition pressure $P = -1$ kbar, which compares favourably with the value of -7 kbar, extrapolated from the experimental phase diagram to T = 0K. The volume collapse associated with the phase transition is calculated to be 24 %. The spin magnetic moment is m = $1.32\mu_B$ at the SIC–LSD minimum. The localized state contributes $1\mu_B$ so that the polarization (mainly of the $5d$ band states) contributes $\sim 0.32\mu_B$.

Analysis of the calculated density of states shows that the strong hybridization effects of the $4f$ electrons are the driving force behind the $\gamma \rightarrow \alpha$ transition in Ce. This leads to an energy lowering of the non-magnetic state below the magnetic state at small volumes. Within this *ab initio* approach the SIC–LSD also provides the mechanism for switching off the $4f$ contribution to cohesion at the phase transition. In Ref. [44] the delocalization-localization energy difference calculated with SIC–LSD was shown to be similar in magnitude to the spin screening energy of the α-phase used in the KVC model, which is the reason for the quite similar predictions of the present model of the $\gamma \rightarrow \alpha$ transition in Ce and the KVC model.

Cerium Monopnictides

The cerium monopnictides CeN, CeP, CeAs, CeSb and CeBi have peculiar properties as a function of applied hydrostatic pressure. At ambient conditions they all crystallize in the rocksalt (B1) crystal structure. With the exception of CeN the lattice constants indicate a trivalent nature of the Ce ions [80]. CeN is closer to a tetravalent Ce configuration, but photoemission experiments reveal significant mixed-valence characteristics [81-83]. Thus, CeN is similar to the collapsed α phase of pure cerium, where the cerium $f-$electron contributes actively to the cohesive properties, while the other Ce pnictides at ambient conditions are more similar to the γ phase of cerium, characterized by 'localized', i.e. atomic-like, $f-$electrons. When pressure is applied discontinuous structural phase transitions occur [80,84-89], again with the exception of CeN, where to the best of our knowledge no pressure experiment has been reported.

The magnetic properties of the cerium pnictides are quite complicated with several antiferromagnetically ordered phases and phase transitions as a function of pressure, temperature and applied magnetic field [85,90,91]. For CeP and CeAs the ground state configuration in zero field and at zero pressure and temperature is the AFI structure, which consists of an antiferromagnetic stacking of (100) planes, within which the Ce spins are ferromagnetically coupled. In CeSb and CeBi the stacking of ferromagnetically ordered (100) planes follows the sequence $(++--)$ (AF1A). The Néel temperatures are quite low, 8–30 K [92], and at high temperatures their susceptibilities show characteristic Curie-Weiss behavior indicative of local moments. Only CeN behaves more as a Pauli enhanced paramagnet. The resistivity of CeN is metallic. In these two respects, CeN is again similar to α-Ce. The resistivity of the other cerium pnictides is semimetallic with extremely low carrier density, of the order of 0.1-3% [92].

Here we present a systematic theoretical study of Ce pnictides concentrating on the structural phase transitions and magnetic structures [45]. We have performed total energy calculations as a function of volume for CeN, CeP, CeAs, CeSb and CeBi for B1 and B2 phases, in the ferromagnetic (F) arrangement of Ce moments, and with the $f-$electron treated as either delocalized (LSD) or localized (SIC–LSD). Our main results for the relevant transition pressures and volumes, together with the experimental data, are summarized in Table 2.

For CeP, our calculation is depicted in Figure 3, which shows total energy curves as a function of specific volume for the 4 different cases. The lowest energy is found in the B1 structure with localized f-electrons. From the common tangent construction, an isostructural phase transition is predicted at a pressure of 71 kbar with a volume collapse of $\Delta V/V_0$ = 8% which agrees well with the experimental values of 55 and

Table 2. Calculated transition pressures for the electronic and structural phase transitions in the cerium pnictides. Also quoted are the specific volumes on the two sides of the transition^f. The notation (d) and (l) refers to calculations with delocalized or localized Ce f−electrons, i.e. tetravalent or trivalent Ce atoms. B2* denotes the distorted B2 structure (see text for discussion).

compound	transition	P_t (kbar)		V_h (a_0^3)		V_l (a_0^3)	
		theo.	expt.	theo.	expt.	theo	expt.
CeN	B1(d) → B2(d)	620	-	148	-	141	-
CeP	B1(l) → B1(d)	71	90^a,55^b	325	308^a	297	298^a
CeP	B1(d) → B2(d)	113	$150(40)^a$	288	285^a	246	247^a
CeAs	B1(l) → B2(d)	114	$140(20)^c$	332	315^c	265	274^c
CeSb	B1(l) → B2*(l)	70	$85(25)^d$	400	398^d	353	354^d
CeSb	B2*(l) → B2*(d)	252	-	311	-	295	-
CeBi	B1(l) → B2*(l)	88	$90(40)^e$	427	399^e	376	360^e
CeBi	B2*(l) → B2*(d)	370	-	317	-	304	-

a: Ref. [84] b: Ref. [85] c: Ref. [86] d: Ref. [88] e: Ref. [89] ^{f}The errorbars in the quoted experimental transition pressures are the present authors' estimates based on the hysteresis loop observed in the experimental PV curves. The quoted volume changes are relative to the equilibrium volume and taken at the average transition pressure.

90 kbar∗ [84,85]. The B2 structure is not favorable for the CeP compound, since the calculated energy is substantially higher than that of the B1 structure. This holds for both localized and delocalized f−electrons. At higher pressures a second phase transition to the B2 structure with delocalized f−electrons is found. The transition pressure is calculated to be 113 kbar and the volume collapse 12%, while experimentally the B1→B2 phase transition occurs at 150±40 kbar* The experimental volume collapse is 11% [84]. Some uncertainty is associated with comparing the total energies calculated for the B1 and B2 phases in the ASA. Therefore we have also investigated the high pressure transition in CeP with the full-potential LMTO method [93]. In this case we have found a transition pressure of 167 kbar, which is somewhat higher, but in good accord with the experimental value.

In comparison with CeP the localized phase of CeAs is slightly more favorable than the delocalized phase, which is caused by the somewhat larger specific volume due to the larger ligand ion. As a consequence the common tangent construction leads to only one phase transition, from the localized B1 structure directly to the delocalized B2 structure. The transition pressure is calculated to be 114 kbar with a 18 % volume collapse. This is in accord with the experimental situation, where a B1→B2 transition is observed at 140±20 kbar, with a volume collapse of 11% [86]. There has been some dispute over the nature of the f−electrons in the high pressure phase [86,87]. The present calculations rule out the presence of localized f−electrons in the B2 phase of CeAs. The theoretical pictures offered by the present work is rather idealized, though. In reality, the ground state of CeAs at a given pressure will be a complicated fluctuating mixture of localized and delocalized f−electrons. The fact that the B1 phase with delocalized f−electrons is very close to the transition line may imply that the B1 phase at pressures just below the transition attains a significant component of delocalized f−electron character, i.e., mixed valent behavior. We take the present calculations as evidence for the high pressure phase having predominantly delocalized f−electrons.

* See Note f at bottom of Table 2.

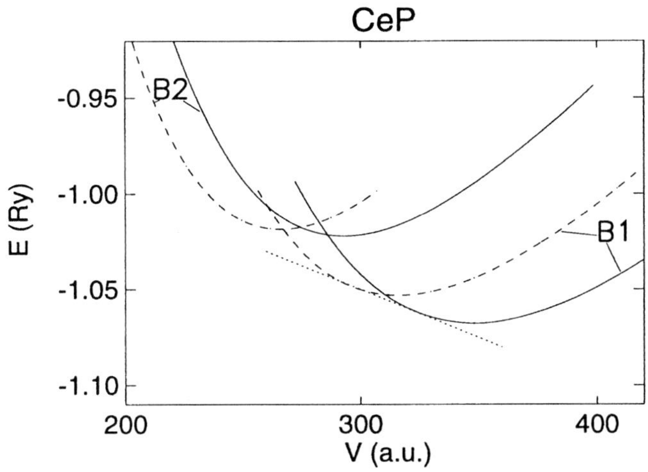

Figure 3. Cohesive energy of CeP (in Ry/formula unit) as a function of atomic volume (in a_0^3/formula unit). Two crystal structures, the B1 and B2, were considered, and each with two different treatments of the Ce f-elctrons. The solid curves correspond to calculations with one localized f−electron per Ce atom, while the dashed curves correspond to itinerant f-electrons. The localization-delocalization phase transition is marked by the common tangent.

For CeSb the phases with localized f−electrons are even further lowered relative to those with delocalized f−electrons. As a consequence the first transition to occur is from B1 to B2, both with localized f−electrons. The transition pressure is found to be 70 kbar, while the experimental B1→B2(distorted) transition occurs at 85±25 kbar, i.e. with increasing pressure the transition is observed at 110 kbar, while on releasing pressure, the reverse transition is seen at 60 kbar. The calculated volume collapse of 11% compares favorably with the 10% volume difference between the two phases seen at 85 kbar in the hysteresis loop [88]. From Table 2, it is seen that a second isostructural B2→B2 transition is predicted to occur at high pressure, at 252 kbar, where the f−electrons are delocalizing. The associated volume collapse is 4%. Experimentally, CeSb has been investigated up to 250 kbar [88], but only one discontinuous phase transition was observed. It seems necessary to investigate CeSb at somewhat higher pressures to clarify whether a second phase transition occurs. From the present work we can not exclude a new structure (i.e., neither B1 nor B2) at the high pressure side of a possible second phase transition.

The trends observed in going from CeP over CeAs to CeSb are seen to continue to CeBi, namely a preference for localized f−electrons in both the B1 and the B2 phase. A transition from B1 to B2 is found at a pressure of 88 kbar with a volume collapse of 9%, which is in perfect accord with the experimentally observed transition at 90±40 with a volume collapse of 8% [89]. As in CeSb, both sides of the transition are characterized by localized f−electrons. Only at a high pressure of $\sim$ 370 kbar is a delocalization transition foreseen, as for CeSb with a good chance of being isostructural.

In contrast to the other cerium pnictides the localization of the f−electrons is never favored in CeN. The equilibrium structure is the B1 structure with a volume of 195 a_0^3. The experimental volume is 212 a_0^3. The slight underestimation of the equilibrium volume may be an indication that the true CeN ground state is of a more complicated structure than that of itinerant f−electrons. The present calculations show, however, that the LDA ground state with itinerant f−electrons is a better representative of the

CeN ground state than the SIC–LSD ground state with fully localized f−electrons. The present work predicts a structural phase transition B1→B2 around 620 kbar.

To address the important question of the magnetic structure we also performed calculations for AFI and AFII antiferromagnetic arrangements in the rocksalt B1 crystal structure for CeP and CeSb. In the calculations with localized f−electrons all the magnetic structures were very close in energy: at the experimental lattice constant the AFI was lower by 0.5 mRy than the AFII and 1.8 mRy than the F structure. On the other hand the magnetic moment was lowest in the F structure: 1.00 μ_B per CeP, compared with 1.07 μ_B in the AFII and 1.04 μ_B in the AFI. In short, the f−electron is more localized in the AFI and AFII structures than in the F structure. In the case of delocalized f−electrons, we found that the ferromagnetic structure was the structure of lowest energy at the experimental lattice constant. However, at high pressure the AFII structure won over the F structure. Therefore, we predict that a change of magnetic structure occurs in the high pressure phase. The AFII structure of CeP is insulating and has a gap of 0.1 eV, whilst the AFI structure is semi-metallic with a density of states at the Fermi level of 0.066 states/Ce and Ry. The electron pockets are situated both at the Γ and X points, whilst the compensating holes are found along the direction starting at the X point and running in parallel to the c-axis. By comparison, for CeSb in the AFI structure the number of states is slightly increased to 0.1 states/Ce and Ry, and more strikingly, due to a reduction of the separation between the Ce d states and the P p states, the holes occur at the Γ point.

In summary, the calculations explain the experimental structural facts as follows: With increasing Z of the ligand, the specific volume at zero pressure increases and the ionicity of the compound decreases. The delocalization of Ce f−electrons is less favorable at large specific volumes, while the B2 structure is favored over the B1 structure with decreasing ionicity. The tetragonal structure observed at high pressure in CeSb and CeBi is almost degenerate with the undistorted B2 structure. Concerning magnetic ordering, the AFI structure is found to be the ground state in both CeP and CeSb. We find an intricate interplay of the charge and magnetization densities: The groundstate is *not* the structure which has the most localized f−electron (AFII). Moreover, our calculations show that the number of carriers and their character is determined by the magnetic structure.

V. CONCLUSIONS

We have demonstrated that the SIC–LSD approach is an *ab initio* electronic structure scheme that describes static Coulomb correlation effects much better than the LSD approximation. It is capable of providing a correct description of the purely electronic localized → delocalized phase transitions in solid materials on an equal footing with the more conventional structural transitions and magnetic phases. This approach goes well beyond the LSD approximation and seems to have a great future concerning applications to heavy fermions, mixed valence states, and in general, strongly correlated systems.

ACKNOWLEDGMENTS

This work has benefited from collaborations within, and has been partially funded by, the Human Capital and Mobility Network on 'Ab-initio (from electronic structure) calculation of complex processes in materials' (contract:ERBCHRXCT930369). A.S.

acknowledges support from the Danish Natural Science Research Council (Grant no. 11-9001-3).

REFERENCES

1. R. O. Jones and O. Gunnarsson, Rev. Mod. Phys. **61**, 689 (1989).
2. W. Pickett, Rev. Mod. Phys. **61**, 433 (1989).
3. N. F. Mott, *Metal-Insulator Transitions* (Taylor and Francis, London, 1974).
4. B. Brandow, Adv. Phys. **26**, 651 (1977); J. Alloys and Compounds, **181**, 377 (1992).
5. K. Terakura, A. R. Williams, T. Oguchi and J. Kübler, Phys. Rev. Lett. **52**, 1830 (1984) ; Phys. Rev. B**30**, 4734 (1984).
6. J. Zaanen, G. A. Sawatzky and J. W. Allen, Phys. Rev. Lett. **55**, 418 (1985).
7. V. I. Anisimov, J. Zaanen and O. K. Andersen, Phys. Rev. B **44**, 943 (1991).
8. B. Johansson, Phil. Mag. **30**, 469 (1974).
9. J. W. Allen and R. M. Martin, Phys. Rev. Lett. **49**, 1106 (1982); J. W. Allen and L. Z. Liu, Phys. Rev. B **46**, 5047 (1992).
10. D. Glötzel, J. Phys. F **8**, L163 (1978); D. Glötzel and R. Podloucky, Physica **102B**, 348 (1980).
11. H. L. Skriver, O. K. Andersen and B. Johansson, Phys. Rev. Lett. **44**, 1230 (1980).
12. J. Hubbard, Proc. R. Soc. London A**276**, 238 (1963); A**277**, 237 (1964); A**281** 401 (1964)
13. V. I. Anisimov, F. Aryasetiawan and A. I. Liechtenstein, J. Phys.: Condens. Matter, **9**, 767 (1997).
14. J. Taylor, *Scattering Theory* (Wiley, New York, 1972).
15. R. D. Cowan, Phys. Rev. **163**, B54 (1967); I. Lindgren, Int. J. Quant. Chem. **5**, 411 (1971); G. W. Bryant and G. D. Mahan, Phys. Rev. B **17**, 1744 (1978); J. P. Perdew, Chem. Phys. Lett. **64**, 127 (1979). Phys. Rev. B **41** 7311 (1990).
16. A. Zunger, J. P. Perdew and G. L. Oliver, Solid State Commun. **34**, 933 (1980).
17. J. P. Perdew and A. Zunger, Phys. Rev. B **23**, 5048 (1981).
18. P. Hohenberg and W. Kohn, Phys. Rev. **136**, B864 (1964): W. Kohn and L. J. Sham, Phys. Rev. A **140**, 1133 (1965).
19. S. H. Vosko, L. Wilk and M. Nusair, Can. J. Phys. **58**,1200 (1980).
20. A. Svane, Phys. Rev. B **51**, 7924 (1995).
21. M. Levy, Proc. Natl. Acad. Sci. USA **76**, 6062 (1979).
22. T. Kotani, Phys. Rev. B **50**, 14816 (1994); T. Kotani, Phys. Rev. Lett. **74**, 2989 (1995).
23. J. G. Harrison, J. Chem. Phys. **78**, 4562 (1983); *ibid.*, **79**, 2265 (1983); M. R. Pederson and C. C. Lin, J. Chem. Phys. **88**, 1807 (1988); S. D. Manoli and M. A. Whitehead, Phys. Rev. A **38**, 3187 (1988); M. M. Rieger and P. Vogl, Phys. Rev. B**52**, 16567 (1995).
24. J. G. Harrison, R. A. Heaton and C. C. Lin, J. Phys. B **16**, 2079 (1983);
25. M. R. Pederson, R. A. Heaton and C. C. Lin, J. Chem. Phys., **80**, 1972 (1984); *ibid.*, **82**, 2688 (1985).
26. Y. Guo and M. A. Whitehead, J. Mol. Struct. (Theochem), **228**, 285 (1991);*ibid.*, **231**, 307 (1991).
27. Y. Ishii, S. Ohnishi and S. Sugano, Phys. Rev. B **33**, 5271 (1986).
28. E. S. Fois, J. I. Penman and P. A. Madden, J. Chem. Phys. **98**, 6352 (1993).
29. J. M. Pacheco and W. Ekardt, Z. Phys. D **24**, 65 (1992).
30. H. Katayama-Yoshida and A. Zunger, Phys. Rev. Lett. **53**, 1256 (1984); Phys. Rev. B **31**, 7877 (1985); *ibid.*, 8317 (1985).
31. R. A. Heaton, J. G. Harrison and C. C. Lin, Phys. Rev. B **31**, 1077 (1985) ; M. R. Pederson and B. M. Klein, Phys. Rev. B **37**, 10319 (1988); K. A. Jackson and C. C. Lin, Phys. Rev. B **38**, 12171 (1988); *ibid.*, **41**, 947 (1990); S. C. Erwin and C. C. Lin, Phys. Rev. B **40**, 1892 (1989).
32. M. J. Puska and R. Nieminen, Phys. Rev. B **43**, 12221 (1991).
33. R. A. Heaton, J. G. Harrison and C. C. Lin, Phys. Rev. B **28**, 5992 (1983); R. A. Heaton and C. C. Lin, J. Phys. C **17**, 1853 (1984); M. R. Pederson and C. C. Lin, Phys. Rev. B **35**, 2273 (1987); S. C. Erwin and C. C. Lin, J. Phys. C **21**, 4285 (1988). F. Gan, Y.-N. Xu, M.-Z. Huang, W. Y. Ching and J. G. Harrison, Phys. Rev. B **45**, 8248 (1992).
34. N. Hamada and S. Ohnishi, Phys. Rev. B **34**, 9042 (1986); Y. Hatsugai and T. Fujiwara, Phys. Rev. B **37**, 1280 (1988).
35. A. Svane and O. Gunnarsson, Phys. Rev. Lett. **65**, 1148 (1990).
36. Z. Szotek, W. M. Temmerman and H. Winter, Phys. Rev. B **47**, 4029 (1993).

37. M. Arai and T. Fujiwara, Phys. Rev. B **51**, 1477 (1995).
38. A. Svane, Phys. Rev. Lett. **68**, 1900 (1992).
39. W. M. Temmerman, Z. Szotek and H. Winter, Phys. Rev. B**47**, 11533 (1993).
40. I. V. Solov'ev, A. I. Liechtenstein and V. A. Gubanov, Sov. Phys. Solid State, **33**, 572 (1991).
41. W. M. Temmerman, Z. Szotek and H. Winter, Phys. Rev. B **47**, 1184 (1993).
42. Z. Szotek, W. M. Temmerman and H. Winter, Phys. Rev. Lett. **72**, 1244 (1994).
43. A. Svane, Phys. Rev. Lett. **72**, 1248 (1994).
44. A. Svane, Phys. Rev. B**53**, 4275 (1996).
45. A. Svane, W. M. Temmerman, Z. Szotek, H. Winter, Solid State Commun. **102**, 473 (1997).
46. A. Svane, J. Trygg, B. Johansson and O. Eriksson, Phys. Rev. B**56**,7143 (1997).
47. A. Svane and O. Gunnarsson, Solid State Commun. **76**, 851 (1990).
48. A. R. Williams and J. Soler, Bull. Amer. Phys. Soc. **32**, 562 (1987).
49. O. K. Andersen, Phys. Rev. B **12**, 3060 (1975);
50. H. L. Skriver, *The LMTO Method* (Springer Verlag, Berlin, 1984).
51. O. K. Andersen and O. Jepsen, Phys. Rev. Lett. **53**, 2571 (1984).
52. O. K. Andersen, O. Jepsen and O. Glötzel, *Canonical description of the band structures of metals*, in *Proceedings of the International School of Physics, Course LXXXIX, Varenna, 1985* edited by F. Bassani, F. Fumi and M. P. Tosi (North- Holland, Amsterdam, 1985), p. 59.
53. O. K. Andersen, Z. Pawlowska and O. Jepsen, Phys. Rev. B **34**, 5253 (1986).
54. R.E. Cohen, W.E. Pickett, and H. Krakauer, Phys. Rev. Lett. **62**, 831 (1990); C. O. Rodriguez, A. I. Lichtenstein, I. I. Mazin, O. Jepsen, O. K. Andersen, and M. Methfessel, Phys. Rev. B **42**, 2692 (1990).
55. I. N. Iskenderov, I. A. Drabkin, L. T. Emelýanova, and Ya. M. Ksendzov, Fiz. Tverd. Tela (Leningrad) **10**, 2573 (1968) [Sov. Phys. Solid State **10**, 2031 (1969)].
56. B. E. F. Fender, A. J. Jacobson, and F. A. Wegwood, J. Chem. Phys. **48**, 990 (1968).
57. A. K. Cheetham and D. A. O. Hope, Phys. Rev. B **27**, 6964 (1983).
58. W. L. Roth, Phys. Rev. **110**, 1333 (1958).
59. J. van Elp, J. L. Wieland, H. Eskes, P. Kuiper, G. A. Sawatzky, F.M.F. de Groot and T.S. Turner, Phys. Rev. B**44**, 6090 (1991).
60. D. C. Khan and R. A. Erickson, Phys. Rev. B **1**, 2243 (1970).
61. D. Herrmann-Ronzaud, P. Burlet, and J. Rossat-Mignod, J. Phys. C **11**, 2123 (1978).
62. A. Fujimori, F. Minami, and S. Sugano, Phys. Rev. B **29**, 5225 (1984); A. Fujimori and F. Minami, Phys. Rev. B **30**,957 (1984).
63. S. Hüfner, J. Osterwalder, T. Riesterer, and F. Hulliger, Solid State Commun. **52**, 793 (1984).
64. H. A. Alperin, J. Phys. Soc. Jpn. Suppl. B **17**, 12 (1962).
65. F. P. Koffyberg and F. A. Benko, J. Appl. Phys. **53**, 1173 (1982).
66. J. B. Forsyth, P. J. Brown, and B. M. Wanklyn, J.Phys. C **21**, 2917 (1988).
67. J. M. Ginder, R. M. Roe, Y. Song, R. P. McCall, J. R. Gaines, E. Ehrenfreund, and A. J. Epstein, Phys. Rev. B **37**, 7506 (1988).
68. D. Vaknin, S. K. Sinha, D. E. Moneton, D. C. Johnson, J. Newsam, C. R. Safinya, and H. E. King, Phys. Rev. Lett. **58**, 2802 (1987).
69. G. A. Sawatzky and J. W. Allen, Phys. Rev. Lett. **53**, 2339 (1984).
70. D. C. Koskenmaki and K. A. Gschneidner, in *Handbook on the Physics and Chemistry of Rare Earths*, edited by K.A. Gschneidner and L. Eyring (North-Holland, Amsterdam, 1978), Vol. 1, Chapter 4.
71. B. Coqblin and A. Blandin, Advan. Phys. **17**, 281 (1968); R. Ramirez and L. M. Falicov, Phys. Rev. B **3**, 1225 (1971).
72. B. Johansson, Phil. Mag. **30**, 469 (1974).
73. D. R. Gustafsson, J. D. McNutt, and R. O. Roelling, Phys. Rev. **183**, 435 (1969).
74. E. Willoud, H. R. Moser, W.-D. Schneider, and Y. Baer, Phys. Rev. B **28**, 7354 (1983).
75. D. Glötzel, J. Phys. F **8**, L163 (1978); D. Glötzel and R. Podloucky, Physica **102B**, 348 (1980).
76. W. E. Pickett, A. J. Freeman, and D. D. Koelling, Physica **102B**, 341 (1980); W. E. Pickett, A. J. Freeman, and D. D. Koelling, Phys. Rev. B **23**, 1266 (1981).
77. O. Eriksson, M. S. S. Brooks, B. Johansson, Phys. Rev. B **41**, 7311 (1990).
78. B. I. Min, H. J. F. Jansen, T. Oguchi, and A. J. Freeman, Phys. Rev. B **34**, 369 (1986).
79. J. M. Wills, O. Eriksson, and A. M. Boring, Phys. Rev. Lett. **67**, 2215 (1991); O. Eriksson, J. M. Wills, and A. M. Boring, Phys. Rev. B **46**, 12981 (1992).
80. A. Jayaraman, W. Lowe, L. D. Longinotti and E. Bucher, Phys. Rev. Lett. **36**, 366 (1976).
81. Y. Baer, R. Hauger, Ch. Zürcher, M. Campagna and G. K. Wertheim, Phys. Rev. B. **18**, 4433 (1978).

82. A. Franciosi, J. H. Weaver, N. Mårtensson and M. Croft, Phys. Rev. B. **24**, 3651 (1981).
83. W. Gudat, R. Rosei, J. H. Weaver, E. Kaldis and F. Hulliger, Solid State Commun. **41**, 37 (1982).
84. I. Vedel, A. M. Redon, J. Rossat-Mignod, O. Vogt and J. M. Leger, J. Phys. C **20**, 3439, (1987).
85. N. Mori, Y. Okayama, H. Takahashi, Y. Haga and T. Suzuki, Physica B **186-188**, 444 (1993).
86. A. Werner, H. D. Hochheimer, R. L. Meng and E. Bucher, Physics Lett. **97A**, 207, (1983).
87. J. M. Leger, I. Vedel, A. M. Redon and J. Rossat-Mignod, J. Magn. Magn. Mater. **63& 64**, 49, (1987).
88. J. M. Leger, D. Ravot and J. Rossat-Mignod, J. Phys. C **17**, 4935, (1984).
89. J. M. Leger, K. Oki, J. Rossat-Mignod and O. Vogt, J. de Physique **46**, 889, (1985).
90. Y. Okayama, Y. Ohara, S. Mituda, H. Takahashi, H. Yoshizawa, T. Osakabe, M. Kohgi, Y. Haga, T. Suzuki and N. Mori, Physica B **186-188**, 531 (1993).
91. F. Hulliger, M. Landolt, H. R. Ott and R. Schmelczer, J. Low Temp. Phys. **20**, 269 (1975).
92. T. Suzuki, Physica B **186-188**, 347 (1993).
93. M. Methfessel, Phys. Rev. B **38**, 1537 (1988); M. Methfessel, C. O. Rodriguez, and O. K. Andersen, Phys. Rev. B **40**, 2009 (1989).

DENSITY FUNCTIONAL CALCULATIONS FOR MOLECULES AND CLUSTERS – Li_n, Li_nO, C_n

R. O. Jones

Institut für Festkörperforschung, Forschungszentrum Jülich
D-52425 Jülich, Germany

INTRODUCTION

Applications of density functional (DF) methods to molecules and atomic clusters have expanded greatly in recent years. The reasons for this are not hard to find: DF calculations using the local spin density (LSD) approximation are free of adjustable parameters and can be applied to systems where accurate calculations of the many-electron wave function are not yet possible. Furthermore, the use of modifications of the LSD approximation to the exchange-correlation energy has resulted in significant improvements in the calculated formation energies of numerous small molecules [1]. This has helped overcome the long-standing reservations of many in the chemical community, and the DF method has found favor in many applications to larger molecules and clusters.

In the present article, we apply DF methods to lithium clusters and their monoxides with up to ten atoms, and give some initial results of a continuing study of carbon clusters with between 14 and 24 atoms. These are areas that show both the advantages and the limitations of DF calculations. Calculations with predictive capability are possible for systems that are difficult to treat with wave function-based methods, and the relative ease with which molecular dynamics (MD) can be combined with DF calculations enables us to study many geometrical configurations and to find isomers that were previously unknown. On the other hand, the different approximations used for the exchange-correlation energy can lead – particularly in the carbon clusters – to significant differences in the energy ordering of the isomers. It is not yet clear whether the non-local modifications to the LSD approximation automatically lead to improved energy differences.

Lithium, the lightest element that is metallic under normal conditions, has served for decades as a model "nearly free electron" system. Recently, the structural and bonding properties of small lithium clusters and of molecules containing Li have received much attention. There is a single s-valence electron in the alkali elements, and the energy surfaces of the clusters show shallower minima and less tendency to directional bonding than found in elements with p- and d-electrons. In fact, there is evidence for pseudorotation in the electron spin resonance (ESR) and two-photon ionization spectra for Li_3 [2,3] and in ESR spectra for Li_5 [4]. The structural flexibility and the absence of a guide to the most stable of the many possible structures is a particular challenge for theoreticians, who also have estimates of the dissociation energies provided by thermochemical (mass-spectrometric) measurements on clusters up to Li_5 [5]. Lithium clusters with up to $\sim$ 40 atoms have been ionized and detected as cations, and their ionization potentials determined up to $n = 26$ [6]. Measurements on the unimolecular dissociation of Li_n^+ clusters [7] have provided information on the cohesive energies and dissociation paths of clusters with $n = 4-42$. Studies of neutral and charged Li clusters using Hartree-Fock (HF) based methods [8] have indicated that planar structures are preferred to $n = 6$, with larger clusters (to Li_9) comprising deformed tetrahedra.

Small molecules containing lithium and other elements have received much attention in recent years, with lithiated molecules containing O, C, or Si showing unusual structures that appear to have more than eight valence electrons occupying bonding orbitals. The first "hyperlithiated" molecule (Li_3O) was identified experimentally [9], and theory has led subsequently to many unexpected structures. There is little doubt that many unusual structures remain to be found.

The elemental clusters that have been studied most intensively in recent years are those of carbon. The detection of clusters up to C_{190} [10] and the explanation of the particular stability of C_{60} [11] were followed by a surge of interest in C_n clusters that has abated little. The original measurements [10] showed a bimodal distribution, with all clusters sizes for $n < 30$ being detected, but only clusters with even n above 30. The latter are usually assumed to comprise hexagons and pentagons ("fullerenes"), but there has been much speculation about the structures of the smaller clusters. It is generally accepted that linear chains are favored up to about ten atoms, where monocyclic rings become more stable. For cluster sizes in the range $n = 10 - 30$, the suggested structures include monocyclic, bicyclic and other rings [12,13], as well as "bowl" and cage structures.

The carbon vapor arising from laser ablation shows the presence of only the ring isomer of C_{20} [14], while calculations with a variety of methods predict that bowl (corannulene) and dodecahedral structures are also stable. The relative stability of these isomers is a matter of some controversy. HF calculations favor the ring structure, as do DF calculations with gradient corrections, but the LSD approximation gives the dodecahedron as the most stable [15]. However, while quantum Monte Carlo calculations [15] favor the bowl structure, coupled-cluster calculations [16] predict that the dodecahedron is the most stable form. Accurate calculations of the many-electron wave function are immensely difficult in clusters of this size. In order to shed more light on the problem, we have undertaken an extensive study of many isomers of C_n clusters with $n \sim 20$. Some of the first results are presented here.

METHOD OF CALCULATION

We describe here the results of density functional (DF) calculations on the energy surfaces of lithium clusters Li_n and their monoxides Li_nO with up to ten atoms (Li_{10}, Li_9O), as well as of carbon clusters C_n with $n = 18, 20$. We have adopted a range of starting structures in all cases and have used both all-electron and pseudopotential calculations. The all-electron calculations use a Gaussian basis set, * and all structures shown below were obtained with this method. In systems where pseudorotation is

*DGauss program from the UniChem package of Cray Research, Inc. The calculations used a triple zeta Gaussian orbital basis with d polarization functions (TZ94+p) and the TZ94 auxiliary basis.

known to occur or where the nature of the most stable isomer is uncertain, we have used a combination of DF calculations with molecular dynamics (MD) † at temperatures between 20-500 K to generate additional geometries and to clarify structural changes. The MD/DF calculations use periodic boundary conditions (PBC) with a (simple cubic) unit cell with lattice constant 20 a.u., and a plane wave basis set with a single point ($k = 0$) in the Brillouin zone and an energy cutoff of 15 a.u. (30 a.u. in systems with O or C). The electron-ion interaction is described by the non-local pseudopotential of Troullier and Martins [17], using the d-component of the potential as the reference local part ("sp-nonlocality").

All molecules were studied with both a local spin density (LSD) and a non-local modification (NLSD) involving the gradient of the density [18]. The MD/DF calculations were all performed with the LSD approximation. For full details of the calculations and the results see Refs [19] (Li_n, Li_nO) and [20] (C_n). We focus here on the structural and binding trends.

LITHIUM CLUSTERS AND THEIR OXIDES

In this section, we outline the results obtained for structures and relative energies of isomers of neutral lithium clusters Li_n to $n = 10$. Energies with respect to the most stable isomer are denoted ΔE, and "bonds" between Li atoms are shown in Fig. 1 if the atoms are separated by less than 3.1 Å.

Li_n: Structures

Li_2: This is the lithium cluster for which the most detailed experimental information is available, and the calculated bond lengths and vibration frequencies of the $^1\Sigma_g^+$ ground state (LSD: 2.80 Å, 325 cm^{-1}; NLSD: 2.80 Å, 322 cm^{-1}) are in reasonable agreement with the measured values (2.673 Å, 351.4 cm^{-1}) [21].

Li_3: The D_{3h} equilateral triangle undergoes a Jahn-Teller distortion to C_{2v} symmetry. There are three equivalent energy wells, and electron spin resonance (ESR) spectra of Li_3 trapped in argon matrices [2] show that the molecule is fluxional, i.e., it undergoes pseudorotation between the different minima. The present calculations support this picture. The energy difference between the most stable structure and the saddle point is minute (LSD: 110 cm^{-1}; NLSD: 165 cm^{-1}) and less than the calculated zero point vibrational energy (320 cm^{-1}). MD/DF calculations on Li_3 at both 20 K and 30 K show pseudorotation clearly.

Li_4, Li_5: HF-based calculations [8] predict that the most stable isomer is the rhombus [singlet, D_{2h}], and this is also found here. These HF-based calculations [8] predict Li_5 to be planar (C_{2v}), with a three-dimensional C_{2v} structure the next most stable. We find two local minima with C_{2v} symmetry, with the pyramidal form more stable by 0.23 eV (NLSD: 0.16 eV). The situation is reminiscent of earlier results on aluminum clusters [22], where planar structures were favored up to Al_4, and planar and three-dimensional isomers had very similar energies in the pentamer.

ESR measurements have been performed on Li clusters deposited at 77 K in an adamantane matrix and annealed to temperatures above 200 K [4]. The spectra show that the Li_n clusters present are fluxional, and the coupling constants indicate that $n > 3$. Pseudopotential DF calculations [23] showed that the isotropic spin population for the planar (C_{2v}) structure of Li_5 agrees with the measured value, and these authors

†CPMD version 3.0 written by J. Hutter, Max-Planck-Institut für Festkörperforschung, Stuttgart, 1996.

proposed a pseudorotation mechanism based on geometry changes within a plane. We have performed MD/DF simulations on Li_5 at 100 K. The results show the ready interchange of the planar and three-dimensional structures, and suggest a simple mechanism – the twisting of the central triangle – for pseudorotation between equivalent trigonal bipyramids via an almost planar form.

Li_6: HF-based calculations [8] show that three structures have similar energies: a planar D_{3h} structure, a quasiplanar pentagonal pyramid (C_{5v}) with one atom slightly above the plane, and a (non-planar) tripyramid (C_{2v}). Both LSD and NLSD calculations give several isomers with comparable energies and favor three-dimensional over planar structures, but there are significant differences from the HF structures. The C_{2v} tripyramid found in the latter distorts here to our most stable form, the symmetric flattened octahedron [D_{4h}, 1(a)]. The planar pentagon [D_{5h} [1(b)] is 0.45 eV higher, followed by the planar D_{3h} structure [1(c), the most stable in the HF calculations] an additional 0.29 eV higher. The symmetries D_{3h}, D_{4h}, and D_{5h} all have twofold representations. In the present cases the highest occupied orbitals are doubly degenerate and fully occupied, so that a Jahn-Teller distortion does not occur.

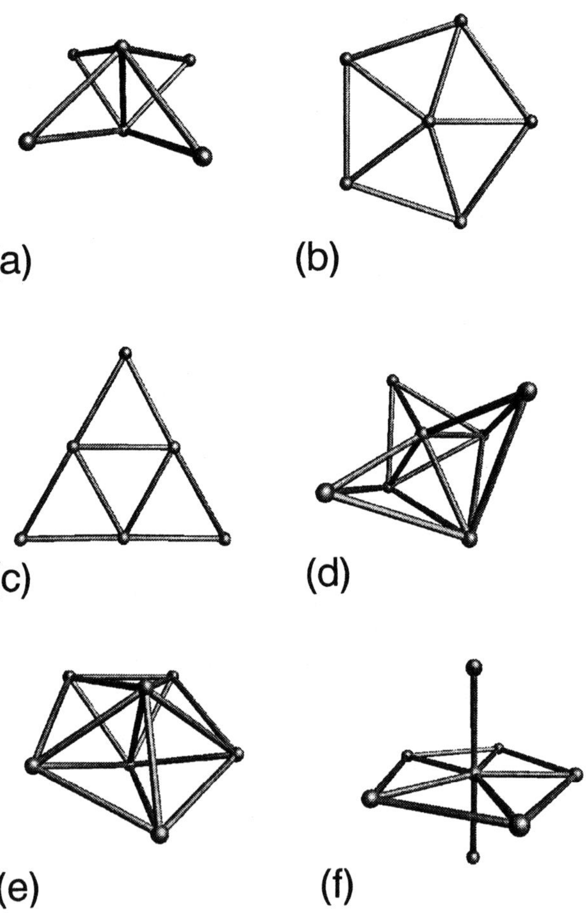

Figure 1. Structures of Li_6 [(a) D_{4h}, (b) D_{5h}, (c) D_{3h}]; Li_7 [(d) C_{3v}, (e) D_{5h}], and Li_8 [(f) D_{5h}].

Li_7, Li_8: The earlier HF-based calculations [8] indicate a most stable isomer of pentagonal bipyramidal form [D_{5h}, 1(e)], followed by a C_{3v} isomer [1(d)] formed by capping three of the faces of a tetrahedron. Similar structures are found in the present work, although the order is reversed [the D_{5h} structure is more stable by 0.10 eV]. Two isomers of Li_8 were considered in the HF-based work [8], a tetrahedron with all four faces capped (T_d) and a related D_{2d} structure. The most stable structure found in the present calculations is, however, a further member in the D_{5h} sequence Li_6 [1(b)] $\rightarrow$ Li_7 [1(e)] $\rightarrow$ Li_8 [1(f)].

Li_9, Li_{10}: The number of isomers increases dramatically with increasing cluster size, so that the identification of the most stable becomes much more difficult. For these clusters we have studied structures derived by adding atoms to stable isomers of smaller clusters, removing them from larger ones, and structures proposed by other authors. We have also used MD/DF simulations to cover larger regions of configuration space and to guide out choice of starting structures for the all-electron calculations. The most stable structures found for Li_9 and Li_{10} are discussed in Ref. [19].

Li_n: Cohesive Energies

The cohesive energy – the energy per atom required to separate the most stable form of a cluster into its constituent atoms – has been estimated for lithium clusters in two ways: from high-temperature gas phase mass spectroscopy [5] and from combining the measured dissociation energies of the Li_n^+ clusters [7] with the ionization potentials [6]. Both sets of results are compared in Fig. 2 with the LSD and NLSD values and with the values extrapolated from MRD-CI calculations for geometries scaled from the HF values [8].

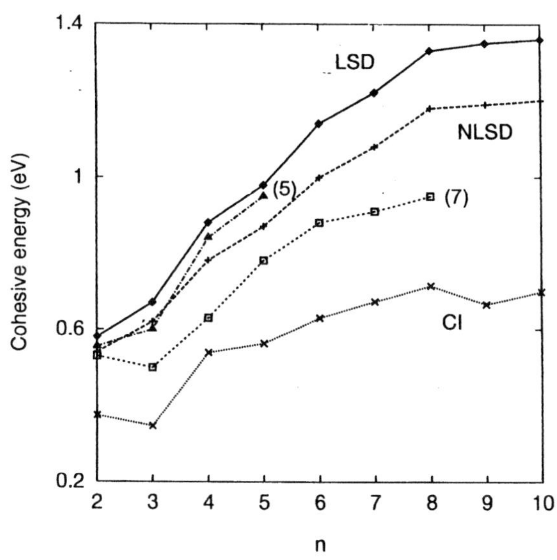

Figure 2. Calculated (LSD, NLSD) and measured [5] and [7] cohesive energies (binding energies per atom) for Li_n clusters to $n = 10$. Also shown are values obtained by extrapolating MRD-CI calculations for scaled HF geometries [8].

The overall agreement between theory and experiment is satisfactory, and the use of the NLSD approximation leads to cohesive energies that are lower than the LSD values. This is a common effect in clusters of elements that form *sp*-bonds, examples being P_n and As_n clusters, where the LSD approximation leads to overestimates in the cohesive energy of ~1 eV [24]. The LSD estimates of the binding energy are much closer to experiment here than in the group 15 clusters, and the effects of gradient corrections are smaller. We see in Fig. 2 that the two experimental techniques give rise to somewhat different results, and it is difficult to say which of the LSD and NLSD estimates is better.

Lithium Cluster Oxides

We now discuss isomers of lithium cluster oxides Li_nO to $n = 9$. We label atom 1 (black) as oxygen, the remaining (grey) atoms are Li. Li-O "bonds" are shown if the atoms are separated by less than 2.3 Å. Oxygen has six valence electrons to one in Li, and the single oxygen atom has a dramatic effect on the structures of small lithium clusters. A comparable effect is evident in the solid state, where the density of *lithium* atoms in the oxide Li_2O is over 80% higher than that in crystalline lithium itself. ‡

LiO, Li_2O: Small lithium monoxide clusters were investigated almost 20 years ago by HF calculations (LiO, LiOLi, LiLiO) [25] and thermochemical measurements (LiO, Li_2O, Li_3O) [9,26]. HF calculations [25] gave a bond length in LiO of 1.71 Å and a vibration frequency of 829 cm^{-1}. The present results [LSD: 1.72 Å, 784 cm^{-1}; NLSD: 1.74 Å, 760 cm^{-1}] follow a common trend for DF calculations to give bonds that are longer than HF values. Compared with the measured dissociation energy (3.49 eV) [21], the present results (LSD: 4.36 eV, NLSD: 3.89 eV) show (3.49 eV) [21], the present results (LSD: 4.36 eV, NLSD: 3.89 eV) show the familiar picture of a substantial LSD overestimate being improved by non-local modifications. HF calculations on the LiOLi molecule [25] gave a linear singlet ground state [$D_{\infty h}$, ${}^1\Sigma_g^+$], with bond length 1.62 Å and vibration frequencies 145 (π_u), 839 (σ_g), and 1091 (σ_u) cm^{-1}. Our bond lengths are slightly longer, but the vibration frequencies are close to the earlier estimates.

The ground state of the LiLiO molecule is a linear triplet state [${}^3\Pi$], characterized by a very long Li-Li bond and a Li-O bond similar to that in the diatomic molecule. This is the simplest example in the present work of a "hypervalent" molecule, where the number of bonds on the central Li atoms appears to be greater than the number of valence electrons.

Li_3O: This was the first "hyperlithiated" molecule to be prepared [9], and it has been studied theoretically by several authors. HF calculations [27,28] indicated that a planar C_{2v} structure was the most stable, and this is supported by recent second order Møller-Plesset (MP2) and complete active space (CAS) SCF calculations [29]. However, the energy surface is very flat, and the barrier against pseudorotation is less than 0.03 eV. The present calculations also give a C_{2v} structure that is weakly perturbed from the equilateral triangle (D_{3h}).

Li_4O and Li_5O: These were studied theoretically by Schleyer and coworkers [27,28] and identified by Wu using high temperature mass spectrometry [30]. The most stable structure in the HF calculations was tetrahedral [T_d], with a Li-O bond of 1.728 Å. The T_d (singlet) state is also the most stable in the present calculations. The Li_5O molecule has several isomers with similar energies separated by low energy barriers. HF calculations [28] were performed for the C_{3v} bipyramid isomer of Li_5O [Fig. 3(b)], which is also a stable structure in the present work.

The most stable isomers of Li_5O show five-fold coordination of the central atom. A discussion of two such molecules (PF_5 and PCl_5) led Berry [31] to propose a pseudorotation mechanism trigonal bipyramid → square pyramid → trigonal bipyramid, in which the isomerization takes place by deformation of the bond angles at the central

‡The density of solid Li is 0.534 g cm^{-3}, that of Li_2O 2.103 g cm^{-3} [*Gmelins Handbuch der anorganischen Chemie. Lithium*, 8. Auflage, edited by E. H. E. Pietsch (Verlag Chemie, Weinheim, 1960), p. 166, p. 260]. The density of Li atoms in Li_2O is then ~0.98 g cm^{-3}. The lattice constant of bcc Li is 3.502 Å (*ibid.*, p. 164), so that the nearest neighbor separation is 3.03 Å.

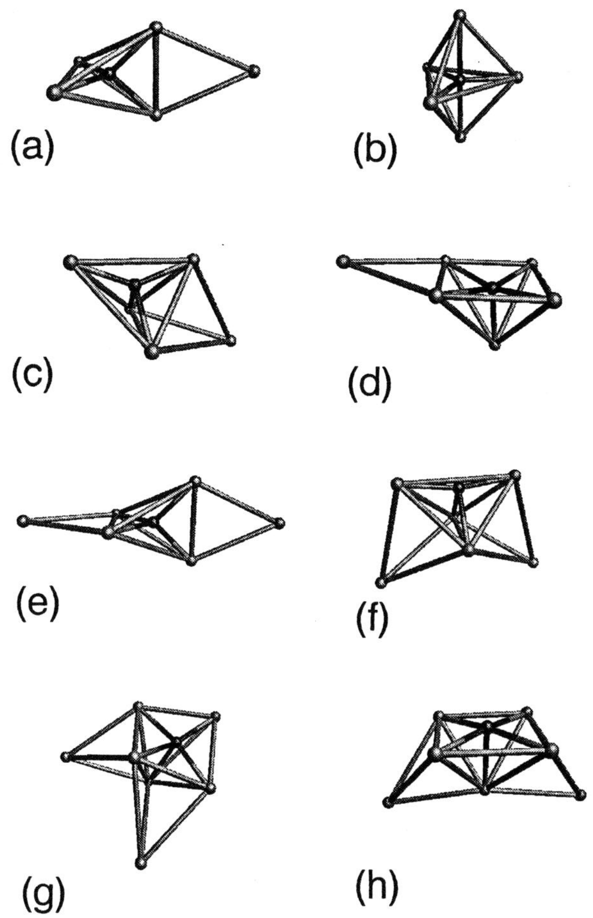

Figure 3. Structures of Li_5O (a-c); Li_6O (d-f); Li_7O (g,h).

atom. In the present calculations the C_s structure [3(c)] is an unstable stationary point in the energy surface only slightly (< 0.05 eV) above the minima noted above. An oxygen atom in a lithium "cage" attached to a Li subcluster is a common pattern in this series.

Li_6O: For this cluster (twelve valence electrons) the first (HF) calculations [28] found a minimum for a D_{3d} distortion of a regular octahedron (O_h). HF and MP2 calculations [32] also found a D_{2d} isomer with two bidentate Li_3 groups [3(e)]. This structure bears an obvious relationship to isomer 3(b) of Li_5O. The most stable isomer in the present LSD calculations is, however, 3(d), followed by 3(e) and 3(f) (C_{2v}), both of which have $\Delta E = 0.07$eV.

A similar picture results from the NLSD calculations, but isomer 3(e) is about 0.04 eV more stable than 3(d), with 3(f) an additional 0.05 eV higher. The energy differences between these structures and those derived from the octahedron are close to the LSD values. The small energy differences between 3(a,b,c) in both sets of calculations indicates that all three are candidates for the most stable isomer.

Li_7O: We have found three stable isomers with structures that are related to the most stable form of Li_5O, and the most stable in both LSD and NLSD calculations is 3(g) (C_s). In the LSD calculations 3(h) (C_{2v}) is 0.14 eV higher, and the NLSD relative energies are very similar. For a discussion of the structures of Li_8O and Li_9O see Ref. [19].

Bonding Trends

For clusters larger than Li_2O there are many novel structures with more than eight valence electrons, so that they appear to have more "bonds" than available electron pairs. Insight into such "hypervalent" molecules has been provided by the groups of Schleyer, Marsden, and others. In a study of Li_nO molecules up to $n = 6$, for example, Schleyer [28] showed that the charge (Mulliken) on the oxygen atom increased little beyond that in Li-O-Li, with the additional electrons contributing primarily to Li-Li bonding. The role of oxygen – with six valence electrons to the one in Li – is crucial in determining the structure.

The oxide clusters Li_nO are generally more compact than their Li_n counterparts and show interesting patterns involving Li_4O (tetrahedral), Li_5O [square (C_{4v}) or rectangular (C_{2v}) pyramids], and Li_6O [approximately octahedral] units. The tetrahedron is the most stable form of Li_4O, and the capped tetrahedra in Li_5O [3(a)] and Li_6O [3(e)] are only slightly higher in energy then the pyramidal structures such as 3(d). Four-fold coordination of the O atom is not favored in larger clusters. If the initial structure of one of the larger clusters (e.g., Li_7O) has a central O atom, this atom moves towards the "surface" as the structure optimization proceeds, leaving behind a Li subcluster [e.g., 3(g)]. This is such a dramatic process that it is natural to speculate that it could occur in much larger clusters or at a surface.

CARBON CLUSTERS

Clusters with around 20 atoms can occur in a great number of isomers. Since relatively few have been studied in the past, we have used a variety of techniques to generate starting structures. Approximately one third of the structures, including those with topologies published previously by other authors, have been obtained from parameterized tight-binding molecular dynamics simulations [33]. Many others have been obtained by removing atoms from larger clusters or adding them to smaller, and multiple ring structures have been assembled from ring geometries already optimized. The use of the MD/DF program to perform simulations at elevated temperatures proved to be decisive in several cases. It would be unwise to pretend that all energetically favorable isomers have been found, as "unexpected" structures occurred with unnerving regularity throughout the work. It does appear, however, that enough isomers have been found to allow some distinct patterns to be identified.

In Figs. 4 and 5 we show the energies relative to the monocyclic rings of a range of isomers of C_{18} and C_{20}, respectively. Results for both the LSD and gradient-corrected (GC) functionals are shown, and the of the plots follows the energetic ordering of the former. Early extended Hückel calculations by Hoffmann [34] indicated that the ring isomers of C_{10}, C_{14} and C_{18} (i.e., C_n with $n = 4q + 2$) were particularly stable relative to the linear chains. A fourfold periodicity is also apparent in the present results for the ring and other structures. Full details will be given in Ref. [20], and I restrict the present discussion to these two cases.

Fig. 4 shows that the monocyclic ring is the most stable isomer of C_{18} in both LSD and GC calculations. The ring is "cumulenic", i.e., all bonds are of equal length. There are several bicyclic (4,5,6,7,8) and "graphitic" (3,9,11) structures. The last of these is a local minimum in the energy surface, but it transforms on annealing to 5 and thence

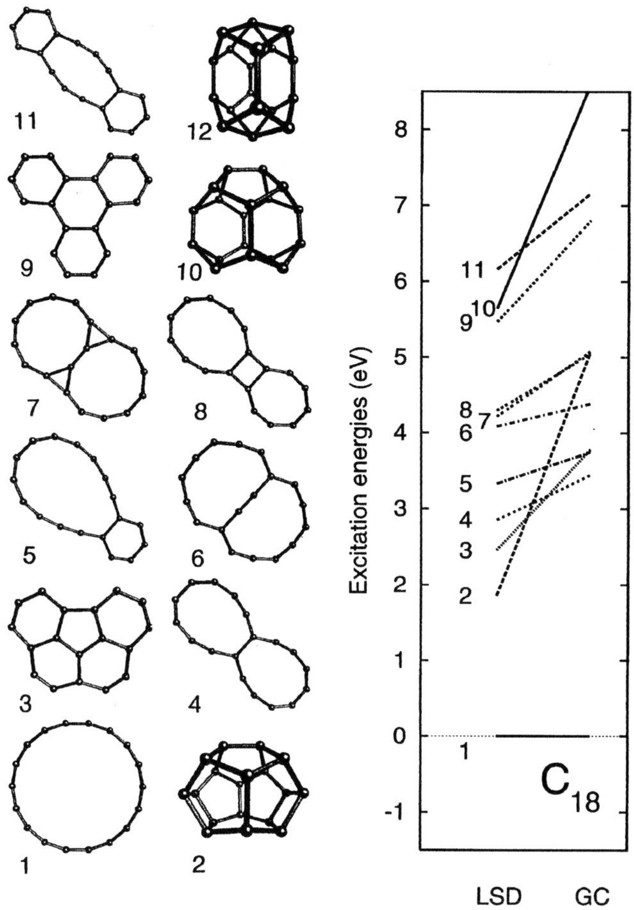

Figure 4. Structures of C_{18}.

to the monocyclic ring. The results for the lower-lying isomers of C_{20} are consistent with the earlier work. Several bicyclic rings are amongst the higher-lying isomers, and there are several cage structures (1,4,6,9,12).

Some general observations are of interest. The most stable cage structures in C_{18} and C_{20} are related by removing two atoms from the latter and allowing the structure to relax. With the exception of $C_{20}(1)$, all cage structures shown incorporate four-membered rings. The monocyclic ring [$C_{20}(3)$] is "polyacetylenic", with bonds of alternating length. The unstable cumulenic form 3′ lies ~1 eV higher in energy. The balance between "strain" and "bond formation" is evident in isomers 2 and 7 of C_{20}, which are both non-planar. The structure $C_{18}(3)$, by contrast, is planar, and could perhaps be described best as a "plate". The occurrence of a single threefold ring in $C_{20}(12)$ is found to be typical of clusters with $n = 4q$, while $C_{18}(7)$ is representative of those with $n = 4q + 2$. Of particular interest is the observation that the calculated energy differences using LSD and GC approximations to the exchange-correlation energy are remarkably similar for the different *families* of structures. If we consider the cage structures, for example, the lines joining the LSD and GC results are almost parallel, and the same applies to the double ring structures. One can see immediately that

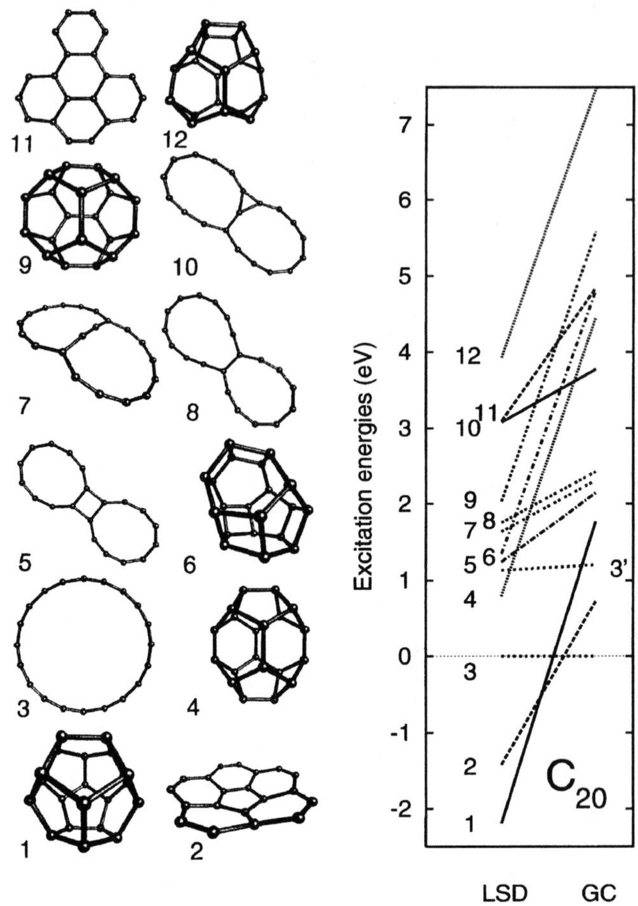

Figure 5. Structures of C_{20}.

there is a qualitative correlation with the number of bonds or the coordination of the constituent atoms. These and other trends will be discussed in detail elsewhere [20].

CONCLUDING REMARKS

We have performed an extensive series of density functional (DF) calculations on lithium clusters and their monoxides with up to ten atoms, and C_n clusters for $n = 14$–24. A variety of starting structures has been used in each, and we have generated many more by combining DF calculations with molecular dynamics. Previously unknown structures have been found in all families.

Lithium shares with the other alkali metals and hydrogen the simplest valence electron configuration, a single s-electron. The energy minima are less pronounced than in many other main-group elements, and the barriers between them are lower. MD/DF simulations show that both Li_3 and Li_5 are fluxional, consistent with the pseudorotation observed in both molecules. The calculated cohesive energies [Fig. 2] show results that are consistent with earlier experience with other elements: overestimates with the LSD approximation to the exchange-correlation energy, improved values with the gradient corrected (NLSD) modification. The lithium cluster monoxides are often "hypervalent", reflecting the delocalized nature of the bonding in these systems.

The first results of an extensive study of carbon clusters also show numerous structures that had not been discussed previously, and also show that LSD and GC approximations lead to distinctly different orderings in the relative energies of the isomers. It is not yet clear which set of results is better, and an answer to this problem will await the availability and analysis of reliable experimental data. We are currently carrying out such a study for C_n^- ions.

Acknowledgments

Parts of this work were performed in collaboration with J. Hutter, A. Lichtenstein, and G. Seifert. I thank C. H. Wu for suggesting the study of molecules containing lithium, and I. Bytheway for helpful discussions at the Workshop.

REFERENCES

1. See, for example, A. D. Becke, J. Chem. Phys. **96**, 2155 (1992); B. G. Johnson, P. M. W. Gill, and J. A. Pople, J. Chem. Phys. **97**, 7846 (1992).
2. D. A. Garland and D. M. Lindsay, J. Chem. Phys. **78**, 2813 (1983).
3. P. Dugourd, J. Chevaleyre, M. Broyer, J. P. Wolf, and L. Wöste, Chem. Phys. Letters **175**, 555 (1990).
4. J. A. Howard, H. A. Joly, R. Jones, P. P. Edwards, R. J. Singer, and D. E. Logan, Chem. Phys. Letters **204**, 128 (1993).
5. C.H. Wu, J. Chem. Phys. **65**, 3181 (1976) [Li_2, Li_3]; J. Phys. Chem. **87**, 1534 (1983) [Li_4]; J. Chem. Phys. **91**, 546 (1989) [Li_5].
6. P. Dugourd, D. Rayane, P. Labastie, B. Vezin, J. Chevaleyre, and M. Broyer, Chem. Phys. Lett. **197**, 433 (1992).
7. C. Bréchignac, H. Busch, P. Cahuzac, and J. Leygnier, J. Chem. Phys. **101**, 6992 (1994).
8. I. Boustani, W. Pewestorf, P. Fantucci, V. Bonačić-Koutecký, and J. Koutecký, Phys. Rev. B **35**, 9437 (1987) [Li_n, Li_n^+]; I. Boustani and J. Koutecký, J. Chem. Phys. **88**, 5657 (1988) [Li_n^-].
9. H. Kudo, C.H. Wu, and H. R. Ihle, J. Nucl. Mater. **78**, 380 (1978); C. H. Wu, H. Kudo, and H. R. Ihle, J. Chem. Phys. **70**, 1815 (1979).
10. E. A. Rohlfing, D. M. Cox, and A. Kaldor, J. Chem. Phys. **81**, 3322 (1984).
11. H. W. Kroto, J. R. Heath, S. C. O'Brien, R. F. Curl, and R. E. Smalley, Nature **318**, 162 (1985).
12. G. von Helden, M.-T. Hsu, N. Gotts, and M. T. Bowers, J. Phys. Chem. **97**, 8182 (1993).
13. H. Handschuh, G. Ganteför, B. Kessler, P. S. Bechthold, and W. Eberhardt, Phys. Rev. Lett. **74**, 1095 (1995).
14. G. von Helden, N. G. Gotts, M. T. Bowers, Nature **363**, 60 (1993); J. Hunter, J. Frye, and M. F. Jarrold, Science **260**, 784 (1993).
15. J. C. Grossman, L. Mitas, and K. Raghavachari, Phys. Rev. Lett. **75**, 3870 (1995).
16. P. R. Taylor, E. Bylaska, J. H. Weare, and R. Kawai, Chem. Phys. Lett. **235**, 558 (1995).
17. N. Troullier and J. L. Martins, Phys. Rev. B **43**, 1993 (1991).
18. The gradient-corrected functional used the modification to the exchange energy of A. D. Becke, Phys. Rev. A **38**, 3098 (1988) and the modified correlation energy functional of J. P. Perdew, Phys. Rev. B **33**, 8822 (1986).
19. R. O. Jones, A. Lichtenstein, and J. Hutter, J. Chem. Phys. **106**, 4566 (1997).
20. R. O. Jones and G. Seifert, Phys. Rev. Lett. **79**, 443 (1997).
21. K. P. Huber and G.Herzberg, *Molecular Spectra and Molecular Structure. IV. Constants of Diatomic Molecules*, (Van Nostrand Reinhold, New York, 1979).
22. R. O. Jones, J. Chem. Phys. **99**, 1194 (1993).
23. R. Kawai, J. F. Tombrello, and J. H. Weare, Phys. Rev. A **49**, 4236 (1994).
24. P. Ballone and R. O. Jones, J. Chem. Phys. **100**, 4941 (1994).
25. D. T. Grow and R. M. Pitzer, J. Chem. Phys. **67**, 4019 (1977).
26. H. Kimura, M. Asano, and K. Kubo, J. Nucl. Mater. **92**, 221 (1980).
27. P. v.R. Schleyer, E.-U. Würthwein, and J. A. Pople, J. Am. Chem. Soc. **104**, 5839 (1982).
28. P. v.R. Schleyer, in *New Horizons in Quantum Chemistry*, edited by P. O. Löwdin and B. Pullman, (Reidel, Dordrecht, 1983), p. 95.

29. M. Gutowski and J. Simons, J. Phys. Chem. **98**, 8326 (1994).
30. C. H. Wu, Chem. Phys. Lett. **139**, 357 (1987).
31. R. S. Berry, J. Chem. Phys. **32**, 933 (1960).
32. J. Ivanic, C. J. Marsden, and D. M. Hassett, J. Chem. Soc., Chem. Commun. **1993**, 822 (1993).
33. D. Porezag, T. Frauenheim, T. Köhler, G. Seifert, and R. Kaschner, Phys. Rev. B **51**, 12947 (1995).
34. R. Hoffmann, Tetrahedron **22**, 521 (1966).

A FUNCTIONAL THEORY OF INTERACTING LOCAL SPINS, SPIN POLARIZED ELECTRONS, AND IONS: HALF METALLIC MAGNETS

A.K. Rajagopal

Naval Research Laboratory
Washington D.C. 20375-5320, USA

INTRODUCTION

Magnetism of manganites such as (La,Ba)MnO_3 has a long history going back to 1950, with the discovery of ferromagnetism and metallic conduction upon doping [1]. Zener [2] proposed to explain these phenomena in terms of a "double exchange" mechanism. This spawned several important developments [3,4,5] leading to new implications of this model. Most recently, the observation of colossal magnetoresistance (CMR) effects [6,7] in these systems has rekindled interest in studying the detailed interplay of magnetism, electronic transport, and structure in these materials. Electronic structure calculations on these systems have been performed [8] and theoretical investigations using modern techniques [9,10,11] have shown the need for self-consistent coupled fields of electrons, localized spins, and ions in this system. In this paper, we offer a functional theory of this type.

We obtain (a) the Pauli-Schrödinger-like equation for the itinerant spin polarized electrons with effective spin-dependent potentials; (b) the Landau-Lifshitz-like equation for the localized spin vector with an effective magnetic field; and (c) the Born-von Karman-like equation for the ion coordinate with effective force equation, with all the "effective" contributions arising from the interactions among all these fields and thus the three equations are coupled to each other and require a self-consistent approach in its full glory. Also the "effective" contributions arise from a single functional of the set of variables mentioned above.

In Section 2, we give a brief account of the formalism. In Section 3, we outline a possible framework for introducing approximation schemes. In Section 4, we give a brief summary and outlook. This is a preliminary account of a collaborative project with Professors S.D. Mahanti and Tom Kaplan of Michigan State University.

FUNCTIONAL FORMULATION OF THE SYSTEM

The Hamiltonian under consideration describes (a) localized spins denoted by the operators, $\hat{\vec{S}}_l$, located on lattice sites, obeying the usual angular momentum commutation rules, interacting among each other via a Heisenberg interaction; (b) interacting itinerant electrons which are spin polarizable, interacting with each other via a two particle interaction (e.g., the Coulomb interaction); (c) interaction of localized spins with the spin density of the itinerant electron system; and (d) possibly also allowing the lattice to vibrate so as to include lattice motion as well. Thus we have here three interacting fields with their Hamiltonian given by:

$$\hat{H} = \hat{H}_S + \hat{H}_e + \hat{H}_{Se} + \hat{H}_i \,, \tag{1}$$

where

$$\begin{aligned}
\hat{H}_S &= \sum_{l \neq l'} J\left(\hat{\vec{R}}_l - \hat{\vec{R}}_{l'}\right) \hat{\vec{S}}_l \cdot \hat{\vec{S}}_{l'} \,, \\
\hat{H}_e &= \hat{T}_e + \hat{U} = \sum_{\sigma} \int d^3r \hat{\psi}^{\dagger}_{\sigma}(\vec{r}) \left(-\frac{\hbar^2}{2m}\nabla^2 + V_c(\vec{r})\right) \hat{\psi}_{\sigma}(\vec{r}) \\
&\quad + \frac{1}{2}\sum_{\sigma,\sigma'} \int\int d^3r d^3r' \hat{\psi}^{\dagger}_{\sigma}(\vec{r}) \hat{\psi}^{\dagger}_{\sigma'}(\vec{r}') U(\vec{r} - \vec{r}') \hat{\psi}_{\sigma'}(\vec{r}') \hat{\psi}_{\sigma}(\vec{r}) \,, \\
\hat{H}_{Se} &= \sum_{l} \int d^3r W\left(\vec{r} - \hat{\vec{R}}_l\right) \hat{\vec{S}}_l \cdot \hat{\vec{s}}(\vec{r}) \,, \\
\hat{H}_i &= \hat{T}_i + \phi = \sum_{l} \frac{\hat{\vec{P}}_l^2}{2M} + \sum_{l \neq l'} \phi\left(\hat{\vec{R}}_l, \hat{\vec{R}}_{l'}\right)
\end{aligned} \tag{2}$$

We introduce external time-dependent fields that couple to these basic entities by means of the external Hamiltonian:

$$\begin{aligned}
\hat{H}^{ext}(t) &= \hat{H}^{ext}_S(t) + \hat{H}^{ext}_e(t) + \hat{H}^{ext}_i(t) \,, \\
\hat{H}^{ext}_S(t) &= \sum_{l} \vec{b}^{ext}_l(t) \cdot \hat{\vec{S}}_l \,, \\
\hat{H}^{ext}_e(t) &= \hat{H}^{ext(1)}_e(t) + \hat{H}^{ext(2)}_e(t) \text{ with} \\
\hat{H}^{ext(1)}_e(t) &= \int d^3r \nu^{ext}(\vec{r}, t) \hat{n}(\vec{r}) \,, \\
\hat{H}^{ext(2)}_e(t) &= \int d^3r \vec{B}^{ext}(\vec{r}, t) \cdot \hat{\vec{s}}(\vec{r}) \,, \text{ and} \\
\hat{H}^{ext}_i(t) &= \sum_{l} \vec{F}^{ext}_l(t) \cdot \hat{\vec{R}}_l \,.
\end{aligned} \tag{3}$$

In the above the itinerant electron density and spin density vector operators are given respectively by

$$\hat{n}(\vec{r}) = \sum_{\sigma} \psi^{\dagger}_{\sigma}(\vec{r}) \psi_{\sigma}(\vec{r}) \,, \quad \hat{\vec{s}}(\vec{r}) = \sum_{\sigma,\sigma'} \psi^{\dagger}_{\sigma}(\vec{r}) \vec{\tau}_{\sigma\sigma'} \psi_{\sigma'}(\vec{r}) \,. \tag{4}$$

Here $\vec{\tau}_{\sigma\sigma'}$ are the matrix components of the standard Pauli spin vector. These operators obey the standard equal time canonical commutation rules (CCR) among themselves: the localized spin operators obey the angular momentum CCR, the electron field operators obey anti commutation rules (ACR) and the ion position and momentum obey the usual CCR. And all of the different types of operators commute among each other at equal times since they represent independent physical quantities.

The equation of motion obeyed by the density matrix, $\hat{\rho}(t)$, of the total system is the usual Neumann equation:

$$i\hbar\frac{d\hat{\rho}(t)}{dt} = \left[\hat{H}(t), \hat{\rho}(t)\right], \text{ with } \hat{H}(t) = \hat{H} + \hat{H}^{ext}(t) , \tag{5}$$

given that at the given initial time, $t = t_0$, the system is in equilibrium, where the density matrix is stationary. The density matrix has unit trace in the Hilbert space of the entire system:

$$tr\hat{\rho}(t) = 1 . \tag{6}$$

The mean value of an arbitrary operator, $\hat{O}(t)$, which may have its own intrinsic time dependence, is defined by

$$\left\langle \hat{O}(t) \right\rangle_t = tr\hat{O}(t)\rho(t) . \tag{7}$$

Then the time rate of change of this mean value is given by, using Eq. (5),

$$i\hbar\frac{d}{dt}\left\langle \hat{O}(t) \right\rangle_t = \left\langle i\hbar\frac{\partial\hat{O}(T)}{\partial t} + \left[\hat{O}(t), \hat{H} + \hat{H}^{ext}(t)\right] \right\rangle_t . \tag{8}$$

Mapping Theorem A (Stationary Case)

In the stationary equilibrium state, $t = t_0$, the system maximizes the von Neumann entropy,

$$S = -tr\hat{\rho}(t_0)\ln\hat{\rho}(t_0) , \tag{9}$$

subject to given mean values of the total Hamitonian without the external fields. We will now prove that the corresponding free energy is a functional of the mean number, n, spin density, $\vec{s}$, of the itinerant electrons, local spin vectors, $\vec{S}_l$, and ion positions, $\vec{R}_l$, in much the same way as in the thermal density functional theory of Mermin-Hohenberg-Kohn. *(Note that the mean values are denoted by the same symbols as for the respective operators but without the carets on their top.)* To prove this, we take the external fields specified in Eq. (3) at $t = t_0$; the coefficients of the respective operators in the external Hamiltonian then serve as Lagrange multipliers in taking care of the constraints. The Lagrange multiplier associated with the mean value of the external field-free Hamiltonian is then the usual temperature, T. Thus, we have the minimum free energy principle,

$$\Omega_{H^{ext}}[\hat{\rho}'] > \Omega_{H^{ext}}[\hat{\rho}] \ , \ \hat{\rho}' \neq \hat{\rho} , \tag{10}$$

where $\hat{\rho}$ minimizes the free energy given by

$$\Omega_{H^{ext}}[\hat{\rho}] = tr\left(\hat{H} - kT\ln\hat{\rho}\right)\hat{\rho} + tr\hat{H}^{ext}\hat{\rho} . \tag{11}$$

The mapping

$$\vec{b}_l \rightarrow \vec{S}_l \ , \ \nu(\vec{r}) \rightarrow n(\vec{r}) \ , \ \vec{B}(\vec{r}) \rightarrow \vec{s}(\vec{r}) \ , \ \vec{F}_l \rightarrow \vec{R}_l \tag{12}$$

is established by considering the Hamiltonian

$$\hat{H}'(t_0) = \hat{H}(t_0) + \hat{H}^{ext'}(t_0) - \hat{H}^{ext}(t_0) , \tag{13}$$

with its density matrix $\hat{\rho}'(t_0)$, with the corresponding mean values of the operators denoted by primes and the associated free energy $\Omega_{H^{ext'}}[\hat{\rho}']$ manipulated by using the above relations to the form

$$\begin{aligned}\Omega_{H^{ext'}}[\hat{\rho}'] &= \Omega_{H^{ext}}[\hat{\rho}'] + tr\left(\hat{H}^{ext'} - \hat{H}^{ext}\right)\hat{\rho}' \\ &> \Omega_{H^{ext}}[\hat{\rho}] + tr\left(\hat{H}^{ext'} - \hat{H}^{ext}\right)\hat{\rho}'. \end{aligned} \tag{14}$$

The last line follows upon using the minimal property of the free energy, Eq. (10). In the same fashion, reversing the primed and unprimed quantities above we obtain

$$\Omega_{H^{ext}}[\hat{\rho}] > \Omega_{H^{ext'}}[\hat{\rho}'] + tr\left(\hat{H}^{ext} - \hat{H}^{ext'}\right)\hat{\rho}\,. \tag{15}$$

Note that the trace in last term is explicitly given by

$$\begin{aligned} tr\left(\hat{H}^{ext'} - \hat{H}^{ext}\right)\hat{\rho}' &= \sum_l \left(\vec{b}_l^{ext'}(t) - \vec{b}_l^{ext}(t)\right)\cdot\vec{S}_l' \\ &+ \int d^3r\left(\nu^{ext'}(\vec{r},t) - \nu^{ext}(\vec{r},t)\right)n'(\vec{r}) \\ &+ \int d^3r\left(\vec{B}^{ext'}(\vec{r},t) - \vec{B}^{ext}(\vec{r},t)\right)\cdot\vec{s}'(\vec{r}) \\ &+ \sum_l \left(\vec{F}_l^{ext'}(t) - \vec{F}_l^{ext}(t)\right)\cdot\vec{R}_l'\,. \end{aligned} \tag{16}$$

Eqs. (14,15) are contradictory if the two density matrices are designed to yield the same average quantities and thus the statement of the 1-1 correspondence is proved for the initial equilibrium condition. Thus we have also shown that the **density matrix** and free energy are functionals of $n(\vec{r})$, $\vec{s}(\vec{r})$, $\vec{S}_l$, and $\vec{R}_l$.

Mapping Theorem B (Nonstationary Case)

We next turn our attention to the mapping theorem when we turn on the time dependence of the fields, so that the 1-1 correspondence may continue to hold. The proof goes along the same lines as originally given by Runge and Gross [12] but extended to the density matrix formalism by Rajagopal and Buot [13]. The method employs Eqs. (7,8) and the Taylor expansions of all the external fields around the initial time, and the various commutation relations. We will not give the details of this demonstration as it is fairly straightforward. This establishes that the time-dependent density matrix continues to be a functional of $n(\vec{r},t)$, $\vec{s}(\vec{r},t)$, $\vec{S}_l(t)$, and $\vec{R}_l(t)$.

Dynamical Equations of Motion

The next step is the derivation of the equations of motion for the various entities appearing in this system. In order to do that, we invoke the "Action Principle", defining the action by

$$B\left[n,\vec{s},\vec{S},\vec{R}\right] = \int_{t_0}^{t_1} dt\; tr\left\{\hat{\Lambda}\left(i\hbar\frac{d}{dt} - \hat{\mathcal{L}}\right)\hat{\rho}(t)\right\}\,. \tag{17}$$

Here $\hat{\Lambda}$ is an auxiliary functional introduced in order to include "backward" evolution in the theory (see Ref. [13]). $\hat{\mathcal{L}}$ is the Liouvillean operator associated with $\hat{H}$, the total Hamiltonian given by Eq. (1). It is defined by

$$\hat{\mathcal{L}}\hat{\rho} = \left[\hat{H},\hat{\rho}\right]_-\,.$$

The upper limit on the time integration is the final time up to which the td (time-dependent) external fields are applied. The stationary action principle is used to obtain the basic coupled equations of motion for the various quantities and the corresponding initial conditions are determined from the free energy functional.

The action functional (17) is a functional of the variables mentioned above and it is stationary with respect to variations of these fields. This feature leads to the equations to be derived below. Again we shall not give the details of the derivation but give here the final result.

(1) The Pauli-Schrödinger-like equation for the electron spinor

$$\begin{aligned}
\underline{\Psi}_\alpha(\vec{r},t) &= \begin{pmatrix} \psi_{\alpha\uparrow}(\vec{r},t) \\ \psi_{\alpha\downarrow}(\vec{r},t) \end{pmatrix} \text{ , } \alpha : \text{quantum number. :} \\
i\hbar\frac{\partial}{\partial t}\underline{\Psi}_\alpha(\vec{r},t) &= \Big\{\Big(-\frac{\hbar^2\nabla^2}{2m} + \nu^{ext}(\vec{r},t) + V_H[n] + V_{xc}\left[n,\vec{s},\vec{S},\vec{R}\right]\Big)\underline{1} \\
&\quad + \Big(\vec{b}^{ext}(\vec{r},t) + \vec{W}_{xc}\left[n,\vec{s},\vec{S},\vec{R}\right]\Big)\cdot\underline{\vec{\tau}}\Big\}\,\underline{\Psi}_\alpha(\vec{r},t) \\
n(\vec{r},t) &= \sum_{\alpha,\sigma}\psi^*_{\alpha\sigma}(\vec{r},t)\psi_{\alpha\sigma}(\vec{r},t)f_{\alpha\sigma\sigma}\ , \\
\vec{s}(\vec{r},t) &= \sum_{\alpha,\sigma,\sigma'}\psi^*_{\alpha\sigma}(\vec{r},t)\vec{\tau}_{\sigma\sigma'}\psi_{\alpha\sigma'}(\vec{r},t)f_{\alpha\sigma\sigma'}
\end{aligned} \tag{18}$$

Here the external and Hartree terms are the usual contributions and the new terms are functional derivatives of the appropriate action functional $\mathcal{B}$ to be discussed further below:

$$\begin{aligned}
V_{xc}\left[n,\vec{s},\vec{S},\vec{R}\right] &= \frac{\delta\mathcal{B}\left[n,\vec{s},\vec{S},\vec{R}\right]}{\delta n(\vec{r},t)}\ ; \\
\vec{W}_{xc}\left[n,\vec{s},\vec{S},\vec{R}\right] &= \frac{\delta\mathcal{B}\left[n,\vec{s},\vec{S},\vec{R}\right]}{\delta\vec{s}(\vec{r},t)}\ .
\end{aligned} \tag{19}$$

(2) The Landau-Lifshitz-like equation for the mean localized spin, $\vec{S}_l(t)$:

$$\frac{d}{dt}\vec{S}_l(t) = \vec{B}_l^{eff}\left[n,\vec{s},\vec{S},\vec{R}\right](t)\times\vec{S}_l(t)\ , \tag{20}$$

where the effective field has the form

$$\vec{B}_l^{eff}\left[n,\vec{s},\vec{S},\vec{R}\right](t) = \vec{b}_l^{ext}(t) + \frac{\delta\mathcal{B}\left[n,\vec{s},\vec{S},\vec{R}\right]}{\delta\vec{S}(t)}\ . \tag{21}$$

(3) The Born-von Kármán-like equation for the mean ion position, $\vec{R}_l(t)$:

$$-\frac{d^2}{dt^2}\vec{R}_l(t) = F_l^{eff}\left[n,\vec{s},\vec{S},\vec{R}\right](t) \tag{22}$$

where the effective force has the form

$$F_l^{eff}\left[n,\vec{s},\vec{S},\vec{R}\right](t) = F_l^{ext}(t) + \frac{\delta\mathcal{B}\left[n,\vec{s},\vec{S},\vec{R}\right]}{\delta\vec{R}(t)}\ . \tag{23}$$

All the above equations have an initial condition given by the corresponding ones obtained from the equilibrium theory.

The functional $\mathcal{B}$ appearing in the above equations is given by

$$\mathcal{B}\left[n,\vec{s},\vec{S},\vec{R}\right] = B_e^s\left[n,\vec{s}\right] + B_S^s\left[\vec{S}\right] + B_i^s\left[\vec{R}\right] - B\left[n,\vec{s},\vec{S},\vec{R}\right]$$
$$-\frac{1}{2}\int\int d^3rd^3r'n(\vec{r},t)n(\vec{r}',t)U(\vec{r}-\vec{r}') \,. \quad (24)$$

It may be noted that $\mathcal{B}\left[n,\vec{s},\vec{S},\vec{R}\right]$ represents the contributions due to interactions and correlations among all the fields appearing in the problem. $B_e^s[n,\vec{s}]$, $B_S^s\left[\vec{S}\right]$, and $B_i^s\left[\vec{R}\right]$ are the corresponding "non-interacting" actions of matter, localized spins, and ions given by:

$$B_e^s[n,\vec{s}] = \int_{t_0}^{t_1} dt\; tr\left\{\hat{\Lambda}\left(i\hbar\frac{d}{dt} - \hat{\mathcal{L}}_e\right)\hat{\rho}_s(t)\right\} \,,$$
$$B_S^s\left[\vec{S}\right] = \int_{t_0}^{t_1} dt\; tr\left\{\hat{\Lambda}\left(i\hbar\frac{d}{dt} - \hat{\mathcal{L}}_S\right)\hat{\rho}_s(t)\right\} \,,$$
$$B_i^s\left[\vec{R}\right] = \int_{t_0}^{t_1} dt\; tr\left\{\hat{\Lambda}\left(i\hbar\frac{d}{dt} - \hat{\mathcal{L}}_i\right)\hat{\rho}_s(t)\right\} \,. \quad (25)$$

Here $\hat{\rho}_s(t)$ is the corresponding density matrix for the noninteracting systems. And, $\hat{\mathcal{L}}_e\hat{\rho}_s = \left[\hat{T}_e,\hat{\rho}_s\right]_-$, $\hat{\mathcal{L}}_S\hat{\rho}_s = \left[\hat{H}_S,\hat{\rho}_s\right]_-$, and $\hat{\mathcal{L}}_i\hat{\rho}_s = \left[\hat{T}_i,\hat{\rho}_s\right]_-$.

These equations have to be solved self-consistently to examine the system behavior. The various approximate results such as adiabatic approximation etc. are derived from the above by making simplified assumptions as to the nature of the density matrix of the interacting system.

AN ALTERNATE FORMULATION: Φ-DERIVABLE THEORY

There is an alternate route to a self-consistent Green functional theory of interacting many body systems. This has the advantage that we can derive from it the transport properties of the system within the same framework. In this theory the central functional is like the action functional introduced above, but is a functional of the spinor (2 × 2) electron Green function,

$$\underline{\mathcal{G}}\left(\vec{r}t;\vec{r}'t'\right) = -\frac{i}{\hbar}tr\left\{\hat{\rho}\left(t_0\right)T\left(\underline{\Psi}\left(\vec{r}t\right)\underline{\Psi}^{\dagger}\left(\vec{r}'t'\right)\right)\right\} \,,$$

the dyadic Green function associated with the local spins,

$$\underline{\underline{\chi}}\left(lt;l't'\right) = tr\left\{\hat{\rho}\left(t_0\right)T\left(\hat{\vec{S}}_l(t)\hat{\vec{S}}_{l'}\left(t'\right)\right)\right\} \,,$$

and the dyadic Green function associated with ion displacements,

$$\underline{\underline{\mathcal{D}}}\left(lt;l't'\right) = -\frac{i}{\hbar}tr\left\{\hat{\rho}\left(t_0\right)T\left(\hat{\vec{R}}_l(t)\hat{\vec{R}}_{l'}\left(t'\right)\right)\right\} \,.$$

The time ordering symbol T is to be taken in the Schwinger-Keldysh sense (i.e., times are path ordered) when dealing with nonequilibrium time dependent formulation.

In the traditional pure-state td-functional theory, one often expresses the stationarity condition equivalently as a one-particle td-Schrödinger-like equations for some newly defined wave functions. This could of course be recast into an effective one-particle Green function equation with the effective potential appearing here as a self-energy functional. In many body theory, Baym [14] developed a powerful formalism called Φ-derivable theory which maintains all the conservation laws. This is based on a rigorous stationary principle involving the one-particle Green function of the system. A few attempts have been made to compare the results of a Φ-derivable approximation with a corresponding density-functional calculation. Here we shall show that a general functional method can be constructed in a manner analogous to the td-functional approach. This has the merit that the self-energy deduced in this way is in general complex unlike in the earlier versions of the td-functional theory.

The Baym functional $W_{\mathcal{U}}[\mathcal{G}]$ is now defined as

$$W_{\mathcal{U}}[\mathcal{G}] = \Phi[\mathcal{G}] - tr\left\{\mathcal{G}_0^{-1}\mathcal{G} - \ln(-\mathcal{G}) - \mathcal{U}\mathcal{G}\right\}. \tag{26}$$

This is **stationary** with respect to variations of the one particle Green function $\mathcal{G}$ provided the exact self energy Σ is given by

$$\frac{\delta\Phi}{\delta\mathcal{G}} = \Sigma\,. \tag{27}$$

Here tr stands for the operation:

$$tr\mathcal{A} = \int dt \int d^3r \sum_i \mathcal{A}_{ii}\left(\vec{r}t;\vec{r}t^+\right) \tag{28}$$

where the time integration is along the real time-line, and the indices i denote the matrix structure of $\mathcal{A}$, when necessary. Here $\mathcal{U}$, a matrix, is a nonlocal external potential. For the initial thermodynamic equilibrium situation, the time integration runs along the line $(t_0, t_0 - i\beta)$, and $\mathcal{G}$ obeys the Martin-Schwinger time periodicity condition [15]. $W_U(\mathcal{G})$ then corresponds to the free energy Ω of the system. It is interesting to point out that in this theory the stationarity of the Baym functional is common to both the nonequilibrium and the equilibrium situations; the difference between the two now arises in the Martin-Schwinger periodicity property of $\mathcal{G}$ for determining the equilibrium Green function.

We should recall that the self energy arises from interactions among the electrons, so that Φ is merely another representation of the interaction contributions. The purpose of the present work is to incorporate this framework into another, very successful formalism in computing the system properties of equally wide ranging magnitude, namely the non-equilibrium time-dependent functional method.

Let us define a universal functional $F[\mathcal{G}]$ to be

$$F[\mathcal{G}] = \Phi[\mathcal{G}] - tr\ \mathcal{G}_0^{-1}\mathcal{G} + tr\ln(-\mathcal{G})\,. \tag{29}$$

Then

$$W_{\mathcal{U}}[\mathcal{G}] = F[\mathcal{G}] + tr\mathcal{U}\mathcal{G}\,. \tag{30}$$

The stationarity property of Eq. (29) is just the equation for the one particle Green function:

$$\frac{\delta F[\mathcal{G}]}{\delta\mathcal{G}} + \mathcal{U} = 0, \text{ or } \mathcal{G}^{-1} = \mathcal{G}_0^{-1} + \mathcal{U} + \frac{\delta\Phi}{\delta\mathcal{G}}\,. \tag{31}$$

Following Kohn, introduce a "non-interacting" system such that

$$\frac{\delta T_S[\mathcal{G}]}{\delta \mathcal{G}} = \mathcal{G}_0^{-1} - \mathcal{G}_S^{-1} , \tag{32}$$

and a one-particle local potential $\mathcal{U}_S$ so that

$$\mathcal{G}_0^{-1} - \mathcal{G}_S^{-1} + \mathcal{U}_S = 0 . \tag{33}$$

By comparing with Eq. (31), and by recalling the introduction, in the Kohn-Sham theory, of a non-interacting system whose properties are denoted by a subscript s, we have introduced the Green function $\mathcal{G}_s$ from which we obtain the same density as for the interacting system. We then write

$$W_{\mathcal{U}}[\mathcal{G}] = T_S[\mathcal{G}] + \tilde{F}[\mathcal{G}] + tr(\mathcal{U}\mathcal{G}) , \tag{34}$$

where

$$\tilde{F}[\mathcal{G}] = F[\mathcal{G}] - T_S[\mathcal{G}] \, , \; T_S[\mathcal{G}] = tr\left(\mathcal{G}_0^{-1}\mathcal{G}_S + \ln(-\mathcal{G}_S)\right) . \tag{35}$$

Then we have for the one-particle equation a Kohn-Sham type equation when the particles are mutually interacting:

$$\frac{\delta T_S[\mathcal{G}]}{\delta \mathcal{G}} + \frac{\delta \tilde{F}[\mathcal{G}]}{\delta \mathcal{G}} + \mathcal{U} = 0 . \tag{36}$$

Combined with Eq. (32) this gives the equation for $\mathcal{G}$ in the form;

$$\begin{aligned} \mathcal{G}_0^{-1} - \mathcal{G}^{-1} + \mathcal{V}_{eff}[\mathcal{G}] &= 0 \, , \text{ where} \\ \mathcal{V}_{eff}[\mathcal{G}] &= \frac{\delta \tilde{F}[\mathcal{G}]}{\delta \mathcal{G}} . \end{aligned} \tag{37}$$

$\mathcal{V}_{eff}$ here is the self-consistent one-electron potential arising from the electron interactions. Models are constructed based on a suitable choice of the functional $\tilde{F}$, for example, Hartree or Hartree-Fock, or other forms depending on the scheme one wishes to adopt. Formally, $\mathcal{V}_{eff}$ is the same as the self-energy functional of the many-body theory but as discussed in Baym's paper approximation schemes are often employed obeying some basic requirements of the formal theory.

The method outlined here folds in the full many body formalism including the finite temperature theory into a Kohn-type scheme. Also it incorporates the Keldysh method when the path-ordered contour on the time path is employed. We will now indicate how this is incorporated to develop another version of a self-consistent theory of half-metallic magnetism discussed earlier.

1) Inclusion of Ionic Motions

We modify the functional as follows to include the ionic motions:

$$W_{\mathcal{U}}[\mathcal{G},\mathcal{D}] = \Phi[\mathcal{G},\mathcal{D}] - tr\left(\mathcal{G}_0^{-1} - \mathcal{U}\right)\mathcal{G} + tr\ln(-\mathcal{G}) - tr\,\mathcal{D}_0^{-1}\mathcal{D} + tr\ln(-\mathcal{D}) . \tag{38}$$

This functional is stationary w.r.t. variations in $\mathcal{G}$ and the phonon Green function $\mathcal{D}$:

$$\begin{aligned} \mathcal{G}^{-1} &= \mathcal{G}_0^{-1} + \mathcal{U} + \frac{\delta \Phi}{\delta \mathcal{G}} , \\ \mathcal{D}^{-1} &= \mathcal{D}_0^{-1} + \frac{\delta \Phi}{\delta \mathcal{D}} . \end{aligned} \tag{39}$$

$\Phi[\mathcal{G},\mathcal{D}]$ now represents the electron-electron and electron-ion interactions, so that

$$\frac{\delta\Phi}{\delta\mathcal{G}} = \Sigma_e[\mathcal{G},\mathcal{D}] \text{ and } \frac{\delta\Phi}{\delta\mathcal{D}} = \Sigma_i[\mathcal{G},\mathcal{D}] \,, \tag{40}$$

Σ_e, Σ_i being the respective self-energy matrices of electrons and ions. Proceeding as before by introducing $T_S[\mathcal{G}]$ and $T_i[\mathcal{D}]$, and introducing $F[\mathcal{G},\mathcal{D}]$, we obtain the effective electron and ion equations of motion:

$$\begin{aligned} W_{\mathcal{U}}[\mathcal{G},\mathcal{D}] &= T_S[\mathcal{G}] + T_i[\mathcal{D}] + \tilde{F}[\mathcal{G},\mathcal{D}] + tr\,\mathcal{U}\mathcal{G} \,, \text{ where} \\ \tilde{F}[\mathcal{G},\mathcal{D}] &= \Phi[\mathcal{G},\mathcal{D}] - tr\left(\mathcal{G}_0^{-1}\mathcal{G} - \ln(-\mathcal{G})\right) \\ &\quad - tr\left(\mathcal{D}_0^{-1}\mathcal{D} - \ln(-\mathcal{D})\right) - T_S[\mathcal{G}] - T_i[\mathcal{D}] \,. \end{aligned} \tag{41}$$

$T_i[\mathcal{D}]$ is defined in a similar manner to $T_S[\mathcal{G}]$ in Eq. (34). We now obtain

$$\begin{aligned} \frac{\delta T_S[\mathcal{G}]}{\delta\mathcal{G}} + \frac{\delta\tilde{F}}{\delta\mathcal{G}} + \mathcal{U} &= 0 \,, \\ \frac{\delta T_i[\mathcal{D}]}{\delta\mathcal{D}} + \frac{\delta\tilde{F}}{\delta\mathcal{D}} &= 0 \,. \end{aligned} \tag{42}$$

These equations are coupled and take the form (following the same procedures as before):

$$\begin{aligned} \mathcal{G}_0^{-1} - \mathcal{G}^{-1} + \mathcal{V}_{\mathit{eff}}[\mathcal{G},\mathcal{D}] &= 0 \,, \text{ where} \\ \mathcal{V}_{\mathit{eff}}[\mathcal{G},\mathcal{D}] &= \mathcal{U} + \frac{\delta\tilde{F}[\mathcal{G},\mathcal{D}]}{\delta\mathcal{G}} \,. \\ &\text{And,} \\ \mathcal{D}_0^{-1} - \mathcal{D}^{-1} + \mathcal{K}_{\mathit{eff}}[\mathcal{G},\mathcal{D}] &= 0 \,, \text{ where} \\ \mathcal{K}_{\mathit{eff}}[\mathcal{G},\mathcal{D}] &= \frac{\delta\tilde{F}[\mathcal{G},\mathcal{D}]}{\delta\mathcal{D}} \,. \end{aligned} \tag{43}$$

$\mathcal{K}_{\mathit{eff}}$ here is the effective "force constant" contributions from both the electrons and the ions of the system, whereas the $\mathcal{V}_{\mathit{eff}}$ is the effective one-electron potential arising from the same system, and the two equations have to be solved self-consistently.

Inclusion of Localized Spin Fields

Here we introduce the dyadic Green function associated with the localized spin field components, which we here denote by the symbol $\underline{\chi}$ with the corresponding self-energy dyadic $\underline{\Sigma}$ so that the corresponding equation determining $\underline{\chi}$ in the presence of other fields is now given by

$$\begin{aligned} \left(\underline{\chi}^{-1}\right)_{ij} &= \left(\underline{\chi}_0^{(em)-1}\right)_{ij} + \left(\underline{\Sigma}_{\mathit{eff}}\right)_{ij} \,, \text{where} \\ \left(\underline{\Sigma}_{\mathit{eff}}\right)_{ij} &= \frac{\delta\tilde{F}[\mathcal{G},\underline{\mathcal{D}},\underline{\chi}]}{\delta\underline{\chi}_{ij}} \,. \end{aligned} \tag{44}$$

The indices i, j refer to cartesian components. The new $\tilde{F}$ is given by

$$\begin{aligned} \tilde{F}[\mathcal{G},\underline{\mathcal{D}},\underline{\chi}] &= \Phi[\mathcal{G},\underline{\mathcal{D}},\underline{\chi}] - tr\left(\mathcal{G}_0^{-1}\mathcal{G} - \ln(-\mathcal{G})\right) \\ &\quad - tr\left(\underline{\mathcal{D}}_0^{-1}\underline{\mathcal{D}} - \ln(-\underline{\mathcal{D}}\right) - tr\left(\underline{\chi}_0^{-1}\underline{\chi} - \ln(-\underline{\chi}\right) \\ &\quad - T_e[\mathcal{G}] - T_i[\underline{\mathcal{D}}] - T_S[\underline{\chi}] \,. \end{aligned} \tag{45}$$

Eq. (43) should be correspondingly supplemented by the above equation and we must now incorporate the localized spin contributions to the electron and ion properties by changing to the new $\tilde{F}$ defined by Eq. (45). We may now point out that the three coupled Green function equations provide us with the self-consistent framework to study the system in detail, including its equilibrium and nonequilibrium properties.

CONCLUDING REMARKS

In conclusion, we have here formulated a self-consistent dynamic theory of mutually interacting species of localized spins, itinerant electrons which are spin polarizable, and ions. Such a theory demands a form for the effective interaction term, Eq. (24). Schemes to determine this in some form must be sought. We have here suggested one in Section 3. Another approach is to use a phenomenological form for the density matrix, such as a Gaussian form for the ions and localized spins, with time-dependent parameters as a variational ansatz. The resulting equations may then be handled by self-consistent techniques. In collaboration with Professors Mahanti and Kaplan, we are trying to set up such schemes. Clearly, this is one of many possible new avenues open to study time-dependent problems.

ACKNOWLEDGMENTS

It is a pleasure to thank Dr. Krystl Hathaway for directing my attention to the problem of magnetism of manganites and stressing the need for a theory of this coupled field problem. I thank Professors S.D. Mahanti and Tom Kaplan of Michigan State University for providing me with almost all the relevant references on this subject and for many friendly discussions. This is a preliminary report of the collaborative work among the three of us. I am thankful to the US National Science Foundation for supporting my trip to Brisbane; to the Office of Naval Research for partial support of the research; and the support and hospitality of Professors John Dobson, Mukunda Das, and Giovanni Vignale.

REFERENCES

1. G.H. Jonker and J.H. Van Santen, Physica **16**, 337 (1950); J.H. Van Santen and G.H. Jonker, *ibid.* **16**, 599 (1950).
2. C. Zener, Phys. Rev. **81**, 440 (1951): *ibid* **82**, 403 (1951).
3. P.W. Anderson and H. Hasegawa, Phys. Rev. **100**, 675 (1955).
4. P.G. de Gennes, Phys. Rev. **118**, 141 (1960).
5. K. Kubo and N. Ohata, J. Phys. Soc. Japan **33**, 21 (1972).
6. A. Urushibara, Y. Moritomo, T. Arima, A. Asamitsu, G. Kido, and Y. Tokura, Phys. Rev. B **51**, 14 103 (1995).
7. A. Asamitsu, Y. Moritomo. Y. Tomioka, T. Arima, and Y. Tokura, Nature (London) **373**, 407 (1995).
8. W.E. Pickett and D.J. Singh, Phys. Rev. B **53**, 1146 (1996).
9. A.J. Mills, P.B. Littlewood, and B.I. Shraiman, Phys. Rev. Lett. **74**, 5144 (1995).
10. H. Röder, J. Zang, and A.R. Bishop, Phys. Rev. Lett. **76**, 1356 (1996).
11. I. Solovyev, N. Hamada, and K. Terakura, Phys. Rev. Lett. **76**, 4825 (1996).
12. E.K.U. Gross, J.F. Dobson, and M. Petersilka, in *Density Functional Theory II*, edited by R.F. Nalewajski, Springer Series *Topics in Current Chemistry* (Springer, New York, 1996). This is a comprehensive review of the density-functional theory of time-dependent phenomena.
13. A.K. Rajagopal and F. Buot, in *Density Functional Theory II*, edited by R.F. Nalewajski, Springer Series *Topics in Current Chemistry* (Springer, New York, 1996). This is a comprehensive

report on the functional theory of coupled fields of electrons, ions, and electromagnetic fields. The theory given in the present paper is another application of the techniques presented in this review article.
14. G. Baym, Phys. Rev. **127**, 1391 (1962).
15. P. C. Martin and J. Schwinger, Phys. Rev. **115**, 1342 (1959).

DENSITY FUNCTIONAL THEORY OF SUPER-PHENOMENA IN CONDENSED SYSTEMS

M.P. Das

Department of Theoretical Physics
Institute of Advanced Studies
The Australian National University
Canberra, ACT 0200, Australia

INTRODUCTION

Density functional theory (DFT) is considered as the standard model for low /very low energy physics. The theory appeared with its full glory when Hohenberg and Kohn [1] proved that the ground state properties of an interacting many body system can be described by a functional of the one-particle density of the system. The theory has a rigorous formal structure with a well-defined energy functional. Unfortunately the theorem does not tell how to construct the functional, hence the functional remains unknown in the exact sense. In view of the complexity of problems of a many-body nature, it is expected to make explicit construction of the functional by suitable approximations.

In the past three decades there have been many interesting developments:

- Generalisation of DFT [2,3] to a variety of physical situations, namely extensions to include spin, finite temperature, relativity, current density, time dependent density etc.
- Invention of new functionals, mainly beyond the local density approximation [4].
- Applications to a variety of systems: atoms, molecules, solids, surfaces etc., both for classical and quantum domains. Difficult problems like van der Waals interactions are now within the reach of DFT.

The preceding chapters of this book have covered most of the current developments on the theory and applications of DFT. We have discussed elsewhere [5] a new application of DFT to vortex matter, appropriate for strongly type II superconductors, e.g. high Tc superconductors. The present chapter discusses a new class of problems of super-phenomena in physics covered well within the realm of DFT. Some important developments appear in the recent literature on Fermi [6] and Bose systems [7] as regards super-phenemena.

Electronic Density Functional Theory: *Recent Progress and New Directions*
Edited by Dobson *et al.*, Plenum Press, New York, 1998

The organisation of this chapter is as follows. In Sec. II we briefly discuss the nature of super-phenomena, within a ground state formulation. This hinges on the question: is there any broken symmetry responsible for bringing a new ground state? The next question, discussed in Section III, is: can a conventional normal state be sustained in presence of a new field ? If there is an anomalous density, it is sustained by a corresponding external potential. The Hohenberg-Kohn theorem is generalised to consider a functional of normal and anomalous densities. Formal Kohn-Sham equations are derived in a variational procedure in Sec. IV. In limiting cases the corresponding Fermi and Bose systems are described by Ginzberg-Landau-Bardeen-Cooper-Schrieffer (GL-BCS) and Gross-Pitaevskii equations. Conclusions are made in Sec. V.

SUPER-PHENOMENA IN THE GROUND STATE

Many philosophical questions have been raised in the past on the nature of the ground state of a quantum system. A likely deeper question can be posed on the observable macroscopic quantum phenomena. Let us consider the following two examples at room temperature, (1) a piece of ordinary metal, which is best described as a Fermi liquid of interacting electrons, subject to an external potential provided by the ions in the system, and (2) a fluid of Bosonic particles , e.g. ^{4}He. In normal conditions these two systems have normal properties as we usually understand with a quasi-particle approach.

When the temperature is brought down close to 0K, so as to quench the kinetic energy of particles completely, one would envisage a frozen state of the particles like atoms in a crystalline environment. In case of electrons a highly correlated state would be possible. This is known as a Wigner crystal and in the case of Bosons, it would be a Bosonic solid. If the quantum particles are of light mass, as in the case of ^{4}He, the zero point motion may be large enough to melt the crystal to a fluid state. Apart from these types of condensed state what else can one envisage?

Experience shows that at low enough temperature the ground state is not really a normal state, rather it is a state dominated by correlations. Effects of correlations are larger than the energy of single particle dynamics. At this point one would like to distinguish the effect of statistics from that of interactions. A pure noninteracting Bose system is understood to have a component of condensation at 0K. Carrying out carefully the effect of interaction, the condensation density may need a modification. This condensation has a likely effect on superfluidity [8].

Similarly for a system of interacting Fermions a Fermion condensate would be formed, which is well known as a GL-BCS condensation in the theory of superconductivity. In both types of quantum system there is a common existence of the so-called "off-diagonal long range order (ODLRO)" [9]. A GL-BCS condensate is sometimes considered as a charged Boson condensate in an approximate description.

While a clear cut formulation of the unique ground state of interacting quantum systems is not available, we wish to discuss some of the underlying basic physics leading to the super-phenomena of the ground state. A multi-component density functional approach would be an appropriate formalism to understand the true ground state properties.

Study of the ground state has been a hypothetical problem, since no real experiment can be performed at 0K. Nevertheless where it has been possible to reach a temperature close enough to 0K, most of the physical systems do exhibit an ordered state dominated by correlations at this low temperature. We propose to identify this state as a state of super-phenomenon. This state possesses either a Fermi or a Bose condensate depending on the type of system.

In our description while the system has come from an elevated temperature to an ultra low temperature there has been a change of an internal symmetry by a spontaneous breakdown. A phase transition has occurred to a super-phase. A broken internal symmetry is a Gauge symmetry. After the system has gone to a broken symmetry state of a new phase, the quasi-particles of the previous normal state are now less relevant. The total number of particles is conserved after the phase transition has taken place, except that there are now two sub-systems, composed of a part of normal quasi-particles and the remaining belonging to the condensates. Note that number conservation is not essential in this theory if one fixes, instead, the chemical potential. It is natural to think of two types of densities as one-body and condensate densities. These two densities are subject to a one-body normal external potential and a symmetry-breaking anomalous potential respectively. In principle, it is possible to show that the ground state can be a super-state. The details depend on the nature of the energy functionals, which we are currently working on.

DFT FOR A SYSTEM WITH CONDENSATES

In the case of two densities ρ_1 and ρ_2 corresponding to two potentials v_1 and v_2, we can prove the corresponding Hohenberg-Kohn theorem for systems possing super-phenomena.

The Hamiltonian is

$$H = T + U + V_1 + V_2 \tag{1}$$

where T and U are the kinetic and interaction energies respectively. If $|\Phi\rangle$ is the ground state we can write the energy functional

$$E_{v1,v2}[\rho_1, \rho_2] = F[\rho_1, \rho_2] + \int dr v_1 \rho_1 + \int\int dr dr' [v_2^* \rho_2 + \text{h.c.}] \tag{2}$$

where $F[\rho_1, \rho_2] = \langle\Phi|T+U|\Phi\rangle$ is an universal functional. Using a *reductio ad absurdum* argument as in Hohenberg and Kohn [1], we can prove that the energy functional is unique in the presence of v_1 and v_2. Hence we can show that v_1 and v_2 have one to one correspondence with ρ_1 and ρ_2 respectively. This means that all observable quantities are functionals of ρ_1 and ρ_2.

Then we can set up two variational equations by varying the energy functional with respect to two densities:

$$\delta E_{v1,v2}[\rho_1, \rho_2]/\delta\rho_1 = 0 \tag{3}$$

and

$$\delta E_{v1,v2}[\rho_1, \rho_2]/\delta\rho_2 = 0. \tag{4}$$

Formally these two coupled equations are to be solved self-consistently. Technically v_1 is a one-body local potential corresponding to the one-body density ρ_1, whereas v_2 is an anomalous potential corresponding to the condensate density ρ_2.

In the case of superconductors we can make an explicit construction of the Hamiltonian (1) by using

$$\rho_1(r) = \langle\phi| \sum_\sigma \psi_\sigma^+(r)\psi_\sigma(r)|\phi\rangle \tag{5}$$

where ψ_σ's are Fermion operators of the normal state and

$$\rho_2(r,r') = \langle\phi|\psi_\sigma(r)\psi_{\sigma'}(r')|\phi\rangle \tag{6}$$

is the anomalous density (pair amplitude) with σ and σ' being up and/or down Fermion spins. Note that V_2 contains a hermitian conjugate term. In their formulation of the theory of superconductivity Oliveira *et al* [6] used v_2 as a mathematical quantity to generate ρ_2.

In the case of Bosons $\rho_1(r)$ is written as in (5) with ψ's (no spin index) as Boson field operators. There is a slight difference in writing ρ_2, it is written as the ground state expectation of $\langle\phi|\psi|\phi\rangle$. Here the condensate wave function couples to the symmetry breaking potential [7]. ρ_2 in either case is an order parameter, supposed to survive below a certain critical temperature. In the ground state it is expected to be nonzero.

With the above definition of order parameters in both the Fermi and Bose cases, we can set up variational equations ((3) and (4)) on a parallel footing. From the knowledge and experience of the normal-state DFT, it will be useful to go to the Kohn-Sham representation of the DFT equations for quasi-particles.

KOHN-SHAM EQUATIONS FOR SYSTEMS POSSESSING CONDENSATES

The interacting system is considered as a fictitious system, described by single particles in a Hartree-like procedure. The usual assumption is that ρ_1 and ρ_2 representing the interacting system are the same as in the noninteracting system. The kinetic energy operator is written as a one-particle operator, so that the Hamiltonian (1) is

$$\begin{aligned} H = & \int dr\psi^+(r)\left[-\hbar^2\Delta/2m + \tilde{v}_1(r) - \mu\right]\psi(r) \\ & + \int\int drdr'\left[\tilde{v}_2^*(r,r')\rho_2(r,r') + \tilde{v}_2(r,r')\rho_2^*(r,r')\right] \end{aligned} \tag{7}$$

1. Fermions:

   The potentials $\tilde{v}_1$ and $\tilde{v}_2$ contain effective contributions from interactions in the following form.

$$\tilde{v}_1 = v_1 + v_{H1} + v_{xc1}\ , \tag{8}$$

$$\tilde{v}_2 = v_2 + v_{H2} + v_{xc2}\ , \tag{9}$$

   where $v_{H1}(r)$ and $v_{H2}(r,r')$ are Hartree potentials of normal and anomalous type respectively. v_{xci} $(i = 1, 2)$ is defined as $[\delta E_{xc}/\delta\rho_i]$, where E_{xc} is a functional of ρ_1 and ρ_2. This functional is defined customarily by subtracting the normal and anomalous Hartree term in addition to the noninteracting kinetic energy.

   The field operators $\psi_\sigma^+(r)$ are spin dependent. By using Bogoliubov-Valatin transformation,

$$\psi_\sigma(r) = \sum_k\left[u_k(r)\gamma_{k\sigma} - v_k^*(r)\gamma_{k-\sigma}^+\right]\ , \tag{10}$$

   we can diagonalize the Hamiltonian (7). The functions $u_k(r)$ and $v_k(r)$ satisfy the coupled eigenvalue equations.

$$\left[-\hbar^2\Delta/2m + \tilde{v}_1(r) - \mu\right]u_k(r) + \int dr'\tilde{v}_2(r,r')v_k(r') = E_k u_k(r) \tag{11}$$

and

$$\left[-\hbar^2\Delta/2m + \tilde{v}_1(r) - \mu\right] v_k(r) + \int dr' \tilde{v}_2^*(r, r') u_k(r) = E_k v_k(r) \tag{12}$$

The operators $\gamma^+_{k\sigma}$ and $\gamma_{k\sigma}$ create and destroy the quasi particle excitations respectively above the superconducting state. They obey usual anticommutation relations as Fermions. The Fermi function $f(E_k)$ comes from the ground state average of $\langle\gamma^+_{k\sigma}\gamma_{k\sigma}\rangle$. For the ground state $f(E_k)$ is a Θ function. Now by using the functions $u_k(r)$ and $v_k(r)$ through (10), (5) and (6), one-particle and condensate densities can be obtained as

$$\rho_1(r) = 2\sum_k \left\{\mid v_k \mid^2 (1 - f(E_k)) + \mid u_k \mid^2 f(E_k)\right\} , \tag{13}$$

and

$$\rho_2(r, r') = \sum_k u_k v_k^* \left(1 - 2f(E_k)\right) . \tag{14}$$

As in the normal Kohn-Sham theory, one needs to solve the whole system of equations self-consistently provided a definite form of $E_{xc}[\rho_1, \rho_2]$ is known.

2. Bosons

The Kohn-Sham method can be implemented for Bosonic systems in a manner similar to the above, keeping in mind the difference that the symmetry breaking field is coupled to the condensate wave function, rather than to the anomalous density. Using the Bogoliubov approximation (valid for a dilute weakly interacting gas) and the Bogoliubov transformation, a clear derivation can be made for the quasi particle states [7,10]. Also we have to solve the variational equation for the condensate wave function (which is known as the nonlinear macroscopic Schrödinger equation or generalised Gross-Pitaevskii equation for Bosons).

CONCLUSIONS

We have chosen the Kohn-Sham procedure for solution of the DF equations. The kinetic energy functional in the Kohn-Sham method is a simple one-body term calculated on the noninteracting basis. The only unknown is the E_{xc} functional. Some trial functionals should be invented for realistic solutions. Before that is done, we take a limiting case by dropping the xc term in the potential, Eqs. (8) and (9). We can produce the BCS gap equation in the Hartree limit [2]:

$$\Delta(k) = W/2 \int dk' \Delta(k') / \sqrt{\{[(k - k')^2 - \mu]^2 + \Delta(k')^2\}}.$$

Here $\Delta(k)$ is the energy gap function which can be expressed in terms of the condensate density. W is the BCS interaction parameter and μ is the chemical potential. A strategy to calculate the xc energy and potential would be a novel project even at the level of LDA. This will provide corrections beyond the BCS interactions.

For Bosons similarly dropping the xc term would produce the nonlinear Gross-Pitaevskii equation

$$\left\{-\hbar^2\Delta/2m + v_1(r) - \mu + \int dr' v(r - r') n_c(r')\right\} \phi(r) + v_2 = 0.$$

Inclusion of further interaction terms would be greatly warranted. Recently we have incorporated a $| \phi |^6$ term in the ground state energy with attractive pair interaction $(v(r-r'))$ to explain ^{7}Li Bose condensates [11].

In summary we have discussed some recent thoughts on the nature of the ground state of quantum systems. We believe that within the realm of DFT one would be able to describe a possible superconducting or superfluid ground state. A correct understanding of the ground state would require a good knowledge of the energy functional, which we currently lack.

I acknowledge many useful discussions with John Dobson, Hardy Gross, Walter Kohn and A. K. Rajagopal.

REFERENCES

1. P. C. Hohenberg and W. Kohn, Phys. Rev. B **136**, 864 (1964).
2. R. M. Dreizler and E. K. U. Gross, *Density Functional Theory* (Springer-Verlag, Berlin, 1990).
3. *Density Functional Theory*, edited by E. K. U. Gross and R. M. Dreizler, (Plenum, New York, 1994).
4. U. von Barth, in *Methods of Electronic Structure Calculations*, edited by V. Kumar *et al* (World Scientific, Singapore, 1995). See also J.P.Perdew in Ref. [3].
5. D. J. C. Jackson and M. P. Das in Ref. [3]; Supercond. Sc. and Tech. **9**, 713 (1996).
6. L. N. Oliveira, E. K. U. Gross and W. Kohn, Phys. Rev. Lett **60**, 2430 (1988).
7. A. Griffin, Canad. J. Phys. **73**, 755 (1995).
8. D. W. Snoke and G. Baym in *Bose-Einstein Condensation*, edited by A. Griffin, D. W. Snoke and S. Stringari (Cambridge University Press, Cambridge, UK, 1995).
9. C. N. Yang, Rev. Mod. Phys. **34**, 694 (1962).
10. A. L. Fetter, Ann. Phys. (NY) **70**, 67 (1972).
11. N. Akhmediev, M. P. Das and A. Vagov (unpublished).

Part V

ABSTRACTS OF POSTER PRESENTATIONS

AB INITIO CALCULATIONS OF MAGNETIC INTERACTIONS IN MAGNETIC METAL ALLOYS VIA THE LSDF APPROACH

M. F. Ling

Department of Physics, Monash University
Clayton, Victoria 3168, Australia

Over the past decade or so, great advances have been made in our understanding of magnetism in metals as a result of the development of *first-principles* electronic structure calculations based on the local spin density functional (LSDF) theory. The author attempts to give a brief review of this aspect of modern theoretical magnetism with special emphasis on the significant role played by the LSDF in obtaining highly reliable ground state properties of magnetic metals and their alloys. On the other hand, some of the shortcomings of LSD will be pointed out and the current efforts to overcome them mentioned. Finally, the important interplay between magnetic and compositional correlations in magnetic metal alloys is discussed in view of some recent results from *first-principles* electronic structure calculations based on LSDF.

UNICHEM AND ELECTRON MOMENTUM SPECTROSCOPY INVESTIGATION INTO THE VALENCE ELECTRONIC STRUCTURE OF TRANS 1,3 BUTADIENE

M.T. Michalewicz[1], D.A. Winkler[2], M.J. Brunger[3], I.E. McCarthy[3] and W. von Niessen[4]

[1]High-Performance Computing and Communications Centre
CSIRO Mathematical and Information Sciences
24th Floor, 150 Lonsdale St.
Melbourne, Victoria 3000, Australia
e-mail: Marek.Michalewicz@hpc.csiro.au.

[2]CSIRO Division of Chemicals and Polymers, Private Bag 10
Rosebank MDC, Clayton, Victoria 3169, Australia

[3]Physics Department, The Flinders University of S.A.
GPO Box 2100, Adelaide, SA 5001, Australia

[4]Institut für Physikalische and Theoretische Chemie
Technische Universität Braunschweig
Braunschweig, Germany

The experimental (e,2e) coincidence spectroscopy, known as electron momentum spectroscopy (EMS), was applied to the trans 1,3 butadiene (C_4H_6) molecule with detailed binding energy spectra and orbital momentum distributions (MDs) being measured. A small selection of this data is presented. We also report on the utilisation of UniChem computational chemistry codes and the Flinders-developed AMOLD program, which together allow us to calculate theoretical MDs for each orbital, to help elucidate the valence electronic structure of butadiene.

In this study we used Kohn-Sham orbitals in the "Target Kohn-Sham Approximation":

$$\varphi_I \simeq \sqrt{S_I^i}\ \psi_i$$

where φ_I is the Dyson orbital, and S_I^i is the spectroscopic factor. This is analogous to the Target Hartree-Fock Approximation (THFA) with the difference that the molecular orbitals ψ_i are obtained from the Density Functional Theory. It turns out that KS orbitals *may be interpreted* as approximate Dyson Orbitals which are basic in the interpretation of the Electron Momentum Spectroscopy experiments [1,2].

We also present results of our many-body Green's function calculation, to the ADC(3) level, for the binding energies and spectroscopic factors of the respective orbitals of C_4H_6. A critical comparison between the experimental and theoretical MDs allows us to determine the *optimum* wavefunction from the basis sets we studied. The determination of this wavefunction then allows us to make further use of the UniChem package to derive butadiene's chemically interesting molecular properties. A summary of these results and comparison of them with the previous results of other workers is presented with the level of agreement typically being good.

REFERENCES

1. P. Duffy, D.P. Chong, M.E. Casida and D.R. Salahub, Phys. Rev. A **50**, 4707 (1994).
2. M.E. Casida, Phys. Rev. A **51**, 2005 (1995).

VAN DER WAALS INTERACTIONS IN DENSITY FUNCTIONAL THEORY

T. C. Choy

Department of Physics, Monash University
Clayton, Victoria 3168, Australia

We shall explore a modification of the Kohn-Sham LDA theory in which the interatomic dipole-dipole interaction is incorporated at the outset by modifying the electron-electron correlation energies belonging to different atoms. By employing a local density ansatz for the exchange correlation hole appropriate to dipole interactions, we have developed an algorithm in which the van der Waals interaction and in principle also higher order multipolar interactions are incorporated self-consistently. In addition to LDA calculations for molecular solids, this theory is a useful basis for the inclusion of dispersion forces in more accurate calculations of the cohesive energy of metals, ionic solids and semiconductors, when appropriate screening functions are included. We demonstrate this feature within the RPA, which recovers the earlier results of Rapcewicz and Ashcroft.

QUANTITATIVE ELECTRON MOMENTUM SPECTROSCOPY OF SOLIDS

A. Kheifets[1], M. Vos[1], E. Weigold[1], and I. E. McCarthy[2]

[1] Research School of Physical Sciences and Engineering
Australian National University, Canberra ACT 0200, Australia

[2] Electronic Structure of Materials Centre, Flinders University
GPO Box 2100, Adelaide, SA 5001, Australia

We determine, both experimentally and theoretically, the spectral momentum density (SMD) of valence electrons in solids

$$\rho(\epsilon, \mathbf{q}) = \sum_j \int d\mathbf{p}\, |C_j(\mathbf{p} + \mathbf{G})|^2 (-\pi)^{-1} Im\, G_j^-(\epsilon, \mathbf{p})\, \delta_{\mathbf{q},\mathbf{p}+\mathbf{G}}. \tag{1}$$

The SMD gives the probability of finding an electron with a given energy ϵ and momentum $\mathbf{q}$. Eq. (1) contains the one-electron momentum space wave function in a given band $C_j(\mathbf{p} + \mathbf{G})$ and the imaginary part of the retarded Green's function $G_j^-(\epsilon, \mathbf{p})$. The Green's function contains the information about collective effects and presents significant interest for the many-body quantum theory of solids.

Experimentally the SMD is determined by the (e,2e) reaction in which an incoming electron knocks out a bound electron and the two outgoing electrons are then detected in time coincidence with fully determined kinematics. This allows one to infer the binding energy and momentum of the target electron *before* the collision by using the laws of energy and momentum conservation [1]. Under conditions of a high incident energy and a large momentum transfer the (e,2e) reaction cross-section is directly proportional to the SMD.

Theoretically we perform band structure calculations using the linear muffin-tin orbital (LMTO) method in the atomic sphere approximation, as described by Skriver [2]. We use the von Barth – Hedin parametrization for the exchange-correlation potential [3]. By performing a Fourier transform we then obtain the one-electron momentum-space wave function $C_j(\mathbf{p}+\mathbf{G})$. The Green's function is calculated using a GWA model for the structureless jellium in which the screening of the Coulomb potential is described within the RPA [4]. The effects of multiple scattering are modeled using Monte Carlo procedures utilizing well known cross sections for elastic and inelastic scattering [5]. Spherical averaging over the irreducible wedge of the Brillouin zone is taken to model the SMD of structurally disordered targets.

These experimental and theoretical procedures have been applied to study SMD in various amorphous and crystalline materials: semiconductors and semimetals (diamond, silicon, germanium, silicon carbide, graphite), metals (aluminium, copper, nickel) and insulators (aluminum and silicon oxides). The experimental SMD in the valence region is described fairly well by the present theoretical model. However, the plasmaron satellites at higher binding energies have much broader structure and are not reproduced exactly by the calculation. The plasmaron is a bound state of a hole coupled with the plasmon collective mode. Unlike the plasmon, the plasmaron carries energy $\simeq 1.5\Omega_p$ rather than Ω_p. This kind of excitation has been predicted theoretically a few decades ago [4] but still there have not been conclusive experimental observations of the plasmaron yet. It is known that GWA-type calculations predict very sharp plasmaron structure. More sophisticated models produce broader plasmaron peaks [6].

REFERENCES

1. M. Vos and I.E. McCarthy, Rev. Mod. Phys. **67**, 713 (1995).
2. H. L. Skriver, *The LMTO Method*, No. 41 in *Springer Series in Solid-State Science* (Springer-Verlag, Berlin, 1984).
3. U. von Barth and L. Hedin, J. Phys. Cond. Matter **5**, 1629 (1972).
4. B.I. Lundqvist, Phys. Kondens. Materie **7**, 117 (1968).
5. M. Vos and M. Bottema, Phys. Rev. B 54 5946 (1996).
6. F. Aryasetiawan, L. Hedin and K. Karlsonn, Phys. Rev. Lett. **77**, 2268 (1996).

PARTICIPANTS

Dr. Ken Adam,
Chemistry Department,
James Cook University,
Townsville, Queensland 4811, Australia

Dr. George Backskay,
Chemistry Department,
University of Sydney,
Sydney, NSW 2006, Australia

Dr. J. Berakdar,
Atomic and Molecular Physics Laboratory,
Research School of Physical Sciences and Engineering,
The Australian National University,
Canberra, ACT 0200, Australia

Dr. Keiron Burke,
Department of Chemistry,
Rutgers University-Camden,
315 Penn St., Camden NJ 08102, USA

A/Prof. Graham S. Chandler,
Department of Chemistry,
The University of Western Australia,
Nedlands, Western Australia 6907,
Australia

Dr. Tuck C. Choy,
Department of Physics,
Monash University,
Clayton, Victoria 3168, Australia

Mr. M. Daniell,
School of Science, Griffith University,
Nathan, Queensland 4111, Australia

Dr. M. P. Das,
Department of Theoretical Physics, IAS,
The Australian National University,
Canberra, ACT 0200, Australia

Mr. B. Dinte,
School of Science, Griffith University,
Nathan, Queensland 4111, Australia

A/Prof. John F. Dobson,
School of Science, Griffith University,
Nathan, Queensland 4111, Australia

Dr. Eberhard Engel,
Institut für Theoretische Physik,
J.W. Göthe Universität Frankfurt,
Robert Mayer Str. 8-10,
60054 Frankfurt am Main, Germany

Dr. Fred Green,
CSIRO Division of Radiophysics,
P.O. Box 76, Epping, NSW 2121, Australia

Mr. L. Hall,
Department of Chemistry,
University of Sydney,
Sydney, NSW 2006, Australia

Ms. Y. Hancock,
Department of Physics,
Monash University,
Clayton, Victoria 3168, Australia

Prof. Olle Heinonen,
Physics Department,
University of Central Florida,
Orlando, Florida 32816, USA

Dr. Dylan Jayatilaka,
Department of Chemistry,
The University of Western Australia,
Nedlands, Western Australia 6907,
Australia

Ms. B. Johnston,
School of Science, Griffith University,
Nathan, Queensland 4111, Australia

Dr. Robert O. Jones,
Forschungszentrum Jülich, IFF,
D-52425 Jülich, Germany

Dr. Anatoli Kheifets,
Electronic Structure of Materials Centre,
School of Physical Sciences,
Flinders University of South Australia,
GPO Box 2100, Adelaide, SA 5001,
Australia

Prof. Walter Kohn,
Physics Department,
University of California at Santa Barbara,
Santa Barbara, California 93106, USA

Prof. David Langreth,
Physics Department, Rutgers University,
PO Box 849, Piscataway, New Jersey 08854,
USA

Mr. H. Le,
School of Science, Griffith University,
Nathan, Queensland 4111, Australia

Prof. Mel Levy,
Department of Chemistry,
Tulane University, New Orleans,
Louisiana 70118-5698, USA

Dr. Michael F. Ling,
Department of Physics,
Monash University,
Clayton, Victoria 3168, Australia

Mr. M. Martoprawiro,
Department of Chemistry,
University of Sydney,
Sydney, NSW 2006, Australia

Dr. Marek Michalewicz,
CSIRO Division of Information Technology,
723 Swanston St.,
Carlton, Victoria 3053, Australia

Dr. Robert Nesbet,
IBM Almaden Research Center,
650 Harry Rd.,
San Jose, California 95120-6099, USA

Dr. I.-K. Oh,
Department of Theoretical Physics, IAS,
The Australian National University,
Canberra, ACT 0200, Australia

Prof. Robert G. Parr,
Department of Chemistry,
University of North Carolina,
Chapel Hill, N. Carolina 27599-3290,
USA

Prof. John P. Perdew,
Physics Department, Tulane University,
New Orleans, Louisiana 70188, USA

Dr. Martin Petersilka,
Institut für Theoretische Physik,
Universität Würzburg,
Am Hubland, D-97074 Würzburg, Germany

Dr. Warren Pickett,
Code 6604, Naval Research Laboratory,
Washington, DC 20375, USA

Dr. Marian W. Radny,
Physics Department, Newcastle University,
Callaghan, New South Wales 2308, Australia

Dr. A. K. Rajagopal,
Code 6860.1 Naval Research Laboratory,
Washington DC 20375-5320, USA

Dr. Andreas Savin,
Laboratoire de Chimie Théorique,
Université Paris VI,
4, Place Jussieu,
F75252 Paris, France

Ms. K. Sendt,
Department of Chemistry,
University of Sydney,
Sydney, NSW 2006, Australia

Mr. W. Shapley,
Department of Chemistry,
University of Sydney,
Sydney, NSW 2006, Australia

A/Prof. Phil V. Smith,
Physics Department, Newcastle University,
Callaghan, NSW 2308, Australia

Dr. Z. Szotek,
Daresbury Laboratory,
Warrington WA4 4AD, UK

Dr. Walter M. Temmerman,
Daresbury Laboratory,
Warrington WA4 4AD, UK

Dr. Jagdish Thakur,
Physics Department,
University of New South Wales,
Kensington, NSW 2052, Australia

Dr. Cyrus J. Umrigar,
Laboratory of Atomic and Solid State Physics,
Cornell University,
Ithaca, New York 14853, USA

Prof. Giovanni Vignale,
Physics Department,
University of Missouri at Columbia,
Columbia, Missouri 65211, USA

Dr. Jun Wang,
School of Science, Griffith University,
Nathan, Queensland 4111, Australia

Dr. Leslie C. Wilson,
Department of Chemistry, Box 5,
Loyola University,
New Orleans, Louisiana 70118, USA

Prof. Weitao Yang,
Chemistry Department, Duke University,
Durham, North Carolina 27706, USA

Prof. Eugene Zaremba,
Physics Department, Queens University,
Kingston, Ontario K7L 3N6, Canada

Index